AF732719

HIGH RESOLUTION FLOW MODELLING IN HYDROLOGY AND GEOMORPHOLOGY

ADVANCES IN HYDROLOGICAL PROCESSES

Advances in Hydrological Processes is a series of books devoted to the publication of selected, original research and peer reviewed papers previously published in the Journal *Hydrological Processes.*

Series Editors M. G. Anderson
N. E. Peters
D. E. Walling

Terrain Analysis and Distributed Modelling in Hydrology
Edited by K. J. Beven and I. D. Moore

Scale Issues in Hydrological Modelling
Edited by J. D. Kalma and M. Sivapalan

Freshwater Ecosystems and Climate Change in North America
A Regional Assessment
Edited by C. E. Cushing

Water Quality Trends and Geochemical Mass Balance
Edited by N. E. Peters, O. P. Bricker and M. M. Kennedy

Distributed Hydrological Modelling
Applications of the TOPMODEL Concept
Edited by K. J. Beven

Glacier Hydrology and Hydrochemistry
Edited by M. Sharp, K. S. Richards and M. Tranter

High Resolution Flow Modelling in Hydrology and Geomorphology
Edited by P. D. Bates and S. N. Lane

Hydrological Applications of GIS
Edited by A. M. Gurnell and D. R. Montgomery

HIGH RESOLUTION FLOW MODELLING IN HYDROLOGY AND GEOMORPHOLOGY

Edited by

P.D. BATES
School of Geographical Sciences, University of Bristol, UK

S. N. LANE
School of Geography, University of Leeds, UK

JOHN WILEY AND SONS
Chichester · New York · Weinheim · Brisbane · Singapore · Toronto

The papers in this volume were originally published in Hydrological Processes Volume 11/2, 11/14, 11/15, 12/8

Baffins Lane, Chichester,
West Sussex PO19 1UD, England
National 01243 779777
International (+44) 1243 779777
e-mail (for orders and customer service enquiries): cs-books @wiley.co.uk
Visit our Home Page on http://www.wiley.co.uk
or http://www.wiley.com

Other Wiley Editorial Offices

John Wiley & Sons, Inc., 605 Third Avenue,
New York, NY 10158-0012, USA

WILEY-VCH Verlag GmbH, Pappelallee 3,
D-69469 Weinheim, Germany

Jacaranda Wiley Ltd, 33 Park Road, Milton,
Queensland 4064, Australia

John Wiley & Sons (Asia) Pte Ltd, 2 Clementi Loop #02–01,
Jin Xing Distripark, Singapore 129809

John Wiley & Sons (Canada) Ltd, 22 Worcester Road,
Rexdale, Ontario M9W 1L1, Canada

Library of Congress Cataloging in Publication Data

High resolution flow modelling in hydrology and geomorphology / edited by P. D. Bates and S. N. Lane
p. cm.—(Advances in hydrological processes)
A selection of papers originally published in the journal Hydrological processes.
Includes bibliographical references.
ISBN 0-471-97875-2 (alk. paper)
1. Groundwater flow—Simulation methods. 2. Geomorphology—Simulation methods.
I. Bates, Paul D. II. Lane, Stuart N. III. Series.
GB1197.7.H54 1999
551.48 — dc21

99-048237

British Library Cataloguing in Publication Data

A catalogue record for this book is available from the British Library

ISBN 0-471-978752

Printed and bound in Great Britain

CONTENTS

INTRODUCTION

STUART N. LANE

School of Geography, University of Leeds, UK

AND

PAUL D. BATES

School of Geographical Sciences, University of Bristol, University Road, Bristol BS8 1SS, UK

Recent research has witnessed increased application of flow models in hydrology and geomorphology, aided by developments in process understanding and representation, boundary condition specification and computational power. This book represents a collection of chapters that reflect this increased application and which seek to convey some of the key themes that are emerging in the use of numerical simulation at high spatial and temporal resolutions for increasing our understanding and prediction of hydrological and geomorphological phenomena. It is possible to envisage a number of key reasons why a concern with models of this type is important:

1. Despite improvements in our ability to measure many environmental phenomena, there are still processes that cannot be understood through either direct measurement or scaled physical models (e.g. Moulin and Ben Slama, Chapter 3). A good example is provided by El-Hames and Richards (Chapter 11) in the case of extreme flash floods in arid regions, where high-magnitude but low-frequency events in relatively inaccessible environments can make direct measurement exceptionally difficulty. Similarly, Darby (Chapter 17) describes how a bank erosion model may be of particular use for supplementing limited empirical data.
2. Point (1) takes on special importance when the behaviour of whole systems is of interest, possibly over large spatial scales (e.g. Bates *et al.*, Chapter 10), and which cannot be readily understood from modelling or measurement of individual components in isolation. Models may be particularly useful in identifying emergent behaviour where we are interested in combinations of processes acting across several scales, processes with a complexity that can only be described through nonlinear equations or where there may be strong process interaction, feedback and system behavioural thresholds operating. In all such situations system behaviour may be highly non-intuitive and models provide a means of quantifying and analysing this complexity within a structured framework. For geomorphologists this is particularly useful in attempting to connect the relatively long time scales of landform development with the complex and transient processes of fluid dynamics which create them. This has not previously been possible and computational fluid dynamics models provide, for the first time, a potential means of bridging this scale gap. A note of caution should be added, however, as models themselves represent truncations of reality as a result of assumptions made regarding process representation or boundary condition specification. Thus, there can be uncertainty as to whether emergent behaviour is a consequence of these assumptions, or a characteristic of the real behaviour of the system being studied. With increased ability to couple different processes together within the same model (e.g. Stewart *et al.*, Chapter 12; Nicholas and Walling, Chapter 16), the possibility of exploring whole system behaviour is increased. This does remain sensitively dependent upon establishing faith in model simulations, and it is hardly surprising that this volume

illustrates a wide range of approaches for assessing the quality of model predictions. Nevertheless use of models to stimulate and support basic process research is a common methodology in disciplines such as meteorology and oceanography, as even a cursory inspection of the major journals in these fields will show. Hydrologists and geomorphologists have been relatively slow to adopt such an approach and the extent to which this type of research will here come to form a part of the scientific mainstream remmains to be seen. The chapter by Stewart *et al.* (Chapter 12) provides an example of a modelling study which is used to assess the adequacy of currently standard field sampling procedures for floodplain contamination studies. In this respect models can provide a numerical 'virtual laboratory' for exploring scientific ideas for reasonableness and conducting numerical experiments aimed at improving process understanding.

3. In addition to identification of emergent system behaviour, high resolution modelling can play an important role in extending the basic understanding that has been obtained from field or laboratory studies. A commonly cited criticism of intensive field investigations is the limited generality of the conclusions that are derived from them. Even if it is possible to measure a particular phenomenon, this may only be possible for a restricted range of boundary conditions, due to the time and effort that it can take to measure a particular case, or due to the difficulties of measuring particular cases (e.g. floodplain inundation), whether because they are extreme and difficult to measure or rarely occur and do not fall within a particular measurement period (e.g. El-Hames and Richards, Chapter 11). Numerical modelling can provide a means of extending a single well-studied case, which is used for model verification, to other situations, by altering boundary condition values. In Chapter 7, Bradbrook *et al.* use this approach to extend a limited number of confluence case-studies to a larger number of different confluence configurations. Although one must question the applicability of a heavily calibrated model beyond the time- and space-specific situation that it was calibrated for, combined field or laboratory calibration with numerical simulation can significantly enhance our understanding of those situations in which particular processes matter. In data-poor environments models may be used prior to field investigation to design instrumentation and provide information on the spatial distribution and accuracy of measurements necessary to define system behaviour.
4. Such studies are becoming possible due to the role of technology, in terms of both computing and data capture, in forcing a change in the research environment. High-power computing applications are now a common currency in a wide range of fields and it seems that we are only just beginning to see the range of possibilities involved. Until relatively recently, increases in computing power meant that data requirements for numerical models far outstripped the capability of field monitoring to supply such information. The chapter by Bates *et al.* (Chapter 10) shows the potential for remote sensing technologies to revise our approach to calibration and validation of two-dimensional (2D) river flow models. We now have the opportunity in an increasing number of areas to use new data capture techniques with high performance in computing to create ever more sophisticated computer representations of the environment. Indeed, technology is now developing areas of research potential faster than the time scale of most scientific projects, particularly if one considers the length of time between conducting a piece of research and seeing that work in print! Nevertheless, models provide a means of taking advantage of technological advances for scientific research and this volume provides an example how technological change has forced methodological change in the environmental sciences, and in particular in geomorphology.
5. Finally, the book contains examples of studies where numerical models are required for their characteristic predictive role. These examples range from bank erosion (e.g. Mosselman, Chapter 18; Darby, Chapter 17) through erosion and deposition studies (e.g. Moulin and Ben Slama, Chapter 3) to floodplain inundation problems (e.g. Bates *et al.*, Chapter 9) and runoff generation (e.g. Singh *et al.*; Jolley and Wheater: El-Hames and Richards, Chapters 13–15).

The aim of this introduction is to set the chapters into the context of recent and future research developments in this area. The book has one unifying theme: the way in which process representation is rapidly developing in terms of the flow modelling techniques and technologies used by both hydrologists and geomorphologists. The book is structured into four sections. These divisions are somewhat artificial, and merely reflect the emphasis that the authors place in their chapters. The first section covers theoretical advances, and seeks to provide an introduction to some basic flow modelling principles as well as some examples of particular aspects of process representations that are currently being developed. In all cases, these chapters take basic processes and explore how they can be better represented in numerical models. The concern of the remaining three sections is with application of flow modelling methods to the areas of hydraulics, hydrology and geomorphology. The second section is largely concerned with two- and three-dimensional modelling of flows over complex topographies, both floodplains and in-channel. The third section has a hydrological emphasis, and uses both the hydraulic models and methods described in the previous section, but extends coverage to different processes, some of which are critical for studies of hillslope hydrology. The final section considers the use of flow modelling in geomorphology, and in particular in sedimentation and bank erosion studies.

The remainder of this introduction is designed to give a flavour of what the book contains through a reflection upon some of the issues that emerge as being of particular interest: (i) process representation; (ii) process coupling; (iii) model dimensionality; (iv) numerical solution; (v) model validation, calibration and parameterization; and (vi) model applications.

PROCESS REPRESENTATION

This book shows how progress is being made in terms of modelling some particularly intractable, but nevertheless critical, geomorphological and hydrological processes. Despite the increasing complexity of simulations performed and the enhanced data visualization techniques used, a model is only as good as the theoretical description that it embodies and the ultimate aim of all modelling research must be to improve basic scientific knowledge. For example, two chapters by Singh, Bengtsson and Westerstrom (Chapter 13) focus upon water movement through and underneath a snowpack. For catchments that are affected by seasonal snowfalls, understanding and predicting water routing during snowmelt is central to forecasting both floods and water supply. Water movement in and under a snowpack is notoriously difficult to measure and these two chapters both adopt a kinematic wave treatment of the problem. They were able to develop analytical solutions that reproduced essential features of snowmelt–runoff hydrographs. Some chapters are particularly ambitious in what they seek to model (e.g. Lane and Richards, Chapter 5) and it is clear that further developments in process representation are required. The book sheds light on a number of issues in this respect.

As is implicit in the term 'high resolution', moves towards distributed modelling at the within-catchment (e.g. Singh, Bengtsson and Westerstrom, Chapter 14) and within-channel scale will result in necessary improvement to basic process representation. This is illustrated by Tchamen and Kahawita (Chapter 2) who note that by simplifying the depth-averaged momentum equations such that there is zero inertia, and combining this with the continuity equation, it is possible to obtain a stable flood routing solution. However, this does not conserve momentum, and hence does not give good distributions of parameters like water depth within the model. Similarly, one of the most important but most intractable of hydraulics questions remains how to close the Reynolds-averaged flow equations (Lane, Chapter 1). There is wide variation in the levels of sophistication or turbulence models used in flow models, and chapters in this volume illustrate how choice of an appropriate turbulence model can be important, but that this is scale-dependent. For instance, Lane and Richards (Chapter 5) found that in applying a depth-averaged model to a multi-thread river channel, model predictions were relatively insensitive to the sophistication of the turbulence closure. Similarly, Hodskinson and Ferguson (Chapter 6) found little change in the correspondence between 3D model predictions of flow in a sharp meander bend and field data when the type of turbulence closure was changed. However, Bradbrook *et al.* (Chapter 7) show that a modification of a second-order turbulence model did significantly improve the correspondence between model predictions and laboratory measurements in a 3D simulation of a laboratory confluence. There are at least three reasons for these apparently contradictory findings. First, some explanation may rest with the dimensionality of the particular application. In a depth-averaged model, required in reach-scale studies due to computational limitations, there is only restricted opportunity for turbulence generation and dissipation within the water column, reflected in the observation by Lane and Richards

(Chapter 5) that the turbulence model only affected predictions close to the channel sidewalls. Second, turbulence models now exist for almost every possible type of flow situation that is of interest. This reflects the need to deal with quite specific flow field problems such as separation in the case of the Bradbrook *et al.* (Chapter 7) study. Both of these points suggest that choice of turbulence model, and process representation more generally, needs to be made both internally (i.e. with reference to other model assumptions) and externally (i.e. with reference to the scale of model application). Third, using changes in correspondence between model predictions and field measurements (e.g. Lane *et al.*, 1999) may not be an effective way of assessing the importance of particular process representations. This is because field measurements may not include those points in the model where a particular process representation does indeed matter. If they are included, any changes in correspondence at those points may be hidden within the variance of the correspondence in the data set as a whole. Moreover, following on from the 'effective parameter' debate commonly cited in hydrology (*cf.* Beven, 1989) as one of the problems with parameterizing distributed models, we may not be able to measure exactly what the model predicts. For example, validation of the modelled velocities of Bates *et al.* (Chapter 9) would require both time and depth averaging of measured instantaneous velocity fields in order to correspond to model predictions. In practice this may be extremely difficult to accomplish. This is a good reason why other forms of assessment of process representation (see below), using sensitivity analysis at the level of the entire set of model predictions, may be required (e.g. Lane *et al.*, 1994; Bates *et al.*, 1996; Bates *et al.*, Chapter 10).

However, turbulence modelling using Reynolds-averaged numerical solutions is only of use up to a point. High-frequency velocity fluctuations may be critical to issues like sediment transport. Central to the work of Nino and Garcia (Chapter 4) is an understanding of the interaction between flow field and particle movement at high frequencies, as a means of understanding grain saltation. These sorts of observations emphasize the need for improved understanding of the interactions between high-frequency flow fluctuations and sediment transport using laboratory and field observations, and use of this understanding to improve parameterization in time-averaged models. Direct numerical simulation remains an impossibility for high Reynolds number, 3D flows, over complex topographies, but recent research has used large eddy simulation (LES) as a possible compromise (Lane *et al.*, 1998; Bradbrook *et al.*, in review). This has seen widespread application to oceanic and atmospheric problems, but as yet has only been used to improve mean flow field predictions in natural river channels (Thomas and Williams, 1995). It is based upon filtering the Navier–Stokes equations at the scale of the local grid-size. At scales coarser than this, the flow field is allowed to evolve. At scales smaller than this, a sub-gridscale turbulence model is used. Whilst the latter retains the problems of other aspects of Reynolds averaging, the more isotropic nature of small-scale turbulent fluctuations permits use of much simpler closure schemes. Hence, Reynolds averaging at sufficiently small scales does not undermine the philosophy of LES, which seeks to obtain a good qualitative and quantitative representation of the flow at the scale of interest, through careful consideration of grid design.

Unsteady flow representation is equally important over longer time-periods due to the need to understand the effects of discharge variation on flow within river channel and floodplain reaches. Both Singh *et al.* (Chapter 13) and El-Hames and Richards (Chapter 11) adopt one-dimensional (1D) approaches for large spatial scales, to help understand flood routing. At smaller scales, 2D approaches are adopted, particularly where the extent of floodplain inundation is of substantive importance. This represents an exceptionally difficult problem to address (Tchamen and Kahawita, Chapter 2) as it will not only require a free surface treatment, but also wetting and drying of the channel and floodplain as the wetted boundary migrates. The challenge results from very shallow flows which can be prone to numerical instability (Tchamen and Kahawite, Chapter 2), particularly as drying occurs. Also, the domain over which the model equations must be solved may be very large compared with the channel itself, so resulting in serious computational demands.

Flood routing (if not flood hydrographs) may be sensitively dependent upon local topographic and roughness effects. These issues are explored in a number of chapters that address the wetting and drying problem (e.g. Bates *et al.*, Moulin and Ben Slama, Nicholas and Walling, Stewart *et al.*, Tchamen and Kahawita Chapter 2). Some of these chapters illustrate just how important field and laboratory observations are for developing our understanding of what processes need to be represented in numerical models. For instance, Nicholas and Walling describe how they had to modify a simplified 2D model of wetting and drying to allow adequate representation of both backwater and recession limb ponding on the floodplain surface.

Sensitive to any question of process representation is the need to balance improvements in physical detail against the associated increase in computational demands (Bates *et al.*, Chapter 10). For instance, Nino and Garcia (Chapter 4) illustrate how careful evaluation of different components of equations describing particle saltation, with respect to experimental results, allows identification of those terms that matter and which must be included (e.g. the lift term), those that matter under certain situations (the Bassett term seems to be critical for sand saltation), those that could matter, but where further theoretical development may be required (e.g. the Magnus force term associated with particle rotation), and those that may be ignored (turbulence effects upon saltation lengths at high particle Reynolds numbers). This research is an exemplary illustration of the application-dependent nature of any model, where model structure and complexity depend on not just the type of sediment being transported, but also on the sediment characteristics themselves. It also emerges in other chapters in this volume. Mosselman found that a model that had previously been successful in sand-bedded channels was much less successful in gravelly channels due to a failure to include the effects of sediment sorting. At a different scale, Jolley and Wheater (Chapter 15) showed that a simple hydrological routing model performed equally as well as a more sophisticated hydrological model based on channel network morphplogy and the 1D hydrodynamic equations for open channel flow.

PROCESS COUPLING

One of the most exciting aspects of process representation is the coupling together of models representing different processes. This is illustrated by Moulin and Ben Slama (Chapter 3) who describe the coupling of a suspended sediment code to a depth-averaged flow model, and use it to explore both sedimentation processes, and the coupling of fine sediment transport to heavy metal adsorption/desorption. The flow model includes wetting and drying, and this provides an important methodological tool for exploring sedimentation processes in systems where inundation extent varies through time. Moulin and Ben Slama show how the wetting–drying problem needs special attention in this situation: solving an advective equation for a conservative tracer over the whole solution domain simultaneously with the flow equations will not prevent the tracer reaching dry areas which the flow has not reached. Their solution is to combine a wetting and drying algorithm with a mask function that only applies the advection equation for the tracer to zones that are predicted as being wet.

Coupling flow processes to sediment transport is particularly advanced in the area of bank erosion (Darby, Chapter 17; Mosselman, Chapter 16), a problem of immense practical importance. The volume includes a number of other examples of process coupling. A particularly novel one is provided by Stewart *et al.* (Chapter 12). Floodplains are now recognized as important sinks for both polluted sediment (e.g. Macklin, 1996) and nutrients (e.g. Haycock and Burt, 1993). Central to understanding the way that floodplains perform this function is knowledge of the spatial patterns of water movement across and into the floodplain, particularly given the difficulty of field measurement of this process. Thus, Stewart *et al.* couple a depth-averaged model for floodplain flows to models for pollution and suspended sediment transport and floodplain infiltration.

DIMENSIONALITY

Dimensionality is a theme that is implicit to many of the chapters herein. In recent years, real progress has been made in the move towards 2D and 3D modelling. This is important as it reflects a growing of geomorphologists with what happens within individual river reaches (Lane and Richards, 1997), as opposed to between them. Implicit to this is the idea that changing model dimensionality also changes the associated scale of enquiry (e.g. Bates *et al.*, Chapter 10). For instance, El-Hames and Richards (Chapter 11) used a 1D model at the catchment scale to simulate flash floods in an arid region. Singh *et al.* (Chapter 14) develop an alternative treatment of the diffusion wave equation for 1D flood routing. Examples of reach-scale applications include the 2D studies of Lane and Richards, Bates *et al.*, Stewart *et al.*, Nicholas and Walling, and

Mosselman (Chapters 5, 9, 12, 16 and 18). At the within-reach scale, Hodskinson and Ferguson (Chapter 6) and Bradbrook *et al.* (Chapter 7) use 3D models. A fundamental reason for this scale–dimensionality relationship is the computational limitations imposed when trying to model larger spatial scales. A further constraint may be the affinity of different process assumptions for particular scales of enquiry. Model assumptions which are reasonable at one scale may break down at another. For example, 2D depth-averaged models may not be suitable for small scales over microtopography as bed gradients may exceed the 10% slope threshold normally quoted for this class of scheme. Similarly, if one reduces the scale of enquiry sufficiently vertical velocities may become important and negate the use of a 2D representation. There is thus a characteristic range of scales over which certain sets of model assumptions can be said to apply. Bates *et al.* (Chapter 9) illustrate just how important scale is in terms of predictions of distributed hydraulic variables in a 2D model of floodplain flow, if not in terms of predicted outflow discharge hydrographs or inundation extents. They found that water depths and velocities changed significantly in response to increases in mesh resolution. A move to three dimensions at the scale of entire reaches is clearly impossible given the need to ensure grid independence using relatively fine mesh resolutions, at such large spatial scales.

The simple relation between dimensionality and scale may however break down, and is a salutary reminder of one of the fundamental problems of modelling natural systems: sub-grid scale parameterization. For instance, Nino and Garcia present results for the saltation of individual sand particles and show that this causes the basic Bagnold approach to sediment transport to be of somewhat more restricted generality. Even with the 3D model used by Bradbrook *et al.*, there remained certain critical processes (e.g. turbulence) which had to be parameterized via a combination of assumptions and empirical observations.

As soon as dimensionality is reduced, new questions arise. For instance, depth averaging of the full 3D form of the Navier–Stokes equations (Lane, Chapter 5) alters the fundamental structure of those equations through the introduction of lateral dispersion terms, and requires model development to allow adequate representation of the effects of those dispersion terms on the redistribution of mass and momentum. Mosselman (Chapter 18) and Lane (Chapter 5) both provide examples of how these dispersion terms might be dealt with, and recent research suggests that this is critical: Lane *et al.* (1999) illustrate how predictions from a depth-averaged model are sensitively dependent upon dispersion terms. As with process representation, this is application specific (e.g. it is of more importance where there is marked streamline curvature).

Dimensionality may become an especially important issue when it comes to model coupling. As Mosselman notes, depth-averaged models will not be effective in predicting lateral channel changes where the banks are especially steep due to the poor accuracy of flow predictions if lateral friction is ignored. Given the strong links between lateral friction and shear stress, and between shear stress and erosion modelling, this remains an area that needs much further research (Darby, Chapter 17).

NUMERICAL SOLUTION

A key theme in high resolution numerical modelling is the development of codes that are numerically stable for the particular applications being studied. El-Hames and Richards (Chapter 11) report on how solution of the mass continuity and momentum equations have to be replaced by a kinematic wave treatment for overland flow routing due to the shallow flows that are involved. Similarly Singh *et al.* (Chapter 8) introduce a kinematic wave model to their mixing model treatment of flood wave diffusion to allow solutions that are more simple, accurate and stable. This reflects a fundamental aim of any modelling approach: to develop solutions that are numerically correct, but also computationally straightforward.

The need to deal carefully with shallow flows for stability reasons is implicit in many of the chapters in this volume, and links directly with process representation. Floodplain wetting and drying is one example. Nicholas and Walling (Chapter 16) describe how it was necessary to simplify the momentum conservation equation to produce a numerical model which would work successfully in situations with complex floodplain topography, the latter being a critical control on the locations of fine sediment deposition. Stewart *et al.* (Chapter 12) use a depth-averaged model with a modified form of the momentum equation to deal with the difficulties of simulating inundation processes where bed slopes are exceptionally low. Both of these examples illustrate the need to balance achieving model stability, against adequate process representation. Similar problems arise when using 3D models, but for a different reason. Three-dimensional treatments of the Navier–Stokes equations are difficult because of the need to determine the position of the free surface, and hence adjust the solution domain accordingly. Bradbrook *et al.* (Chapter 7) get around this problem by using a porosity correction, where flux through the surface layer of cells is increased where pressure on the surface suggests that there should be water surface super-elevation and reduced where there should be depression. This represents an effective change in cell height, and allows the free surface to be incorporated into the solution, without encountering the numerical stability problems that may arise with other treatments of the free surface.

However, not all numerical stability issues need to be dealt with through process representation: one critical alternative is to give consideration to numerical solution methods. Tchamen and Kahawita (Chapter 2) show how by giving close attention to numerical solution issues, it is possible to deal with some of the numerical instability problems that result from wetting and drying. They use linearization with switching between explicit and implicit time integration to produce solutions that are stable whilst still dealing effectively with associated source terms. Numerical solution may also be central to allowing efficient simulation, minimizing the amount of computational effort required to achieve desired results. This may be especially important for high resolution modelling over complex topographies where mesh complexity, density and spatial extent are all high. Stewart *et al.* (Chapter 12), for instance, describe how a modified Galerkin treatment of certain terms in the shallow water continuity equation can avoid the need for mesh refinement to achieve stable solutions, particularly in areas where associated increases in mesh resolution are not required. Additionally, use of element-by-element techniques to store the matrices of the linear system avoids the need to assemble these at each time step and thus saves both computational time and computer memory.

VALIDATION, CALIBRATION AND PARAMETERIZATION

Many of the chapters included in this volume attempt some form of model validation, albeit to varying degrees of success. Singh *et al.* (Chapter 8) use predicted outflow hydrographs to evaluate three different treatments of the 1D diffusion wave equation for flood routing. Nino and Garcia (Chapter 4) obtained good results when comparing laboratory-measured saltation trajectories with an associated saltation model for sand grains. Mosselman (Chapter 18) compares predictions of observed river bank recession rates to those modelled using a depth-averaged flow model coupled to models for bank failure and sediment transport. Hodskinson and Ferguson (Chapter 6) compared field measurements of two components of velocity with 3D velocity predictions for flow in meander bends. Bradbrook *et al.* (Chapter 7) compared laboratory measurements with model predictions of 3D flow in a simple confluence geometry.

A number of chapters attempt to apply high resolution models to quite complex situations, perhaps one of the main rationales behind high resolution modelling, but this also makes these sorts of models increasingly dependent upon boundary condition representation if model predictions are to be reasonable (Bates *et al.*; Nicholas and Walling; Chapters 10 and 16). Hodskinson and Ferguson (Chapter 6) applied a 3D model to flow in sharp meander bends, where flow separation has commonly been observed. This involved a complex bed morphology and flow structure, involving a deep scour pool on the outer bend of the meander and a flow separation zone on the inner bend. Comparison with independently acquired field measurements was particularly encouraging, and illustrates the potential for this sort of modelling even in quite complex morphological situations. Nicholas and Walling (Chapter 16) considered flow inundation and suspended sediment deposition over a complex floodplain topography, and attribute sensible model performance to a good representation of floodplain topography in the finite difference scheme. As they note, this represents a real challenge for boundary condition specification, as it is rare for traditional sources of floodplain topographic information (e.g. contour maps) to provide sufficient information of floodplain surface micro-topography. These sorts of studies all provide an important reminder that model predictions may only ever be as good as the input data used to derive them. For instance, El-Hames and Richards (Chapter 11) describe how a major difference between observed and simulated discharge hydrographs was due to assumptions that had to be made about the spatial distribution of input precipitation data.

In the presence of divergence between model predictions and independent acquired data, and provided some confidence may be placed in the latter, the modeller must choose between: (i) process representation; (ii) boundary condition specification; and (iii) model parameterization; as the source of error. If (i) is inadequate, there will be limits on the utility of improving boundary condition specification. Similarly, if either (i) or (ii) are inadequate, changing model parameterization may be able to produce the right results, but not necessarily for the right reasons. An example of the latter is provided by Lane *et al.* (1999) who show how a depth-averaged model of flow in a confluence can produce predictions of depth-averaged velocity that are as good as those estimated from the output of a 3D model, but only with a value of bed roughness that is 50% higher than that measured in the field. Any situation where a model is producing the right results for the wrong reasons will limit the model's utility for prediction, as there is no guarantee that the calibration will hold beyond the specific situation that the calibration has been developed for. For this reason, simply considering model outputs may be insufficient, and as models increase in complexity, it is important to extend validation to individual components associated with each part of the model (e.g. Stewart *et al.*, Chapter 12). This is necessary as if there is divergence between model predictions and verification measurements it is important to establish exactly which part of the model is associated with most of the divergence to allow research efforts to be directed accordingly. A number of chapters illustrate this, Mosselman notes three reasons for poor model predictions of bank erosion rates: incorrect bank erosion process specification; a failure to represent bank accretion processes; and poor representation of bed topographical change. Identification of these was made possible by the availability of different sorts of verification data, including water depth and bank migration information. El-Hames and Richards (Chapter 11) validate separately both the hillslope and channel routing components of their 1D model of flash flood generation. Stewart *et al.* (Chapter 12) consider the flow, sediment and solute transport and infiltration components of their model of floodplain flow and transport processes.

Internal model validation does not just involve individual components of a model, but may also incorporate different spatial zones within a model's domain, and this can be equally important in identifying where a model has poor process representation. An example of this is provided by floodplain inundation studies, where it is critical to assess predictions of the position of the wetted boundary with independent data. Bates *et al.* (Chapter 10) recognize the difficulty of acquiring such data from direct field measurement, and therefore used remotely sensed information. They found a good correspondence between model predictions and the satellite data in terms of large-scale features, but found that this was less the case in terms of small-scale features. The main reason for this was the poor resolution of available input topographical data, which limited the model's ability to resolve these features. However, it was also due to the resolution of the satellite imagery which was greater than that of the model. Nevertheless, despite the development of such methods, data constraints restricted calibration and validation procedures for 2D schemes to those techniques developed for the previous generation of 1D models. These were based on the analysis of bulk flows and were clearly of limited suitability for distributed schemes. Increasingly, the potential now exists to develop distributed high resolution modelling driven by distributed high resolution data, although this must be tempered with the realization that for spatially distributed nonlinear models apparently simple external perturbations to the system may have effects that could not be predicted *a priori.* The chapter by Bates *et al.* (Chapter 9) provides an example of this effect where the 2D modelling discussed above is shown to have a spatially complex and nonlinear sensitivity to simple, uniform changes in topography, friction and grid resolution. This has clear implications for newly emergent distributed model calibration and validation techniques that we are only just beginning to explore.

Whilst comparison of model predictions with independently acquired data must be critical to any modelling effort, this is not always possible. One of the major reasons for this is a lack of available data sets suitable for the problems being investigated, which may itself be one of the reasons for adopting a modelling approach. Similarly, if there is divergence between model results and field data, then it is necessary to ask *which* is wrong. It is traditional to assume that the model results are incorrect, but verification of Computational Fluid Dynamics (CFD) code is not straightforward both because verification data may well be in error

(Lane *et al.*, 1999) but also because of differences between the nature of model predictions and independent observations. This is illustrated by Bradbrook *et al.* (Chapter 7) who compare model predictions of turbulent kinetic energy (TKE) with those from a measuring device such as an acoustic Doppler velocimeter (ADV). This is difficult because of the sensitivity of ADV estimates of TKE to the design of the instrument and decisions that are made after data collection as to the most appropriate data filtering option. Even assuming that it is possible to locate and orient a measurement instrument within the same local coordinate system as that which is being used to drive the model (Lane *et al.*, 1998), problems will arise when the instrument measuring volume is different in size from the grid size associated with model predictions. Whilst an inability to undertake some form of indepeendent comparison may raise philosophical uncertainties over whether or not a model's predictions may be relied upon, it is important to remember that there are other means by which a model's predictive ability may be assessed (e.g. Howes and Anderson, 1988).

One of the most basic checks of any model performance is mass conservation which is particularly useful for spotting basic model coding errors (Darby, Chapter 17). Stewart *et al.* (Chapter 12) show that checks on mass conservation are clearly critical where there is variable input and output of water, sediment or pollutants. Of equal importance are analytical or benchmark solutions (Darby, Chapter 17) for direct comparison with numerical solutions as well as comparisons of model predictions with laboratory results. However, when a model is applied to field situations, it is not uncommon for a model that is conserving mass and which has performed adequately in both benchmark and laboratory tests to produce different results from those that are measured, and this represents a particularly difficult situation. As Darby (Chapter 17) notes, there are three conditions that may result in this situation: (i) inadequate model structure; (ii) poor representation of initial boundary conditions or model parameters; and (iii) ptoblems with the numerical solution. Carefully constructed sensitivity analyses (e.g. Bates *et al.*; Lane and Richards; Chapters 9 and 5) may be particularly useful in such situations and can allow: (i) assessment that, in response to representative variation of model input parameter values and boundary conditions, theoretically realistic model behaviour is experienced (e.g. Howes and Anderson, 1988); (ii) assessment that a model is sufficiently sensitive to represent the actual variation in the prototype system (Howes and Anderson, 1988); (iii) identification of those input parameters and boundary conditions to which the model is most sensitive (Young *et al.*, 1971; McCuen, 1973; Beven, 1979; Howes and Anderson, 1988) and which therefore must be given detailed attention in terms of their acquisition, quality and associated error; (iv) improvement in model representation through limiting the sensitivity of a model component (McCuen, 1973) or by optimizing of a set of parameter values in the context of specific model behaviour (Decoursey and Snyder, 1969; McCuen, 1973; Rogers *et al.*, 1985; Calver, 1988; Beven and Binley, 1992); and (v) assignment of the likely magnitude of error in a model prediction that arises from a particularly specification, through combining an estimate of the error in parameter specification with the sensitivity of a particular prediction to that parameter (McCuen, 1973).

A standard means of model calibration involves (iv) above and was used by Stewart *et al.* (Chapter 12) who used roughness as a calibration parameter to obtain a good fit between observed and predicted flood hydrographs for floodplain flow models. Bates *et al.* (Chapter 9) provide an illustration of (iii) above. They found that bulk flow and inundation extent predictions were relatively insensitive to both friction parameterization and the level of microtopographical complexity included in the model. However, in terms of distributed hydraulic variables, both inundation depths and velocities showed complex behaviour in response to friction change in particular, and topographic complexity change to a lesser extent. These findings are useful as they show that what the model will be used for will determine the extent to which internal validation needs to be given careful consideration. For discharge hydrograph predictions, lower mesh and topographic complexity may be perfectly adequate. For distributed predictions within the model (e.g. Stewart *et al.*; Chapter 12), more attention will need to be given to topographic representation and mesh resolution. Lane and Richards use sensitivity analysis for the purpose of (i), showing how a depth-averaged rigid-lid model for flow in a gravel-bed stream displayed theoretically realistic behaviour. They also used it for (v) above, to produce surfaces of prediction confidence, which described the likely uncertainty in model predictions that arise from uncertain boundary condition specification, notably bed roughness. As Bates *et al.* (Chapter 9) and Lane and Richards (Chapter 5) show, if sensitivity analysis is applied to high-resolution models in two and three-dimensions, that it must be spatially distributed to capture the real nature of model sensitivity.

Equally important is the realization of results that are either intuitively sensible or which accord with previous results from field studies of different situations, particularly where verification data are not available for the study in question. In the case of Stewart *et al.* (Chapter 12) the deposition of suspended sediment was found to be more complex due to strong dependencies upon local topography and floodplain hydraulics. Pollutant deposition was much less complex, and related directly to infiltration patterns. Both of these suggest sensible behaviour in terms of what we know about what controls sediment and pollutant depositional processes.

MODEL APPLICATIONS

This volume is not simply about numerical model development. At the core of most chapters is an attempt to use numerical models to further understanding of particular research issues. Once Hodskinson and Ferguson have established that their model is capable of producing a good representation of flow processes in a meander bend, they use it to explore those morphological situations in which flow separation occurs in meander bend apices. They were able to confirm suggestions that outer curvature was a key influence upon flow separation, but the exact nature of the relationship was dependent upon the form of the inner bank point bar and the nature of curvature in the upstream bend. Bradbrook *et al.* use a similar approach, albeit for labratory-style confluences. Physical models of laboratory confluences are restricted in the range of boundary conditions that may be investigated due to both the time that it takes to set up and measure in a laboratory study, and limitations in the range of boundary conditions that may be imposed. Bradbrook *et al.* study a single confluence configuration, but then adjust both the momentum ratio and the level of discordance between the two channels. They were able to confirm laboratory observations (e.g. Biron *et al.*, 1993) that bed discordance may be a critical aspect of confluence dynamics, but that its role is conditioned by the tributary momentum ratio: strong secondary circulation was observed provided the depth differential was 25% or more, and as long as the velocity in the shallower tributary was at least as fast as in the deeper channel.

Coupled models are also often associated with particular practical applications. For instance, Stewart *et al.* use model predictions to provide basic information for refining data-sampling strategies for assessing the deposition of heavy metals on floodplains. Moulin and Ben Slama used their coupled flow and sediment transport model to evaluate the effect of a power station intake upon sedimentation processes and a point discharge of heavy metal laden sediment upon deposition in a 400 m reach of river.

CONCLUSIONS: RESEARCH CHALLENGES

What does emerge from the chapters in this book are a set of critical research challenges, some of which urgently need consideration, and many of which interconnect the themes identified above. Turbulence remains one of the most intractable questions for high resolution flow modelling. Research into natural channels at small scales has only recently begun to explore models that might represent the effects of turbulence-driven secondary circulation. Similarly, for situations where there is any form of shear, and consequent implications for mass and momentum transfer (such as at the boundary between in-channel and floodplain flows), 3D investigation, combined with exploration of turbulence model effects, are required to help improve the parameterization of 2D approaches. The move to three dimensions, with more sophisticated turbulence modeling, increases dependence upon specification of necessary boundary conditions. Without the latter, the extent to which real process representation can be achieved will be limited. Concurrently, whilst higher resolution modelling, with better process and boundary condition representation, may improve model predictions, this needs to be judged against questions of practicality and feasibility.

One aspect that this book does not address in detail, but which will no doubt become increasingly important in the immediate future, are formal attempts to deal with parameter and boundary condition uncertainty in models of this kind. If questions of practicality and feasibility mean that process or boundary condition representation must be restricted, it is important that some attempt is made to express the uncertainty in model predictions that results. Whilst uncertainty analysis is well established in hydrology, this is much less the case in terms of floodplain and in-channel studies, with some exceptions such as Romanowicz *et al.* (1996) and Aronica *et al.* (1998). This is not surprising, as there are many fewer parameters that have to be dealt with in 2D and 3D hydraulic models, the computational demands of these models have traditionally limited the ability to conduct large numbers of numerical simulations and only recently has progress been made in some areas of basic process representation. The latter is important as the information gained from an uncertainty analysis will be strongly dependent upon the processes that are included in the model and, as this volume shows, significant improvements in model performance may be achieved by improvements in process representation *per se.* However, the growth of high resolution approaches does require a greater resolution of boundary condition specification, which may not always be possible, and it is in this area where future research must begin to determine the causes and magnitudes of uncertainties in model predictions.

A related issue is model validation. Some journals (e.g. American Society of Mechanical Engineers, *Journal of Fluids Engineering*) have set basic standards which papers concerned with numerical modelling must meet, often in the form of an editorial policy statement. These were primarily designed to guide numerical solution issues, notably associated with truncation and convergence errors (ASME, 1993) but have been extended to include validation issues more generally. For instance, it is now necessary to specify the reliability that can be placed in validation data, and methods have been developed that allow comparison of model predictions and validation data, where uncertainty is associated with both. Coleman and Stern (1997) show how a poor level of correspondence between model predictions and validation data may actually be quite an encouraging level of fit when uncertainties in numerical solution and validation data are combined together.

A further area of concern is the extent to which CFD models, originally developed in mathematical or engineering disciplines, can be treated by environmental scientists as 'black boxes'. Model code and numerical solvers are now freely or cheaply available, and the computing power with which to undertake significant simulations can now be found on most academics' desk tops. However, there are perhaps good reasons why we too should get involved in model development. Ultimately, modelling is just another tool to enable us to conduct science. Established codes merely reflect established theory, and once the potential stemming from new applications to environmental problems has been exhausted, the black box approach may leave a deskilled academic community unable to conduct fundamental process investigations. A continued reliance on agencies outside hydrology and geomorphology to generate our basic scientific tools could eventually lead to a two-tier system of research. This will require a significant change in the training of environmental science researchers. Fortunately, a number of chapters in this issue illustrate that model development is beginning to be addressed. However, if interdisciplinary comparisons are made, the relative lack of development of high reslution flow modelling techniques in hydrology and geomorphology is a major problem.

Finally, the relative balance and focus of the chapters in the book illustrate the differential uptake of these techniques between the two disciplines covered. Geomorphology, particularly in Europe, has rapidly adopted CFD techniques as a standard methodology over the last decade. This reflects the subject areas where it was initially adopted and the journals in which the research was disseminated. Uptake of high resolution CFD techniques in hydrology on the other hand has been less marked, and it is instructive to note that the state of the art in distributed catchment modelling is still represented by the European Hydrological System (Système Hydrologique Européen, SHE) model which was developed in the 1980s (Abbott *et al.*, 1986). In hydrology the use of numerical modelling has become largely stalled following criticisms of the SHE-type approach starting with Beven (1989) and an apparent lack of willingness to adopt the uncertainty analysis framework that Beven's work suggested. This has not been true of the closely related field of groundwater modelling, where distributed numerical schemes with stochastic uncertainty are increasingly a standard technique. Considerable scope for the adoption of high resolution numerical modelling would therefore seem to exist in catchment hydrology and it is hoped that this volume may give impetus to the exploration of this potential.

REFERENCES

Abott, M. B., Bathurst, J. C., Cunge, J. A., O'Connell, P. E., and Rasmussen, J. 1986. 'An introduction to the European Hydrological System — Système Hydrologique Européen, 'SHE', 2. Structure of a physicaly based, distributed modelling system, *Journal of Hydrology*, **87**, 61–77.

Aronica, G., Hankin, B., and Beven, K. 1998. 'Uncertainty and equifinality in calibrating distributed roughness coefficients in a flood propagation model with limited data', *Advances in Water Resources*, **22**, 349–365.

ASME (American Society of Mechanical Engineers). 1993. '*Journal of Fluids Engineering* editorial policy statement on the control of numerical accuracy', *ASME Transactions, Journal of Fluids Engineering*, **115**, 340.

Bates, P. D., Anderson, M. G., Price, D. A., Hardy, R. J., and Smith, C. N. 1996. 'Analysis and development of hydraulic models for floodplain flows', Chapter 7 in Anderson, M. G., Walling, D. E., and Bates, P. D. (Eds), *Floodplain Processes*, Wiley, Chichester, pp. 215–254.

Beven, K. J. 1979. 'A sensitivity analysis of the Penman–Monteith actual evapotranspiration estimates', *Journal of Hydrology*, **44**, 169–190.

Beven, K. J. 1989. 'Changing ideas in hydrology: The case of physically based distributed models, *Journal of Hydrology*, **105**, 79–102.

Beven, K. J. and Binley, A. 1992. 'The future of distributed models: model calibration and prediction uncertainty', *Hydrological Processes*, **6**, 279–298.

Biron, P., De Serres, B., Roy, A. G., and Best, J. L. 1993. 'Shear layer turbulence at an unequal depth confluence' in Clifford, N. J., French, J. R., and Hardisty, J. (Eds), *Turbulence: Perspectives on Flow and Sediment Transport*. Wiley, Chichester, pp. 197–214.

Calver, A. 1988. 'Calibration, sensitivity and verification of an physically-based rainfall–runoff model', *Journal of Hydrology*, **103**, 103–115.

Coleman, H. W. and Stern, F. W. 1997, 'Uncertainties and CFD code validation, *ASME Transactions, Journal of Fluids Engineering*, **119**, 795–803.

Decoursey, D. G. and Snyder, W. M. 1969. 'Computer-oriented method of optimising hydrologic model parameters', *Journal of Hydrology*, **9**, 34–56.

Haycock, N. E. and Burt, T. P. 1993. 'Role of floodplain sediments in reducing the nitrate concentration of subsurface run-off: a case study in the Cotswolds, UK', *Hydrological Processes*, **7**, 287–295.

Howes, S. and Anderson, M. G. 1988. 'Computer simulation in geomorphology, in Anderson, M. G. (Ed.), *Modelling Geomorphological Systems*, Wiley, Chichester, pp. 421–440.

Lane, S. N. and Richards, K. S. 1997. 'Linking river channel form and process: time, space and causality revisited, *Earth Processes and Landforms*, **22**, 249–260.

Lane, S. N., Richards, K. S., and Chandler, J. H. 1994. 'Distributed sensitivity analysis in modelling environmental systems', *Proceedings of the Royal Society*, Series A, **447**, 49–63.

Lane, S. N., Biron, P. M., Bradbrook, K. F., Butler, J. B., Chandler, J. H., Crowell, M. D., McLelland, S. J., Richards, K. S., and Roy, A. G. 1998. 'Integrated three-dimensional measurement of river channel topography and flow processes using acoustic doppler velocimetry', *Earth Surface Processes and Landforms*, **23**, 1247–1267.

Lane, S. N., Bradbrook, K. F., Richards, K. S., Biron, P. A., and Roy, A. G. 1999. 'The application of computational fluid dynamics to natural river channels: three-dimensional versus two-dimensional approaches', forthcoming in *Geomorphology*.

McCuen, R. H. 1973. 'The role of sensitivity analysis in hydrologic modelling', *Journal of Hydrology*, **18**, 37–53.

Macklin, M. G. 1996. 'Fluxes and storage of sediment-associated heavy metals in floodplain systems: assessment and river basin management issues at a time of rapid environmental change', Chapter 13 in Anderson, M. G., Walling, D. E., and Bates, P. D. (Eds), *Floodplain Processes*. Wiley, Chichester, pp. 441–460.

Rogers, C. C. M., Beven, K. J., Morris, E. M., and Anderson, M. G. 1985. 'Sensitivity analysis, calibration and predictive uncertainty of the Institute of Hydrology Distributed Model', *Journal of Hydrology*, **81**, 179–191.

Romanowicz, R., Beven, K. J., and Tawn, J. 1996. 'Bayesian calibration of Flood inundation models', Chapter 10 in Anderson, M. G., Walling, D. E., and Bates, P. D. (Eds), *Floodplain Processes*, Wiley, Chichester, pp. 333–361.

Thomas, T. G. and Williams, J. J. R. 1995. 'Large eddy simulation of a symmetrical trapezoidal channel at a Reynolds number of 430,000', *Journal of Hydraulic Research*, **33**, 825–842.

Young, G. K., Tseng, M. T., and Taylor, R. S. 1971. 'Estuary water temperature sensitivity to meteorological conditions, *Water Resources Research*, **7**, 1173–1181.

SECTION I

THEORETICAL ADVANCES

1

HYDRAULIC MODELLING IN HYDROLOGY AND GEOMORPHOLOGY: A REVIEW OF HIGH RESOLUTION APPROACHES

STUART N. LANE*

Department of Geography, University of Cambridge, Downing Place, Cambridge CB2 3EN, UK

ABSTRACT

This paper will introduce the basic principles associated with hydraulic modelling of surface waters for geomorphological and hydrological purposes and illustrate how these have been applied to specific problems. The basic principles governing fluid flow are derived from the principles of conservation of mass and momentum. In the case of the shallow flow problems that typify most geomorphological and hydrological contexts, these equations involve some modifications: (i) as the boundary layer is likely to extend throughout the flow depth, it is possible to assume a hydrostatic pressure distribution; (ii) special conditions need to be determined for both the bottom and water surface, including the possibility of horizontal gradients of atmospheric pressure for large-scale applications, and wind stress; and (iii) it is generally permissible to ignore the Coriolis terms. Direct application of the resultant equations is complicated by the need to Reynolds-average, which introduces additional terms but no additional equations. These terms have to be determined through empirical or semi-empirical transport equations, usually termed turbulence models. Current applications of these equations to geomorphological and hydrological applications are reviewed. Applications to river channels have generally not made use of the full three-dimensional form of the governing equations, and have either been one-dimensional, or, more commonly, two-dimensional. The latter involves depth averaging of the governing equations but requires parameterization of the effects of secondary circulation upon the transport of momentum. This has emphasized secondary circulation generated by curvature of the depth-averaged streamlines, but has yet to address secondary circulation associated with topographic discordance at river channel confluences or diffluences or owing to anisotropic turbulence. Applications to unsteady flows require special attention to be given to the effects of spatial and temporal variation in the depth of inundation, and the associated treatments are reviewed.

KEY WORDS computational fluid dynamics; rivers; hydraulic modelling; floodplains; turbulence modelling

INTRODUCTION

Recent years have witnessed a considerable increase in the use of numerical models of flow in geomorphology and hydrology. The benefits of numerical modelling in this context are clear: numerical methods may allow both improved understanding and improved simulation of key processes, providing two- or three-dimensional prediction fields, which can provide considerable insight into the distribution of processes within units of the landscape. Indeed, concomitant with these benefits is the growth of more powerful computer capacity, and the availability of a growing range of commercial and academic

* Correspondence to: Stuart N. Lane, Department of Geography, University of Cambridge, Downing Place, Cambridge, CB2 3EN, UK

computational fluid dynamics software. This is reflected in geomorphological and hydrological applications, where increasingly sophisticated modelling methods are employed, incorporating progressively more physical detail.

Given the extensive research in computational fluid dynamics, it is unlikely that basic development of such methods by geomorphologists and hydrologists will be either practical or necessary. However, there is scope for further research in the application of these tools to specific geomorphological and hydrological problems, whose uniqueness is a product of the interaction of the operation of the fundamental physics underlying these models, with time- and space-specific boundary conditions (Richards, 1996). However, if geomorphologists and hydrologists are to use these models effectively, often without specific training in computational fluid dynamics methods, it is critical that they are aware of the physical basis of the models they use, and of the inherent limitations of the assumptions made during model development. Thus, the aim of this paper is to establish a context for the remaining papers in this issue through a review that links the basic mathematical principles fundamental to numerical modelling of open channel flow processes, with the large literature that describes their application to particular geomorphological and hydrological problems. The material presented is confined to that which is relevant to the free surface flow of water, and explicitly aims to review the state of *current* research. Three key themes are identified as being relevant to geomorphological and hydrological problems: (i) how to describe and model the effects of turbulence on the transport of momentum; (ii) the relative importance of solutions of different dimensionality; and (iii) the frequent need to consider unsteady rather than steady flow problems.

BASIC PRINCIPLES

Derivation of the basic equations that govern shallow open channel flow begins with the assumption that a fluid can be considered as a continuum (Tritton, 1988), where the details of the molecular structure only appear in constituent equations that recognize the effects of molecular processes on the overall flow characteristics. Although Newton's laws of motion provide a basis for the study of motion of particles or rigid bodies, they need modification for the application to the continuum that represents a fluid (Euler, 1707–1783). The Eulerian transformation of Newton's laws begins by considering a unit of fluid of some arbitrary volume (the Euler cut principle, after Tritton, 1988). Through time, this fluid unit will move through space, and the constitution of the unit will itself change.

Using these principles it is possible to obtain two key equations (White, 1974) for a single fluid unit from Newton's laws of motion: (i) the law of conservation of mass for an incompressible fluid in Eulerian form

$$\frac{\partial u}{\partial x} + \frac{\partial v}{\partial y} + \frac{\partial w}{\partial z} = 0 \tag{1}$$

where u, v and w are the components of velocity in the x, y (planform) and z (vertical) directions respectively; and (ii) the Navier–Stokes momentum equations for an incompressible fluid

$$\frac{\partial}{\partial t}(\rho u) + \frac{\partial}{\partial x}(\rho u^2) + \frac{\partial}{\partial y}(\rho uv) + \frac{\partial}{\partial z}(\rho uw) - 2\rho u\zeta \sin\Phi + \frac{\partial p}{\partial x} - \frac{\partial \tau_{xx}}{\partial x} - \frac{\partial \tau_{xy}}{\partial y} - \frac{\partial_{xz}}{\partial z} = 0 \tag{2a}$$

$$\frac{\partial}{\partial t}(\rho v) + \frac{\partial}{\partial x}(\rho uv) + \frac{\partial}{\partial y}(\rho v^2) + \frac{\partial}{\partial z}(\rho vw) - 2\rho v\zeta \sin\Phi + \frac{\partial p}{\partial y} - \frac{\partial \tau_{xy}}{\partial x} - \frac{\partial \tau_{yy}}{\partial y} - \frac{\partial \tau_{yz}}{\partial z} = 0 \tag{2b}$$

$$\frac{\partial}{\partial t}(\rho w) + \frac{\partial}{\partial x}(\rho uw) + \frac{\partial}{\partial y}(\rho vw) + \frac{\partial}{\partial z}(\rho w^2) + \rho g + \frac{\partial p}{\partial z} - \frac{\partial \tau_{xz}}{\partial x} - \frac{\partial \tau_{yz}}{\partial y} - \frac{\partial \tau_{zz}}{\partial z} = 0 \tag{2c}$$

where ρ is the density of water, ζ is the angular rotation of the earth, Φ is latitude, p is pressure, g is the acceleration due to gravity,

$$\frac{\tau_{ij}}{\rho} = \mu\left(\frac{\delta v_j}{\delta x_i} + \frac{\delta v_i}{\delta x_j}\right)$$

and μ is the coefficient of viscosity of for a Newtonian fluid (the molecular viscosity).

For the shallow open channel flow problems in which geomorphologists are commonly interested, these equations need some modifications and additions. First, in the case of shallow water, it is generally assumed that the vertical scale is much smaller than the horizontal scale and the boundary layer extends throughout the water depth. This allows Equation (2c) to be simplified to the hydrostatic pressure distribution [see Rodi *et al.* (1981) or Vreugdenhill (1994) for justification]

$$\frac{\partial p}{\partial z} = -\rho g \tag{3}$$

Assuming density is constant with depth, and integrating over depth, Equation (3) becomes:

$$p = \rho g(h + z_{\mathrm{b}}) \tag{4}$$

where h is the water depth and z_{b} is the bottom elevation. This allows the pressure gradients in Equations (2a) and (2b) to be redefined, and, assuming no horizontal variation in density, Equations (2a) and (2b) become

$$\frac{\partial}{\partial t}(u) + \frac{\partial}{\partial x}(u^2) + \frac{\partial}{\partial y}(uv) + \frac{\partial}{\partial z}(uw) - 2u\Omega \sin\Phi + g\frac{\partial h}{\partial x} + g\frac{\partial z_{\mathrm{b}}}{\partial x} + \frac{1}{\rho}\frac{\partial \tau_{xx}}{\partial x} - \frac{1}{\rho}\frac{\partial \tau_{xy}}{\partial y} - \frac{1}{\rho}\frac{\partial \tau_{xz}}{\partial z} = 0 \tag{5a}$$

$$\frac{\partial}{\partial t}(v) + \frac{\partial}{\partial x}(uv) + \frac{\partial}{\partial y}(v^2) + \frac{\partial}{\partial z}(vw) - 2v\Omega \sin\Phi + g\frac{\partial h}{\partial y} + g\frac{\partial z_{\mathrm{b}}}{\partial y} + \frac{1}{\rho}\frac{\partial \tau_{xy}}{\partial x} - \frac{1}{\rho}\frac{\partial \tau_{yy}}{\partial y} - \frac{1}{\rho}\frac{\partial \tau_{yz}}{\partial z} = 0 \tag{5b}$$

Secondly, it is necessary to recognize that there will be special conditions at both the (free) water surface and the bottom, since water particles will be unable to cross either boundary. At the bottom, assuming that it is solid, the normal velocity components to the bottom must disappear, that is

$$u\frac{\partial z_{\mathrm{b}}}{\partial x} + v\frac{\partial z_{\mathrm{b}}}{\partial y} - w = 0 \tag{6}$$

where z_{b} is the bottom elevation. If it is also assumed that at the bottom there is a no-slip condition, u and v must also be zero. This assumption is important because it requires determination of a bottom stress, which by definition is determined from the properties of the flow field above and around the point of interest.

At the surface, the normal velocity component must also vanish, but this can be achieved by movement of the water surface. This introduces a time derivative

$$\frac{\partial h}{\partial t} + u\frac{\partial h}{\partial x} + v\frac{\partial h}{\partial y} - w = 0 \tag{7}$$

In practice, at the water surface, it may be necessary to introduce a number of additional modifications. First, if the equations are being applied to problems over which there might be spatial variation in atmospheric pressure (p_{a}), generally large-scale simulations, it may be necessary to introduce additional horizontal pressure gradient terms into Equations (5a) and (5b): $\partial p_{\mathrm{a}}/\partial x$ and $\partial p_{\mathrm{a}}/\partial y$, respectively. In the case of geomorphological applications concerned with river channels and floodplains, these can be ignored. In the case of large-scale floodplain flow simulations, it will be necessary to introduce wind stress (τ_{wi}) terms at the water surface. These are determined by empirical relationships and exist in a wide variety of forms, such as

$$\tau_{wi} = \rho_{air} c_f W_i^2 \tag{8}$$

where c_f is drag coefficient and W_i is the wind velocity in the i direction.

Finally, and following from the general concern of geomorphologists with small-scale applications to river channel systems, it is generally acceptable to ignore the terms associated with the Coriolis effect in Equations (2) and (5) ($-2\rho u\Omega \sin \Phi$ and $-2\rho v\Omega \sin \Phi$).

REYNOLDS AVERAGING AND TURBULENCE CLOSURE

It is possible to show that shallow open channel flow satisfies, at every point in space, Equations (1) and (5), modified in the vicinity of boundaries by Equations (6)–(8) and horizontal atmospheric pressure gradient terms where necessary. It follows that it only remains to specify the appropriate boundary conditions (such as the channel topography) to be able to model a turbulent river flow. However, direct application of these equations results in a computational problem. To solve the equations with sufficient accuracy a three-dimensional computational grid is required. This must be large enough to cover the area of interest but with spacings smaller than the smallest turbulent motion. The computations must be unsteady, utilizing a time step smaller than that associated with the fastest eddies. The number of grid nodes required shows a Reynolds number dependence (Younis, 1992). Thus, Emmons (1970) calculated that for the relatively simple problem of steady flow in a pipe with a Reynolds number of 10^7, 10^{22} numerical operations would be required. For such a simple problem, and assuming a computer operating at 160 MHz, this would take just under 2 million years.

In computational fluid dynamics, therefore, the engineer uses one of two solutions to this problem: (i) a statistical theory of turbulence making use of empirical correlation between observations of the structure of turbulent flow; or (ii) or a semi-empirical analysis of the *effects* of turbulent motions on the *mean* flow properties. The latter is the most commonly used treatment of the turbulence problem and involves a technique called Reynolds (1895) averaging. This assumes that each variable (v_i) can be split into a slowly varying mean value ($\overline{v_i}$) and a random variation v_i') about it

$$v_i = \overline{v_i} + v_i' \tag{9}$$

Reynolds averaging and depth averaging have no effect on the structure of Equation (1), the conservation of mass equation, but introduce a number of new terms into Equations (5), the momentum equations. These are called the turbulent or Reynolds shear stresses ($\overline{v_i v_j}$) and stress terms in Equation (5) are redefined as

$$\frac{\tau_{ij}}{\rho} = \mu\left(\frac{\delta v_j}{\delta x_i} + \frac{\delta v_i}{\delta x_j}\right) - \overline{v_i v_j} \tag{10}$$

The Reynolds shear stresses represent the transport of momentum that can be attributed to turbulence; they can never be zero in a turbulent flow. In the turbulent flows associated with natural channels, they tend to be much larger than their molecular (laminar) counterparts (Rodi, 1980). Their introduction adds a total of four new terms (two in each equation), which can only be defined by (unavailable) knowledge of the turbulent structure because no additional equations have been introduced to solve for these. The Reynolds-averaged equations do not close (Olson and Wright, 1990), and there are no further physical laws to resolve this dilemma (White, 1974). As there is no direct way of obtaining the Reynolds shear stresses, the only alternative is to model their consequences using flow parameters that are either known or knowable

(Younis, 1992); such a model generally consists of a set of differential equations that allow the Reynolds-averaged Navier–Stokes equations to become closed.

Thus far, in the application of hydraulic models to geomorphological and hydrological problems, most turbulent models have been based on a narrow range of options, based on a Boussinesq (1877) assumption, which uses a mixture of physical reasoning and dimensional analysis (Dreissler, 1977). An analogy is made between molecular motion, which leads to Stokes' viscosity law in laminar flow, and turbulent motion (Rodi, 1980). Boussinesq proposed that the Reynolds shear stresses may be modelled as being proportional to the mean rates of strain, the proportionality coefficient (ν_{ij}) then defining the eddy viscosity coefficient (Younis, 1992):

$$-\overline{v_i v_j} = \nu_{ij}\left(\frac{\partial v_j}{\partial x_i} + \frac{\partial v_i}{\partial x_j}\right) \tag{11}$$

Turbulent eddies are thought of as lumps of fluid that, like molecules, collide and exchange momentum (Rodi, 1980). Molecular viscosity is proportional to the average velocity and the mean free path of the molecules. Hence the eddy viscosity (ν_{ij}) is considered as proportional to a velocity, and a length that characterize the fluctuating motion. Whereas the molecular viscosity is a property of the fluid, the eddy viscosity is a property of the flow, and it may not be regarded as a constant (Olson and Wright, 1990). If this were the case, the velocity profile for turbulent flow would be identical to that for laminar flow (Olson and Wright, 1990) and the flow, by definition, would not be turbulent.

In principle (Corrsin, 1957; Bradshaw, 1969), the analogy between molecular and turbulent motion cannot be correct because the turbulent eddies are not rigid bodies and their identity changes, through time. Nevertheless, most current modelling strategies in alluvial channels make use of the Boussinesq (1877) approximation. The simplest applications of the Boussinesq approximation have involved a 'zero-equation' turbulence model (e.g. Nelson and Smith, 1989a,b; Shimizu *et al.*, 1990; Bates, 1992; Bridge, 1992), which either specifies a constant eddy viscosity or determines the eddy viscosity using a mixing length hypothesis. The former is difficult since the eddy viscosity concept has a poor physical basis and the viscosity is therefore difficult to measure in the field. The latter approach is based on Prandtl (1925); here, the eddy viscosity is thought to depend upon an average fluctuating velocity and a mixing length. It is possible to specify simple empirical formulae for the mixing length. For instance, Schlichting (1955) proposed that for the zone of flow near to the boundary, the mixing length would be the product of water depth and Von Karman's constant and this has seen widespread application to river channel hydraulics problems, and particularly field monitoring. For instance, it provides the basis of the 'law of the wall' model (Carson, 1971) which is frequently used to measure the bed shear stress within alluvial channels (e.g. Ashworth and Ferguson, 1986).

The main problem with such zero-equation models is that they implicitly assume that turbulence is dissipated where it is generated (Rodi, 1980). They do not account for the transport of turbulence by the mean flow and, in particular, for situations where the state of turbulence at a point is influenced significantly by turbulence generation elsewhere in the flow. One- and two-equation models overcome this limitation by accounting for the transport of turbulence quantities through solving additional differential transport equations. Rodi (1980) notes that if the velocity fluctuations associated with turbulence are to be characterized by a single scale, the physically most meaningful is the square root of the kinetic energy of the turbulent motion per unit mass ($\sqrt{k}$). This kinetic energy will mainly be contained within the large-scale velocity fluctuations and hence $\sqrt{k}$ is a velocity scale for the large-scale turbulent motion. The eddy viscosity is determined from

$$\nu_{ij} = c'_{\mu}\sqrt{k_{ij}}\,l_{ij} \tag{12}$$

where c'_{μ} is an empirical constant determined from experiment and l_{ij} is a length scale. Kolmogorov (1942) and Prandtl (1945) both argued that it may be possible to determine the distribution of k by solving a transport equation for this quantity of the form

$$\frac{\partial k}{\partial t} + v_i \frac{\delta k}{\delta x_i} = \frac{\delta}{\delta x_i}\left(\overline{v_i'\left(\frac{v_i'v_j'}{2} + \frac{P}{\rho}\right)}\right) - \overline{v_i'v_j'}\frac{\delta v_i}{\delta x_j} - v_{ij}\left(\overline{\frac{\delta v_i}{\delta x_j}\frac{\delta v_i}{\delta x_j}}\right) \tag{13}$$

rate of change of k	convective transport owing to mean motion	diffusive transport owing to velocity and pressure fluctuations	transfer of k from the mean to the turbulent motion through shear	transfer of k into the internal energy of the fluid through viscous dissipation

Comparison with the zero-equation models is useful; if the rate of change, convective transport and diffusive transport terms in Equation (13) are zero then the problem simplifies to a zero-equation form. Equation (13) *per se* is of no use (Rodi, 1980) because it introduces yet more unknown correlations in the diffusive transport and viscous dissipation terms. In practice, the diffusive transport is assumed to be proportional to the gradient of k, and the viscous dissipation to both k, and a length scale

$$\left(\overline{v_i'\left(\frac{v_i'v_j'}{2} + \frac{P}{\rho}\right)}\right) = \frac{v_{ij}}{\sigma_k}\frac{\delta k}{\delta x_i} \tag{14}$$

$$v_{ij}\left(\overline{\frac{\delta v_i}{\delta x_j}\frac{\delta v_i}{\delta x_j}}\right) = C_D \frac{k^{3/2}}{l_{ij}} \tag{15}$$

Equation (15) is logical, as the rate of dissipation is generally governed by the large-scale motion, characterized by k, and l, even though the dissipation takes place at the smallest eddies (Rodi, 1980). The empirical constants (σ_k, c_μ' and C_D) have been determined experimentally (Launder and Spalding, 1972).

One-equation models are a significant improvement over zero-equation models in that they endeavour to account for convective and diffusive transport of the turbulent velocity scale. Their main problem, however, is calculation of the length scale (l_{ij}). Two-equation models argue that the length scale is also subject to transport processes and can be modelled in a similar way to the velocity scale (Rodi, 1980). The length scale would be expected to show some spatial variation; eddy size will change with transport downstream; dissipation will destroy the small eddies and therefore effectively increase the eddy size; and vortex shedding connected with the energy cascade will also reduce the eddy size. It is possible to specify an exact transport equation for l_{ij}, derived from the Navier–Stokes equations (Tennekes and Lumley, 1972), but this equation contains complex correlations whose behaviour is little known and for which drastic model assumptions must be introduced in order to make the equations tractable (Rodi, 1980). However, the equation has a similar structure to the k, equation, Equation (13), containing terms that represent the rate of change of l_{ij}, convection, diffusion, generation of vorticity owing to vortex stretching across the energy cascade and viscous destruction of vorticity. The diffusion term is modelled using the gradient assumption, see Equation (14), but the generation and destruction terms require a more empirical analysis in which the two are treated together (Rodi, 1971). The resultant equation for the length scale, for flow in shallow water situations (i.e. ignoring buoyancy) is

$$\frac{\delta l_{ij}}{\delta t} + v_i \frac{\delta l_{ij}}{\delta x_i} = \frac{\delta}{\delta xi}\left(\frac{v_{ij}}{\sigma_1}\frac{\delta l_{ij}}{\delta x_i}\right) - C_{1\varepsilon}\frac{l_{ij}}{k}(P+G) - C_{2\varepsilon}\frac{l_{ij}^2}{k} \tag{16}$$

rate of change of l_{ij}	convective transport owing to mean motion	diffusive transport owing to velocity and pressure fluctuations	generation of l_{ij} minus destruction of l_{ij}

where G is a buoyancy production term.

Table I. Values of empirical constants in a two-equation κ–ε model, as determined from grid turbulence experiments

c'_u	σ_k	σ_1	$C_{1\varepsilon}$
0·09	1·0	1·3	1·44

Definition of equations for both k and l_{ij} leads to one final problem with respect to the two-equation model; choice of the appropriate constants (σ_1, $C_{1\varepsilon}$, $C_{2\varepsilon}$). As for the one-equation situation, this is done empirically and Launder and Spalding (1974), after computer optimization, suggest the values shown in Table I.

Application of these more sophisticated models to geomorphological and hydrological problems is less common, although Lane *et al.* (1995) and Lane and Richards (1998, this issue) made use of a two-equation code in applying a hydraulic model to a braided channel. The model has been extensively tested, and numerous modifications have been developed for particular applications. For instance, Bradbrook *et al.* (1998, this issue) and Hodkinson and Ferguson (1998, this issue) make use of a modified two-equation code, based upon renormalized group (RNG) theory, for the special cases of separated flows over a bed step and in the puter bend of a meander, respectively. Bradbrook *et al.* found that the RNG modification resulted in significant improvement in the correspondence between model predictions and laboratory observations.

This aside, there remain serious problems with the two-equation model, notably the model's dependence upon the validity of the Boussinesq approximation, and its inability to predict the occurrence of turbulence driven secondary circulation (see below). One alternative is to avoid the Boussinesq approximation by determining the Reynolds stresses indirectly or directly from their own, individual, differential transport equations (Younis, 1992). There are two main approaches to this. The first, the algebraic stress model, simplifies the transport equations for individual Reynolds stresses to algebraic expressions (e.g. Naot and Rodi, 1982) which are then determined as functions of the production, convection, diffusion and dissipation of turbulent kinetic energy. Although such a treatment is able to describe some of the types of motion that models involving a Boussinesq approximation fail to represent, the algebraic expressions are determined using eddy–viscosity relationships that are more characteristic of the two-equation models described above, and which contain additional empirical terms that are not a consequence of rigorous derivation, but are intended to account for the influence of turbulence gradients on the secondary motion (Cokljat and Younis, 1995). The second approach, the Reynolds stress model, derives the Reynolds stresses directly from their own differential transport equations. This has the advantage that some of the terms in these equations are exact, and, fortuitously, it is these terms that tend to be most influential (Younis, 1992). However, these models are more computationally demanding and as yet have only been applied to extremely simple problems in rectangular ducts and channels (e.g. Malin and Younis, 1990) and, most recently, compound regular channels (Cokljat and Younis, 1995).

These observations must be put into context. For example, the American Society of Civil Engineers Task Committee on Turbulence Models in Computational Hydraulics (ASCE, 1988), noted that much of the development of more sophisticated turbulence closures was proceeding without due consideration to the extent to which such sophistication was required, particularly with reference to natural alluvial channels. For instance, although the two-equation model described above may reduce the need for guesswork or experience in the specification of a uniform scalar eddy viscosity, it may be the case that other uncertainties associated with model application (e.g. determination of boundary conditions) are more significant. A study by Tingsanchali and Maheswaran (1990) found only small differences between flow field predictions with (i) different values of a uniform scalar eddy viscosity, and (ii) with a two-equation modelled viscosity.

Furthermore, it is almost certainly the case that turbulence has an important effect on the transport of momentum, and therefore on mean flow properties in the contexts relevant to the geomorphologist (e.g. Cokljat and Younis, 1995). The problem is not the importance of turbulence, but rather the way in which models based on the Boussinesq approximation, which has dominated current geomorphological and hydrological applications, describe the effects of turbulence on the mean flow properties. It is this that stimulates further interest in the algebraic stress and Reynolds stress turbulence models, most notably in critical areas of flow, such as the channel–floodplain transition and zones of flow convergence and divergence.

DIMENSIONALITY, DEPTH AVERAGING AND SECONDARY CIRCULATION

One-dimensional treatment of the governing equations has a long history in both hydraulic engineering, geomorphology and hydrology, and such models are still being used in certain circumstances (e.g. Bhallamudi and Chaudhury, 1991; Correia *et al.*, 1992; Niekerk *et al.*, 1992) often at quite small spatial scales (e.g. Keller and Florsheim, 1993; Carling and Wood, 1994). However, two-dimensional, depth-averaged models have been extensively described in the engineering literature for some time (e.g. Rodi *et al.*, 1981) and are increasingly being used in geomorphological problems (e.g. Dietrich and Smith, 1983; Odgaard and Bergs, 1988; Nelson and Smith, 1989a,b; Bridge, 1992; Bridge and Gable, 1992; Bates and Anderson, 1993).

Two-dimensional treatments begin by depth averaging Equations (1) and (5), with the aim of calculating the horizontal distribution of the depth-average velocity components, $\bar{U}$ and $\bar{V}$ (Rodi *et al.*, 1981), where capital letters indicate depth-averaged quantities and overbars indicate time-averaged quantities. These are defined by

$$V_i = \frac{1}{h - z_b}\int_{z_b}^{h} v_i \, \partial z \tag{17}$$

Applying Equation (17) to Equation (1), the depth-averaged form of the law of conservation of mass becomes;

$$\frac{\partial h}{\partial t} + \frac{\partial}{\partial x}[(h - z_b)\bar{U}] + \frac{\partial}{\partial y}[(h - z_b)\bar{V}] = 0 \tag{18}$$

The depth-averaged forms of the Navier–Stokes momentum equations become, for each of the *x*- and *y*-components, and ignoring the Coriolis terms

x-component momentum equation

$$\begin{aligned}\frac{\partial}{\partial t}(\bar{U}) + \frac{\partial}{\partial x}(\bar{U}^2) + \frac{\partial}{\partial y}(\bar{U}\bar{V}) = &- g\frac{\partial h}{\partial x} - g\frac{\partial z_b}{\partial x} + \frac{1}{\rho}\frac{\partial}{\partial x}(\overline{\tau_{xx}}) + \frac{1}{\rho(h - z_b)}\frac{\partial}{\partial y}(\overline{\tau_{xy}}) - \frac{\tau_{bx}}{\rho(h - z_b)} \\ &+ \frac{1}{\rho}\frac{\partial}{\partial x}\int_{z_o}^{h}\rho(\bar{u} - \bar{U})^2 \, \partial z + \frac{1}{\rho(h - z_b)}\frac{\partial}{\partial y}\int_{z_o}^{h}\rho(\bar{u} - \bar{U})(\bar{v} - \bar{V}) \, \partial z\end{aligned} \tag{19}$$

y-component momentum equation

$$\begin{aligned}\frac{\partial}{\partial t}(\bar{V}) + \frac{\partial}{\partial x}(\bar{U}\bar{V}) + \frac{\partial}{\partial y}(V^2) = -g\frac{\partial h}{\partial y} - g\frac{\partial z_b}{\partial y} + \frac{1}{\rho}\frac{\partial}{\partial x}(\overline{\tau_{xy}}) + \frac{1}{\rho(h - z_b)}\frac{\partial}{\partial y}(\overline{\tau_{yy}}) \\ - \frac{\tau_{by}}{\rho(h - z_b)} + \frac{1}{\rho(h - z_b)}\frac{\partial}{\partial x}\int_{z_o}^{h}\rho(\bar{u} - \bar{U})(\bar{v} - \bar{V}) \, \partial z + \frac{1}{\rho(h - z_b)}\frac{\partial}{\partial y}\int_{z_o}^{h}\rho(\bar{v} - V)^2 \, \partial z\end{aligned} \tag{20}$$

Equations (19) and (20) may be simplified further. They are specified in their elliptical form. If downstream events cannot influence the flow upstream, apart from the water level, the equations can be simplified to their parabolic form. In this situation, the $\partial/\partial x$ terms are much smaller than the $\partial/\partial y$ terms and $\bar{V}$ is much less than $\bar{U}$. Thus, transport of momentum by dispersion in the x-direction is minimal (Rodi *et al.*, 1981) and the associated terms can be ignored.

Depth averaging results in the introduction of two new groups of terms: (i) the bottom stresses (τ_{bx} and τ_{by}); and (ii) the dispersion terms [the last two terms in Equations (19) and (20)].

Bottom stresses

In turbulent flow, the bottom stresses are assumed to be a quadratic function of the depth-averaged velocity

$$\overline{\tau_{bi}} = C_f \rho \overline{V_i}(\overline{V_i^2} + \overline{V_j^2})^{1/2} \tag{21}$$

where C_f is a friction coefficient, the Darcy–Weisbach friction factor. Langbein (1966) provides justification for this by using a random walk model for particles of water. He was able to show that channel resistance (head loss per unit area) varies as the square of velocity. Flume studies (Leopold *et al.*, 1960) also suggest a linear relationship between boundary shear stress and the square of velocity. Equation (21) is used extensively in depth-averaged models in river research (e.g. Dietrich and Whiting, 1989; Nelson and Smith, 1989a,b; Shimizu and Itakura, 1989; Struiksma and Crosato, 1989; Shimizu *et al.*, 1990; Tingsanchali and Mahewswaran, 1990; Yeh and Kennedy, 1993).

However, it is important to investigate the assumption behind Equation (21). Its origins can be traced to Chezy (1775), and it is based upon an assumption that there is uniform flow. As application of these equations generally involves discretization of the channel or floodplain, this is not a problem, as the assumption is limited to each grid cell, and does not apply to the channel as a whole, where convergence and divergence of the depth-averaged streamlines at the channel scale will cause spatial variation between cells in the magnitude of bottom stress, secondary circulation and turbulence transfer of momentum. The key issue, however, is the slope of the relationship, which defines the channel roughness [essentially C_f in Equation (21)]. Leopold *et al.* (1960) found that the magnitude of the roughness coefficient increased as one moved from straight channels to sinuous channels, different roughness components being additive in subcritical flow. Use of the square law relationship can only be justified if there is an acceptable description of the roughness coefficient.

There are a number of possibilities for expression of the roughness coefficient. A number of approaches have expressed it in terms of Manning's n.

$$C_f = gn^2 h^{-1/3} \tag{22}$$

Uniform values of Manning's n have been used by Shimizu and Itakura (1989), Tingsanchali and Mahewswaran (1990) and Shimizu *et al.* (1990). However, most notably in coarse-grained channels, C_f will be spatially variable, reflecting spatial variation in grain size and sedimentological structure. An alternative method (e.g. Dietrich and Whiting, 1989) involves specification of a roughness height, but both n and roughness height specifications demand information on grain size. Thus, how the roughness coefficient is determined is less important than the quality of information used during determination. There may or may not be some attempt to correct estimates based on grain size for the increase in roughness associated with sedimentological structures, sediment transport or the form of drag associated with reach-scale topography, such that friction is often used as one of the key calibration parameters. In general (e.g. Dietrich and Whiting, 1989; Nelson and Smith, 1989b; Struiksma and Crosato, 1989), spatial variations in the roughness coefficient have tended to be ignored, and a characteristic grain size for a reach is specified. This is clearly an area in which further research is required.

Dispersion terms and secondary circulation

The dispersion terms in Equations (19) and (20) are a product of vertical non-uniformity in the velocity field (which is inevitable because the velocities tend to fall to zero at the channel bottom) and represent deviations from the depth-averaged velocities within a vertical profile. Rodi (1980) and Rodi *et al.* (1981) note that the physical meaning of these dispersion terms is similar to the turbulent stress terms (see above), in that both represent gradients of transport of momentum. From measurements of depth-averaged quantities it is usually not possible to distinguish between the turbulent and dispersion contributions to the momentum transport. As with the Reynolds shear stresses, no additional equations arise during the depth-averaging process and determination of these terms requires either knowledge of the secondary flow field or a model of their effects on the mean flow properties.

Initial research suggested that the dispersion terms would be minimal in boundary layer-type flows (e.g. McGuirk and Rodi, 1978; Rastogi and Rodi, 1978). However, this is contrary to field evidence from natural channels which indicates that the dispersion terms can be important, associated with helical secondary flow owing to either channel curvature (e.g. meander bends; Hey and Thorne, 1975) or to the junction of two channels (Smith, 1974; Church and Jones, 1982; Ashmore *et al.*, 1992). If no secondary flow correction is made, then depth-averaged models will tend to constrain the largest velocities to lie near the inside of channel bends (Bernard and Schneider, 1992) instead of predicting gradual velocity migration to the outside, as observed in the bends of many natural channels (e.g. Dietrich and Whiting, 1989).

As with the treatment of turbulence closure, most analyses undertake a semi-empirical or empirical treatment of the dispersion terms to model their effects on the mean transport of momentum and hence the depth-averaged flow velocities. Two key methods exist. The first is more analytical (e.g. Johannesson and Parker, 1989a), and combines the two-dimensional forms of the mass and momentum equations with an empirical shape function for the expected distribution of *primary* velocity through the vertical to provide a credible representation of secondary flow for small curvature problems (Bernard and Schneider, 1992). This method is often described as taking the moment of momentum and is used extensively (e.g. Jin and Steffer, 1993; Yeh and Kennedy, 1993; Johannesson and Parker, 1989a,b,c; Nelson and Smith, 1989b). For the central region of a wide channel (Johannesson and Parker, 1989b)

$$\begin{aligned} \bar{u} &= \bar{U}(x, y)T(\zeta) \\ \bar{v} &= \bar{V}(x, y)T(\zeta) + v(x, y, \zeta) \end{aligned} \tag{23}$$

where v is the transverse helical flow, ζ is the ratio of elevation above the bed to total flow depth $[z/(h - z_b)]$, $T(\zeta)$ is dimensionless velocity shape function, and the x- and y-directions are defined as the streamwise and transverse directions, respectively. The depth averages of $\bar{u}$ and $\bar{v}$ are equal to $\bar{U}$ and $\bar{V}$, and therefore

$$\int_0^1 T(\zeta)\, \mathrm{d}\zeta = 1 \quad \text{and} \quad \int_0^1 \bar{v}(\zeta)\, \mathrm{d}\zeta = 0 \tag{24}$$

This allows the dispersion terms in the depth-averaged equations to be expressed as a function of T, which in turn is some function of the shape of the vertical variation of the primary flow (e.g. Engelund, 1974). For linear order in curvature, the transverse momentum balance reduces to a state that allows immediate solution of v, and which can then be applied to the depth-averaged equations for streamwise momentum and mass conservation.

However, Bernard (1989) notes that in situations with irregular lateral boundaries, non-uniform depth or non-uniform curvature, both the suitability and the applicability of the Johannesson and Parker model is questionable. The method requires specification of the depth-averaged mass and momentum equations in terms of the streamwise and transverse components, which is difficult in the presence of irregularly curved channels.

Thus, the second approach is based upon qualitative arguments concerning the interaction of lateral curvature, bottom friction and vertical non-uniformity, to develop a model that reproduces the depth-averaged influence of the secondary flow. In essence, this method *models* the z-dependence of the streamwise and out-of-plane velocities, and the effects of centrifugal and frictional forces on a vertical column of water. The dispersion terms in Equations (19) and (20) are replaced by a term that represents the shear stress associated with secondary flow. The latter is determined from the streamwise vorticity, which includes empirically derived terms for vorticity production, dissipation, diffusion and vortex stretching. Bernard (1992) begins with the assumption that secondary shear stresses may be responsible for a significant proportion of momentum transport whenever the depth-averaged streamlines are curved. Denoting the dispersion terms in Equation (19) and (20) with the term $\boldsymbol{S}$, this can be defined approximately as;

$$\boldsymbol{S} = \frac{1}{\rho}\frac{\bar{V}}{\left(\overline{V_i^2} + \overline{V_j^2}\right)^{1/2}}\left[\frac{1}{h - z_\mathrm{b}}\frac{\overline{V_j}\,\partial h\tau_\mathrm{s}/\partial x - \overline{V_i}\partial h\tau_\mathrm{s}/\partial y}{\left(\overline{V_i^2} + \overline{V_j^2}\right)^{1/2}} + \frac{\tau_\mathrm{s}}{2r}\right] \tag{25}$$

where τ_s is the shear stress generated by the secondary circulation and r is the streamline radius of curvature which is related directly to the velocity derivatives and is defined by

$$r = \frac{\left(\overline{V_i^2} + \overline{V_j^2}\right)^{3/2}}{\overline{V_i}\,\overline{V_j}(\partial \overline{V_j}/\partial y - \partial \overline{V_i}/\partial x) + \overline{V_i^2}\partial \overline{V_j}/\partial x - \overline{V_j^2}\,\partial \overline{V_i}/\partial y} \tag{26}$$

The secondary stress (τ_s) is then expressed through

$$\tau_\mathrm{s} = \rho h\Omega\sqrt{C_\mathrm{f}}|\overline{V_i}| \tag{27}$$

where $\Omega = C_2\omega_\mathrm{s}/12$, ω_s is the streamwise vorticity and C_2 is an empirically derived constant. This leaves ω_s as the only value to be modelled to determine τ_s. As with the turbulence terms, the streamwise vorticity is dealt with in terms of production, transport and dissipation. Likewise, it makes use of semi-empirical principles and parameters, which must be adjusted to make model predictions fit experimental data (Bernard and Schneider, 1992). Following Johannesson and Parker (1989a), Bernard and Schneider (1992) recommend the use of

$$\frac{\partial\Omega}{\partial t} = \underbrace{\frac{A_\mathrm{s}\sqrt{C_\mathrm{f}}(\overline{V_i^2} + \overline{V_j^2})^{1/2}}{rh}}_{\text{vorticity production}} - \underbrace{D_\mathrm{s}\sqrt{C_\mathrm{f}}\Omega\frac{(\overline{V_i^2} + \overline{V_j^2})^{1/2}}{h}}_{\text{vorticity dissipation}} + \underbrace{\frac{\nu}{h}\nabla.(h\nabla\Omega)}_{\text{vortex diffusion owing to turbulence}} + \underbrace{\bar{V}.\nabla\Omega}_{\text{vortex advection}} \tag{28}$$

where A_s is an empirical coefficient that determines the rate of vorticity production and D_s is an empirical coefficient that determines the rate of vorticity dissipation. Bernard (1992) and Bernard and Schneider (1992) provide a full derivation of these terms.

The qualitative arguments that lead to the structure of Equation (28) need mention. The production and dissipation of streamwise vorticity are proportional to both the magnitude of the bottom friction and the depth-averaged velocity, and inversely proportional to the water depth, which one would expect. A radius of curvature term is introduced for the production term, as this is central to the generation of streamwise vorticity; where the smaller the radius of curvature, the greater the vorticity production. The vorticity dissipation term represents the functional argument that the greater the magnitude of vorticity, the greater the dissipation, although in the derivation of Equation (28) this term arises from the inclusion of a radial shear stress term at the bottom of the water column (Bernard and Schneider, 1992). Experimental evidence suggests the empirical coefficient A_s has the value 5·0 and D_s has the value 0·5. Two extra terms are added such that the term Ω is truly proportional to vorticity. The first is associated with lateral diffusion of the streamwise vorticity owing to turbulence. The second is associated with the advection of vorticity which is assumed to be much greater than the stretching term (which is not included). Evidence suggests that this is the case (Bernard and Schneider, 1992) provided the magnitude of the primary flow velocity is larger than that associated with secondary motion. Strictly speaking, Equation (28) is only valid when $r \gg h$. Otherwise, the depth is comparable to the radius of curvature and the amount of vorticity generation becomes too large. This has resulted in a correction to the generation term through a self-limiting function, where rh is replaced by $rh(1 + 9h^2/r^2)$.

Evaluation of the depth-averaging process

The acceptability of depth averaging needs to be considered with respect to the relative importance of these vertical motions. These may be small scale and related to the nature of the channel boundary, particularly microtopographic features (e.g. Hassan and Reid, 1991; Clifford *et al.*, 1992), or large in scale and associated with planform channel curvature (e.g. Dietrich and Whiting, 1989; Johannesson and Parker, 1989a) or changes in bed elevation (e.g. Best, 1986; Ashmore *et al.*, 1992). Rodi *et al.* (1981) note that even in the presence of such three-dimensional effects, in many rivers the depth-averaging method may be sufficiently accurate for practical purposes, particularly if the width to depth ratio is high and changes in bed elevation are not rapid (Bhallamudi and Chaudhury, 1991).

A number of models have made no correction for the effects of secondary circulation (e.g. Odgaard, 1989; Shimizu and Itakura, 1989; Struiksma and Crosato, 1989). However, several authors have described the influence of secondary circulation on momentum transfer both in the downstream (e.g. Odgaard and Bergs, 1988) and the cross-stream (e.g. Dietrich, 1987; Johannesson and Parker, 1989b; Nelson and Smith, 1989b; Shimizu *et al.*, 1990) directions. This is particularly the case in meandering channels, where the convective transport of the primary flow momentum by the secondary flow leads to a significant outward redistribution of primary flow velocity (Dietrich and Whiting, 1989; Nelson and Smith, 1989b).

Most of the treatment of secondary circulation has been through the analytical treatment described by Johannesson and Parker (1989b). Odgaard and Bergs (1988), Bridge (1992) and Bridge and Gabel (1992) all include a convective acceleration term associated with change in momentum of the downstream flow in the downstream direction. However, Dietrich and Smith (1983) add to this a convective acceleration term associated with change in momentum of the downstream flow in the cross-stream direction. However, there has been substantial disagreement as to the relevance of these terms. Central to the debate is the role of irregularity in streamline curvature. Researchers who have found the change in momentum of the downstream flow in the cross-stream direction to be small have generally been studying regular meander bends, either mathematically [e.g. Yeh and Kennedy (1993) used sine-generated curves] or experimentally [e.g. Odgaard and Berg (1988) used laboratory flumes]. However, as Jin and Steffer (1993) note, bends that have constant radius of curvature are rarely found in the field. Once channel irregularity is incorporated, and particularly downstream changes in the radius of curvature of bar forms (e.g. Nelson and Smith, 1989a,b; Dietrich and Whiting, 1989), the cross-stream term becomes important.

Regardless of the relative importance of these terms, these methods of treating the dispersion terms created by depth averaging suffer from a number of problems. First, the analytical treatments of the dispersion terms require prior specification of the local radius of curvature and often have to assume a uniform bend shape, which is necessary even if the change in momentum of the downstream flow in the cross-stream direction is considered. This is not always easy. For instance, Bridge and Gabel (1992) found it difficult to define the tangent to the centreline where the banks are not always parallel. Secondly, and more importantly, it is possible to question the relevance of the assumed causes of secondary flow in certain parts of channels, notably confluences and diffluences, where field and laboratory evidence suggest that secondary circulation is associated with processes additional to those represented in the corrections described above. First, Best (1986), Ashmore *et al.* (1992) and Biron *et al.* (1993) all describe the importance of topographic discordance as a secondary flow-generating mechanism. In addition to planform streamline curvature, a helical circulation arises from horizontal separation vortices formed in the lee of avalanche faces at the entrance to confluences. These are essentially associated with streamline curvature in the x–z plane. These problems could be more completely addressed through a three-dimensional treatment of the problem. Such models are available (e.g. Shimizu *et al.*, 1990) but, owing to their computational demands, have seen little application. The way to proceed would seem to be based upon using three-dimensional models to improve understanding of secondary circulation effects, and from this to investigate the way in which such three-dimensional effects may be parameterized in two-dimensional models. This was the method used by Shimizu *et al.* (1990) for meanders in which the result was a velocity profile treatment of the effects of secondary circulation, in a form not dissimilar to that of Johannesson and Parker (1989a). Confluences and diffluences await such a treatment. Second, helical motion is not just produced by topographic or water surface effects but also by turbulent shear in the flow. Nezu *et al.* (1993) describe turbulence-driven secondary circulation, owing to anisotropy in the magnitude of turbulent intensities between the vertical and cross-stream components or vertical and downstream components. They found that the importance of this anisotropy was dependent upon channel morphology, and that it was less developed if the width–depth ratio was greater than 5. However, field research (Biron *et al.*, 1993) has illustrated the importance of the shear zone in channel confluences in generating exactly the same effect. This is a clear instance of the need to use the more sophisticated algebraic stress and Reynolds stress turbulence models, as these have the ability to model turbulence-driven secondary circulation (e.g. Naot and Rodi, 1982; Cokljat and Younis, 1995).

STEADY AND UNSTEADY FLOWS

The assumption of a steady state may be both implicit and explicit. It is implicit to the Reynolds averaging procedure, Equations (9) and (10), in which the effects of instantaneous velocity fluctuations on momentum transfer and therefore average flow parameters are modelled, although this does not provide information on the velocity fluctuations themselves. These fluctuations represent continual local fluid accelerations and decelerations which may be critical for sediment transport (cf. Kalinske, 1947; Bridge and Bennett, 1992; Lapointe, 1992; Nelson *et al.*, 1993), implying that the assumption of a steady state in this sense is not appropriate. However, at the scale of river and floodplain reaches this is necessary for computational reasons, and, at this scale, turbulence processes associated with the erosion and deposition of individual particles may cancel each other, both across space and through time, particularly when sediment transport rates are high.

The assumption of steady state is also explicit at larger time-scales. In the case of steady flow, or unsteady flow where there is no change in the position of the free surface, it may be possible to make a 'rigid lid' approximation. This involves specification of an artificial, fixed surface, and the introduction into the momentum equation of a non-zero pressure term, which indicates the level that the water surface would have attained if it were free (Vreugdenhill, 1994). The hydrostatic pressure distribution, Equation (4), becomes

$$p = p_s + \rho g(h_s + z_b) \tag{29}$$

where subscript s indicates the fixed surface and p_s is the pressure term consequent upon the rigid lid approximation. In the case of the depth-averaged mass equation, Equation (18), this causes the $\partial/\partial t$ term to disappear. In the case of the depth-averaged forms of the x- and y-momentum equations, Equations (19) and (20), the water surface gradient terms are replaced by the gradients of the water surface pressure term in the x- and y-directions, respectively. These terms are evaluated during solution, essentially by determining the distribution of surface pressure required to conserve mass (e.g. Bernard, 1992). In the case of steady flows, a rigid lid approximation is only acceptable if spatial variation in water surface elevation is small compared with depth (Vreugdenhill, 1994). In the case of unsteady flow, it is acceptable if both the spatial and temporal variation in water surface elevation is small, and if the local Froude number is less than 0·5, or 0·7 if variations in the width and depth of flow are sufficiently gentle (Bernard, 1992). In either case, use of a rigid lid model is conditional upon accurate knowledge of water surface elevation (see Lane and Richards, 1998, this issue). If information on changes in water surface elevation through time is available, it may be possible to simulate unsteady flow as a series of steady states within which the discharge is assumed to be constant. This assumption is used frequently in modelling studies of natural channels (e.g. Andrews and Nelson, 1989; Nelson and Smith, 1989a,b; Struiksma and Crosato, 1989).

In the majority of circumstances, the geomorphologist is dealing with unsteady flow, and may be interested in predicting change in water surface elevation and spatial patterns of inundation through time rather than relying upon its pre-specification. The classic example is floodplain flow, where both the means of predicting and the kinds of prediction required have resulted in a need to model hydraulically unsteady flow. Overbank flows result in a discontinuity in the stage–discharge curve (e.g. Knight, 1989) as a result of both the sudden change in cross-sectional shape and the shear layer that exists between the channel and the floodplain, and which indicates momentum transfer and therefore a reduction in conveyance capacity. The nature of this discontinuity is difficult to predict, as it depends upon knowledge of the dimensions and orientation of the shear layer, which depends upon the exact geometrical configuration of, and the relative roughness differential between, the channel and floodplain cross-section. Likewise, floodplain protection studies require the specification of spatiotemporal variation in floodplain inundation that more conventional stage–discharge analyses are unable to predict (Gee *et al.*, 1990). Improving the downstream routing of floodplain flows using two-dimensional models might even result in better hydrograph prediction (Baird *et al.*, 1992), through better representation of floodplain storage attenuation effects.

Flood routing models have therefore been developed and applied for application to unsteady flow (e.g. King and Roig, 1988; Gee *et al.*, 1990; Baird *et al.*, 1992; Bates *et al.*, 1992; Bates and Anderson, 1993; Jin and Kranenburg, 1993). Traditional approaches have involved the computation of backwater curves (e.g. HEC-2 developed by the US Army Corps of Engineers). The main problems with these are: (i) that they are often one-dimensional, assuming that the water surface is horizontal in the transverse direction; and (ii) that they are often only quasi-unsteady, where the peak hydrograph flow is used to compute flood elevations; they cannot therefore deal with the differential inundation patterns that result from varied hydrograph rises. More sophisticated treatments involve use of the unsteady form of the Navier–Stokes equations, and the simulation of temporal variation in the spatial domain of the solution, where the exact position of the boundary is not known *a priori*, but must be determined during solution as its position is determined by the state of the fluid at the boundary (Lynch and Gray, 1980; France, 1981).

Solutions to this can be divided into four broad groups: (i) spatially fixed, excluding partially wet elements; (ii) spatially fixed, but including partially wet elements with a special treatment; (iii) deforming boundary; and (iv) boundary dividing. The first approach simply involves specifying a 'wet' criterion and a 'dry' criterion and removing or adding grid cells or elements from or to the solution domain accordingly, and correcting for the special case of partially wet or partially dry elements through some form of boundary treatment (e.g. Laura and Wang, 1984; Kawahara and Umetsu, 1986). Most importantly, this method *per se* fails to conserve both mass and momentum (Holz and Nitsche, 1982). Although, with simple changes in the extent of inundation, this approach is computationally efficient (France, 1981), if the change in the inundation extent is large then additional computational nodes will need to be carried along with each solution step (Akanbi and Katopodes, 1988) and this is computationally demanding (Lynch and Gray, 1980; Gopalakrishnan and Tung, 1983). In most geomorphological and hydrological applications, the pattern of inundation is likely to be complex, if the modeller has little *a priori* expectation as to the way the boundary will move, the spatial extent of boundary migration may be large and a high resolution is required in the entire spatial domain through which the boundary will move.

Mass conservation can be achieved during both the wetting and drying process using a fixed grid (e.g. Holz and Nitsche, 1982; Zhao *et al.*, 1994). For instance, Holz and Nitsche (1982) note that in the case of an element that becomes partially dry during the previous computational step, it is possible to extrapolate the velocities, then on the element boundary, to neighbouring and completely wetted elements, and to use these values for prescribing mass flow into these elements. This is the first step of an iterative procedure that includes assessing whether the calculated water level on the boundary of the wetted element agrees with that of the remaining water volume on the partially dry element. If not, the mass flow must be re-estimated.

Again, this method suffers from the large computational time, and, despite conserving mass, it does not conserve momentum.

Approaches that simply drop partially wet elements are also likely to be computationally unstable, as large changes in the lateral extent of the free surface may result in spurious velocities on partially wet areas, where gravity terms in the momentum equations dominate the solution (Bates *et al.*, 1995). The second approach involves keeping partially wet elements in the solution and cancelling g in the momentum equations to avoid spurious velocities. This does not affect the scheme mass conservation as the mass continuity equation is unaltered.

The third solution method involves grid deformation across the spatial domain such that nodal points at the perimeter of the boundary are allowed to move with the boundary, the grid deforming behind the boundary as the boundary nodes migrate (e.g. Katopodes and Strelkoff, 1978; Lynch and Gray, 1980; Gopalakrishnan and Tung, 1983; Akanbi and Katopodes, 1988). This allows the number of computational nodes necessary to produce a certain accuracy in the calculation of flow conditions to remain constant with time (Katopodes and Strelkoff, 1978) and has the advantage of being computationally more efficient than approaches employing a spatially fixed grid, although solution convergence rates can be slow (King and Roig, 1988). It also suffers from the need to ensure that boundary nodes actually follow the boundary and the need to account accurately for the effects of grid deformation on model predictions (Lynch and Gray, 1980). Particularly problematic in geomorphological and hydrological problems is that, as a result of grid deformation during solution, the grid becomes disconnected from the spatial domain during flood recession and must be reconstructed during the rising stage (Holz and Nitsche, 1982). The importance of a close coupling of the computational grid to the boundary topographical characteristics has been noted (e.g. Bates *et al.*, 1992), and, as most geomorphological and hydrological applications involve complex changes of topography and other boundary conditions across space, deforming grids may be inappropriate unless the deformation also checks for topographic effects. Nevertheless, there may remain some potential for grid deformation that is closely coupled with spatial variation in boundary conditions, owing to the computational advantage therein.

The final solution involves dividing the boundary of the spatial domain in the vicinity of the moving boundary. This can be achieved both explicitly (e.g. Holz and Nitsche, 1982) and implicitly (e.g. King and Roig, 1988). Holz and Nitsche (1982) developed a simple method of creating a set of subelements in the vicinity of the flow boundary. This involved tentatively assuming a position for the flow boundary and dividing the spatial domain in this region into a number of smaller elements. An iterative procedure was then used to deform the subelement mesh geometry in order to accommodate the corrected position of the free surface. A more recent approach subdivides elements implicitly in the vicinity of the moving boundary through a pseudo-porosity treatment (e.g. King and Roig, 1988; Gee *et al.*, 1990; Bates *et al.*, 1992; Bates and Anderson, 1993), in which elements make a transition between fully dry, partially wet and fully wet according to flow boundary position. This is achieved through the use of a domain coefficient that describes the proportion of water in an element that is available for flow. The flow boundary is approximated in partially wet elements by using the domain coefficient to scale the simulated elemental water volume to the actual elemental water volume for each time step. With some small modifications to ensure solution stability, the results have been encouraging (e.g. Bates and Anderson, 1993; Anderson and Bates, 1994). Most notably, this allows a smooth transition between the fully dry and the fully wet state, with large node spacings should a high level of computational efficiency be required.

Extending the wetting and drying process to three dimensions will be even more challenging. Most approaches assume a fixed number of grid cells in the vertical. As an element dries, and water depth tends to zero, grid cells will become progressively smaller, so causing the solution to be unstable. In such situations, the only viable solution may be simply to remove partially wet elements from the solution (P. D. Bates, personal communication).

Two other issues need to be mentioned when dealing with unsteady channel–floodplain flows. First, close attention should be given to division of the spatial domain (either into a finite difference grid or a finite element surface). This will require a finer grid in any zones where there may be sudden changes in boundary conditions such as topography (e.g. Bates *et al.*, 1992) or roughness (e.g. Holz and Nitsche, 1982). Secondly, specification of the most appropriate time step also needs careful consideration and will need to be defined by both the rate of change of discharge and the numerical schemes used for solution. The faster the rate of increase or decrease of discharge, the shorter the time step will need to be. This may require variable time step lengths to avoid computational inefficiency, with longer time steps for periods of slower rates of change of discharge and vice versa. In practice, time steps may be directed by the numerical scheme being used, and notably the Courant number. Effective coverage of these issues would require too much space, but they are considered by Press *et al.* (1992). Explicit solution schemes are forward looking with new values of a variable expressed as a function of only its existing values. They require the time step to be judged with reference to the Courant number, which should always be less than one

$$\frac{|\nu|\,\Delta t}{\Delta x} \leqslant 1 \tag{30}$$

where ν is the rate at which a property moves through a spatial domain that is discretized with a grid spacing of x and a time step t. In essence, the Courant number constrains the domain of influence of discretized points. Provided the Courant condition is met, by choice of appropriate values of both x and t, then only adjacent points in time or space can be used to determine new values, and the solution is stable. For this reason, there will be a close relationship between spatial discretization and the time step used, which necessarily means that smaller discretizations will require shorter time steps, and therefore run the risk of increasing computational inefficiency. This problem may be partly overcome with implicit schemes, in which the new value of a variable is judged as a function of its new value. This has been shown (e.g. Press *et al.*, 1992) to produce more stable solutions even with larger time steps, and hence Courant numbers.

CONCLUSION

It is perhaps important to finish this review on a cautious note. It is all too easy to get excited by the potential generality of models (cf. Bathurst, 1993), and notably by the apparent reliability of a tool for the study of natural systems that is not fraught with the problems of both field and laboratory observation. Rhoads (1994) claims that 'holistic modelling' of complex, non-linear systems allows the investigator to explore the relative influence of contingent initial conditions *and* general physical or chemical laws on predicted patterns of system behaviour. Both Rhoads and Richards (1990) argue that it is through experimentation with models that we might see emergence in the form of synergistic qualities, structures or properties that cannot be predicted from relationships between isolated components of a system. This is an appealing argument and acceptable up to a point. However, models in themselves are truncations of reality, and may be equally reductionist in that their predictions will be conditioned by: (i) incomplete knowledge of the 'whole' system; (ii) the extent to which the 'correct' initial conditions are established in their full spatial form at the start of the system; (iii) the assumptions used by the modeller during model formulation (e.g. whether or not one assumes a rigid lid, which will constrain model predictive ability); and (iv) the dependence of model predictions on parameterizations which may have a weak physical basis [e.g. the viscosity term introduced by Reynolds averaging in Equation (11)]. Indeed, if a model's effectiveness is assessed by reference to its predictive ability alone, it becomes very difficult to distinguish between which of the many possible model assumptions are responsible for the empirical inadequacy, or whether the model structure is an incomplete representation of reality. Furthermore, even if a model predicts successfully, this is no guarantee that it will hold beyond the specific situation to which it was applied. The 'whole', the requisite assumptions and the most appropriate parameterization may be very different in different field contexts and at different points in time. This point is reinforced if parameterization or optimization is based upon empirical observations at particular points in space and time. Thus, there is a particularly difficult problem if a model is being used to fulfil what are often perceived to be its two key roles of prediction and simulation. It is unlikely that a single model would suffice for every situation that the geomorphologist or hydrologist encounters, and it is important that model performance is evaluated with respect to specific field problems as they are encountered.

Many of the issues discussed in this review are scale dependent and their relative importance varies spatially within a model application as well as between the particular locations to which models are applied. Implicit to much of the review presented, is the need for physical realism in the numerical simulation of flow processes. It is common for engineers to judge physical realism through a quantitative assessment of predictive ability. Application of such an approach to geomorphological and hydrological problems may not always be appropriate. In the environmental sciences, where field measurement is critical for both boundary condition specification and model evaluation, obtaining a measurement that is both relevant and accurate may be difficult. To judge model performance with respect to predictive ability alone must recognize that measurements may be just as much in error as the models themselves, and model *performance*, judged through checks on mass conservation and sensitivity analysis, as well as model predictions must be considered (see Bates *et al.*, 1998, this issue).

Where predictive ability is poor, or where substantial uncertainty may exist with respect to both the model structure and the boundary conditions required to allow its application to a specific problem, it may be tempting to argue that ever more sophisticated models will be required to achieve an acceptable set of predictions. This embodies some of the discussion above. However, one must be cautious of such an approach. The uncertainty associated with boundary condition specification may be much greater than that associated with poor description of the physical process itself. This does not mean that including better physical representation is a futile activity, as it is only with such incorporation that it becomes possible to assess just how important that representation might be. In situations where the nature of the boundary condition is central to the operation of the process that the improved representation is trying to describe (e.g. turbulence), any modelling attempt must balance the advantages that come from such improvements, with the problems of determining the boundary conditions required to allow effective model application.

ACKNOWLEDGEMENTS

Paul Bates, Kate Bradbrook, Anthony Reeves and Keith Richards provided helpful comments on an earlier draft of this paper. This research is supported by NERC Grant GR3/9715 (to S.N.L., Professor K. S. Richards and Dr J. H. Chandler).

REFERENCES

Akanbi, A. and Katopodes, N. D. 1988. 'Model for flood propagation on initially dry land', *ASCE J. Hydraul. Engng*, **114**, 689–719.

Anderson, M. G. and Bates, P. D. 1994. 'Initial testing of a two-dimensional finite element model for floodplain inundation', *Proc. R. Soc. London, Ser. A*, **444**, 149–159.

Andrews, E. D. and Nelson, J. M., 1989. 'Topographic response of a bar in the Green River, Utah, to variation in discharge', in Ikeda, S. and Parker, G. (Eds), *River Meandering*, American Geophysical Union Monograph, 12. pp. 463–485.

ASCE, 1988. 'Turbulence modelling of surface water flow and transport: part V', *ASCE J. Hydraul. Engng*, **114**, 1052–1073.

Ashmore, P. E., Ferguson, R. I., Prestegaard, K. L., Ashworth, P. J., and Paola, C. 1992. 'Secondary flow in anabranch confluences of a braided gravel-bed stream', *Earth Surf. Process. Landf.*, **17**, 299–311.

Ashworth, P. J. and Ferguson, R. I. 1986. 'Interrelationships of channel processes, changes and sediments in a proglacial braided river', *Geograf. Annal.*, **68A**, 361–371.

Baird, L., Gee, D. M., and Anderson, M. G. 1992. 'Ungauged catchment modelling II. Utilization of hydraulic models for validation', *Catena*, **19**, 33–42.

Bates, P. D. 1992. 'Finite element modelling of floodplain inundation', *PhD Thesis,* University of Bristol.

Bates, P. D. and Anderson, M. G. 1993. 'A two-dimensional finite element model for river flow inundation', *Proc. R. Soc. London, Ser. A*, **440**, 481–491.

Bates, P. D., Anderson, M. G., Baird, L., Walling, D. E., and Simm, D. 1992. 'Modelling floodplain flows using a two-dimensional finite element model. *Earth Surf. Process. Landf.*, **17**, 575–588.

Bates, P. D., Anderson, M. G., and Hervouet, J. M., 1995. 'Initial comparison of two 2-dimensional finite-element codes for river flood simulation', *Proc. Inst. Civil Engrs, Wat. Maritime Energy*, **112**, 238–248.

Bates, P. D. 1998. 'Investigating two-dimensional finite element predictions of floodplain inundation using fractal generated topography', *Hydrol. Process.*, **12**, 1257–1277 (this issue).

Bathurst, J. C. 1993. 'The third postgraduate symposium', *Circulation*, **15**, 2–5.

Bernard, R. S. 1989. 'Explicit numerical algorithm for modelling incompressible approach flow', *Technical Report HY-5*. US Army Corps Engineers Waterways Experiment Research Station, Vicksburg, Mississippi.

Bernard, R. S. 1992. 'STREMR: numerical model for depth-averaged incompressible flow', *Technical Report HY-105*. US Army Corps Engineers Waterways Experiment Research Station, Vicksburg, Mississippi.

Bernard, R. S. and Schneider, M. L. 1992. Depth-averaged numerical modelling for curved channels', *Technical Report HL-92-9*. US Army Corps Engineers, Waterways Experiment Research Station, Vicksburg, Mississippi.

Best, J. L., 1986. 'The morphology of river channel confluences', *Progr. Phys. Geog.* **10**, 157–174.

Bhallamudi, S. M. and Chaudhury, M. H. 1991. 'Numerical modelling of aggradation and degradation in alluvial channels. *ASCE J. Hydraul. Engng*, **119**, 1145–1164.

Biron, P., De Serres, B., Roy, A. G., and Best, J. L. 1993. 'Shear layer turbulence at an unequal depth confluence', in Clifford, N. J., French, J. R., and Hardisty, J. (Eds), *Turbulence: Perspective on Flow and Sediment Transport*. Wiley, Chichester. pp. 197–214.

Boussinesq, J. 1877. *Essai sur la théorie des eaux courantes*. Mémoires présentés pars divers savants à l'Académie des Sciences, Paris.

Bradbrook, K. F., Biron, P. M., Lane, S. N., Richards, K. S., and Roy, A. G. 1998. 'Investigation of controls on secondary circulation in a simple confluence geometry using a three-dimensional numerical model', *Hydrol. Process.*, **12**, 1371–1396 (this issue).

Bradshaw, S. 1969. 'The use of transport equations for Reynolds stresses', *Proceedings of the Boeing Symposium on Turbulence, Seattle, June 1969*.

Bridge, J. S. 1992. 'A revised model for water flow, sediment transport, bed topography and grainsize sorting in natural river bends', *Wat. Resour. Res.*, **28**, 999–1013.

Bridge, J. S. and Bennett, S. J. 1992. 'A model for the entrainment and transport of sediment grains of mixed sizes, shapes and densities', *Wat. Resour. Res.*, **28**, 337–363.

Bridge, J. S. and Gabel, S. L. 1992. 'Flow and sediment dynamics in a low sinuosity river: Calamus River, Nebraska Sandhills', *Sedimentology*, **29**, 499–550.

Carling, P. A. and Wood, N. 1994. 'Simulation of flow over pool-riffle topography: a consideration of the velocity reversal hypothesis', *Earth Surf. Process. Landf.*, **19**, 319–332.

Carson, M. A. 1971. *The Mechanics of Erosion*. Pion, London. 174 pp.

Chezy, A. de, 1775. 'Memoire sur la vitesse de l'eau conduite dans une régole', MS reprinted in *Ann. Ponts Chaussées*, **60**, 1921.

Church, M. and Jones, D. 1982. 'Channel bars in braided rivers', in Hey, R. D., Bathurst, J. C., and Thorne, C. R. (Eds), *Gravel-Bed Rivers*. Wiley, Chichester. pp. 291–324.

Clifford, N. J., Robert, A., and Richards, K. S. 1992. 'Estimation of flow resistance in gravel-bedded rivers: a physical explanation of the multiplier of roughness length', *Earth Surf. Process. Landf.*, **17**, 111–126.

Cokljat, D. and Younis, B. A. 1995. 'Second order closure study of open channel flows', *ASCE J. Hydraul. Engng*, **112**, 94–107.

Correia, L. R. P., Krishnappan, G., and Graf, W. H. 1992. 'Fully coupled unsteady mobile boundary flow model', *ASCE J. Hydraul. Engng*, **118**, 476–494.

Corrsin, S. 1957. 'Some current problems in turbulent shear flow', in National Academy of Sciences (Ed.), *Naval Hydrodynamics*, Publication 515. National Academy of Sciences, Washington, D.C.

Dietrich, W. E. 1987. 'Mechanics of flow and sediment transport in river bends', in Richards, K. S. (Ed.), *River Channels: Environment and Process*, IBG Spl. Publ., 18. Blackwell, Oxford. pp. 179–227.

Dietrich, W. E. and Smith, J. D. 1983. 'Influence of the point bar on flow through curved channels', *Wat. Resour. Res.*, **19**, 1173–1192.

Dietrich, W. E. and Whiting, P. J. 1989. 'Boundary shear stress and sediment transport in river meanders of sand and gravel', in Ikeda, S. and Parker, G. (Eds), *River Meandering*, American Geophysical Union Monograph, 12. pp. 1–50.

Dreissler, R. G. 1977. 'Turbulence processes and simple closure schemes', in Frost, W. and Moulden, T. H. (Eds), *Handbook of Turbulence*, Vol. 1, *Fundamentals and Applications*. Plenum, New York. pp. 165–186.

Emmons, H. W. 1970. *Ann. Rev. Fluid Mech.*, **2**, 15–36.

Engelund, F. 1974. 'Flow and bed topography in channel bends', *ASCE J. Hydraul. Div.*, **100**, 1631–1648.

France, P. W. 1981. 'A simple technique for the analysis of free surface flow problem', *Adv. Wat. Resour.*, **4**, 20–22.

Gee, D. M., Anderson, M. G., and Baird, L. 1990. 'Large-scale floodplain modelling', *Earth Surf. Process. Landf.*, **15**, 513–523.

Gopalakrishnan, T. C. and Tung, C. C. 1983. 'Numerical analysis of a moving boundary problem in coastal hydrodynamics', *Int. J. Num. Meth. Fluids*, **3**, 179–200.

Hassan, M. and Reid, I. 1991. 'The influence of microform bed roughness elements on flow and sediment transport in gravel-bed rivers', *Earth Surf. Process. Landf.*, **15**, 739–750.

Hey, R. D. and Thorne, C. R. 1975. 'Secondary flows in river channels', *Area*, **7**, 191–195.

Hodkinson, A. and Ferguson, R. I. 1998. 'Numerical modelling of separated flow in river beds: model testing and experimental investigation of geometric controls on the extent of flow separation at the concave bank', *Hydrol. Process.*, **12**, 1323–1338 (this issue).

Holz, K.-P. and Nitsche, G. 1982. 'Tidal wave analysis for estuaries with intertidal flats', *Adv. Wat. Resour.*, **5**, 142–148.

Jin, X. and Kranenburg, C. 1993. 'Quasi-three-dimensional numerical modelling of shallow-water circulation', *ASCE J. Hydraul. Engng*, **119**, 458–472.

Jin, Y.-C. and Steffer, P. M. 1993. 'Predicting flow in curved open channels by depth-averaged method', *ASCE J. Hydraul. Engng*, **119**, 109–123.

Johannesson, H. and Parker, G. 1989a. 'Linear theory of river meanders', in Parker G. (Ed.), *River Meandering*, American Geophysical Union Monograph, 12. pp. 181–213.

Johannesson, H. and Parker, G. 1989b. 'Velocity redistribution in meandering rivers', *ASCE J. Hydraul. Engng*, **115**, 1019–1039.
Johannesson, H. and Parker, G. 1989c. 'Secondary flow in mildly sinuous channels', *ASCE J. Hydraul. Engng*, **115**, 289–308.
Kalinske, A. A. 1947. 'Movement of sediment as bedload in rivers', *Trans. Am. Geophys. Union*, **28**, 615–620.
Katopodes, N. and Strelkoff, T. 1978. 'Computing two-dimensional dam-break flood waves', *ASCE J. Hydraul. Div.*, **104**, 1269–1288.
Kawahara, M. and Umetsu, T. 1986. 'Finite element method for moving boundary problems in river flow', *Int. J. Num. Meth. Fluids*, **6**, 365–386.
Keller, E. A. and Florsheim, J. L. 1993. 'Velocity-reversal hypothesis — a model approach', *Earth Surf. Process. Landf.*, **18**, 733–740.
King, I. P. and Roig, L. C. 1988. 'Combined modelling of one- and two-dimensional flow systems', in Niki, K. and Kawahara, M. (Eds), *Computational Models in Flow Analysis*. pp. 711–718.
Knight, D. W. 1989. 'Hydraulics of flood channels', in Beven, K. and Carling, P. A. (Eds), *Floods: Hydrological, Sedimentological and Geomorphological Implications*. Wiley, Chichester, pp. 83–105.
Kolmogorov, A. N. 1942. 'Equations of turbulent motion of an incompressible fluid', *Akad. Nank. SSR., Seriafizicheska*, **1–2**, 56–58.
Lane, S. N. and Richards, K. S. 1998. 'High resolution, two-dimensional spatial modelling of flow processes in a multi-thread channel', *Hydrol. Process.*, **12**, 1279–1298 (this issue).
Lane, S. N., Richards, K. S., and Chandler, J. H. 1995. 'Within-reach spatial patterns of process and channel adjustment', in Hickin, E. J. (Ed.), *River Geomorphology*. Wiley, Chichester, pp. 105–130.
Langbein, W. B. 1966. 'A random walk model of hydraulic friction', *Bull. Int. Assoc. Sci. Hydrol.*, **11**, 5–9.
Lapointe, M. 1992. 'Burst-like sediment suspension events in a sand-bedded river', *Earth Surf. Process. Landf.*, **17**, 253–270.
Launder, B. E. and Spalding, D. 1972. *Lectures in Mathematical Models of Turbulence*. Academic Press, New York.
Launder, B. E. and Spalding, D. 1974. 'The numerical computation of turbulent flows', *Comp. Meth. Appl. Mech. Engng*, **3**, 269–289.
Laura, R. A. and Wang, J. D. 1984. 'Two-dimensional flood routing on steep slopes', *ASCE J. Hydraul. Engng*, **110**, 1121–1135.
Leopold, L. B., Bagnold, R. A., Wolman, M. G., and Brush, L. M. 1960. 'Flow resistance in sinuous or irregular channels', *USGS Professional Paper*, **282D**, 111–135.
Leschziner, M. A. and Rodi, W. 1979. 'Calculation of strongly curved open channel flow', *ASCE J. Hydraul. Engng*, **105**, 1297–1314.
Lynch, D. R. and Gray, W. G. 1980. 'Finite element simulation of flow in deforming regions', *J. Comput. Phys.*, **36**, 135–153.
Malin, M. R. and Younis, B. A. 1990. 'Calculation of turbulent buoyant plumes with a Reynolds stress and heat-flux transport closure', *Int. J. Heat Mass Trans.*, **33**, 2247–2264.
McGuirk, J. and Rodi, W. 1978. 'A depth-averaged mathematical model for the near field of side discharges into open channel flow', *J. Fluid Mech.*, **86**, 861–871.
Naot, D. and Rodi, W. 1982. 'Calculation of secondary currents in channel flow', *ASCE J. Hydraul. Div.*, **108**, 948–968.
Nelson, J. M. and Smith, J. D. 1989a. 'Flow in meandering channels with natural topography', in Parker, G. (Ed.), *River Meandering*, American Geophysical Union Monograph, 12. pp. 69–102.
Nelson, J. M. and Smith, J. D. 1989b. 'Evolution and stability of erodible channel beds', in Parker, G. (Ed.), *River Meandering*, American Geophysical Union Monograph, 12, pp. 321–377.
Nelson, J. M. Shreve, R. L., and McLean, S. R. 1993. 'Near-bed turbulence structure and bedload transport: implications for bedform mechanics', *3rd International Geomorphology Conference Hamilton Canada, 23–29 August, 1993.*
Nezu, I., Tominaga, A., and Nakagawa, H. 1993. 'Field measurements of secondary currents in straight rivers', *ASCE J. Hydraul. Engng*, **119**, 598–614.
Niekerk, A. Van, Vogel, K. R., Slingerland, R. L., and Bridge, J. S. 1992. 'Routing of heterogeneous sediment over mobile bed; model development', *ASCE J. Hydraul. Engng*, **118**, 246–263.
Odgaard, A. J. 1989. 'River meander model; 1. Development', *ASCE J. Hydraul. Engng*, **115**, 1433–1450.
Odgaard, A. J. and Bergs, M. A. 1988. 'Flow processes in a curved alluvial channel', *Wat. Resour. Res.*, **24**, 45–56.
Olson, R. M. and Wright, S. J. 1990. *Essentials of Engineering Fluid Mechanics*. Harper Row, New York. 638 pp.
Prandtl, L. 1925. 'Liber die ausgebildete turbulenz', *Z. Angew. Math. Mech.*, **5**, 136–139.
Prandtl, L. 1945. 'Uber ein neues formelsystem für die ausgebildete turbulenz', Translated as *Jet Propulsion Lab. Publ.*, **13**, 1952.
Press, W. H., Teukolsky, S. A., Vetterling, W. T., and Flannery, B. P. 1992. *Numerical Recipes in FORTRAN*, 2nd edn. Cambridge University Press, Cambridge.
Rastogi, A. and Rodi, W. 1978. 'Predictions of heat and mass transfer in open channels. *ASCE J. Hydraul. Div.*, **104**, 397–420.
Reynolds, O., 1895. 'On the dynamical theory of incompressible viscous fluids and the determination of the criterion', *Phil. Trans. R. Soc.*, **186A**, 123–164.
Rhoads, B. L. 1994. 'On being a real geomorphologist', *Earth Surf. Landf.*, **19**, 269–272.
Richards, K. S. 1990. 'Real geomorphology', *Earth Surf. Process. Landf.*, **15**, 195–197.
Richards, K. S. 1996. 'Samples and cases: generalisation and explanation in geomorphology', in Rhoads, B. L. and Thorn, C. E. (Eds), *The Scientific Basis of Geomorphology*, Binghamton Symposium in Geomorphology, Wiley, Chichester.
Rodi, W. 1971. 'On the equation governing the rate of turbulent energy dissipation', *Imp. Coll. Mech. Engng Report*, **TWF/TN/A/14**.
Rodi, W., 1980. *Turbulence Models and their Application in Hydraulics*. IAHR, Delft. 104 pp.
Rodi, W., Pavlovic, R. N., and Srivatsa, S. K. 1981. 'Prediction of flow and pollutant spreading in rivers', in Fischer, H. B. (Ed.), *Transport Models for Inland and Coastal Waters*. Academic Press, New York. 542 pp.
Schlichting, H. 1955. *Boundary Layer Theory*. McGraw-Hill, New York. 817 pp.
Shimizu, Y. and Itakura, T. 1989. 'Calculation of bed variation in alluvial channels', *ASCE J. Hydraul. Engng*, **115**, 367–384.
Shimizu, Y., Yamaguchi, H., and Itakura, T. 1990. 'Three-dimensional computation of flow and bed deformation', *ASCE J. Hydraul. Engng*, **116**, 1090–1106.
Smith, N. D. 1974. 'Sedimentology and bar formation in the Upper Kicking Horse River, a braided outwash stream', *J. Geol.*, **82**, 205–223.

Struiksma, N. and Crosato, A. 1989. 'Analysis of a two-dimensional bed topography model for rivers', in Parker, G. (Ed.), *River Meandering*, American Geophysical Union Monograph, 12. pp. 153–180.
Tennekes, H. and Lumley, J. L. 1972. *A First Course in Turbulence*, Massachusetts Institute of Technology Press, Massachusetts.
Tingsanchali, T. and Maheswaran, S. 1990. 'Two-dimensional depth-averaged flow computation near groynes', *ASCE J. Hydraul. Engng*, **116**, 103–125.
Tritton, D. J. 1988. *Physical Fluid Dynamics*. Oxford University Press, Oxford. 519 pp.
Vreugdenhill, C. B. 1994. *Numerical Methods for Shallow Water Flow*. Kluwer Academic Publishers, Dordrecht, Netherlands.
White, F. M. 1974. *Viscous Fluid Flow*. McGraw-Hill, New York. 725 pp.
Yeh, K.-C. and Kennedy, J. F. 1993. 'Moment model of non-uniform channel-bend flow; 1. Fixed bed', *ASCE J. Hydraul. Engng*, **119**, 776–795.
Younis, B. A. 1992. *Is Turbulence Modelling of any Use?* AIRH Conference. Institution of Civil Engineers, London, April 1992.
Zhao, D. H., Shen, H. W., Tabios, III, G. Q., Lai, J. S., and Tan, W. Y. 1994. 'Finite-volume two-dimensional unsteady-flow model for river basins', *ASCE J. Hydraul. Engng*, **120**, 863–883.

2

MODELLING WETTING AND DRYING EFFECTS OVER COMPLEX TOPOGRAPHY

G. W. TCHAMEN* AND R. A. KAHAWITA

Section Hydraulique, Département de génie civil, École polytechnique de Montréal, C.P. 6079 succ. Centre-ville, Montréal (Québec), H3C 3A7, Canada

ABSTRACT

The numerical simulation of free surface flows that alternately flood and dry out over complex topography is a formidable task. The model equation set generally used for this purpose is the two-dimensional (2D) shallow water wave model (SWWM). Simplified forms of this system such as the zero inertia model (ZIM) can accommodate specific situations like slowly evolving floods over gentle slopes. Classical numerical techniques, such as finite differences (FD) and finite elements (FE), have been used for their integration over the last 20–30 years. Most of these schemes experience some kind of instability and usually fail when some particular domain under specific flow conditions is treated. The numerical instability generally manifests itself in the form of an unphysical negative depth that subsequently causes a run-time error at the computation of the celerity and/or the friction slope. The origins of this behaviour are diverse and may be generally attributed to:

1. The use of a scheme that is inappropriate for such complex flow conditions (mixed regimes).
2. Improper treatment of a friction source term or a large local curvature in topography.
3. Mishandling of a cell that is partially wet/dry.

In this paper, a tentative attempt has been made to gain a better understanding of the genesis of the instabilities, their implications and the limits to the proposed solutions. Frequently, the enforcement of robustness is made at the expense of accuracy. The need for a positive scheme, that is, a scheme that always predicts positive depths when run within the constraints of some practical stability limits, is fundamental. It is shown here how a carefully chosen scheme (in this case, an adaptation of the solver to the SWWM) can preserve positive values of water depth under both explicit and implicit time integration, high velocities and complex topography that may include dry areas. However, the treatment of the source terms: friction, Coriolis and particularly the bathymetry, are also of prime importance and must not be overlooked. Linearization with a combination of switching between explicit–implicit integration can overcome the 'stiffness' of the friction and Coriolis terms and provide stable numerical integration. The treatment of the bathymetry source term is much more delicate. For cells undergoing a transient wet–dry process, the imposition of zero velocity stabilizes most of the approximations. However, this artificial zero velocity condition can be the cause of considerable error, especially when fast moving fronts are involved. Besides these difficulties linked with the internal position of the front within a cell versus the limited resolution of a numerical grid, it appears that the second derivative that defines whether the bed is locally convex or concave is a key indicator for stability. A convex bottom may lead to unbounded solutions. It appears that this behaviour is not linked to the numerics (numerical scheme) but rather to the mathematical theory of the SWWM. These concerns about stability have taken precedence, until now, over the crucial and related question of accuracy, especially near a moving front, and how these possible inaccuracies at the leading edge may affect the solution at interior points within the domain.

This paper presents an in depth, fully two-dimensional space analysis of the aforementioned problem that has not been addressed before. The purpose of the present communication is not to propose what could be

* Correspondence to: G. W. Tchamen, Section Hydraulique, Département de génie civil, École Polytechnique de Montréal, C.P. 6079 succ. Centre-ville, Montréal, Québec H3C 3A7, Canada.

viewed as a 'final solution', but rather to provide some key considerations that may reveal the ingredients and insight necessary for the development of accurate and robust solutions in the future.

KEY WORDS numerical simulation; free surface flows; Riemann solvers; numerical instability; finite volumes; dry bed computations

INTRODUCTION

The governing equations of the two-dimensional (2D) SWWM are

$$\frac{\partial H}{\partial t} + \nabla \cdot (\vec{V}h) = 0$$

$$\frac{\partial (uh)}{\partial t} + \nabla \cdot [u(\vec{V}h)] + \frac{\partial (gh^2/2)}{\partial x} = \nabla \cdot [K_d \cdot \nabla (uh)] - gh\frac{\partial Zb}{\partial x} - g\frac{N^2\sqrt{u^2+v^2}}{h^{4/3}}uh + R \cdot vh \tag{1}$$

$$\frac{\partial (vh)}{\partial t} + \nabla \cdot [v(\vec{V}h)] + \frac{\partial (gh^2/2)}{\partial y} = \nabla \cdot [K_d \cdot \nabla (vh)] - gh\frac{\partial Zb}{\partial y} - g\frac{N^2\sqrt{u^2+v^2}}{h^{4/3}}vh - R \cdot uh$$

(I) (II) (III) (IV) (V) (VI) (VII)

where

t	time
x, y	longitudinal and transverse coordinates of the horizontal plane
$Zb(x, y)$	bottom topography elevation with respect to a reference plane
$H(x, y, t)$	Free surface elevation with respect to the same reference used to define Zb
h	water depth, $h = H(x, y, t) - Zb(x, y)$
$\vec{V} = u \cdot \vec{1}_x + v \cdot \vec{1}_y$	velocity vector with u and v being the x and y components, respectively
g	gravitational acceleration
K_d	diffusion (turbulent) coefficient tensor
N	Manning friction factor
R	Coriolis factor

This model equation set has been used for a long time to represent the water flow over a 2D horizontal plane (xy) where vertical accelerations are considered negligible. Many practical flows in coastal areas, rivers, channel and floodplain systems may be adequately described by the above equation set since these flows verify the underlying hypothesis of a hydrostatic pressure distribution. Term I is the unsteady term representing local time variation. Term II is the convective variation. Pressure and diffusive contributions are only present in the velocity equations and are constituted by terms III and IV, respectively. Gravity effects owing to bed slope are embodied into the source term V. Friction shear with the bed and Coriolis acceleration resulting from the earth's rotation are described by terms VI and VII, respectively.

This paper concentrates on modelling situations where some part of the domain, which is initially dry, is subsequently flooded during the computation, or, conversely, initially wet areas are dried out by the retreat of the water front. These two actions may take place simultaneously, each on different sides of the same domain. Practically, these phenomena are present in floods resulting from natural storms, catastrophic failures of dams or dikes, flows over tidal flats, etc. Recently, we have also been made aware of the use of the wetting and drying SWWM to model the spreading of industrial liquid fuel or the resulting flammable cloud of heavy gas (liquid natural gas or liquid propane) for environmental safety (Brewer *et al.*, 1982).

With a time-scale of a few hours to several days, floods originating from natural storms and resulting in flows over floodplains of gentle slope are generally the slowest phenomena among the situations cited above. Reduced forms, usually neglecting the convective term II, of the SWWM [Equation (1)] have been used to model this specific situation (Di Giammarco and Todini, 1994). For flows caused by a dike failure, this approximation is rarely justifiable since local accelerations are large while a majority of dams are built on sloping slides usually followed by valleys with a complex topography. The complete form of the System (1) must consequently be considered for this category of flow, whose time-scale is of the order of several hours. The typical time-scale for the spreading of a liquid fuel or a flammable heavy gas usually takes place within the space of a few minutes to a maximum of about one hour, the time of safe dilution. The modelling of this last category also requires the complete System (1) to be used, augmented with some appropriate mass source term to account for evaporation and liquid spill discharge. A critical, non-exhaustive review of the propositions for modelling each of the above cited categories is presented in the next section.

The third section consists of specifying the framework of a finite volume technique using Riemann solvers that has been retained by the authors to explore the subject. The discretization discussed there is based, without loss of generality, on an underlying unstructured triangular mesh with elemental integration peformed on the associated median dual cell. A 'depth-positive' scheme derived from an adaptation of the Van Leer scheme for gas dynamics (van Leer, 1982) is presented. Numerical integration of the source terms, a subject that is usually omitted in Computational Fluid Dynamics (CFD) publications, is then discussed. Source terms include the bed friction, Coriolis acceleration and bottom slope. Linearization coupled with explicit–implicit switches appear to be the appropriate ingredient to achieve stability.

The fourth section is devoted to the special problems related to wetting and drying and how it may be treated within the proposed framework. It is shown that the adaptation of the van Leer scheme always predicts positive depth for both explicit and implicit time integration. Depth positivity is also analysed for the zero inertia model (ZIM). It is shown that the boundedness of the solution is dependent on the curvature of the bed topography. The same dependence on bottom curvature is shown for the SWWM velocity equation through the use of a theoretically derived necessary condition for the existence of the solution to the Riemann problem for the SWWM. Stability of the treatment for the closed boundaries, a purely two-dimensional space problem, is also tackled in relation to the same constraint imposed for the existence of a solution. The reasons why the popular Roe scheme may be unsuited to the modelling of wetting and drying are also exhibited in relation to the same constraint.

Elements of a more accurate mesh adaptation algorithm expected in the future are suggested in the fifth section. It appears that the preceding statement regarding the existence of a solution converts naturally into a reasonably sound basis for a mesh adaptation criterion.

Illustrative test cases of the proposed techniques are presented through figures and discussed in the sixth section. It includes smooth wave, bore and dam break run-up over adverse as well as descending slopes. The final section is devoted to some concluding remarks.

REVIEW OF THE LITERATURE

Zero inertia model

The case of a flood that is principally an overland flow on a plain of gentle slope or flows typically encountered in irrigation systems are some cases in which the convective acceleration may be neglected since water surface gradients and friction effects are dominant. Xanthopoulos and Koutitas (1976) considered a flood with characteristic velocity of order $U = 1$ m/s, characteristic depth $D = 1$ m, horizontal distance $L = 1000$ m, gravitational acceleration $g = 10$ m/s^2. For this case, the inertial acceleration (U^2/L) is estimated to be of the order of 10^{-3} and is negligible compared with the surface slope (g^*S_0) and friction

$(C^{-2}*U^2/D)$ terms, which are of the order of 10^{-2}. The Saint-Venant system reduces, in this case, to the zero inertia model:

$$\frac{\partial H}{\partial t} + \nabla \cdot (\vec{V}h) = 0 \tag{2}$$

$$\frac{\partial H}{\partial x} + \left(\frac{N^2 \cdot \sqrt{u^2 + v^2}}{h^{4/3}} \right) u = 0 \tag{3}$$

$$\frac{\partial H}{\partial y} + \left(\frac{N^2 \cdot \sqrt{u^2 + v^2}}{h^{4/3}} \right) v = 0 \tag{4}$$

The method adopted by Di Giammarco and Todini (1994) and Di Giammarco *et al.* (1994) is to insert the velocity expressions, deduced from the zero inertia momentum equations (3) and (4), into the continuity relation [Equation (2)] and obtain a diffusive parabolic equation for the depth

$$\frac{\partial H}{\partial t} = \frac{\partial}{\partial x}\left[k \frac{\partial H}{\partial x} \right] + \frac{\partial}{\partial y}\left[k \frac{\partial H}{\partial y} \right] \tag{5}$$

where

$$k = N^{-1} \cdot \varphi(\nabla H) \cdot h^{5/3} \text{ with } \varphi(\nabla H) = \left[\left(\frac{\partial H}{\partial x} \right)^2 + \left(\frac{\partial H}{\partial y} \right)^2 \right]^{-1/4} \tag{6}$$

with the diffusion coefficient k depending on the friction parameter. It will be explained why a stable numerical approximation to the zero inertia system is easily attainable compared with the full Saint-Venant system. Xanthopoulos and Koutitas (1976) performed a comparison of their ZIM results against some laboratory experimental results of dam break flows. They noted that as the initial head behind the dam increases, the fit between experimental and numerical results becomes less satisfactory. It is highly probable that this behaviour is a consequence of neglecting the convective terms, since a large initial head behind the dam would induce large downstream velocities and accelerations after dam failure.

It appears that mass conservation is the only law that is rigorously respected in the ZIM, with propagation of the front being mainly dictated by the surface gradient, while being tempered by friction effects. Simulations performed with the ZIM employed a fixed Eulerian grid size of the order of 1000 m. It is clear that in these situations the resolution attainable for the position of the front and associated velocities is crude. In fact, a scenario of this type would typically correspond to flooding from a natural storm over a large plain, where the main interest would be to model the extent of the flooded areas, with knowledge of water depths being only incidental.

Full shallow water wave model

Simulations of more rapidly varied events, such as dam break flows over steep topographic valleys, require the inclusion of inertia terms and thus the full solutions to the Saint-Venant system (SWWM). Numerical approximations to the SWWM implemented for drying and wetting processes have been proposed over fixed Eulerian or Lagrangian grids by Holsters (1976), Katopodes and Strelkoff (1978), Sakkas and Strelkoff (1973) and Zhang *et al.* (1992a), among others.

The Lagrangian grid is a natural alternative when one is faced with a moving boundary phenomenon. The implementation of a Lagrangian algorithm is feasible in one space dimension (Vasiliev, 1970; Di Monaco and Molinaro, 1984). However, the extension to two space dimensions is not a trivial task and this is probably why the published FE models for the full Saint-Venant system (Lynch and Gray, 1980; Akanbi and Katopodes, 1988) do not contain any practical applications. The criteria that should dictate the node movements are not simple. The Lagrangian algorithm may be summarized by the use of a modified SWWM where the fluxes are adapted to take into account node movement:

$$F_N^{LG} = \begin{bmatrix} h\cdot(\vec{V}-\vec{V}_G)\cdot\vec{N} \\ uh\cdot(\vec{V}-\vec{V}_G)\cdot\vec{N}+g\dfrac{h^2}{2}\cdot N_x \\ vh\cdot(\vec{V}-\vec{V}_G)\cdot\vec{N}+g\dfrac{h^2}{2}\cdot N_y \end{bmatrix} = F_N^{Euler} - F_N^{GRID}$$

where

$$F_N^{Euler} = F_N = \begin{bmatrix} h\cdot(\vec{V}\cdot\vec{N}) \\ uh\cdot(\vec{V}\cdot\vec{N})+g\dfrac{h^2}{2}\cdot N_x \\ vh\cdot(\vec{V}\cdot\vec{N})+g\dfrac{h^2}{2}\cdot N_y \end{bmatrix} F_N^{GRID} = (\vec{V}_G\cdot\vec{N})\cdot U = (\vec{V}_G\cdot\vec{N})\begin{bmatrix} h \\ uh \\ vh \end{bmatrix}$$

$\vec{V}_G$ is the velocity at which the grid nodes are moved. Its value may be determined locally (pointwise), as was done by Akanbi and Katopodes (1988), where $\vec{V}_G(x, y) = Fct[\vec{V}(x, y)]$, or globally, by a node relocation scheme according to a global grid velocity (Vasiliev *et al.*, 1970) or another mesh adaptation criterion. The modified flux, F_N^{LG}, preserves conservation over a moving reference frame.

Applications of the 2D SWWM over a Eulerian grid are numerous: for example, Holsters (1976), Rajar (1978), Falconer and Chen (1991) and Zhang *et al.* (1992a) in the FD context, and Kawahara and Umetsu (1986), Goutal (1987), Holz and Leister (1988), Zhang *et al.* (1992b), Bates and Anderson (1993) and many others using FE techniques. The proceedings of the conference on 'Modelling of Floods Over Initially Dry Areas' (Molinaro, 1994), which is devoted to the present subject, contains a number of communications that show real case applications. Most of these models use some steps of the following algorithm to handle wetting and drying areas.

(WD1) Identification of all wet nodes: depth $h > \varepsilon$, ε being the limiting minimum depth above which the node is declared wet. This step involves, for some FE models, other variants on this classification. For example, a dry element is an element in which all nodes are dry, an intermediate element is one in which at least one node is dry, while, finally, wet elements are those which have all their nodes wet.

(WD2) Application of the 'classical' scheme over all wet elements. Some models run the classical scheme over the whole grid, including wet, intermediate and dry elements. For these models some post-corrections are usually necessary.

(WD3) Use of an '*ad hoc*' scheme for evaluating the depth at all dry nodes. Use, for example, a convex mean of the following type (where the index 'nb' stands for 'neighbour'):

$$h_P = \alpha_P \cdot h_P + \Sigma_{nb} a_{nb} \cdot h_{nb} \tag{7}$$

with $\alpha_P > 0$, $a_{nb} > 0$ (convexity), and $\alpha_P + \Sigma_{nb} a_{nb} = 1$ (consistency with globally constant solution). Usually, only directly adjacent nodes that are linked to the reference node, indexed 'P', through a mesh segment, are used in the above summation, subject to the condition that the water level at this neighbouring node is greater than the bottom topography between the two nodes: $(z_{nb} + h_{nb}) > z_P$. Such convex interpolation schemes are not always globally conservative when combined with the application of the classical scheme over a wet element. Certain models apply conservative corrections. Conservation of mass constitutes the principal weakness of these algorithms.

(WD4) Application of a modified momentum equation over dry and intermediate elements. These modifications vary from, directly setting the velocity to zero, to application of the momentum equation without a bottom (friction) slope or a modified bottom (friction) slope term. It is almost a general rule that the momentum equation is never rigorously applied inside a fixed Eulerian grid model.

Meselhe and Holly (1993) expressed precisely the underlying philosophy of the solutions proposed up to now: 'In simulating a dry period, numerical accuracy during the dry period itself is not the primary consideration. Rather, the primary goal is to achieve accuracy and robustness before and after the dry period. Therefore, although the dry period itself is of no practical importance, its integration in the simulation of irrigation systems is extremely important.' The time may now have come when modellers will go a step further and ask for both robustness and accuracy. Before detailing our personal views on this subject, let us summarize this brief review of literature by specifying the problems to be solved.

Definition of the problems

Physics. The physics of the problem has already been described. In a practical sense, a river that has flowed along its traditional 'depression' or bed, will occasionally rise and flood surrounding plains, which may be flat or sloping, populated, forested, etc. These episodes are followed invariably by a period in which the flood subsides, water surface elevations decrease and a drying out of the flooded areas is initiated. This physically observed behaviour is tied to a moving boundary phenomenon which we would like to model.

Numerics. The challenges involved in the numerical representation of this problem are numerous. First, it is not possible in advance to delimit the space that will be affected (wetted) by the flows. There exist two principal methods that may be implemented to solve the problem.

1. A Lagrangian approach where a grid is continuously generated, modified and adapted to closely match the boundaries of the wetted region. This leads to complex algorithms for node displacements over a 2D complex topographic terrain. The feasibility of such an approach has, up to now, not been proven by a real case application.

2. Use of a fixed Eulerian grid. This approach supposes that the modeller is able to make an 'educated guess' about the region that will be affected by the flow field and consequently develop a mesh system that is most appropriate to the case being modelled. Most of the models reviewed here (including that of the authors) belong to this category. The numerical difficulties encountered for this class of schemes are identified below.

(NP1) The full Saint-Venant system is highly non-linear and simulation of wetting and drying areas involves mixed sub- and supercritical regimes. Traditional hydraulic schemes have historically suffered from instabilities when dealing with such flows. Depending on the numerical integration scheme employed, when the flow depth is zero, all other flow variables are set to zero, or the scheme is applied over the whole domain whether or not the depth is positive.

(NP2) The second category of numerical difficulty is associated with the use of a fixed Eulerian grid and may be designated 'the small depth anomaly'. This problem usually occurs within cells that are just partially wet. To make this point clear, let us explain how this is related to the assumption made regarding the shape of the topographic variable. Figure 1a,b shows that partial wetting of a cell can only happen in the case of a linearly varying bathymetry; for a 'staircase' bathymetry (Figure 1a), a cell is either completely dry or wet. In both cases, numerical instabilities usually originate in these 'frontal' cells in the form of a negative depth and/or too large velocity since the manner in which to perform a rigorous discretization of the basic conservation laws over the cell is not readily apparent. These unphysical negative depths may also originate from the use of a non-monotone scheme or from round-off errors in a region with small depth.

(a) Staircase representation of the bathymetry. Every cell is either wet or dry.

(b) Continuous representation of the bathymetry. Near the front, a partially wet cell may exist.

Figure 1. Possible representation of the bathymetry

(NP3) The third category of problems is related to the velocity prediction inside a cell that is partially wet. Since a methodology to apply the entire momentum equations over these cells does not present itself, it is essential to apply an appropriate control over the velocity in these cells, since velocity values that are too high may cause instabilities. This raises the question of the implications for accuracy of the proposed control, which has to be addressed.

Note that all these categories of anomalies are not completely independent. Unphysical negative depths may originate from the use of an improper scheme in a supercritical flow region, or an excessively large velocity may appear after division by a previously evaluated depth, which may be too small.

APPLICATION OF RIEMANN SOLVERS TO THE NUMERICAL MODELLING OF THE SWM

The numerical technique retained here for the integration of the SWWM set of equations is the finite volume technique. The system (SW1) is first written in matrix notation which is more suitable for the demonstration of the finite volume integral summations

$$U_t + \nabla \cdot [\overrightarrow{F(U)} - \overrightarrow{G(U)}] = S \tag{8}$$

where

$$U = [H, uh, vh]^t = [H, \vec{V}h]^t$$

is the vector of conserved variables,

$$\overrightarrow{F(U)} = \left[\vec{V}h, \vec{V} \otimes \vec{V}h + g\frac{h^2}{2} \cdot \bar{I}\right] = \begin{bmatrix} uh & vh \\ u(uh) + g\dfrac{h^2}{2} & u(vh) \\ v(uh) & v(vh) + g\dfrac{h^2}{2} \end{bmatrix}$$

is the convection pressure tensor,

$$\overrightarrow{G(U)} = [0, K_d \cdot \nabla(\vec{V}h)] = \begin{bmatrix} 0 & 0 \\ \dfrac{\partial(uh)}{\partial x} & \dfrac{\partial(uh)}{\partial y} \\ \dfrac{\partial(vh)}{\partial x} & \dfrac{\partial(vh)}{\partial y} \end{bmatrix}$$

the diffusion tensor, and

$$S(U) = \begin{bmatrix} 0 & 0 \\ -\left(g\dfrac{\partial Zb}{\partial x}\right)h - \left(g\dfrac{N^2\cdot\sqrt{u^2+v^2}}{h^{4/3}}\right)\cdot uh & R\cdot vh \\ -R\cdot uh & -\left(g\dfrac{\partial Zb}{\partial y}\right)h - \left(g\dfrac{N^2\cdot\sqrt{u^2+v^2}}{h^{4/3}}\right)\cdot vh \end{bmatrix}$$

is a global source term including bottom slope, bed friction and Coriolis acceleration:

$$S(U) = \begin{bmatrix} 0 & 0 & 0 \\ S_{bx} = -\left(g\dfrac{\partial Zb}{\partial x}\right) & C_{\mathrm{f}} = -\left(g\dfrac{N^2\cdot\sqrt{u^2+v^2}}{h^{4/3}}\right) & R \\ S_{by} = -\left(g\dfrac{\partial Zb}{\partial y}\right) & -R & C_{\mathrm{f}} = -\left(g\dfrac{N^2\cdot\sqrt{u^2+v^2}}{h^{4/3}}\right) \end{bmatrix}$$

$$\cdot \begin{bmatrix} H \\ uh \\ vh \end{bmatrix} + \begin{bmatrix} 0 \\ \dfrac{\partial(Zb^2/2)}{\partial x} \\ \dfrac{\partial(Zb^2/2)}{\partial y} \end{bmatrix}$$

i.e.

$$S(U) = MS\cdot U + Q \tag{9}$$

Among the numerical schemes for the integration of convection-dominated flow problems, Riemann solvers have become probably the most popular during the last decade. They constitute a class of finite volume schemes for which a strong relation with the physics is maintained during the numerics. The underlying methodology was pioneered by Godunov (1959): problem-dependent variables are represented at the discrete level by a set of grid cell integral means, so that the distribution is a supposed piecewise constant over a cell:

$$U_P^n = \frac{1}{Mes(\Omega_P)}\int_{\Omega_P} U(x, y, t^n)\, ds$$

Here, $Mes(\Omega_P)$ refers to 'measure of', in this case the surface area of the node control volume. This spatial distribution step is called projection, and is followed by a time evolution step that updates the problem variables from time t^n to time t^{n+1}. This is achieved by integrating the system (SW2) over each elementary control volume Ω_P, which is the median cell constructed as a dual partition of an underlying triangular mesh of FE type illustrated in Figure 2:

$$\int_{t^n}^{t^{n+1}}\left(\int_{\Omega_P} U_t(x, y, t)\, \mathrm{d}s\right)\mathrm{d}t = \int_{\Omega_P}\left(\int_{t^n}^{t^{n+1}} U_t(x, y, t)\, \mathrm{d}t\right)\mathrm{d}s = \int_{\Omega_P}[U(x, y, t^{n+1}) - U(x, y, t^n)]\, ds$$

$$= Mes(\Omega_P)\Delta_t U_P^n = U_P^{n+1} - U_P^n = \int_{t^n}^{t^{n+1}}\left\{\int_{\Omega_P} S[U(x, y, t)]\, \mathrm{d}s\right\}\mathrm{d}t - \int_{t^n}^{t^{n+1}}\left[\oint_{P1\,P2\,P3\,P4\,P5\,P6\,P7\,P1}(F(U)\cdot\vec{N})\, \mathrm{d}l\right] dt$$

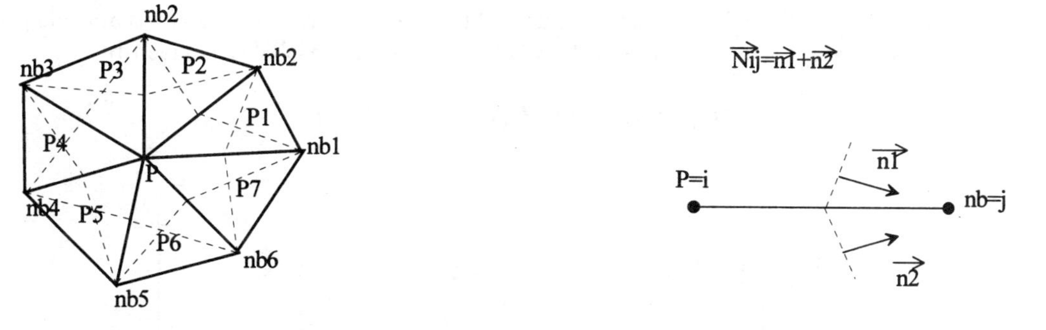

(a) Median cell: control volume associated to node P (P1P2P3P4P5P6P7P1)

(b) Normal of the boundary between nodes P and nb

Figure 2. Elementary median cell control volume and boundary normals

i.e.

$$U_P^{n+1} = U_P^n + \frac{1}{Mes(\Omega_P)}\left(\int_{t^n}^{t^{n+1}}\left\{\int_{\Omega_P} S[U(x,y,t)]\,\mathrm{d}s\right\}\mathrm{d}t - \int_{t^n}^{t^{n+1}}\left\{\sum_i\left[\int_{Pi}^{Pi+1}(F(U)\cdot\vec{N})\,\mathrm{d}l\right]\right\}\mathrm{d}t\right) \tag{10}$$

To perform this operation, the numerical fluxes at the interfaces between cells are needed. The Riemann problem designates precisely the model initial value problem with discontinuous piecewise constant initial distributions for the hyperbolic system:

$$U_t + \nabla \cdot F(U) = 0 \tag{11}$$

with initial distribution

$$\begin{aligned} U(x,y,t=0) &= U_L \quad \text{for} \quad (x, -x_\mathrm{I}, y - y_\mathrm{I})\cdot\vec{N} < 0 \\ U(x,y,t=0) &= U_R \quad \text{for} \quad (x, -x_\mathrm{I}, y - y_\mathrm{I})\cdot\vec{N} > 0 \end{aligned} \tag{12}$$

where $(x_\mathrm{I}, y_\mathrm{I})$ and $\vec{N}$ define, respectively, the position of the interface and its unit normal vector directed from the state at the left (U_L) toward the state at the right (U_R). It is pertinent to specify here how this one-dimensional Godunov algorithm is used within a 2D space context. First, recall the hypothesis:

- (HP1) The vector of the conserved variables $U = (H, uh, vh)^T$ is a supposed constant within a control volume and equal to the real distributed integral mean value. (refer to figure 3)
- (HP2) Zb, the variable representing the bottom elevation is not a 'problem variable' to be determined, but is supplied as part of the input. It is assumed, as in most FE schemes, that Zb varies linearly.

The implication of these two hypotheses is that the water depth, $h = H - Zb$, is not constant over the cell, so that at the interface between two control volumes, the water depth must be evaluated as

$$h_L = H_L - \frac{Zb_L + Zb_R}{2}$$
$$h_R = H_R - \frac{Zb_L + Zb_R}{2}$$

Various Riemann solvers differ in the way in which they estimate the solution (or the flux) of the model problem [Equations (11) and (12)] (Toro, 1992; Tchamen, 1996). The scheme proposed by van Leer for gas dynamics is adapted here for the SWWM because, as will be shown later, it has some favourable properties; in particular the positive depth constraint. Van Leer's flux for the 2D SWWM is given by

$$F_N^{VL}(U_L, U_R) = F_N^+(U_L) + F_N^-(U_R) \tag{13}$$

where

$$F_N^+(U) = \left[\begin{pmatrix} F_N(U) \\ F_N^+(1) = \dfrac{1}{4}hc(Fr_N + 1)^2 \\ F_N^+(2) = F_N^+(1)\left[u - \dfrac{Nx}{2}(V_N - 2c)\right] \\ F_N^+(3) = F_N^+(1)\left[v - \dfrac{Ny}{2}(V_N - 2c)\right] \\ 0 \end{pmatrix} \begin{matrix} Fr_N > 1 \\ \\ |Fr_N| < 1 \\ \\ Fr_N < -1 \end{matrix}\right] \tag{14}$$

$$F_N^-(U) = \left[\begin{pmatrix} 0 \\ F_N^-(1) = -\dfrac{1}{4}hc(Fr_N - 1)^2 \\ F_N^-(2) = F_N^-(1)\left[u - \dfrac{Nx}{2}(V_N + 2c)\right] \\ F_N^-(3) = F_N^-(1)\left[v - \dfrac{Ny}{2}(V_N + 2c)\right] \\ F_N(U) \end{pmatrix} \begin{matrix} Fr_N > 1 \\ \\ |Fr_N| < 1 \\ \\ Fr_N < -1 \end{matrix}\right] \tag{15}$$

with $V_N = \vec{V} \cdot \vec{N} = (u, v) \cdot (Nx, Ny)$, $c = \sqrt{gh}$ and $Fr_N = V_N/c$ designating, respectively, the normal component of the velocity, c the celerity of a gravity wave, and Fr_N the 'Froude number in the normal direction'.

Source term integration

The presentation of the treatment of the source terms will be made in the spirit of a 'splitting process' approach, i.e. disregarding all other terms but the unsteady derivative

$$U_t = S(U) \tag{16}$$

For a linear source term in the scalar case [i.e. $S(U) = ds \cdot U + q$], the solution to Equation (16) is

$$U(t) = U(0) \cdot \exp(ds \cdot t) + \frac{q}{ds} \cdot [\exp(ds \cdot t) - 1] \tag{17}$$

and the generalization to the complete system is traditionally achieved through an eigenmode decomposition

$$U(t) = \sum_{k=1}^{k=p} \left\{ U_k^0 \cdot \exp(ds_k \cdot t) + \frac{q_k}{ds_k} \cdot [\exp(ds_k \cdot t) - 1] \right\} \cdot VS_k \tag{18}$$

where ds_k and VS_k $(k = 1, \ldots, p)$ are, respectively, the eigenvalues and the associated eigenvectors of the matrix S. U_k^0 and q_k are the coefficients of the decomposition of the initial solution $U(0)$ and of the constant component Q of the source term, respectively, decomposed over the basis of the VS_k. This classical solution forms the basis of the integration of the friction and bed slope source term described in Tchamen (1998). However, in the present case a much simpler solution with closed accuracy has been retained.

Friction, Coriolis and bed slope terms. The friction, Coriolis and topography source terms for the SWWM may be written in the linearized form $S(U) = MS \cdot U + Q$ [Equation (9)] where

$$MS = \begin{bmatrix} 0 & 0 & 0 \\ S_{bx} = -g\dfrac{\partial z}{\partial x} & C_f = -g\dfrac{N^2\sqrt{u^2+v^2}}{h^{4/3}} & R \\ S_{by} = -g\dfrac{\partial z}{\partial y} & -R & C_f = -g\dfrac{N^2\sqrt{u^2+v^2}}{h^{4/3}} \end{bmatrix} \quad \text{and} \quad Q = \begin{bmatrix} 0 \\ \dfrac{\partial(Zb^2/2)}{\partial x} \\ \dfrac{\partial(Zb^2/2)}{\partial y} \end{bmatrix}$$

Integration of Equation (16) is performed by means of a backward Euler scheme

$$U^{n+1} = [I - \Delta t \cdot MS]^{-1} \cdot [U^n + \Delta t \cdot Q] \tag{19}$$

which is preferred here to the forward Euler scheme otherwise used:

$$U^{n+1} = [I + \Delta t \cdot MS] \cdot U^n + \Delta t \cdot Q \tag{20}$$

This choice is motivated by numerical stability considerations. The amplification matrix $[I + \Delta t \cdot MS]$ of Equation (19) has eigenvalues 1 and $1 + \Delta t \cdot (C_f \pm i \cdot F)$, whereas Equation (20) has eigenvalues 1 and $1/1 - \Delta t \cdot (C_f \pm i \cdot F)$. A Fourier stability analysis would reveal that the backward Euler is unconditionally stable since all the eigenvalues of its amplification matrix are less than or equal to 1. On the other hand, the same Fourier stability condition of eigenvalues being less than 1 applied to the forward Euler scheme would imply an upper bound on the allowable time-step. Another argument in favour of the use of the backward Euler for the integration of the source term is provided by the Taylor series development around zero:

$$\exp(ds \cdot \Delta t) = 1 + \sum_{n=1}^{n=\infty} \frac{(ds \cdot \Delta t)^n}{n!} \tag{21}$$

for the exact solution, Equation (17), and

$$\frac{1}{1 - ds \cdot \Delta t} = 1 + \sum_{n=1}^{n=\infty} (ds \cdot \Delta t)^n \tag{22}$$

for the backward Euler scheme which 'are much closer' to each other than to

$$1 + ds \cdot \Delta t \tag{23}$$

which is only a first-order approximation of the two preceding developments. Note that Equations (21) and (22), although they have the same limit at 0 and $+\infty$, do not generally tend exactly to the same limit. Equation (21) tends to 0 more rapidly than Equation (22) does when S_f tends to $+\infty$.

It has to be pointed out that, strictly speaking, the friction source term is non-linear, and the linearization performed before deriving this analysis is tantamount to 'freezing' the term

$$C_f = -g\frac{N^2\sqrt{u^2+v^2}}{h^{4/3}}$$

at the initial time-step.

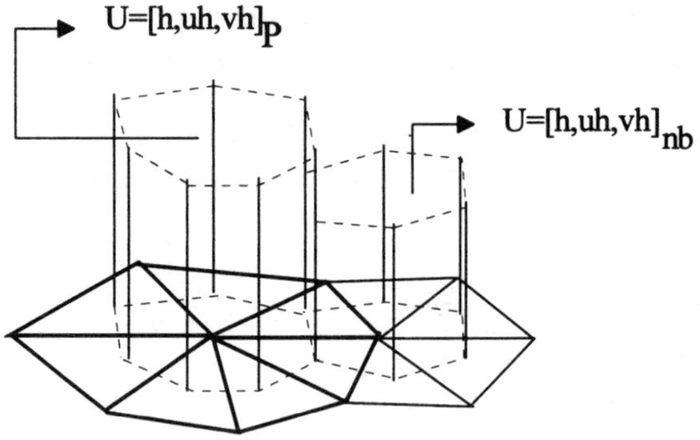

Figure 3. Schematization of the piecewise constant distribution of the dependent variables over the elementary control volumes used for the local Riemann problem

The following section will show how the previously identified problems are tackled in the context of the Van Leer Riemann solver.

'OUR SOLUTION TO SPECIAL PROBLEMS'

Mixed sub/supercritical flows

Traditional hydraulic schemes, such as the Preissman scheme, which is presented in detail in Cunge *et al.* (1980), suffer from instabilities when dealing with mixed sub/supercritical flows. In our opinion, this problem arises from the fact that the stencil of the scheme (i.e. the set of points used to discretize the differential laws at a particular node), remains unchanged when the flow changes from one regime to the other (sub- to supercritical or vice versa). This contradicts what is known from the physics, which is also confirmed by the mathematical properties of the governing system. It is explained in classical hydraulic textbooks (for example, Stoker, 1957; Henderson, 1966; Abbott, 1979) that the eigenvalues of the Jacobian matrix $\vec{V}\cdot\vec{N}-c$, $\vec{V}\cdot\vec{N}$, $\vec{V}\cdot\vec{N}+c$ represent, in fact, the magnitude and direction of the velocity at which the information related to the family of three elementary waves travel through the domain in the direction defined by the unit vector $\vec{N}$. The subcritical regime corresponds to the case where the three eigenvalues are of mixed signs. Conversely, the case where all the eigenvalues are of the same sign defines the supercritical regime. In accordance with this theoretical basis, an appropriate scheme for the simulation of mixed regime flows should adapt its stencil in the supercritical regime in a way that causes it to line up along the upstream region (from where the information originates). This may explain, at least in part, some of the difficulties experienced by the Preissman scheme (Cunge *et al.*, 1980) when applied to mixed flow regimes. It is also noteworthy that when the inertial terms are dropped from the momentum equations, the eigenvalues are always of mixed sign (i.e. subcritical regime), and stabilizing techniques for the simulation of wetting and drying are feasible (Meselhe and Holly, 1993).

An examination of the proposed van Leer scheme of integration shows that when $MIN(Fr_N^{\mathrm{L}}, Fr_N^{\mathrm{R}}) \geqslant 1$ then $F_N^{\mathrm{LR}}(U_{\mathrm{L}}, U_{\mathrm{R}}) = F_N(U_{\mathrm{L}})$, and, when $MAX(Fr_N^{\mathrm{L}}, Fr_N^{\mathrm{R}}) \leqslant -1$ then $F_N^{\mathrm{LR}}(U_{\mathrm{L}}, U_{\mathrm{R}}) = F_N(U_{\mathrm{R}})$.

It appears that Van Leer's scheme respects the theoretical behaviour at the time of switching from one regime to another. Note that since $Fr_N = V_N/c$, the limits $Fr_N = \pm 1$ correspond, respectively, to the upper and the lower values for which the eigenvalues are all of the same sign, i.e. the flow enters the supercritical regime. In fact, all the schemes that belong to the Riemann solver category share this property, which is one of the reasons for their robust behaviour through mixed regimes and highly convective flows. This partially explains their popularity over the last decade.

Positive depth schemes

In recent years, interest in so called 'positive schemes' has been expressed in the computational fluid dynamics (CFD) literature (Eindfelt *et al.*, 1991; Struijs *et al.*, 1991; Hundsdorfer *et al.*, 1995). This does not correspond exactly to what is being referred to here as a 'positive depth scheme'. An attempt to clarify the similarities and differences between these two concepts will be made here. Consider the one-dimensional linear advection equation for a physical property $c(x, t)$ (concentration, for example)

$$\frac{\partial c}{\partial t} + u\frac{\partial c}{\partial x} = 0 \tag{24}$$

A scheme is said to be *positive* if it always predicts a positive value for c_i^{n+1} in the solution of Equation (24) when the time-step is within the stability bounds. Note that this requirement is physically motivated when c represents a quantity such as mass, density or concentration. In practice, a sufficient condition is used: *a scheme is positive* if its update formula may be written in the form

$$c_i^{n+1} = \sum_{k=-m}^{k=p} (a_k \cdot c_{i+k}^n) = H(c_{-m}^n, \ldots, c_{i-1}^n, c_i^n, c_{i+1}^n, \ldots, c_p^n) \tag{25}$$

with

$$a_k \geqslant 0 \tag{26}$$

Note that if Equations (25) and (26) are satisfied, then when c_i^n are non-negative, then the predicted updates, c_i^{n+1}, are also non-negative. It ensures that for quantities positive by nature, coherent initial values guarantee positivity of all predicted future values.

A scheme is said to be *monotone* if the condition

$$\frac{\partial H}{\partial c_k^n}(c_{-m}^n, \ldots, c_{i-1}^n, c_i^n, c_{i+1}^n, \ldots, c_p^n) \geqslant 0$$

$$k = -m, \ldots, i-1, i, i+1, \ldots, p$$

is verified. For a linear scheme, i.e.

$$\frac{\partial H}{\partial c_k^n} = a_k$$

in this case, a_k is independent of the c_j for $k, j = -m$ to p; this condition is the same as the positivity constraint (Equations (25) and (26)).

The '*total variation*' of c_i^n, defined by $TV(c^n) = \Sigma_i |c_i^n - c_{i-1}^n|$, satisfies $TV(c^{n+1}) \leqslant TV(c^n)$. This property is called *total variation diminishing* (TVD). It is easy to verify that a monotone scheme satisfies the TVD property.

1. By writing the difference $c_i^{n+1} - c_{i-1}^{n+1}$ using Equation (25), there results

$$c_i^{n+1} - c_{i-1}^{n+1} = \sum_{k=-m}^{k=p} [a_k \cdot (c_{i+k}^n - c_{i-1+k}^n)]$$

2. $TV(c^{n+1}) = \Sigma_i |c_i^{n+1} - c_{i-1}^{n+1}| = \Sigma_i |\Sigma_{k=-m}^{k=p} [a_k \cdot (c_{i+k}^n - c_{i-1+k}^n)]| \leqslant \Sigma_i \Sigma_{k=-m}^{k=p} [a_k \cdot |c_{i+k}^n - c_{i-1+k}^n|]$, since the a_k are non-negative (Constraint (26)).
3. For consistency, with the uniform solution, the a_k needs to satisfy $\Sigma_{k=-m}^{k=p} a_k = 1$.

From the three points elaborated above, it follows that

$$TV(c^{n+1}) \leqslant \sum_i \sum_{k=-m}^{k=p} [a_k \cdot |c_{i+k}^n - c_{i-1+k}^n|] = \sum_j \sum_{i=j-m}^{i=j+p} [a_{j+i} \cdot |c_j^n - c_{j-1}^n|] = \sum_j \left[|c_j^n - c_{j-1}^n| \sum_{i=j-m}^{i=j+p} a_{j+i} \right]$$

i.e.

$$TV(c^{n+1}) \leqslant \sum_i |c_i^n - c_{i-1}^n| = TV(c^n)$$

All these concepts, the latest being introduced in the CFD community to analyse the stability and the consistency of schemes, are recalled here to show the relationships between themselves and with the notion of 'depth positivity'.

The next example shows how the above concepts are related to the stability. Recall two classical finite difference integration schemes applied to Equation (24), the first being

$$c_i^{n+1} = \left(1 - \frac{u \cdot \Delta t}{\Delta x}\right) c_i^n + \frac{u \cdot \Delta t}{\Delta x} c_{i-1}^n \tag{27}$$

which is the classical 'first-order upwind' scheme, and

$$c_i^{n+1} = c_i^n + \frac{u \cdot \Delta t}{\Delta x} c_{i-1}^n - \frac{u \cdot \Delta t}{\Delta x} c_{i+1}^n \tag{28}$$

the 'forward–time centred space' (FTCS) scheme.

The requirement for a scheme to predict a positive value for c_i^{n+1}, assuming positive initial data $[\ldots c_{i-1}^n, c_i^n, c_{i+1}^n, \ldots]$ is equivalent to requiring that the update of a monotone function will be a monotone function. It is a remarkable fact that if the requirement of positivity is applied to the two preceding schemes, it appears that the first-order upwind is positive if, and only if, the Courant-Friedrichs-Lewy (CFL) condition

$$\frac{u \cdot \Delta t}{\Delta x} \leqslant 1 \tag{29}$$

is satisfied, while it appears that the FTCS scheme is unable to satisfy the positivity constraint. These two results are in exact concordance with what would be obtained from a linearized Fourier stability analysis. This fact illustrates that the notions of 'linear stability in the Fourier sense', 'positivity' and 'monotonicity' are all equivalent in the linear case. These notions will not be developed further here. The reader interested in a more detailed presentation is referred to the book by Hirsch (1990) where these concepts are extensively defined and illustrated.

Extension of these properties to a system of equations is not automatic. This is particularly true for the positivity, which is the property that now concerns us. To illustrate this, let us consider the following one-dimensional version of the linearized SWWM without the source term

$$\frac{\partial h}{\partial t} + u_0 \frac{\partial h}{\partial x} + h_0 \frac{\partial u}{\partial x} = 0$$

$$\frac{\partial u}{\partial t} + g \frac{\partial h}{\partial x} + u_0 \frac{\partial u}{\partial x} = 0$$

which may be written in matrix form as

$$\frac{\partial U}{\partial t} + A \cdot \frac{\partial U}{\partial x} = 0 \tag{30}$$

where

$$U = \begin{bmatrix} h \\ u \end{bmatrix} \quad \text{and} \quad A = \begin{bmatrix} u_0 & h_0 \\ g & u_0 \end{bmatrix} \tag{31}$$

A natural way to extend the first-order upwind scheme to the coupled system [Equations (30) and (31)] is to apply the scalar scheme [Equation (27)] to a decoupled transform of Equations (30) and (31). Fortunately for this case, the decoupling of Equations (30) and (31) is easily performed through the operation of Equation (30) with the matrix of the left eigenvectors of the matrix A:

$$R^{-1} \cdot \left(\frac{\partial U}{\partial t} + A \cdot \frac{\partial U}{\partial x} \right) = 0 \tag{32}$$

where

$$A = R \cdot D \cdot R^{-1} = (2gc_0)^{-1} \cdot \begin{bmatrix} -c_0 & c_0 \\ g & g \end{bmatrix} \cdot \begin{bmatrix} u_0 - c_0 & \\ & u_0 + c_0 \end{bmatrix} \cdot \begin{bmatrix} -g & c_0 \\ g & c_0 \end{bmatrix}$$

Equations (30) and (31) produce the following pair of uncoupled linear equations:

$$\frac{\partial(u \cdot c_0 - g \cdot h)}{\partial t} + (u_0 - c_0) \frac{\partial(u \cdot c_0 - g \cdot h)}{\partial x} = 0 \tag{33}$$

$$\frac{\partial(u \cdot c_0 + g \cdot h)}{\partial t} + (u_0 + c_0) \frac{\partial(u \cdot c_0 + g \cdot h)}{\partial x} = 0 \tag{34}$$

Let us assume, without loss of generality, that $u_0 \geqslant 0$; application of the first-order upwind scheme would then produce

$$(u \cdot c_0 - g \cdot h)_i^{n+1} = \left[1 + (u_0 - c_0) \frac{\Delta t}{\Delta x} \right] \cdot (u \cdot c_0 - g \cdot h)_i^n - (u_0 - c_0) \cdot \frac{\Delta t}{\Delta x} \cdot (u \cdot c_0 - g \cdot h)_{i+1}^n \tag{35}$$

$$(u \cdot c_0 - g \cdot h)_i^{n+1} = \left[1 - (u_0 - c_0) \frac{\Delta t}{\Delta x} \right] \cdot (u \cdot c_0 - g \cdot h)_i^n + (u_0 - c_0) \cdot \frac{\Delta t}{\Delta x} \cdot (u \cdot c_0 - g \cdot h)_{i-1}^n \tag{36}$$

which correspond respectively to the subcritical $[(u_0 - c_0) \leqslant 0]$ and the supercritical $[(u_0 - c_0) \geqslant 0]$ discretization of Equation (33). The second equation (34) would produce

$$(u \cdot c_0 + g \cdot h)_i^{n+1} = \left[1 - (u_0 + c_0) \frac{\Delta t}{\Delta x} \right] \cdot (u \cdot c_0 + g \cdot h)_i^n + (u_0 + c_0) \cdot \frac{\Delta t}{\Delta x} \cdot (u \cdot c_0 + g \cdot h)_{i-1}^n \tag{37}$$

Depth relations are deduced from the above discretizations by subtracting Equation (35) from Equation (37) in the subcritical case

$$\begin{aligned} (2g \cdot h)_i^{n+1} &= \left[(u_0 + c_0) \cdot \frac{\Delta t}{\Delta x} \right] \cdot (g \cdot h)_{i-1}^n + \left(2 - 2c_0 \frac{\Delta t}{\Delta x} \right) \cdot (g \cdot h)_i^n + \left[(c_0 - u_0) \cdot \frac{\Delta t}{\Delta x} \right] \cdot (g \cdot h)_{i+1}^n \\ &+ \left[(u_0 + c_0) c_0 \cdot \frac{\Delta t}{\Delta x} \right] \cdot u_{i-1}^n - \left(2c_0 \cdot u_0 \frac{\Delta t}{\Delta x} \right) \cdot u_i^n + \left[(u_0 - c_0) \cdot c_0 \frac{\Delta t}{\Delta x} \right] \cdot u_{i+1}^n \end{aligned} \tag{38}$$

and subtracting Equation (36) from Equation (37) for the supercritical case, to produce

$$(2g \cdot h)_i^{n+1} = \left(2u_0 \cdot \frac{\Delta t}{\Delta x} \right) \cdot (g \cdot h)_{i-1}^n + \left(2 - 2u_0 \frac{\Delta t}{\Delta x} \right) \cdot (g \cdot h)_i^n + \left(2c_0^2 \cdot \frac{\Delta t}{\Delta x} \right) \cdot (u_{i-1}^n - u_i^n) \tag{39}$$

It appears that all the coefficients of the h_k^n $(k = i-1,\ i,\ i+1)$ in Equations (38) and (39) are positive as soon as the Courant condition, $(u_0 + c_0) \cdot \Delta t/\Delta x \leqslant 1$, is satisfied. But the 'source term' of these equations are not automatically positive. In Equation (39), for example, the sign of the source term is determined by the sign of the difference $(u_{i-1}^n - u_i^n)$. Consequently, the positivity of the predicted depth (h_i^{n+1}), is subject to a relationship of the form

$$\alpha \cdot (u_{i-1}^n - u_i^n) + \beta \cdot \bar{c}_{i-1}^n + \gamma \cdot \bar{c}_i^n \geqslant 0 \tag{40}$$

with $[(\alpha,\ \beta,\ \gamma) \geqslant 0]$. An interesting consequence of this analysis is that adaptation of the Roe scheme to the SWWM (Glaister, 1988) may predict negative depth since this scheme involves a linearization of the type in Equations (30) and (31), where

$$h_0 = \frac{h_{i-1} + h_i}{2} \quad \text{and} \quad u_0 = \frac{u_{i-1}\sqrt{h_{i-1}} + u_i\sqrt{h_i}}{\sqrt{h_{i-1}} + \sqrt{h_i}}$$

Later on, in the section on velocity control, it will be shown that a necessary condition for the existence of a solution to the Riemann problem in Equation (45) has the form of Equation (40).

The important conclusion of the preceding relationship is that extension of the first-order upwind scheme to a coupled system does not automatically produce a depth-positive scheme. Most of the schemes that appear in the literature as 'positive', 'monotone' or 'TVD' have only been verified for the scalar transport equation (24). For our purpose, we are interested in the subset of this class that always ensures a positive depth when applied to the full non-linear coupled SWWM. A sufficient condition that this requirement be met is that the scheme for the depth relationship may be written in the following form:

$$h_i = q + \sum_{nb} a_{nb} \cdot h_{nb} \tag{41}$$

with the coefficients q and a_{nb} being non-negative $[(q,\ a_{nb}) \geqslant 0]$. No assumption of linearity, i.e. q and a_{nb} being constant, is made, so that non-linear schemes may also be encompassed. Equation (41) also applies to non-linear schemes. The 'positive depth scheme' (PDS) is fairly close to the concept of the 'positive coefficients rule' presented by Patankar (1980).

It is possible to verify that the Van Leer scheme defined earlier belongs to the family of PDS by inserting the Van Leer fluxes [Equations (13)–(15)] in the update formula, Equation (10). The depth component of this relationship reduces to

$$h_P^{n+1} = \left[1 - \frac{\Delta t}{Mes(\Omega_P)} \cdot \sum_{k=1}^{k=Max\_nb} (CVF_N^+)_P \cdot l_{Pk}\right] \cdot h_P^n - \frac{\Delta t}{Mes(\Omega_P)} \cdot \sum_{k=1}^{k=Max\_nb} [(CVF_N^-)_k \cdot l_{Pk}] \cdot h_k^n$$

where

$$CVF_N^+ = \begin{bmatrix} V_N & \text{when} & Fr_N \geqslant -1 \\ \frac{c}{4}(Fr_N + 1)^2 & \text{when} & -1 \leqslant Fr_N \leqslant 1 \\ 0 & \text{when} & Fr_N \leqslant -1 \end{bmatrix} \quad \text{and}$$

$$CVF_N^- = \begin{bmatrix} 0 & \text{when} & Fr_N \geqslant -1 \\ -\frac{c}{4}(Fr_N - 1)^2 & \text{when} & -1 \leqslant Fr_N \leqslant 1 \\ V_N & \text{when} & Fr_N \leqslant -1 \end{bmatrix}$$

It appears that this scheme always predicts positive depth provided that the CFL-type condition

$$\Delta t \leqslant \frac{Mes(\Omega_P)}{[Max(V_N, c)]_P \cdot \sum_{nb=1}^{nb=Max\_nb} (l_{Pnb})} \leqslant \frac{Mes(\Omega_P)}{\sum_{nb=1}^{nb=Max\_nb} (CVF_N^+) \cdot (l_{Pnb})}$$

is satisfied. In the Appendix, an extension of the Van Leer scheme to implicit time integration, together with the condition that must be fulfilled to keep depth positive, is presented.

To close this section, it will be shown through analysis of the zero inertia model that although 'positive depth' is a desirable property, it is not sufficient to guarantee stable bounded solutions. Positive depth is much easier to realize in the ZIM than in the full SWWM. Integration of Equation (5) over a control volume of the type shown in Figure 2a, together with use of the divergence theorem, produces

$$Mes(\Omega_P) \cdot \int \frac{\partial H}{\partial t} \mathrm{d}t = Mes(\Omega_P) \Delta_t H_P^n = \int_{t^n}^{t^{n+1}} \left[\oint_{\partial \Omega_P} \left(\frac{\partial H}{\partial x} N_x + \frac{\partial H}{\partial y} N_y \right) dl \right] d$$

A simple approximation to the gradient in the normal direction of the type

$$\left(\frac{\partial H}{\partial x} N_x + \frac{\partial H}{\partial y} N_y \right)_{Pnb} = \left(\frac{\partial H}{\partial n} \right)_{Pnb} \approx \frac{H_P - H_{nb}}{\| Pnb \|}$$

would give

$$H_P^{n+1} = \left(1 - \frac{\Delta t}{Mes(\Omega_P)} \sum_{nb=1}^{nb=Max\_nb} k_{nb} \cdot l_{nb} \right) H_P^n + \frac{\Delta t}{Mes(\Omega_P)} \cdot \sum_{nb=1}^{nb=Max\_nb} H_{nb}^n$$

for explicit time integration. All coefficients would be positive provided that

$$\Delta t \leqslant \frac{Mes(\Omega_P)}{\sum_{nb=1}^{nb=Max\_nb} k_{nb} \cdot l_n}$$

Implicit time integration on the other hand would produce

$$\left(1 + \frac{\Delta t}{Mes(\Omega_P)} \sum_{nb=1}^{nb=Max\_nb} k_{nb} \cdot l_{nb} \right) H_P^{n+1} = H_P^n + \frac{\Delta t}{Mes(\Omega_P)} \cdot \sum_{nb=1}^{nb=Max\_nb} H_{nb}^{n+1}$$

which is unconditionally positive. Note that in the preceding relationships, the dependent variable is H, the free surface elevation, and not h, the water depth. It is, of course, possible to switch to a relationship for the water depth by inserting the relation $h(x, y, t) = H(x, y, t) - Zb(x, y)$ in the original equation (5):

$$\frac{\partial h}{\partial t} = \frac{\partial}{\partial x}\left[k \frac{\partial h}{\partial x} \right] + \frac{\partial}{\partial y}\left[k \frac{\partial h}{\partial y} \right] + \frac{\partial}{\partial x}\left[k \frac{\partial Zb}{\partial x} \right] + \frac{\partial}{\partial y}\left[k \frac{\partial Zb}{\partial y} \right] \tag{42}$$

which has exactly the same form as Equation (5) plus an additional term

$$\frac{\partial}{\partial x}\left[k \frac{\partial Zb}{\partial x} \right] + \frac{\partial}{\partial y}\left[k \frac{\partial Zb}{\partial y} \right] \tag{43}$$

After recalling the definition of $k = N^{-1} \cdot \varphi(\nabla H) \cdot h^{5/3}$, Equations (38) and (39) may be expanded into

$$\left[N^{-1} \cdot \varphi(\nabla H) \cdot h^{2/3}\left(\frac{\partial Zb}{\partial x}, \frac{\partial Zb}{\partial y}\right)\right] \cdot \nabla h + [(\Delta Zb) \cdot N^{-1} \cdot \varphi(\nabla H) \cdot h^{2/3}] \cdot h \tag{44}$$

This last equation (44) includes a convection related to the gradient of h which may be inserted into Equation (42), while preserving the positive depth property by upwind discretization. The second part of Equation (44) is written like a linearized source term of degree one in h. The sign of the source term is determined by the term

$$\Delta Zb = \frac{\partial^2 Zb}{\partial x^2} + \frac{\partial^2 Zb}{\partial y^2}$$

which determines the bottom curvature. The result is that a concave bottom will correspond to a negative source term, while a convex bottom corresponds to a positive one. To maintain depth positivity with respect to such a term, a simple idea is the approximation

$$(\Delta Zb)h = \frac{1 + Sgn(\Delta Zb)}{2} h^n + \frac{1 - Sgn(\Delta Zb)}{2} h^{n+1}$$

In the section on source term integration it was recalled that such a source term has a solution whose behaviour follows $h(t) = h(0) \cdot \exp(at)$ [Equation (17)], for a source term of the form $S(h) = a \cdot h$. It appears that a concave bed leads to a bounded solution, whereas a convex one may produce unbounded depths in the solution of Equation (42). This behaviour is still present in the full SWWM and will be demonstrated through a velocity analysis.

Velocity control

In the general case, an analysis of the velocity field predicted by a numerical model is quite complex since no arguments analogous to the 'positive depth' may be appealed to. However, close analysis of the particular case of pure one-dimensional flow, for which physical and theoretical arguments suggest a uniform sign for the velocity and where a numerical time-explicit integration of the bed friction source term exhibits a change in the sign of the velocity near important source terms, leads to the conclusions expressed above about the integration of such terms.

While performing a theoretical derivation of the Riemann problem for the SWWM, the present authors (Tchamen, 1996) found a necessary condition for the existence of a solution

$$(V_N)_R - 2c_R \leqslant (V_N)_L + 2c_L \tag{45}$$

It has been found through numerical experimentation that most solutions to the Riemann solvers (Godunov, 1959; Roe, 1981; Osher, 1982) have displayed instabilities (mainly negative depth) when the condition in Equation (45) was violated. Practically, it appears that a 'too strong' bottom curvature may be one of the main reasons for the violation of Equation (45). For a convex bottom,

$$\left(\frac{\mathrm{d}Zb}{\mathrm{d}x}\right)_R < \left(\frac{\mathrm{d}Zb}{\mathrm{d}x}\right)_L$$

and so the velocity increase owing to the bottom slope is more important for the downstream cell, indexed 'R', than for the upstream cell, 'L':

$$-g \cdot \left(\frac{\mathrm{d}Zb}{\mathrm{d}x}\right)_L \cdot \Delta t < -g \cdot \left(\frac{\mathrm{d}Zb}{\mathrm{d}x}\right)_R \cdot \Delta t$$

It is clear that a sufficiently large time-step, combined with a strong local curvature of a convex shaped bed $[d^2Zb/dx^2 < 0]$, may subsequently lead to violation of Equation (41). In conjunction with the increase of $(V_N)_R$ due to the bottom curvature, a very small depth $[h_R(c_R)]$ will drive the solution to further violate Equation (41). This behaviour generally takes place in a partially dry frontal cell for the following reason:

- $h_R \sim H_R \sim \Delta H_R = (H_R^{n+1} - H_R^n) \sim \Delta t \cdot (\Sigma V_N h) \cdot |\partial\Omega_R|/Mes(\Omega_R)$ [see Equation (10)], which is inversely proportional to the size $[Mes(\Omega_R)/|\Omega_R|]$ of the mesh.
- $u_R \sim \Delta u_R = (u_R^{n+1} - u_R^n) \sim \Delta(uh)_R/h_R \sim \Delta t \cdot (\Sigma u \cdot V_N h + g(h^2/2)) \cdot |\partial\Omega_R|/h_R \cdot Mes(\Omega_R) \sim (\Sigma u \cdot V_N h + g(h^2/2))/(\Sigma V_N h)$ is not directly (in a single computational step) dependent on the mesh.

These two observations explain how the depth can become infinitely small at a computational node simply because of the size of the related mesh, while the velocity remains almost unaffected. Note that this behaviour is also linked to the use of the conservation form of the SWWM for the discretization.

Another practical situation where violation of Equation (45) weakens the stability of the solution is the case of a closed downstream boundary in supercritical flows. Figure 4 illustrates this particular situation. A problem that has begun to surface in the recent literature deals with the so-called 'two-dimensional dam-break' (Alcrudo and Garcia-Navarro, 1993; Tchamen, 1992; Ambrosi, 1995) and is illustrated in Figure 5.

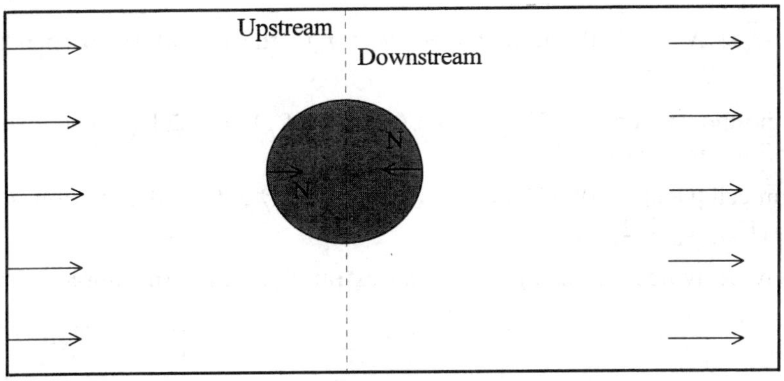

Figure 4. Sketch of an immersed solid closed boundary

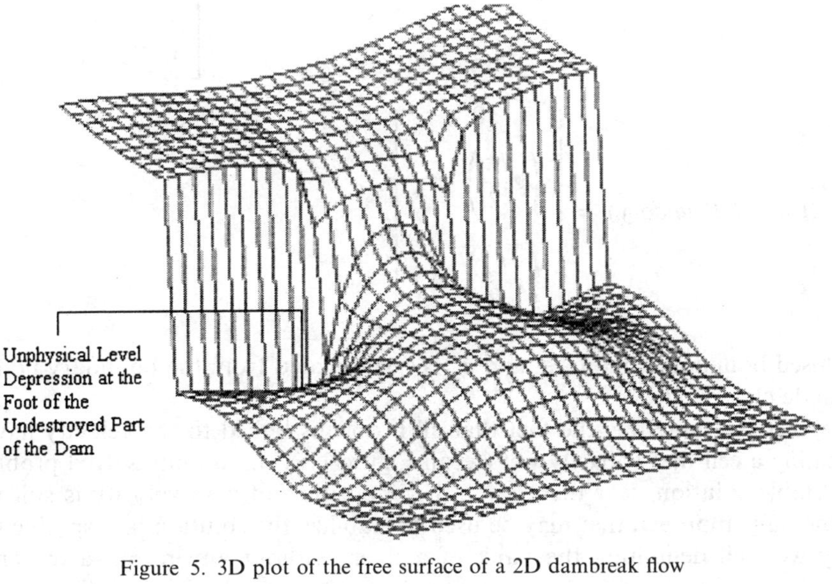

Figure 5. 3D plot of the free surface of a 2D dambreak flow

Downstream, at the foot of the undestroyed part of the dam, non-physical free surface depressions are clearly visible. Personal experience from numerous experiments (Tchamen, 1992) has shown that if the initial depth downstream of the dam is not sufficiently large, negative depths tend to occur at these areas of free surface depression. This is possibly the reason why no case of a dambreak with an initially dry downstream bed has ever been presented in the published literature, especially for any model using the Roe Riemann solver (Alcrudo and Garcia-Navarro, 1993; Tchamen, 1992; Ambrosi, 1995). An explanation for this behaviour may be hidden in the following observations.

- One of the most popular ways of dealing with some types of boundary conditions is to use the 'mirror cell' concept.
- In this context, assuming that the normal between the boundary cell indexed 'L' and the dummy cell indexed 'R' is directed outside the flow domain, the variables in the dummy cell are defined by

$$h_R = h_L$$
$$(\vec{V} \cdot \vec{N})_R = -(\vec{V} \cdot \vec{N})_L$$
$$(\vec{V} \cdot \vec{T})_R = (\vec{V} \cdot \vec{T})_L$$

where $\vec{T} = -N_y \cdot \vec{l}_x + N_x \cdot \vec{l}_y$ is the unit tangent vector to the boundary (perpendicular to the normal vector $\vec{N} = N_x \cdot \vec{l}_x + N_y \cdot \vec{l}_y$).

- For upstream boundary cells $[(V_N)_L > 0]$, $(V_N)_R - (V_N)_L = -2(V_N)_L < 0$, so the condition in Equation (45) will never be violated.
- For a downstream cell $[(V_N)_L < 0]$, $(V_N)_R - (V_N)_L = -2(V_N)_L > 0$, and Equation (45) will be violated as soon as $(Fr_N)_L = (V_N)_L / c_L > 2$.

This difficulty may be avoided by using a different extrapolation for the closed boundary flux

$$\mathit{Flux}(\mathit{Closed\ Boundary}) = F_N[U_L(\mathit{Boundary\ node}), U_R] = \begin{bmatrix} V_N h = 0 \\ u \cdot V_N h + \left(g\dfrac{h^2}{2}\right) \cdot N_x \\ v \cdot V_N h + \left(g\dfrac{h^2}{2}\right) \cdot N_y \end{bmatrix}$$

$$\mathit{Flux}(\mathit{Closed\ Boundary}) = \begin{bmatrix} 0 \\ \left(g\dfrac{h_L^2}{2}\right) \cdot N_x \\ \left(g\dfrac{h_L^2}{2}\right) \cdot N_y \end{bmatrix}$$

$V_N = 0$ is the closed boundary condition, and h_L is extrapolated from the boundary node mesh centroid to the boundary node closed side.

It is necessary here to underline the fact that all problems linked to the velocity have not been solved. Specifically, treating a cell that is partly wet (recall Figure 1) is still an unresolved problem.

To obtain a stable solution near these cells, the imposition of zero velocity is still used. To date, this appears to be the only approach that may be used to stabilize the solution in a specific situation, such as a stable, partially wet cell delimiting the bank of a river, without any recourse to some kind of subcell resolution. However, this artifice, which seemed appropriate in the preceding case, may be a source of inaccuracy for the case of a propagating front. Obviously, for this situation, an imposed zero front velocity will result in an artificial slowing of the propagation velocity.

The next section exposes some elements that may be useful for the construction of more accurate solutions, especially near moving fronts.

ELEMENTS OF AN ADAPTIVE GRID FOR A MOVING BOUNDARY ALGORITHM

The adaptive grid is a concept close to but different to that of a Langrangian grid. A Langrangian grid algorithm generally moves nodes with a local value of the velocity of one determining parameter (flow velocity or a wave speed such as a shock wave speed). The adaptive grid concept has a wider meaning. It includes methods that allow nodes to be added or deleted from the original mesh, nodes to be displaced or even allows simultaneous combinations of the preceding two operations. This algorithm usually proceeds with the following steps.

1. An evaluation of the error distribution is made over an initial mesh. It presupposes that the choice of the variable indicative of the accuracy has been made before. The error here is only an estimate since the exact solution is not known. One of the most sophisticated methodologies to estimate the error is the one now used by some authors (Peraire *et al.*, 1987; Hetu, 1991). The error is the difference between a linearly varying computed solution and an assumed exact quadratic solution. Less advanced techniques only use the gradient of an indicative parameter.

2. Modify the initial grid in order to match a desired error distribution. A uniform distribution of the error $A_i \cdot E_i = C$, where A_i and E_i are the mesh size and local estimate of the error, respectively, and C is a constant, appears to produce the most accurate solution for elliptical equations. This is currently the most popular error distribution used in practice. As indicated above, mesh modifications are made using three approaches.

(a) *Mesh refinement/derefinement*. Nodes are inserted into an element, or at the vertdle of each segment, for which the error is too large. The inverse procedure, i.e. deleting nodes for which the errors in the surrounding elements are acceptable, is also implemented. This overall procedure is repeated until the desired error distribution is obtained.

(b) *Mesh movement*. In this technique, the number of nodes and elements remain fixed, only node positions are updated in order to attain the desired error distribution. This approach is closest to the Lagrangian approach for which node movements are dictated by the local velocities.

(c) *Interpolation with remeshing*. This method may be considered as being a union of the two preceding techniques, since it may include an increase or decrease in the number of nodes, as well as node displacements. In fact, these operations are performed by generating a completely new mesh at each step of the adaptive process. The new mesh is generated in such a way as to reach the desired error distribution.

At present, no adaptative grid algorithm has been implemented for the SWWM by us. However, the main ingredients for such a project have been defined and are available.

1. *Error indicative parameter*. In the context of compressible gas dynamics, some authors (Peraire *et al.*, 1987) use the Mach number as the indicative parameter in error evaluation. It is known that the Froude number is the hydraulic counterpart of the Mach number. Another more convincing argument in favour of the Froude number is suggested by the constraint $(V_N)_R - 2c_R \leqslant (V_N)_L + 2c_L$ [Equation (45)] which is equivalent to a Froude number constraint $(\overline{Fr}_N) - (\overline{Fr}_N)_L \leqslant$ where $\overline{Fr} = V_N/(c_L + c_R)/2$ is a Froude number evaluated using a mean celerity $c_L + c_R/2$. This criterion on the Froude number applies naturally on a segment by segment adaptation. Every segment on which this constraint is violated would be split by

inserting a node in the middle, until the criterion is verified over the whole domain. Alternatively, a remeshing technique that satisfies this constraint is also feasible.

2. A crucial region where concern for the accuracy of results is primordial is situated near wetting and drying fronts. It has already been pointed out how this question of accuracy arises in a cell that is partially wet. In an adaptive algorithm that involves a complete remeshing procedure, it may be appropriate to locate the position of the front exactly. Such procedures used to date (Akanbi and Katopodes, 1988) update the front position by assuming the initial velocity to be the front velocity at the previous time-step. Clearly, important source term effects, such as the influence of gravity from the bed slope and retarding of the flow owing to bed shear, are not accounted for in this procedure. Logically, inclusion of these effects should produce more accurate solutions. Referring to Dressler (1958), it is possible to estimate the front movement by using the following relationship:

$$w(t^n + \Delta t) = \left[w(t^n) - \frac{S_b}{C_f} \right] (-g \cdot C_f \cdot \Delta t) + \frac{S_b}{C_f} \tag{46}$$

$$x(t^n + \Delta t) = x(t^n) - \frac{1}{g \cdot C_f} \left[w(t^n) - \frac{S_b}{C_f} \right] (-g \cdot C_f \cdot \Delta t) + \frac{S_b}{C_f} \cdot \Delta t \tag{47}$$

where $x(t)$ and $w(t)$ are, respectively, the position of the front and its propagation speed. $S_b = \partial Z_b / \partial x$ is the bed slope, and the bed friction has been linearized by

$$S_f = \frac{N^2 \cdot |w| \cdot w}{h^{4/3}} = \frac{N^2 \cdot |(V_N) + 2\sqrt{gh}|}{h^{4/3}} \cdot w = C_f \cdot w$$

The initial front velocity is $w(t^n) = V_N + 2\sqrt{gh}$. If the channel may be treated as being frictionless ($C_f = 0$), the following relationships are exactly the ones suggested by Dressler (1958)

$$w(t^n + \Delta t) = g \cdot S_b \cdot \Delta t + w(t) = g \cdot S_b \cdot \Delta t + [V_N + 2\sqrt{gh}] \tag{48}$$

$$x(t^n + \Delta t) = x(t^n) + \frac{g \cdot S_b}{2} \cdot (\Delta t)^2 + [V_N + 2\sqrt{gh}] \cdot \Delta t \tag{49}$$

Equations (46)–(49) are sufficient to update the front location while accounting for bed topography and friction effects.

ILLUSTRATIVE TEST CASES

Four test cases are presented. The first case computes the drop in level after the recession of a hypothetical flood over a river (in this case, the Red River in Vietnam was used). The results show the appearance of some dry areas after nearly an hour. Clearly, it is difficult to assess the accuracy of these predictions. Two of the three remaining cases possess exact solutions and treat the 'run up' and descent of a wave over a frictionless sloping beach. For these last two cases, the exact solution is compared with:

(a) solution 'A': zero velocity is imposed at the transition cells.

(b) solution 'C': the full numerical SWWM is used over the whole domain, including the momentum equation over the transition cell.

The denomination solution 'A' and solution 'C' are used here to be consistent with the notation in Figure 9 which was presented in a previous paper (Tchamen and Kahawita, 1994).

Figure 6a shows the very irregular bed topography of the Red River. It was assumed that a flood had submerged the river bringing the water surface elevation to 14 m. This level was taken as the initial

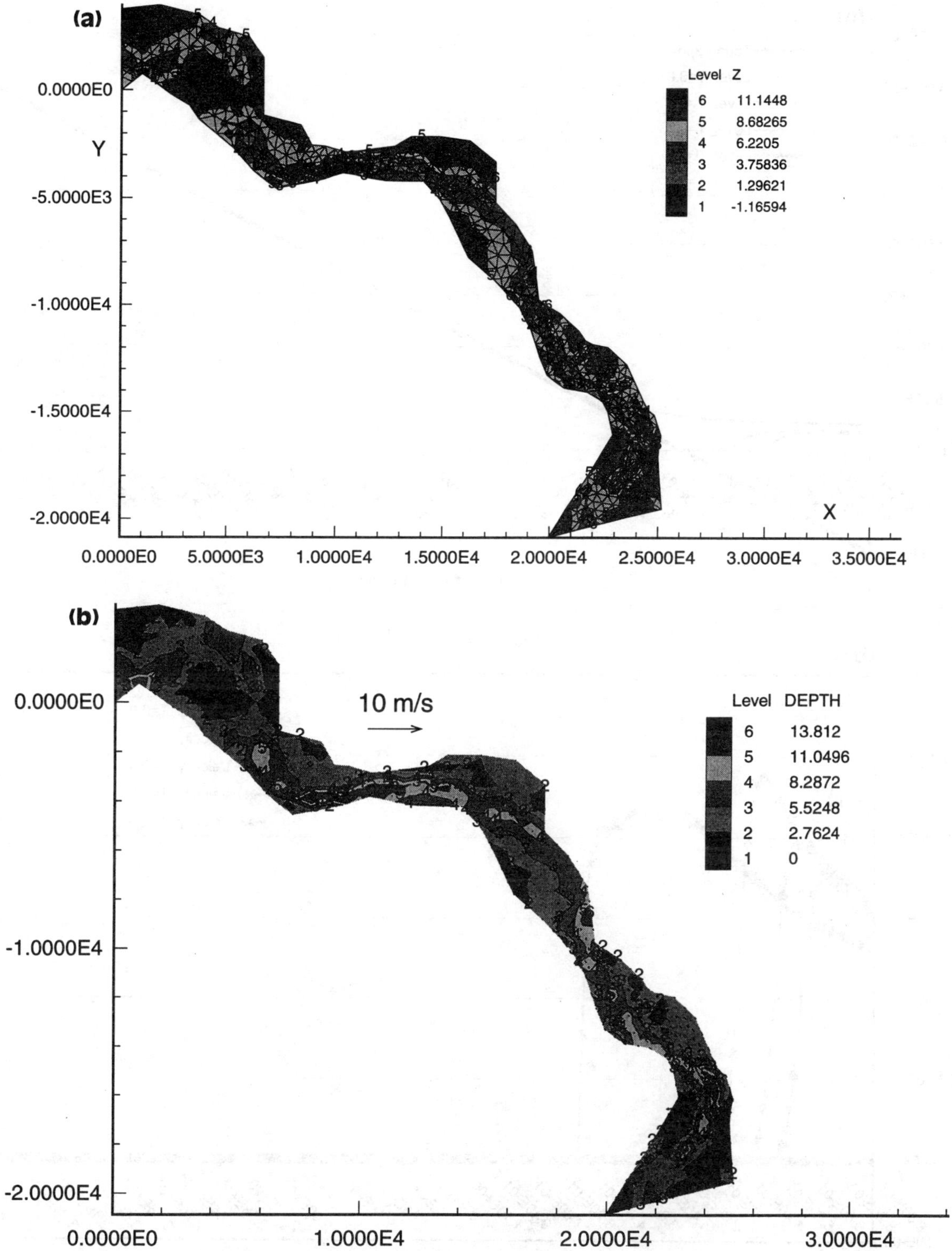

Figure 6. (a) Bottom isocontours plotted over the computational triangular mesh. (b) Velocity field plotted over depth isocontours taken nearly 4400 seconds after the beginning of the drop in level

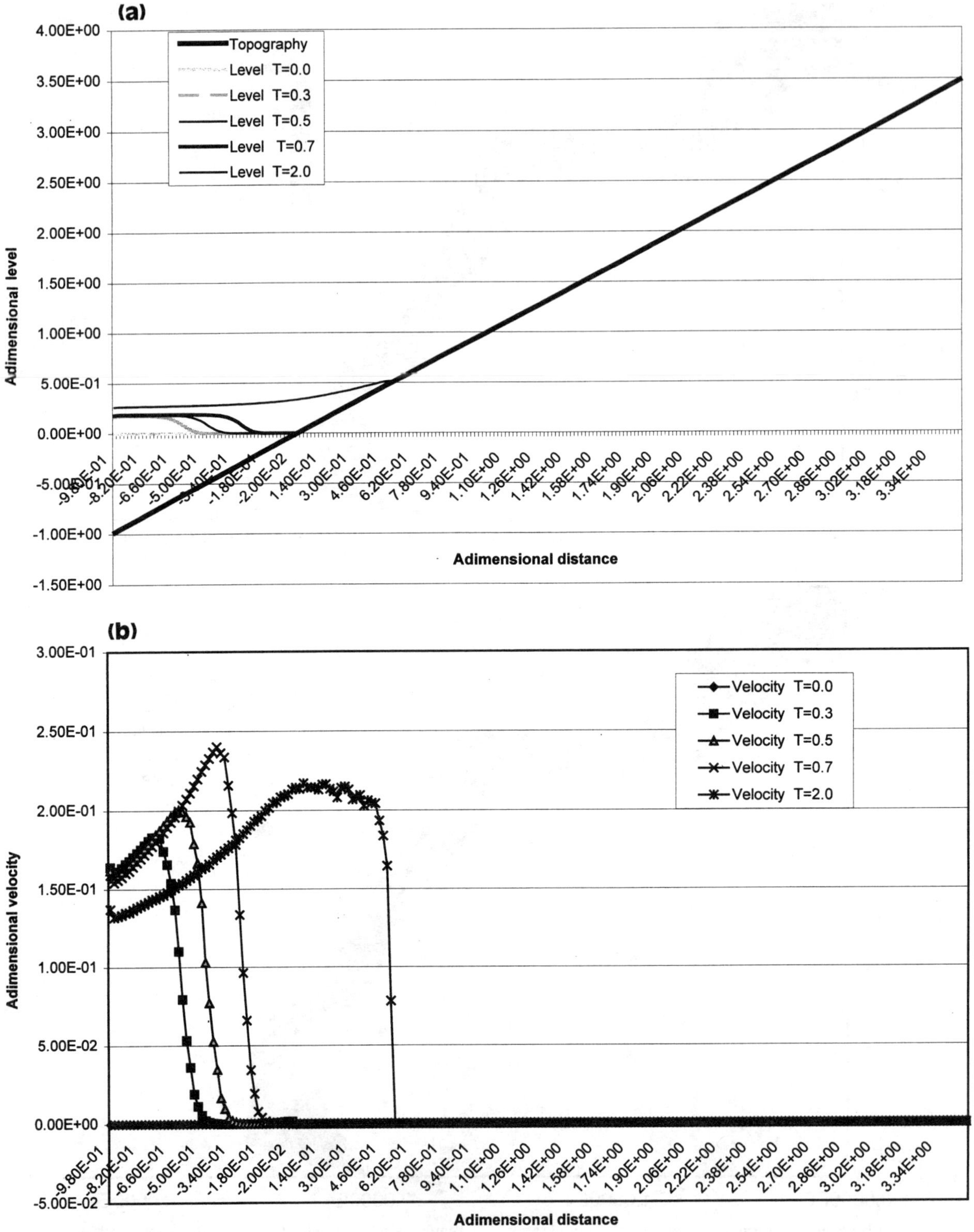

Figure 7. Uniform bore on a beach: (a) evolution of level profiles; (b) evolution of velocity profiles

condition, with a zero velocity imposed over the whole of the computational domain. Figure 6b presents a superposition of the velocity field over water depth contours after nearly one hour. The results are stable and appear plausible. However, before using such results in a forecast situation it would be desirable to conduct a field validation using data from a measured flood event.

The case of a 'uniform' bore over a beach, similar to the one presented by Hibberd and Peregrine (1979) or Kobayashi *et al.* (1987), reproduced with the present model is presented in Figure 7. The upstream boundary for this case is assumed to represent the seaward limit from where the incoming bore originates. The strength of the initial bore was taken to be 1·6 as in Hibberd and Peregrine (1979). The boundary condition at the seaward limit is imposed on the Riemann invariant associated with the incoming wave $\alpha_N = V_N - 2c = V_N^0 - 2c^0$, where V_N^0 and c^0 are normal velocity and celerity and at the initial time associated with the incoming bore. The results shown in Figure 7 demonstrate that the present model has the capacity to handle, in a stable fashion, both dry bed and discontinuous flow conditions.

The solution to the 'run up' of a wave on a sloping beach presented by Carrier and Greenspan (1958) has been used for the third test. The initial condition is given by a one parameter level profile

$$\eta = \varepsilon\left[1 - \frac{5a^3}{2(a^2+\sigma^2)^{3/2}} + \frac{3a^5}{2(a^2+\sigma^2)^{5/2}}\right] \tag{50}$$

where

$$\sigma^2 = \eta - x \tag{51}$$

and the parameters are ε and $a = 1{\cdot}5\sqrt{1 + 0{\cdot}9\varepsilon}$. In Equations (50) and (51), η and x represent the non-dimensional level and distance. The notation of η for the level is in accordance with the paper of Carrier and Greenspan (1958) and Figure 7 is drawn from this reference. The numerical simulation is performed with the ε parameter fixed at 0·2, a constant bed slope of $\alpha = 0{\cdot}01$ and a reference length of $L_0 = 1000$ m. This produces a reference time of $T = \sqrt{L_0/\alpha \cdot g} \approx 101$ s, and a reference velocity of $V_0 = \sqrt{g \cdot \alpha \cdot L_0} \approx 9{\cdot}9$ m/s. The curves in Figure 8 are in dimensionless quantities. Comparison of each exact curve (Carrier and Greenspan, 1958) to its corresponding numerically simulated profile [Figure 8a and 8b] shows that the adaptation of the Van Leer scheme to the SWWM reproduces the exact evolution quite well.

The fourth case is the simulation of an instantaneous release of water running down a frictionless sloping plane. The exact solution to this problem is given by Dressler (1958). The maximum depth behind the dam before the release is 10 m. The results presented in Figure 9 show a significant departure of the zero velocity in the transition element solution (Solution 'A') from the exact solution and the full SWWM numerical solution (solution 'C').

CONCLUSIONS

In this paper, an attempt has been made to portray the general context in which wetting and drying phenomena occur in reality and list the different challenges to overcome for their successful numerical simulation. The difficulties involved for the correct mesh treatment (especially a fixed Eulerian mesh), which is not fully wetted, address the assumption made for the bathymetry profile between the selected computational nodes. It is important, for stability as well as for accuracy purposes, to define a coherent set of assumptions about the flow variables and the bottom.

The first-order Van Leer–Riemann solver has been used, together with an assumption of locally constant surface water level and linear bathymetry; thus, a locally linear water depth variation is the variable used for the evaluation of fluxes, while the surface level is the globally conserved variable for the equation of mass.

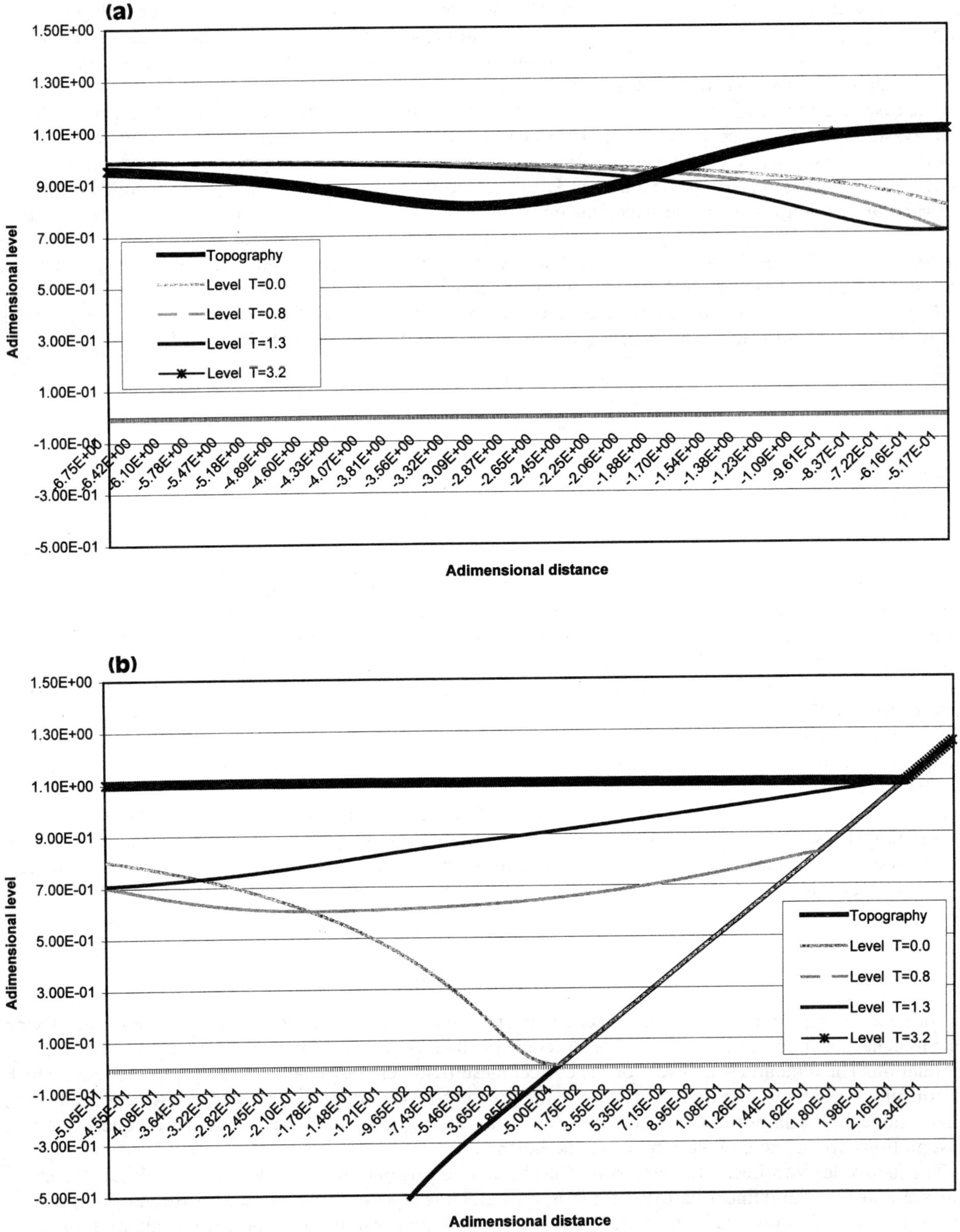
(a)
1.50E+00
1.30E+00
1.10E+00
9.00E-01
7.00E-01
5.00E-01
3.00E-01
1.00E-01
-1.00E-01
-3.00E-01
-5.00E-01
Adimensional level
Topography
Level T=0.0
Level T=0.8
Level T=1.3
Level T=3.2
-6.75E+00
-6.42E+00
-6.10E+00
-5.78E+00
-5.47E+00
-5.18E+00
-4.89E+00
-4.60E+00
-4.33E+00
-4.07E+00
-3.81E+00
-3.56E+00
-3.32E+00
-3.09E+00
-2.87E+00
-2.65E+00
-2.45E+00
-2.25E+00
-2.06E+00
-1.88E+00
-1.70E+00
-1.54E+00
-1.38E+00
-1.23E+00
-1.09E+00
-9.61E-01
-8.37E-01
-7.22E-01
-6.16E-01
-5.17E-01
Adimensional distance
(b)
1.50E+00
1.30E+00
1.10E+00
9.00E-01
7.00E-01
5.00E-01
3.00E-01
1.00E-01
-1.00E-01
-3.00E-01
-5.00E-01
Adimensional level
Topography
Level T=0.0
Level T=0.8
Level T=1.3
Level T=3.2
-5.05E-01
-4.55E-01
-4.08E-01
-3.64E-01
-3.22E-01
-2.82E-01
-2.45E-01
-2.10E-01
-1.78E-01
-1.48E-01
-1.21E-01
-9.65E-02
-7.43E-02
-5.46E-02
-3.65E-02
-5.00E-04
1.75E-02
3.55E-02
5.35E-02
7.15E-02
8.95E-02
1.08E-01
1.26E-01
1.44E-01
1.62E-01
1.80E-01
1.98E-01
2.16E-01
2.34E-01
Adimensional distance

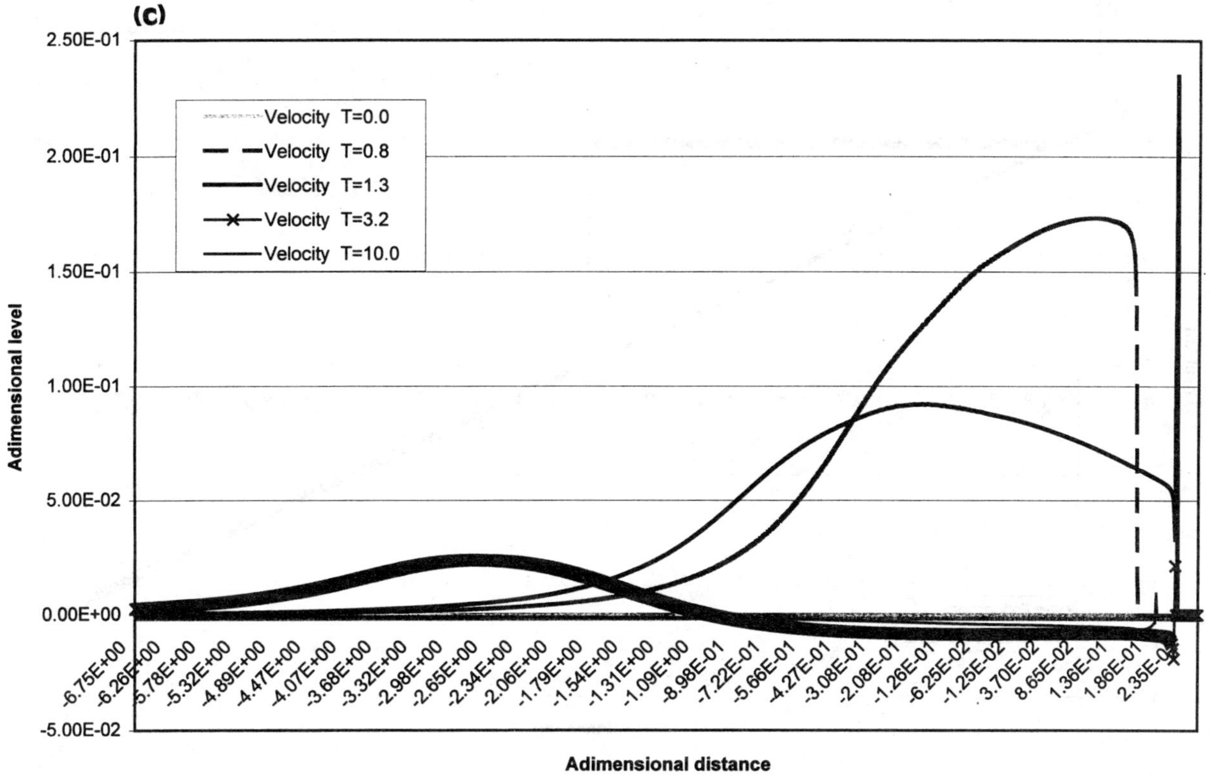

Figure 8. (a), (b) Evolution of level profiles for the climbing non-breaking wave test. (c) Evolution of velocity profiles for the climbing non-breaking wave test (Carrier and Greenspan, 1958)

From this study, some determining factors for the selection of a numerical scheme to simulate flows over domains including dry areas have been exhibited. The guarantee of depth positivity is a primary requirement with obvious physical motivation. The capacity of the scheme to adapt its integration stencil as a function of the local flow regime is also inspired from physical evidence, and is a decisive factor required to obtain a stable solution in flows with mixed regimes (subcritical/supercritical). The Van Leer scheme has been retained because it fulfils these conditions. However, satisfaction of these requirements does not ensure stable solutions. Appropriate treatment is needed in the integration of source terms, especially for transition cells between wet and dry regions. A too large bottom curvature may produce numerical instabilities that are sometimes linked to theoretically unbounded mathematical solutions. Imposition of a zero velocity in transition cells (solution 'A') appears to be a sufficiently robust artifice to obtain stable solutions for different numerical techniques applied to the SWMM. Test 1 presented above illustrates this fact for finite volume Riemann solvers. Some finite element models such as those of Kawahara and Umetsu (1986) and Bates and Anderson (1993), among others, have also used this artifice with success. The finite difference model of Falconer and Chen (1991) has also retained this technique of stabilizing the solution in simulations involving dry areas. The present scheme exhibits excellent robustness when treating the run up of a bore flow on a partially sloping beach (Test 2). Tests 3 and 4 revealed that, while the numerical solution 'A' can be quite accurate with a slowly moving wave over an adverse slope, it may exhibit dramatic errors in the case of a rapidly moving wave descending along a sloping beach.

At the present time, it is the authors' opinion that a dynamic mesh adaptive technique has the potential to improve the quality of the numerical simulation of wetting and drying over a complex topography.

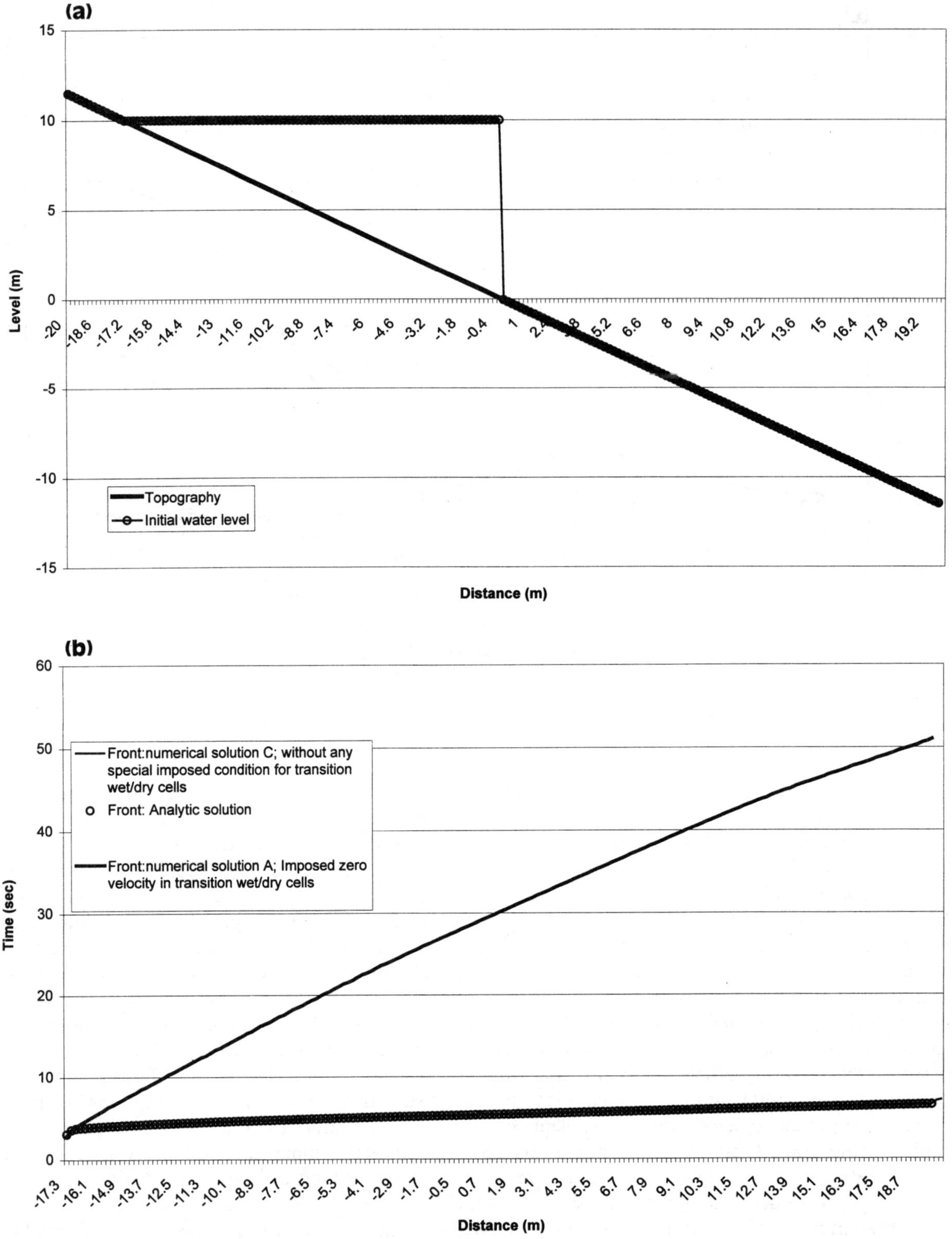

Figure 9. Dambreak over a sloping beach: (a) initial water surface profile; (b) position of the flooding front as a function of time

APPENDIX: IMPLICIT VERSION OF THE VAN LEER'S SCHEME AND DEPTH POSITIVITY

One way to build an implicit version of the Van Leer scheme is to estimate the surface flux of convection and pressure by a weighted formula between the two time limits

$$\int_{t^n}^{t^{n+1}} F_N^{LR}(U_L, U_R) \cdot dt = \Delta t[\theta \cdot F_N^{LR}(U_L^{n+1}, U_R^{n+1}) + (1-\theta) \cdot F_N^{LR}(U_L^{n+1}, U_R^{n+1})] \quad \text{(A1)}$$

where $0 \leqslant \theta \leqslant 1$ is the implicit weighting factor. The time $(n+1)$ component of this flux is approximated by a first-order Taylor approximation in time

$$F_N^{LR}(U_L^{n+1}, U_R^{n+1}) = F_N^{LR}(U_L^n, U_R^n) + \left[\frac{\partial F_N^{LR}}{\partial U}(U_L^n, U_R^n)\right] \cdot \Delta_t U \quad \text{(A2)}$$

where

$$\left[\frac{\partial F_N^{LR}}{\partial U}(U_L^n, U_R^n)\right] \cdot \Delta_t U = \left[\frac{\partial F_N^+}{\partial U}(U_L^n)\right] \cdot (U_L^{n+1} - U_L^n) + \left[\frac{\partial F_N^-}{\partial U}(U_R^n)\right] \cdot (U_R^{n+1} - U_R^n) \quad \text{(A3)}$$

which follows Steve (1988). Expressions for the coefficients of the matrices

$$A_N^+ = \left[\frac{\partial F_N^+}{\partial U}(U^n)\right] \quad \text{and} \quad A_N^- = \left[\frac{\partial F_N^-}{\partial U}(U^n)\right]$$

have been derived, but what is given below are expressions from slightly different approximations:

$$A_N^+ = \left[\frac{\partial F_N}{\partial U}(U^n)\right]^+ = RV \cdot D^+ \cdot RV^{-1}$$

and

$$A_N^- = \left[\frac{\partial F_N}{\partial U}(U^n)\right]^- = RV \cdot D^- \cdot RV^{-1}$$

which are usually called Steger–Warming split matrices, with RV being the right eigenvector matrix and D the diagonal matrix of eigenvalues. This last choice was made to obtain an adequate way to control depth positivity, but sacrificing, at the same time, the exact Van Leer flux Jacobians, which would have ensured a 'Newton' convergence for steady-state problems. Insertion of Equation (A2) into the general update Equation (10) produces

$$\left[\frac{Mes(\Omega_P)}{\Delta t} \cdot I + \sum_{nb=1}^{nb=Max\_nb} A_N^+(U_P^n) \cdot l_{Pnb}\right] \cdot \Delta_t U_P + \sum_{nb=1}^{nb=Max\_nb} (A_N^-(U_{nb}^n) \cdot l_{Pnb}) \cdot \Delta_t U_{nb}$$

$$= - \sum_{nb=1}^{nb=Max\_nb} [F_N^+(U_P^n) + F_N^-(U_{nb}^n)] \cdot l_{Pnb} + Mes(\Omega_P) \cdot \int_{t^n}^{t^{n+1}} S[U(t)] \, dt \quad \text{(A4)}$$

Relation (A4) may be expanded into U_P and U_{nb} by using $\Delta_t U = U^{n+1} - U^n$:

$$\begin{aligned}
&\left[\frac{Mes(\Omega_P)}{\Delta t}\cdot I + \sum_{nb=1}^{nb=Max_nb} A_N^+(U_P^n)\cdot l_{Pnb}\right]\cdot U_P^{n+1} = Mes(\Omega_P)\cdot\int_{t^n}^{t^{n+1}} S[U(t)]\,dt\\
&+\left(\left[\frac{Mes(\Omega_P)}{\Delta t}\cdot I + \sum_{nb=1}^{nb=Max_nb} A_N^+(U_P^n)\cdot l_{Pnb}\right]\cdot U_P^n - \sum_{nb=1}^{nb=Max_nb}[F_N^+(U_P^n)]\cdot l_{Pnb}\right)\\
&+\left(\left[\sum_{nb=1}^{nb=Max_nb} A_N^-(U_{nb}^n)\cdot l_{Pnb}\right]\cdot U_{nb}^n - \sum_{nb=1}^{nb=Max_nb}[F_N^-(U_{nb}^n)]\cdot l_{Pnb}\right)\\
&-\left(\left[\sum_{nb=1}^{nb=Max_nb} A_N^-(U_{nb}^n)\cdot l_{Pnb}\right]\cdot U_{nb}^{n+1}\right)
\end{aligned}\tag{A5}$$

To analyse depth positivity, only the first component of this vector equation is considered. Two necessary conditions have been derived:

$$\left(\frac{Mes(\Omega_P)}{\Delta t} + \frac{c_P}{4}\left[\sum_{nb}(2\theta - 1 - Fr_N^P)(1 + Fr_N^P)\cdot l_{Pnb}\right]\right) \geqslant 0 \tag{A6}$$

from the coefficient of h_P^n, and

$$\left(\sum_{nb}(1 + Fr_N^{nb})\left[\frac{(1 - Fr_N^{nb}) - 2\theta}{4}\right]\cdot l_{Pnb}\right) \geqslant 0 \tag{A7}$$

from h_{nb}^n. Coefficients of h_P^{n+1} and h_{nb}^{n+1} are unconditionably non-negative. To have the sum involved in Equation (A7) positive, it is sufficient to impose on each individual term the requirement that they all be positive. Equation (A7) produces the constraint

$$\theta \leqslant Min_{P=1,Max\_nb\_Node}\left[\frac{1 - Fr_N^P}{2}\right] \tag{A8}$$

and a sufficient relation to satisfy Equation (A6) is simply

$$\Delta t \leqslant \begin{cases} \dfrac{Mes(\Omega_P)}{\dfrac{c_P}{4}\left[\sum_{nb}(1 + Fr_N^P - 2\theta)(1 + Fr_N^P)\cdot l_{Pnb}\right]} = \Delta t_{Min}^P & \text{if } \Delta t_{Min}^P > 0\\ \infty & \text{if } \Delta t_{Min}^P \leqslant 0\end{cases} \tag{A9}$$

Recall that Equations (A8) and (A9) are sufficient conditions for depth positivity, which are not, however, automatically necessary. This means that, in practical calculations, constraint on the time-step, like Equation (A9), may be relaxed while still leading to stable convergent solutions. Finally, note that the two equations (A8) and (A9) look 'reasonable', since for highly convective flows (Fr_N close to 1), the implicit weight θ will be driven by Equation (A8) near 0, i.e. towards an explicit scheme. It is well known that implicit schemes are efficient and accurate when the magnitude of the time-step is in agreement with the real wave propagation speeds dictated by the physics.

The Steger–Warming split coefficient matrix is given by

$$A_N^*(1,1) = (\lambda 1^* - \lambda 3^*)\cdot V_N + (\lambda 1^* + \lambda 3^*)\cdot c$$

$$A_N^*(1,2) = (\lambda 3^* - \lambda 1^*)\cdot N_x$$

$$A_N^*(1,3) = (\lambda 3^* - \lambda 1^*)\cdot N_y$$

$$A_N^*(2,1) = (u - c\cdot N_x)\cdot(V_N + c)\cdot\lambda 1^* + (2V_t\cdot c\cdot N_y)\cdot\lambda 2^* + (u + c\cdot N_x)\cdot(c - V_N)\cdot\lambda 3^*$$

$$A_N^*(2,2) = (c\cdot N_x^2 - u\cdot N_x)\cdot\lambda 1^* + (2c\cdot N_y^2)\cdot\lambda 2^* + (c\cdot N_x^2 + u\cdot N_x)\cdot\lambda 3^*$$

$$A_N^*(2,3) = (c\cdot N_x\cdot Ny - u\cdot N_y)\cdot\lambda 1^* - (2c\cdot N_x\cdot Ny)\cdot\lambda 2^* + (c\cdot N_x\cdot Ny + u\cdot N_y)\cdot\lambda 3^*$$

$$A_N^*(3,1) = (v - c\cdot N_y)\cdot(V_N + c)\cdot\lambda 1^* - (2V_t\cdot c\cdot N_x)\cdot\lambda 2^* + (v + c\cdot N_y)\cdot(c - V_N)\cdot\lambda 3^*$$

$$A_N^*(3,2) = (c\cdot N_x\cdot N_y - v\cdot N_x)\cdot\lambda 1^* - (2c\cdot N_x\cdot N_y)\cdot\lambda 2^* + (c\cdot N_x\cdot N_y + v\cdot N_x)\cdot\lambda 3^*$$

$$A_N^*(3,3) = (c\cdot N_y^2 - v\cdot N_y)\cdot\lambda 1^* + (2c\cdot N_x^2)\cdot\lambda 2^* + (c\cdot N_y^2 + v\cdot N_y)\cdot\lambda 3^*$$

where $\lambda 1 = V_N - c$, $\lambda 2 = V_N$ and $\lambda 3 = V_N + c$ are the eigenvalues of the Jacobian matrix $A_N = \partial F_N/\partial U$. The signed eigenvalues ($* = \pm$) are simply defined as

$$\lambda^+ = \frac{\lambda + |\lambda|}{2} \quad \text{and} \quad \lambda^- = \frac{\lambda - |\lambda|}{2}$$

REFERENCES

Abbott, M. B. 1979. *Computational Hydraulics*, Pitman, London.

Akanbi, A. A. and Katopodes, N. D. 1988. 'Model for flood propagation on initially dry land', *J. Hydraul. Engng*, **114**, 689–706.

Alcrudo, F. and Garcia-Navarro, P. 1993. 'A high-resolution Godunov-type scheme in finite volumes for the 2D shallow-water equations', *International Journal for Numerical Methods in Fluids* **16**, 489–505.

Ambrosi, D. 1995. 'Approximation of shallow water equations by Roe's Riemann solver', *Int. J. Num. Meth. Fluids*, **20**, 157–168.

Bates, P. D. and Anderson, M. G. 1993. 'A two-dimensional finite-element model for river flow inundation', *Proc. R. Soc. Lond. A*, **440**, 481–491.

Brewer, G. D., Wittlin, G., Versaw, E. F., Parmley, R., Cima, R. and Walther, E. G. 1982. 'Assessment of crash fire hazard of LH2-fueled aircraft', *NASA CR-165525*. NASA Lewis Research Centre, Cleveland, Ohio.

Carrier, G. F. and Greenspan, H. P. 1958. 'Water waves of finite amplitude on a sloping beach', *J. Fluid Mech.*, **4**, 97–109.

Cunge, J. A., Holly, F. M. and Verwey, A. 1980. *Practical Aspects of River Computational Hydraulics*. Pitman, London.

Di Giammarco, P. and Todini, E. 1994. 'A control volume finite element method for the solution of 2-D overland flow problems', *Proceedings of the Modelling of Flood Propagation over Initially Dry Area Conference*. Milan, 1994.

Di Giammarco, P., Todini, E., Consuegra, D., Joerin, F. and Vitalini, F. 1994. 'Combining a 2-D flood plain model with GIS for flood delineation and damage assessment', *Proceedings of the Modelling of Flood Propagation over Initially Dry Area Conference*. Milan.

Di Monaco, A. and Molinaro, P. 1984. 'Lagrangian finite element model of dam-break wave on dry bed versus experimental data', *Hydrosoft'84 Hydraulic Engineering Software, Proceedings of the International Conference*, Yugoslavia, September 1984, pp. 2-111–2-120.

Dressler, R. F. 1958. 'Unsteady non-linear waves in sloping channels', *Proc. R. Soc. London Ser. A*, **247**, 186–198.

Einfeldt, C. D., Roe, P. L. and Sorgreen, B. 1991. 'On Godunov-type methods', *J. Comp. Phys.*, **92**, 273–295.

Falconer, R. A. and Chen, Y. 1991. 'An improved representation of flooding and drying and wind stress effects in a two-dimensional tidal numerical model', *Proc. Instn. Civ. Engrs, Part 2*, **91**, 659–678.

Glaister, P. 1988. 'Approximate Riemann solutions of the shallow water equations', *J. Hydraul. Res.*, **26**, 293–306.

Godunov, S. K. 1959. 'A difference scheme for numerical computation of discontinuous solutions of equation', *Math. Sbornik*, **47**, 271–306 [in Russian].

Goutal, N. 1987. 'Résolution des équations de Saint-Venant en régime transcritique par une méthode d'éléments finis: application aux bancs découvrants', *Bulletin de la D.E.R. Série C, EDF. ou Thèse de Doctorat*, Université de Paris VI.

Henderson, F. M. 1966. *Open Channel Flow*. MacMillan, New York.

Hétu, J. F. 1991. *Méthode d'éléments finis adaptatives pour les écoulements visqueux*', PhD Thesis, École Polytechnique de Montréal.

Hibberd, S. and Peregrine, D. H. 1979. 'Surf and run-up on a beach: a uniform bore', *J. Fluid Mech.*, **95**, 323–345.

Hirsch, C. 1990. *Numerical Computation of Internal and External Flows*, Volume 2. Wiley, Chichester.

Holsters, H. 1976. 'Mathematical modelling of tidal flats', in *Mathematical Modelling of Estuarial Flows*. Springer, pp. 172–177.

Holz, K. P. and Leister, K. 1988. 'Accurate waterline calculation in the Wadden Sea', *Computational Metheds in Subsurface Hydrology*, Springer, Berlin, pp. 77–83.
Hundsdorfer, W., Koren, B., Van Loon, M. and Verwer, K. 1995. 'A positive finite-difference advection scheme', *J. Comp. Phys.*, **117**, 35–46.
Katopodes, N. and Strelkoff, T. 1978. 'Computing two-dimensional dam-break flood waves', *J. Hydraul. Div., ASCE*, **104**, 1269–1287.
Kawahara, M. and Umetsu, T. 1986. 'Finite element method for moving boundary problems in river flow', *Int. J. Num. Meth. Fluids*, **6**, 365–386.
Kobayashi, N., Otta, A. K. and Roy, I. 1987. 'Wave reflection and run-up on rough slopes', *J. Waterway Port Coastal Ocean Engng*, **113**, 282–298.
Lynch, D. R. and Gray, W. G. 1980. 'Finite element simulation of flow in deforming regions', *J. Comp. Phys.*, **36**, 135–153.
Meselhe, E. A. and Holly, F. M. 1993. 'Simulation of unsteady flow in irrigation canals with a dry bed', *J. Hydraul. Engng*, **119**, 1021–1039.
Molinaro, P. (ed.) 1994. *Proceedings of the Modelling of Flood Propagation over Initially Dry Area Conference*, Milan, Italy, 1994.
Osher, S. 1982. 'Upwind difference schemes for hyperbolic systems of conservation laws', *Math. Comp.*, **38**, 339–374.
Patankar, S. V. 1980. *Numerical Heat Transfer*. Hemisphere, New York.
Peraire, J., Vadahti, M., Morgan, K. and Zienkiewicz, O. C. 1987. 'Adaptive remeshing for compressible flows', *J. Comp. Phys.*, **72**, 449–466.
Rajar, R. 1978. 'Mathematical simulation of dam-break flow', *J. Hydraul. Div., ASCE*, **104**, 1011–1026.
Roe, P. L. 1981. 'Approximate Riemann solvers, parameter vectors and difference schemes', *J. Comput. Phys.*, **43**, 357–372.
Sakkas, J. G. and Strelkoff, T. 1973. 'Dam-break flood in prismatic dry channel', *Journal of the Hydraulic division, ASCE*, 100, No. HY12, pp. 2195–2216.
Steve, H. 1988. 'Schémas implicites linéarisés décentrés pour la résolution des équations d'Euler en plusieurs dimensions', PhD Thesis, Université de Provence, France.
Stoker, J. J. 1957. *Water Waves*. Wiley Interscience: New York.
Struijs, R., Deconinck, H. and Roe, P. 1991. 'Fluctuation splitting schemes for the 2D Euler equations', *von Karman Institute for Fluid Dynamics, Lecture Series, Computational Fluid Dynamics*, Vol. 1991-01, 42 pages.
Tchamen, G. W. 1992. 'Une étude expérimentale de l'onde de crue consécutive à une rupture de barrage', *MSc Thesis*, École Polytechnique de Montréal.
Tchamen, G. W. 1998. 'On the use of Riemann solvers for the modeling of shallow water equations', *PhD Thesis*, École Polytechnique de Montréal.
Tchamen, G. W. and Kahawita, R. 1994. 'The numerical simulation of wetting and drying areas using Riemann solvers', *Proceedings of the Modelling of Flood Propagation over Initially Dry Area Conference*. Milan, pp. 127–139, 1994.
Toro, E. F. 1992. 'Riemann problems and the WAF method for solving the two-dimensional shallow water equations', *Phil. Trans. R. Soc. Lond. A*, **338**, 43–68.
Van Leer, B. 1982. 'Flux-vector splitting for the Euler equations', in *Lecture Notes in Phys.*, **170**, 507.
Vasiliev, O. F. 1970. 'Numerical solution of the non-linear problems of unsteady flows in open channels', *Proceedings, Second International Conference on Numerical Methods in Fluid Dynamics*, Berkeley, California.
Xanthopoulos, Th. and Koutitas, Ch. 1976. 'Numerical simulation of a two-dimensional flood wave propagation due to dam failure', *J. Hydraul. Res.*, **14**, 321–331.
Zhang, B. N., Bouttes, F. and Dhatt, G. 1992b. 'A shallow water finite element model for moving fronts', *Comp. Meth. Subsurf. Hydrol.*, 91–93; or 'Finite element modeling of fluids', *von Karman Institute for Fluid Dynamics, Lecture Series, Computational Fluid Dynamics*, Vol. 1992-04, 133 pages.
Zhang, H., Youssef, H., Long, N. D. and Kahawita, R. 1992a. 'A 1-D numerical model applied to dam-break flows on dry beds', *J. Hydraul. Res.*, **30**, 211–222.

3

THE TWO-DIMENSIONAL TRANSPORT MODULE SUBIEF. APPLICATIONS TO SEDIMENT TRANSPORT AND WATER QUALITY PROCESSES

C. MOULIN AND E. BEN SLAMA*

EDF, Direction des Etudes et Recherches, Laboratoire National d'Hydraulique, 6, Quai Watier 78401 Chatou Cedex, France

ABSTRACT

This paper describes a new numerical method used within the SUBIEF software to compute the transport of diluted or suspended tracers in a two-dimensional *depth-averaged free surface flow* for domains where the flood extent varies with time. The *transport equations* and the numerical choices for time and space discretization in finite elements are described prior to developing two alternative formulations for dealing with partially dry elements. Finally, we depict two typical applications of the software: the simulation of *sedimentation* in the intake of a power plant qualitatively validated by comparing numerical results obtained from SUBIEF with those from a physical model, and a simulation of *heavy metal* adsorption/desorption on to sediment particles in a fluvial flow which are subsequently deposited and resuspended.

KEY WORDS sediment transport; water quality; finite elements; floodplains; heavy metals

INTRODUCTION

Sediment transport processes are a major concern in many environmental studies (Fischer, 1979). Here, in contrast to pure advection processes, one must also account for sediment deposition and erosion processes, which occur when currents reach threshold values. Numerical models of this phenomenon therefore have to provide accurate estimates of both hydrodynamic variables and these critical thresholds in order to estimate properly the behaviour of transported sediments over complex topography. While such codes have been shown to be reasonably successful where the domain is fully inundated, simulations involving the dynamic wetting and drying of the computational domain create an additional series of numerical problems that have yet to be adequately resolved. In this paper we attempt to identify these problems and define a new numerical algorithm capable of simulating sediment transport in floodplains and estuaries where the position of the domain boundary dynamically changes through the simulation.

These algorithms have been implemented within the TELEMAC system, developed by Laboratoire National d'Hydraulique (LNH) to deal with free surface flow problems. These codes are based on finite element techniques, and include modules that deal with two- or three-dimensional hydrodynamics, bedload or suspended sediment transport and waves. Here the choice of unstructured grids allows spatial variation in mesh characteristics, making it possible to build large-scale models (up to hundreds of square kilometres) that include the detail of small-scale features (<1 m) where necessary.

* Correspondence to: E. Ben Slama, EDF, Direction des Etudes et Recherches, Laboratoire National d'Hydraulique, 6, Quai Watier, 78401 Chatou Cedex, France

In particular we are here concerned with the SUBIEF module, which is used to compute the transport of one or more tracers. At first, the hydrodynamics are computed on a finite element mesh at each time step by the TELEMAC-2D module. The results consist of depth-integrated velocity fields and water depths that are stored in a file that acts as an input for SUBIEF. The tracers can either be dissolved or suspended in the water column, and SUBIEF includes both erosion and deposition processes. As the velocities computed by TELEMAC-2D are integrated over the depth, the concentrations are also assumed to be vertically constant in SUBIEF. This code has been widely used at LNH to study sedimentation in settling basins and outfalls. It has also begun to be used in water quality modelling. In this case, a unique software interface has been developed, which we describe here for the first time, that allows users to take some source terms into account, either by adding an existing water quality module, or by writing their own model.

As the range of applications to which SUBIEF is applied has increased it has become increasingly necessary to deal with complex flows over domains that incorporate dry areas. Studies of flows in estuaries provide an example of this complexity, with computational cases that combine a number of important modelling challenges: large computational domains, long time-scales (several tides) and complex topographies (Marcos *et al.*, 1995). For such situations, the choice of numerical method is of great importance to achieve good quality results, as standard numerical schemes can often lead to build up of mass and sediment concentration in dry areas. In this paper we therefore describe the development of a new numerical scheme that attempts to overcome this limitation of previous methods. To illustrate the capability of the developed model we then present two applications of SUBIEF. First we consider a real case study of sedimentation at a power plant intake. Results of SUBIEF are compared with those of a scale model, showing to what extent SUBIEF can be used for predictive computations that include a complex geometry. Secondly, the utility of the developed scheme for water quality modelling is illustrated by a simulation of heavy metal adsorption/desorption to sediments following a metal-polluted industrial release. These sediments are subsequently advected, deposited and resuspended by a hypothetical fluvial flow.

DISCRETIZING THE TRANSPORT EQUATION FOR DYNAMIC BOUNDARY PROBLEMS

We now describe the development of a mass conservative transport equation for problems involving dry areas within the computational domain.

Choice of the starting formulation: the problem of dry areas

A two-dimensional (2D) vertically integrated tracer transport equation can be developed from the mass conservation formulation. If the conservative variable is the tracer mass, M, then

$$M = hc \tag{1}$$

where h is the water depth and c is the tracer concentration.

The depth-averaged conservation equation then becomes

$$\frac{\partial M}{\partial t} + \mathrm{div}(M\vec{u}) = \mathrm{div}(\bar{\bar{K}} \cdot h\, \overrightarrow{\mathrm{grad}}\; c) + S_M \tag{2}$$

where u is the depth-averaged velocity, K is the dispersion tensor and S_M are source terms for the tracer mass.

Equation (2) provides the best basis on which to build a mass-conservative discretization of the transport equation. However, the choice of the variable M is not a good one for applications, such as floodplains or tidal flats, where the extent of dry areas varies in time. Indeed, one is often more interested in concentration than in mass. In an analogous fashion, predictions in thermal applications are usually based on temperature rather than heat. Thus, when M is obtained, we then have to compute

$$c = \frac{M}{h} \tag{3}$$

In dry areas of the mesh where h tends to zero this operation will obviously be impossible. A further immediate consequence could be a loss of monotonicity in the concentrations computation. As an example, in the case of a plume with a 30 °C temperature at the outfall, we could still get a temperature of 50 °C in the far field if the plume gets close to dry areas. Moreover, from a mathematical point of view, M cannot be described with linear space interpolations when h and c are already linearly interpolated (we should use, for $M = hc$, quadratic functions instead). If M and h were linearly interpolated, then c [computed with Equation (3)] would not belong to the finite element approximation subspace (it could not be a linear function), and monotonicity would hardly be ensured with a mass conservation constraint. Giving a minimum value for h to achieve monotonicity would also not give a general method and has not therefore been used within SUBIEF.

Using a non-conservative formulation as the basis of the model, we use Equation (1) to split Equation (2) into

$$c\frac{\partial h}{\partial t} + h\frac{\partial c}{\partial t} + h\vec{u}\ \overrightarrow{\text{grad}}\ c + c\ \text{div}(h\vec{u}) = \text{div}(\bar{\bar{K}} \cdot h\ \overrightarrow{\text{grad}}\ c) + S_M \tag{4}$$

As (h, u) obeys the continuity equation

$$\frac{\partial h}{\partial t} + \text{div}(h\vec{u}) = S_h \tag{5}$$

where S_h are source terms for h, we get

$$\frac{\partial c}{\partial t} + \vec{u}\ \overrightarrow{\text{grad}}\ c = \frac{1}{h}\text{div}(\bar{\bar{K}} \cdot h\ \overrightarrow{\text{grad}}\ c) + \frac{S_M + cS_h}{h} \tag{6}$$

This method leads to two important properties of the numerical scheme. First, Equation (6) still contains some divisions by h. However, their effect is minimal when applied to the dispersion and source terms. Indeed, the dispersion term usually has a stabilizing effect on the numerical solution. Moreover, the laws that are used to compute source terms in most cases have no physical meaning when computed for very small water depths. (For example, Krone's deposition law assumes $h \gg 1$ mm.) Dividing by h would not raise any disagreement with the physics of the problem (i.e. would not generate any divisions by zero) if we assumed that Equation (6) only has to be solved where there is water ($h > 0$). However, from a numerical point of view, the problem of dry zones remains. These choices will be discussed in the section on dry elements. Secondly, we see that, *at a continuous level*, Equation (6) is theoretically mass conservative. However, we have to derive a discrete formulation of Equation (6) from a discretization of Equation (1) in order to obtain a numerical proof of the tracer mass conservation. This will be considered in the following section.

A mass conservative discretization of Equation (6)

From the following time discretization of Equation (2)

$$\frac{h^{n+1}c^{n+1} - h^n c^n}{DT} + \text{div}(h^n c\vec{u}) = \text{div}(\bar{\bar{K}}h \cdot \overrightarrow{\text{grad}}\ c) = c_i S_h \tag{7}$$

it is necessary to search for a discretization of c that would enable the use of the conservative continuity discretization used in TELEMAC-2D

$$\frac{h^{n+1} - h^n}{DT} + \mathrm{div}(h^n \vec{u}) = S_h \tag{8}$$

with

$$\vec{u} = \theta_u u^{n+1} + (1 - \theta_u) u^n \tag{9}$$

where θ_u is a weighting coefficient.
Hence, Equation (7) can be written as

$$c\left[\frac{h^{n+1} - h^n}{DT} + \mathrm{div}(h^n \vec{u}) - S(h)\right] - c\left[\frac{h^{n+1} - h^n}{DT}\right] + \frac{h^{n+1}c^{n+1} - h^n c^n}{DT} + h^n \vec{u} \cdot \overrightarrow{\mathrm{grad}}\; c$$
$$= \mathrm{div}(\bar{\bar{K}}h \cdot \overrightarrow{\mathrm{grad}}\; c) + c_i S_h - c S_h \tag{10}$$

which can be simplified using Equation (8) to

$$h^{n+1}\frac{c^{n+1} - c}{DT} + h^n \frac{c - c^n}{DT} + h^n \vec{u}\; \overrightarrow{\mathrm{grad}}\; c = \mathrm{div}(\bar{\bar{K}}h \cdot \overrightarrow{\mathrm{grad}}\; c) + (c_i - c) S_h \tag{11}$$

It is of great importance to note from Equation (8) that DT is the time-step used to compute hydrodynamics, which should theoretically be the same for SUBIEF. There are two cases:

(i) In the general case, taking $c = c^{n+1}$ leads to a classical discretization of $\partial c/\partial t$

$$h^n\left[\frac{c^{n+1} - c^n}{DT} + \vec{u}\; \overrightarrow{\mathrm{grad}}\; c^{n+1}\right] = \mathrm{div}(\bar{\bar{K}}h \cdot \overrightarrow{\mathrm{grad}}\; c^{n+1}) + (c_i - c) S_h \tag{12}$$

This discretization leads to a fully implicit scheme in c.

(ii) For the particular case of a steady flow $h^{n+1} = h^n$, $\vec{u}^{n+1} = \vec{u}^n$ and Equation (11) becomes

$$h^n\left[\frac{c^{n+1} - c^n}{DT} + \vec{u}\; \overrightarrow{\mathrm{grad}}\; c\right] = \mathrm{div}(\bar{\bar{K}}h \cdot \overrightarrow{\mathrm{grad}}\; c) + (c_i - c) S_h \tag{13}$$

regardless of the choice of c. In this particular case, it is thus possible to use a classical time discretization of c

$$c = \theta_c c^{n+1} + (1 - \theta_c) c^n \quad \text{with} \quad 0{\cdot}5 < \theta_c < 1{\cdot}0 \tag{14}$$

where θ_c is a further weighting coefficient.

Advection schemes

The first advection scheme to be implemented in SUBIEF was the method of characteristics (Hervouet, 1986) with a linear interpolation. The link with finite elements was ensured by fractional steps, first by computing the advection term, and then by solving the remaining terms of the equation using finite element techniques. This technique was widely used in the past at LNH as it ensures monotonicity. However, as a solution, it tends to be highly diffusive, and no theoretical proof of the tracer mass conservation can be obtained. As we pointed out in the Introduction, it is possible to get a theoretically mass conservative method by using a pure finite element scheme, without fractional steps. To achieve this the choice of discretization for the advection term $\vec{u}\ \overrightarrow{\text{grad}}\ c$ is

$$\vec{u} \cdot \overrightarrow{\text{grad}}\ c[\theta_c c^{n+1} + (1 - \theta_c)c^n] \tag{15}$$

and the weighting function used to stabilize this term is

$$\psi_i = \varphi_i + \frac{DT}{2}\vec{u} \cdot \overrightarrow{\text{grad}}\ \varphi_i \tag{16}$$

where φ_i is a Galerkin linear weighting function.

We thus choose a SUPG-like weighting function for the advection term, the upwind coefficient being the Courant number (Brooks and Hughes, 1982). This approach ensures stabilization of the gradient term by adding an artificial diffusion in the flow direction only. One way to illustrate the effect of this method is to imagine that any spurious numerical oscillation generated by steep concentration gradients will be spatially diffused, and kept within their creation area. This term therefore preserves mass conservation. As there is an implicit time discretization of the advection term, the SUPG weighting function will be found (as with any coefficient of c^{n+1}) in the matrix of the final system. It can be shown that this function does not affect the conditioning of the matrix (Hervouet and Moulin, 1993). Finally, the numerical diffusion of this scheme is, for most of the time, lower than the 'physical' dispersion of the flow $[\bar{\bar{K}}]$ computed in an anisotropic way with an Elder-like $\alpha u^* h$ formulation (Fischer, 1979).

DEALING WITH DRY ELEMENTS

A finite element formulation of Equation (6) can be summarized as

$$\int_\Omega \left[\frac{\partial c}{\partial t} + \vec{u}\ \overrightarrow{\text{grad}}\ c = \frac{1}{h}\text{div}(\bar{\bar{K}} \cdot h\ \overrightarrow{\text{grad}}\ c) + \frac{S_M - cS_h}{h}\right]\varphi_i\ \text{d}\Omega \tag{17}$$

where Ω is the computational domain.

A consideration of this equation gives at least two ways of dealing with dry elements. The first solves Equation (17) over the whole finite element domain. To do this it is necessary to compute the $1/h$ terms, using criteria such as

$$\frac{1}{h} = 0 \text{ if } h < \varepsilon, \quad \text{computing } \frac{1}{h} \text{ if } h > \varepsilon \tag{18}$$

where ε is some small value of h.

This method raises two problems: (i) defining (choosing) ε, and (ii) solving Equation (17) over the whole domain will not prevent the tracer concentration reaching dry areas, especially in the case of a highly diffusive advection scheme. The hydrodynamic computations have to provide an exact local mass conservation, especially on elements that are partially wet, or Equation (8) will not be satisfied. In this case, Equation (8) acts as an extra source term in the tracer transport Equation (11). A practical consequence of this poor local hydrodynamic resolution is usually an artificial tracer concentration build up on the dry or partially dry elements.

The second approach solves Equation (4) over a subdomain defined by the wet elements. This approach seems interesting because it avoids any divisions by zero when $h = 0$, since Equation (17) does not have to be solved where there is no water. This method would obviously not allow concentrations to appear in dry areas, but the question of determining whether an element is dry or wet still remains. It can be solved just by defining a 'mask' calculated for each element 'iel':

mask(iel) = 0 if the element is dry
mask(iel) = 1 if the element is wet and solving

Thus

$$\sum_{\text{iel}} \int_{\text{iel}} \text{mask(iel)}\left[\frac{\partial c}{\partial t} + \vec{u}\ \overrightarrow{\text{grad}}\ c = \frac{1}{h}\text{div}(\bar{\bar{K}} \cdot h\ \overrightarrow{\text{grad}}\ c) + \frac{S_M - cS_h}{h}\right]\varphi_i\ \text{d}\Omega \tag{19}$$

This solution is consistent with the element by element solution techniques implemented in SUBIEF in which system matrices are built for each element, as shown in Equation (19).

SOURCE TERMS

To complete the model it is merely necessary to impose source terms specific to the application being considered. For most fluvial applications the source terms S_M included in Equation (6) can be point sources, erosion and deposition or terms describing chemical reactions that are used to develop water quality simulations. These are now described in detail.

Point sources

In the particular case where S_M is given by an imposed concentration c_i of tracer within the water discharged to be diluted, we have $S_M = c_i S_h$ and Equation (6) becomes

$$\frac{\partial c}{\partial t} + \vec{u}\ \overrightarrow{\text{grad}}\ c = \frac{1}{h}\text{div}(\bar{\bar{K}} \cdot h\ \overrightarrow{\text{grad}}\ c) + \frac{c_i - c}{h} S_h \tag{20}$$

Deposition and erosion of sediments

Deposition and erosion fluxes are calculated with the Krone (Krone, 1962) and Partheniades (Partheniades, 1965) formulae, respectively

$$Q_\text{d} = W_c c\left[1 - \left(\frac{u^*}{u_\text{d}^*}\right)^2\right] \tag{21}$$

$$Q_\text{e} = M\left[\left(\frac{u^*}{u_\text{e}^*}\right)^2 - 1\right] \tag{22}$$

where Q_d is the deposition flux in kg m^{-2} s^{-1}, Q_e is the erosion flux in kg m^{-2} s^{-1}, W_c is the settling velocity in ms^{-1}, M is the Partheniades constant, u_d^* and u_e^* are the critical settling and erosion velocities and u^* is the friction velocity at the bottom.

Hence, deposition will occur only below some critical threshold of velocity u_d^* and erosion above some other, higher threshold, u_e^*. Bottom evolution in m (i.e. bottom level variation) is then equal to

$$E = \frac{Q_\text{d} - Q_\text{e}}{C_\text{sf}} DT \tag{23}$$

For each time step DT, where C_sf is the concentration of the settled layer (g l^{-1}).

SUBIEF is uncoupled from TELEMAC-2D, so that there is no interaction between the sediment and the hydrodynamics. It is thus possible to perform different SUBIEF computations with one given result of TELEMAC-2D (one hydrodynamic field), assuming that deposition or erosion computed by SUBIEF does lead to significant change in the bottom topography. Practically, 2D depth-averaged hydrodynamic models are less sensitive to topography than one might think, and we assume that the hydrodynamics remain valid if, at each node of the mesh, the time-cumulated evolution is less than 5% of the water depth. If the bottom evolution computed by SUBIEF exceeds this threshold, another evaluation of the hydrodynamic field has to be made in order to proceed with SUBIEF computations in a satisfactory manner.

Other source terms

Other source terms (including chemical reactions) may be used to build water quality models. These models are defined by the nature of the advected variables and the reactions between them (see section on Modelling heavy metals). They are introduced into SUBIEF by means of a water quality file. This file consists of an algebraic definition of source terms, such as

$$\mathrm{d}[O2]/\mathrm{d}t = -\{k1\}^*[O2] + \{k2\}^*[O2]\ \ldots \tag{24}$$

Where $\{k1\}$ and $\{k2\}$ are parameters, and $[O2]$ is a water quality state variable. The content of this file is analysed by a software interpreter, translated into FORTRAN and compiled when SUBIEF is run. Users can thus define their own water quality models without any knowledge of computer programming.

APPLICATIONS

In order to demonstrate the capabilities of the developed model we describe here two typical applications conducted at Laboratoire National d'Hydraulique.

The Civaux power plant intake

The Civaux power plant is located on the River Vienne, a tributary of the River Loire, in France. As shown on Figure 1, a weir was built across the river to maintain a minimum water level. The water intake for the power plant is located just upstream of their weir (distance ≈15 m). At the beginning of the modelling study, the power plant was still being built but the intake had been completed and the pumping station was ready to function.

Initially, the inflow (screened by grids) was located on the downstream side of the intake structure (Figure 1). This geometry created a large recirculation zone between the intake and the weir, encouraging deposition. Observations made on the site allowed the thickness of the deposited sediment layer to be estimated. This was found to be almost 2 m, even before pumping commenced. In the Civaux area, the mean discharge of the River Vienne is about 70 $m^3\ s^{-1}$, while the discharge of the one and ten years return period flows are 300 and 1000 $m^3\ s^{-1}$, respectively. The river at this site is known to be quite stable with no noticeable erosion or deposition in spite of significant seasonal sediment loads. Thus, the observed deposition appeared to be mainly a result of the new pumping station geometry. A redesign of the intake was necessary to reduce deposition, as this would affect the long-term performance of the plant, and for this a lateral, rather than downstream, intake was proposed (see Figure 2). A model was therefore required in order to estimate the performance of this new design over a range of typical discharges and to determine the likely benefit.

Numerical simulations of the hydrodynamics at this site using a one-dimensional code suggested that sediment was transported at low discharges (*c.* 35 $m^3\ s^{-1}$) and that the non-dimensional shear stress τ^*

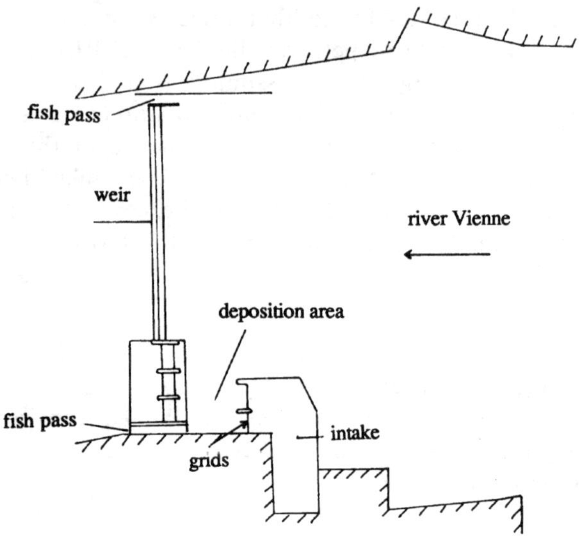

Figure 1. General view of the intake in its original, downstream-facing configuration

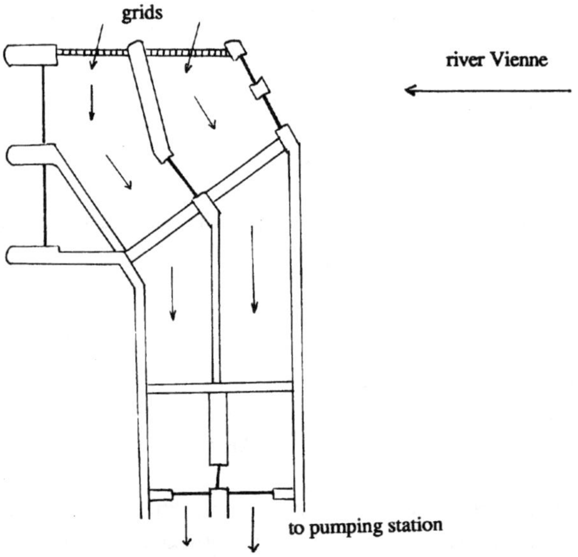

Figure 2. Detailed view of the redesigned intake with lateral inflow

varied from 0·4 (discharge equal to 60 $m^3 s^{-1}$) to 2·5 Nm^{-2} (discharge equal to 400 $m^3 s^{-1}$). The average sediment grain size was 0·3 mm so that transport included both bedload and suspended load. Furthermore, the velocity field near the grids is undoubtedly 3D. The Civaux power plant intake is thus a relatively complex case. A numerical resolution of the whole process (3D suspended and bedload transport) is not yet possible and would require a prohibitive amount of data for parameterization and validation. On the other hand, neither could these processes be entirely studied with a physical model. This would require scaling laws to be respected for each parameter (Froude number, settling velocity, critical bed load shear stress), which is impossible. It was thus decided to study bedload transport with a physical model, while a 2D numerical model based on SUBIEF would allow simulation of suspended transport. Both of these models were

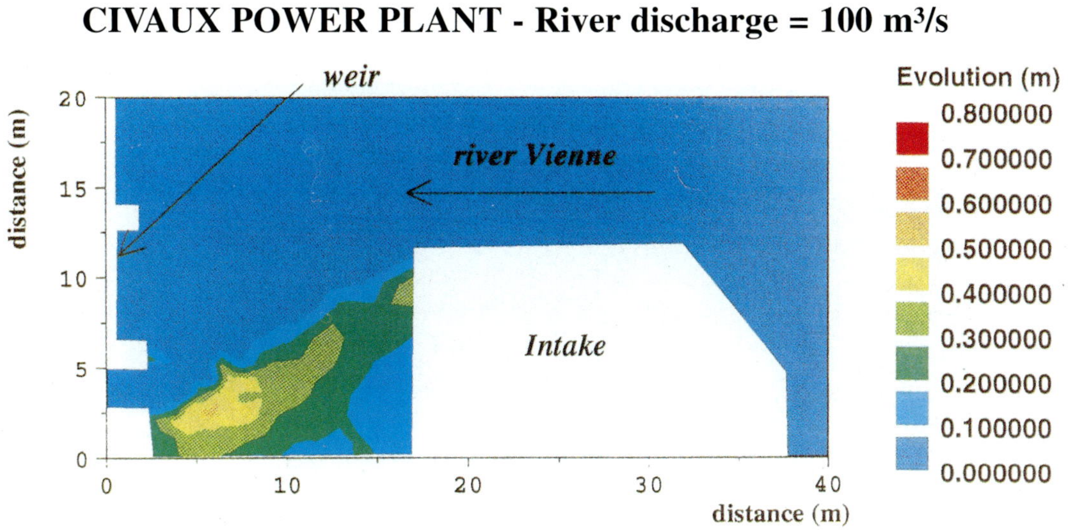

Figure 3. Results of a numerical simulation with SUBIEF for the downstream-facing intake showing the thickness of the deposited suspended sediment after 10 days

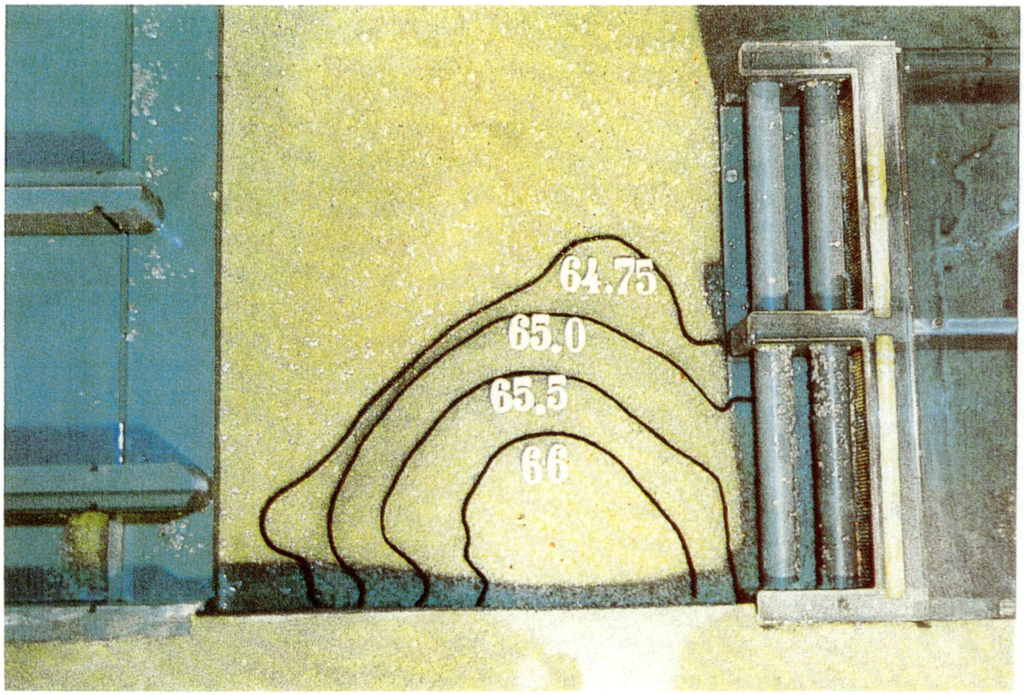

Figure 4. Final result of a simulation with the physical model for the downstream-facing intake showing the thickness of the deposited bed load sediment

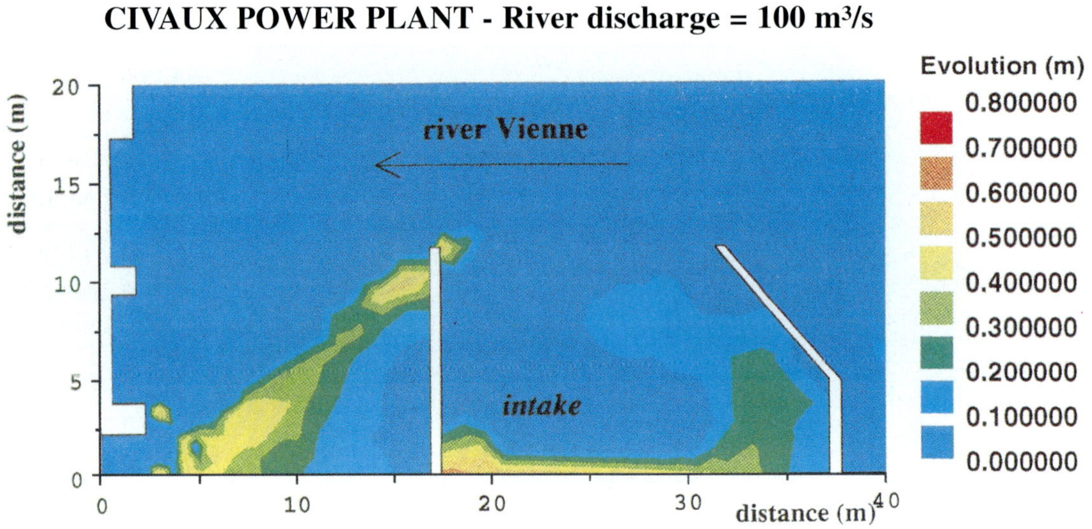

Figure 7. Results of a numerical simulation with SUBIEF for the lateral-facing intake showing the thickness of the deposited suspended sediment after 10 days

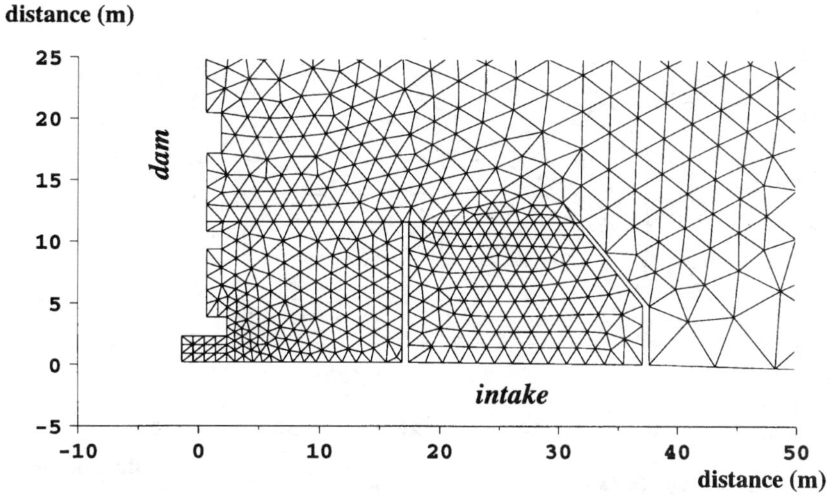

Figure 5. Finite element mesh configured for the lateral intake design

designed to provide the finest scale resolution velocity field possible in order to ensure a good reproduction of the transport phenomena (at least qualitatively).

A SUBIEF computation was made for the initial intake configuration using parameters obtained after calibration against *in situ* flow and sediment data taken before the downstream-facing intake became operational (no pumping). Here, the settling velocity, W_c, was taken to be 0·04 ms^{-1}, assuming a mean particle diameter of 0·3 mm, with a critical settling velocity, u_d^*, of 0·02 ms^{-1}. The initial concentration in the domain, c, for various steady-state simulations, was taken as 0·03 kg m^{-3} for discharge = 60 $m^3 s^{-1}$, 0·15 kg m^{-3} for discharge = 100 $m^3 s^{-1}$ and 0·25 kg m^{-3} for discharge = 200 $m^3 s^{-1}$, respectively. These values were taken from the Engelund Hansen and Van Rijn formulae (Van Rijn, 1989) for a uniform particle diameter of 0·3 mm. Finally, the dispersion coefficient, D, was taken as 0·1 $m^2 s^{-1}$.

After calibration, a check was made that the TELEMAC-2D hydrodynamic model provided similar velocity fields to those measured on the scale model using a propeller current meter and coloured tracers for lines of current. Results from the numerical and physical models (Figures 3 and 4) show that for a river discharge of 100 $m^3 s^{-1}$ both achieve a qualitatively reasonable simulation of deposition in the recirculation. At this discharge, with the intake in the initial, downstream-facing configuration, the simulations indicate that bedload transport dominates over suspended load.

Having ascertained that the models performed reasonably with the initial configuration, steady-state simulations using identical parameters were performed over a range of discharges (100, 200 and 400 $m^3 s^{-1}$) to assess the behaviour of the intake redesign. Figures 5 and 6 show the mesh and the velocity field computed with TELEMAC-2D, while Figure 7 shows the resulting deposition of suspended sediment predicted by the numerical model. Two recirculation zones can be observed, one located downstream, which was observed in the initial configuration, and a large recirculation zone inside the intake itself. With the lateral intake, however, deposition was lower than in the initial configuration, to the extent that pumping operations would not be affected. In the new recirculation zone inside the intake some suspended sediment deposition also occurred. The SUBIEF computation indicated that after 10 days of simulation there is a sediment thickness inside the intake of between 0·15 and 0·9 m. This amount of predicted deposition was much less than that which occurred with the original, downstream-facing configuration, and the lateral intake design (Figure 2) was therefore selected for this site.

This application therefore demonstrates the ability of the developed model to simulate complex flow and sediment transport fields. While absolute validation using field or physical model data was not possible, the correspondence obtained between these data sets provides good evidence that the model behaviour is reasonable.

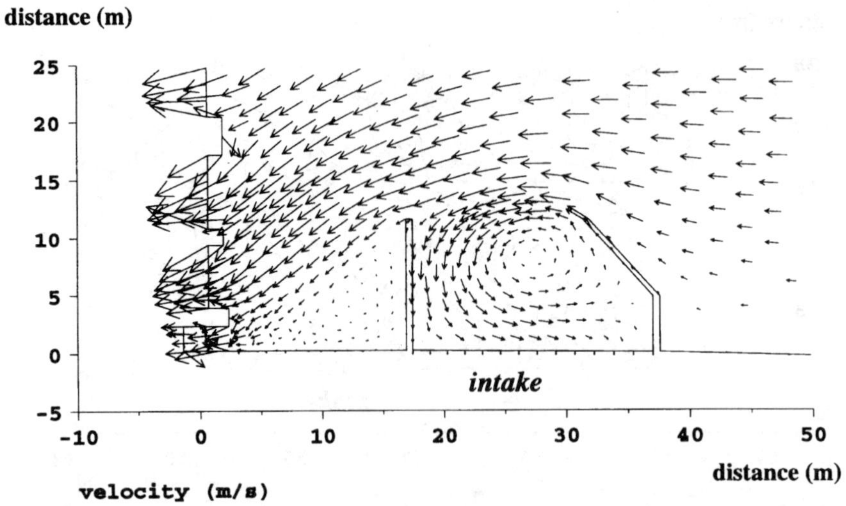

Figure 6. Velocity field computed with TELEMAC-2D for the lateral take design

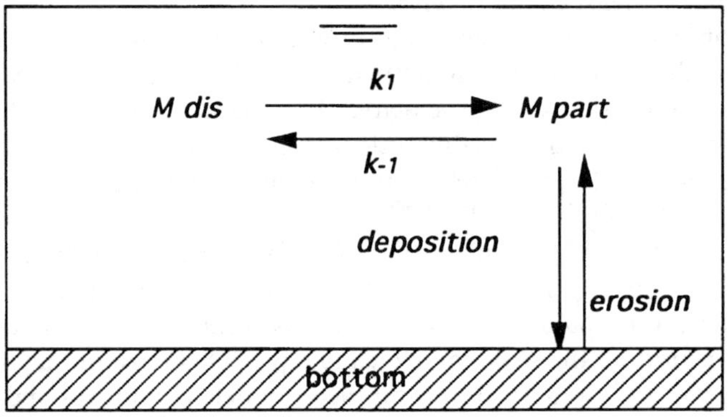

Figure 8. A first approach to the modelling of heavy metals. The dissolved metals, M_{dis}, are adsorbed on to sediments, while the particulate fraction, M_{part}, can settle on the bottom and be eroded

Modelling heavy metals

The industrial release of heavy metals in rivers is another common environmental concern. We present here a basic module to study the deposition of heavy metals with SUBIEF. This module is applied to a hypothetical 400 m long reach of river carrying suspended sediment, within which a point source release of dissolved metals occurs. These metals are adsorbed by the sediments in the water column, leading to a 'solid' or particulate metal fraction. As the sediments settle in the river bed, the particulate fraction of metals is 'trapped' in the sediment layer, until it is resuspended by erosion processes (see Figure 8). Practically, deposition fluxes are usually low, with 1 g m^{-2} day^{-1} being a good order of magnitude estimate. However, settling processes occur over long time-scales, with the peak being reached in summer during low water events. Important accumulated amounts of metal can thus be trapped within the river bed sediment layer. Erosion typically occurs in autumn or winter with the first important floods, and sediments accumulated during several months are often resuspended almost instantaneously. Significant dissolved concentrations of metals can thus be reached. The case presented here is hypothetical; however, it is again possible to assess the general behaviour of the model.

We assume here that there is only one grain size with a single settling velocity. Sediments will thus be modelled in SUBIEF by a single tracer, denoted SS, though it would of course be possible to take several grain sizes into account by using several tracers. This tracer obeys the equation

$$\frac{\partial[SS]}{\partial t} + \vec{u}\,\overrightarrow{\text{grad}}[SS] = \frac{1}{h}\text{div}(\bar{\bar{K}} \cdot h\,\overrightarrow{\text{grad}}[SS]) - \frac{Q_{\text{d}}[SS]}{h} \tag{25}$$

where $[SS]$ is the concentration of SS.

The adsorption processes are modelled with a first-order kinetics reversible reaction between the dissolved and particulate fractions M_{dis} and M_{part}

$$\frac{\partial[M_{\text{dis}}]}{\partial t} + \vec{u}\,\overrightarrow{\text{grad}}[M_{\text{dis}}] = \frac{1}{h}\text{div}(\bar{\bar{K}} \cdot h\,\overrightarrow{\text{grad}}[M_{\text{dis}}]) + k_{-1}([M_{\text{part}}] - k_{\text{d}}[SS][M_{\text{dis}}]) + \frac{M_{\text{dis}}]_i - [M_{\text{dis}}]}{h}S_h \tag{26}$$

$$\frac{\partial[M_{\text{part}}]}{\partial t} + \vec{u}\,\overrightarrow{\text{grad}}[M_{\text{part}}] = \frac{1}{h}\text{div}(\bar{\bar{K}} \cdot h\,\overrightarrow{\text{grad}}[M_{\text{part}}]) + k_{-1}([M_{\text{part}}] - k_{\text{d}}[SS][M_{\text{dis}}]) + \frac{Q_{\text{d}}[M_{\text{part}}]}{h} \tag{27}$$

where $[M_{\text{dis}}]$ and $[M_{\text{part}}]$ are the concentrations of M_{dis} and M_{part}, respectively, in µg l^{-1}, $[M_{\text{dis}}]_i$ is the source concentration of M_{dis} in µg l^{-1}, k_{-1} is the kinetics coefficient in s^{-1} and k_{d} is the sharing coefficient between M_{dis} and M_{part} in lg^{-1}.

The domain for the application consisted of a small river reach (400 m long) separated into two parts. Two bridge piles stand in each part of the reach. The mesh contains 6037 triangular elements, and is refined behind the bridge piles (Figure 9). The water depth computed by TELEMAC-2D is shown on Figure 10. The total discharge in the river was 150 $m^3\ s^{-1}$.

The main parameters for SUBIEF are shown in Table I.

Figures 11 and 12 show, respectively, the dissolved metals concentrations $[M_{\text{dis}}]$ and the daily deposition flux of particulate metalsl $[Q_{\text{d}}(M_{\text{part}})]$ for a steady-state simulation. Deposition of particulate metals can only

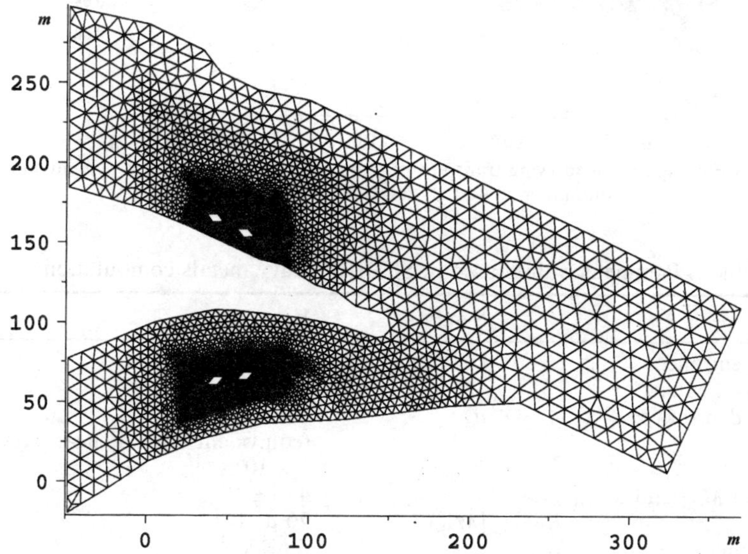

Figure 9. Mesh of the hypothetical computational domain

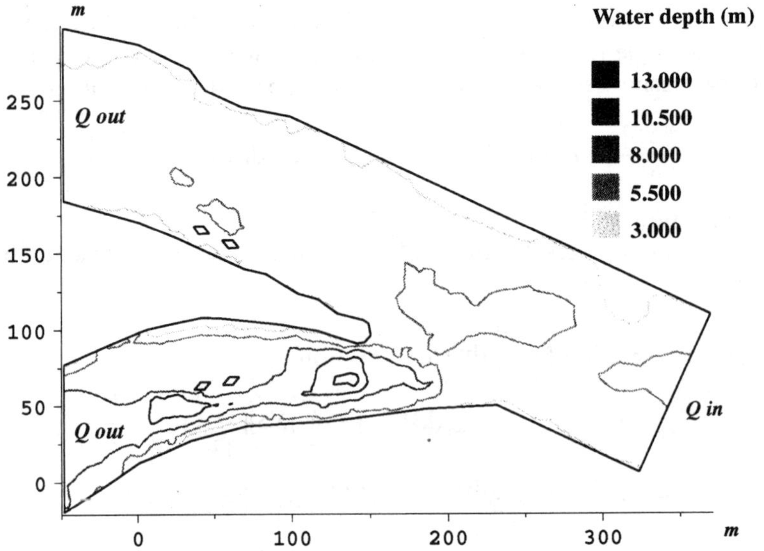

Figure 10. Water depths computed by TELEMAC-2D. The inflow discharge is 150 $m^3 s^{-1}$

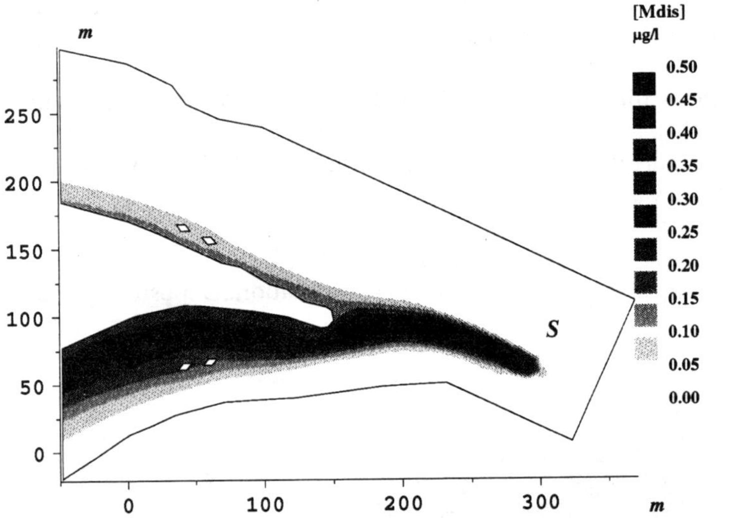

Figure 11. Results of a SUBIEF computation showing transport of dissolved metals, M_{dis}, from a point source release of 5 $m^3 s^{-1}$ containing 20 μg l^{-1} dissolved metals solution

Table I. Parameter values used in SUBIEF heavy metals computation

Name	Symbol	Value
Imposed concentration at entrance	$[SS]_{in}$	100 mg l^{-1}
Settling velocities for SS and Mpart	W_c	1×10^3 ms^{-1}
Critical shear velocity for deposition	u_d^{*c}	0·003 ms^{-1} (equivalent critical shear stress equals 0·1 Nm^{-2})
Kinetics coefficient	k_{-1}	1×10^6 s^{-1}
Sharing coefficient between M_{dis} and M_{part}	k_d	45 l g^{-1}
Source concentration of M_{dis}	$[M_{dis}]_i$	20 μg l^{-1}
Water discharge of the source	—	5 $m^3 s^{-1}$

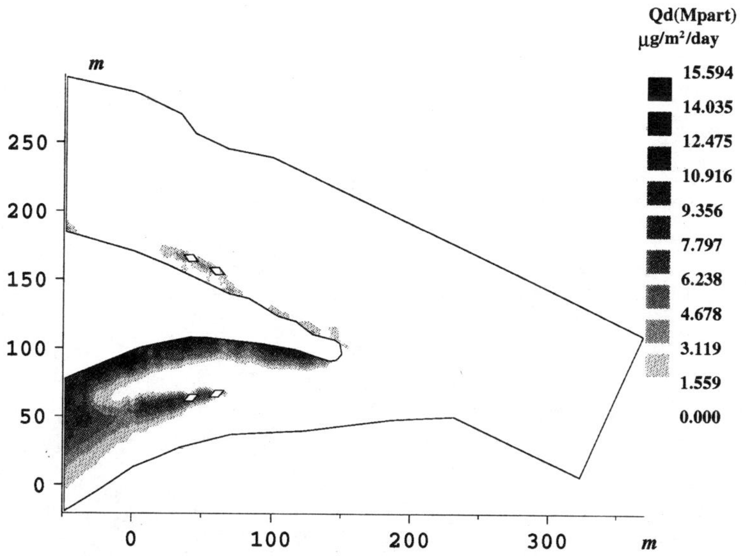

Figure 12. Results of a SUBIEF computation showing the daily deposition flux of particulate-associated metals from a point source release of 5 m^3 s^{-1} containing 20 µg l^{-1} dissolved metals solution

occur when the shear stress is lower than the critical shear stress for deposition. The adsorption reaction also needs some time to make the particulate fraction of metals appear in significant amounts. The combination of these two facts determines whether or not there is deposition of heavy metals. This occurs behind the bridge piles and close to the inner side of the reach as one would expect from a consideration of the velocity field.

CONCLUDING REMARKS

We have presented the principles of a 2D transport module called SUBIEF. This software has already been used as an industrial product in many studies involving dilution aspects or suspended sediment transport. The possibility given to users to associate their own set of chemical reactions to transport–dispersion processes makes SUBIEF an interesting research tool for the modelling of environmental processes. Future improvements will include new erosion laws and alternative advection schemes, as well as parallelization and external coupling with other software. The SUBIEF software is also being extended to geochemical transport–dispersion applications. It will therefore provide a tool for EDF to study the effect of nuclear waste repositories.

REFERENCES

Brooks, A. N. and Hughes, T. J. R. 1982. 'Streamline upwind Petrov–Galerkin formulations for convection dominated flows with particular emphasis on the Navier–Stokes equations', *Comp. Meth. Appl. Mech. Engng*, **32**, 199–259.

Fischer, H. B. 1979. *Mixing in Inland and Coastal Waters*. Academic Press, San Diego. 483 pp.

Hervouet, J.-M. 1986. 'Carac. Module de convection en éléments finis, par la méthode des caractéristiques', *EDF Report no. HE-41/86/21, EDF/LNH, 6 quai Watier*, 78401 Chatou Cedex.

Hervouet, J.-M. and Moulin, C. 1993. 'Nouveaux schémas de convection dans TELEMAC2D-version 2.3. Apport de la methods SUPG', *EDF Report no. 94NJ00023, EDF/LNH, 6 quai Watier*, 78401 Chatou Cedex. 45 pp.

Krone, R. B. 1962. 'Flume studies of the transport of sediment in estuarial shoaling processes: final report', HESER Laboratory, University of Berkeley, California.

Partheniades, E. 1965. 'Erosion and deposition of cohesive soils', *ASCE Hydraul. Div.*, **91**(HY1), 105–138.

Van Rijn, L. 1989. *Handbook of Sediment Transport by Currents and Waves*. Delft Hydraulics, Delft.

4

USING LAGRANGIAN PARTICLE SALTATION OBSERVATIONS FOR BEDLOAN SEDIMENT TRANSPORT MODELLING

YARKO NIÑO[1] AND MARCELO GARCÍA*

Hydrosystems Laboratory, Dept of Civil Engng, University of Illinois at Urbana-Champaign, 250 N. Mathews, Urbana, IL 61901, USA

ABSTRACT

A Lagrangian model for the saltation of sand in water is proposed. Simulated saltation trajectories neglecting particle rotation and turbulence effects compare fairly well with experimental observations. The model for particle motion is coupled with a stochastic model for particle collision with the bed, such that a number of realizations of the saltation process can be simulated numerically. Model predictions of mean values and standard deviations of saltation height, length and streamwise particle velocity agree fairly well with experimental observations. Model predictions of the dynamic friction coefficient are also in good agreement with experimental observations, but they underestimate the value of 0·63 proposed by Bagnold for this coefficient. The saltation model is applied to the estimation of bedload transport rates of sand using a Bagnoldean formulation. Modelled values of the bedload transport rates overestimate those predicted by commonly used bedload formulae, which appears to be a consequence of problems in the definition of the dynamic friction coefficient. These results seem to indicate a few problems with the Bagnoldean formulation, particularly regarding the continuum assumption for the bedload layer, which would be valid only for very high particle concentrations and small particle diameters, and also regarding the evaluation of the shear stress exerted on the bed by the saltating particles.

KEY WORDS Lagrangian model; saltation; particle motion; stochastic collision; Bagnold formulation; bedload sediment transport

INTRODUCTION

Saltation is described as the unsuspended transport of particles over a granular bed by a fluid flow, in the form of consecutive hops within the near-bed region. It is governed mainly by the action of hydrodynamic forces that carry the particles through the flow, the downward pull of gravity and the collision of the particles with the bed, which transfers their streamwise momentum into upward momentum, thus sustaining the saltation motion (Niño and García, 1995). This differs from an earlier definition given by Bagnold (1973) who assumed that the only upward impulses exerted on the saltating particles were those resulting from the impact of particles with the bed. Thus, Bagnold neglected the effect of hydrodynamic lift and vertical impulses owing to flow turbulence, which have been shown to play an essential role in the saltation phenomenon (e.g. Leeder, 1979a; Bridge and Dominic, 1984; Bridge and Bennett, 1992; Niño *et al.*, 1994; Niño and García, 1995).

Two recent experimental studies on saltation of gravel and sand by Niño *et al.* (1994) and Niño and García (1997) have given detailed information on the physics of particle saltation. In particular, they have provided a

* Correspondence to: Marcelo García, Hydrosystems Laboratory, Department of Civil Engineering, University of Illinois at Urbana-Champaign, 250 N. Mathews, Urbana, IL 61801, USA.

[1] Current Address: Department of Civil Engineering, University of Chile, Santiago, Chile.

Contract grant sponsors: NSF and Petroleum Research Fund (ACS).
Contract grant numbers: CTS-9210211 and PRF 24328-G2

description of the particle collision with the bed, allowing calibration of a stochastic model for this phenomenon, and have also provided statistics for the geometric and kinematic properties of the saltation trajectories.

Modelling of particle saltation in water has been attempted by several researchers (e.g. Tsuchiya, 1969; Reizes, 1978; Hayashi and Ozaki, 1980; Murphy and Hooshiari, 1982; Van Rijn, 1984; Wiberg and Smith, 1985; Sekine and Kikkawa, 1992; Lee and Hsu, 1994; Niño and García, 1994). The most recent attempts are those of Niño and García (1994), regarding the saltation of gravel, and Lee and Hsu (1994), regarding the saltation of sand. These models consist of a Lagrangian equation governing particle motion, which is coupled with assumptions regarding initial conditions for the integration of saltating particle trajectories. Niño and García (1994) used a stochastic approach to estimate those initial conditions based on the observations by Niño *et al.* (1994), whereas Lee and Hsu (1994) relied on a deterministic estimation of such conditions.

Niño and García's (1994) model predicted mean values and standard deviations of saltation characteristics of gravel-size particles, such as saltation length, height and streamwise particle velocity, as well as the dynamic friction coefficient, that are in good agreement with the experimental results of Niño *et al.* (1994). On the other hand, Lee and Hsu (1994) calibrated the angular velocity of sand particles during saltation, which controls the amount of lift given by the Magnus force resulting from particle rotation, in order for their model to predict adequately their own observations of saltation. However, their angular velocities were of rather large magnitudes when compared with the direct measurements of this velocity by Niño and García (1995). It was speculated by Niño and García that this may be a consequence of the rather large values of the saltation length and heights measured by Lee and Hsu compared with other available data (e.g. Abbott and Francis, 1977; Niño *et al.*, 1994; Niño and García, 1997). However, it may also be related to the uncertainties in the estimation of the Magnus force, for which Lee and Hsu relied on Rubinow and Keller's (1961) analysis, which is valid only for vanishingly small values of the particle Reynolds number.

Niño and García (1994) used predictions of their model for gravel saltation to estimate bedload transport rates. They invoked a Bagnoldean formulation, such as that proposed by Bridge and Dominic (1984), that relates bedload transport rates to characteristics of saltation such as the mean streamwise particle velocity and the dynamic friction coefficient. The bedload predictions obtained by Niño and García tended to overestimate observed values of bedload transport rates of gravel material, which indicated that Bagnold's hypotheses used in the derivation of the model were not applicable to the transport of gravel-size particles.

Herein, the model for gravel saltation proposed by Niño and García (1994) is applied to the saltation of fine sand, and model predictions are compared with the experimental results of Niño and García (1997). The saltation model is applied to estimate bedload transport rates using a Bagnoldean formulation. Model predictions are then compared with typical empirical and semi-empirical bedload formulae, and Bagnold's hypotheses are discussed in the light of this.

EQUATION FOR SEDIMENT PARTICLE SALTATION IN A TURBULENT BOUNDARY LAYER

The present analysis is based on the instantaneous equations for particle motion in an unbounded fluid flow proposed by Mei *et al.* (1991) (see also Maxey and Riley, 1983), which were modified by Niño and García (1994) in order to analyse the case of particle saltation in a turbulent boundary layer.

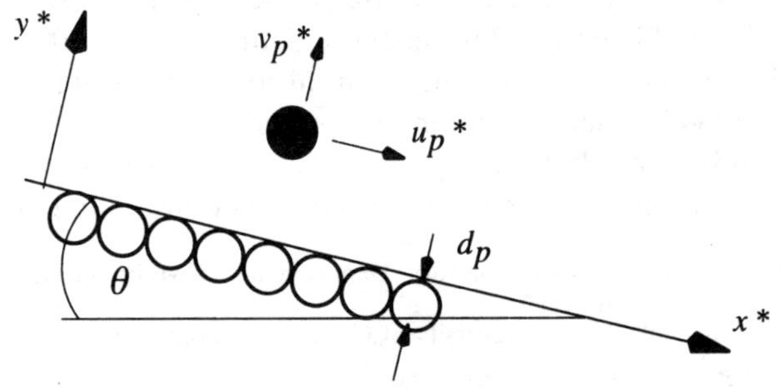

Figure 1. Definition sketch for saltating particle

The situation illustrated in Figure 1 is considered, where a particle is saltating along a bed formed by similar particles, and inclined to an angle θ with respect to the horizontal. The coordinates in the streamwise and bed-normal directions are denoted by x^* and y^*, respectively. Following Niño and García (1994) a system of dimensionless equations governing particle velocity components in the streamwise and bed-normal directions is obtained, such that

$$\frac{du_p}{dt} = -\frac{3}{4}\alpha C_D \mid u_R \mid (u_p - u_f) + \alpha C_m \frac{du_f}{dy} v_p + \frac{\alpha \sin\theta}{\tau_*} + 9\frac{\alpha}{\sqrt{R_p}\tau_*^{1/4}} \frac{1}{\sqrt{\pi}} \int_0^t \frac{d}{d\tau}(u_f - u_p)\frac{d\tau}{\sqrt{t-\tau}} \tag{1a}$$

$$\frac{dv_p}{dt} = -\frac{3}{4}\alpha C_D \mid u_R \mid v_p + \frac{3}{4}\alpha C_L(\mid u_R \mid_T^2 - \mid u_R \mid_B^2) - \frac{\alpha \cos\theta}{\tau_*} - 9\frac{\alpha}{\sqrt{R_p}\tau_*^{1/4}} \frac{1}{\sqrt{\pi}} \int_0^t \frac{d}{d\tau}(v_p)\frac{d\tau}{\sqrt{t-\tau}} + \frac{3}{4}\alpha \mid u_R \mid \left(S - \frac{1}{2}\frac{du_f}{dy}\right) \tag{1b}$$

where the variables have been made dimensionless using the particle diameter, d_p, as a length scale, the flow shear velocity, u_*, as a velocity scale and the ratio d_p/u_* as a time scale. The terms in the right-hand side of Equation (1a) represent the effects of non-linear drag, added mass, gravitational acceleration and Basset history force, respectively. The terms in the right-hand side of Equation (1b) represent the effects of non-linear drag, lift force owing to the non-uniformity of the mean flow velocity profile, gravitational acceleration, Basset history force and Magnus force owing to particle rotation, respectively.

In Equations (1a) and (1b), u_p and v_p denote the dimensionless particle velocity components in the streamwise and bed-normal directions, respectively; $u_f = u_f(y)$ denotes the dimensionless mean flow velocity (which has a component in the streamwise direction only); y denotes the dimensionless bed-normal coordinate; $\mid u_R \mid$ denotes the dimensionless magnitude of the particle slip velocity evaluated at the particle centroid, defined as

$$\mid u_R \mid^2 = [u_p - u_f(y_p)]^2 + v_p^2 \tag{2}$$

with y_p denoting the dimensionless bed-normal coordinate of the particle centroid, and $\mid u_R \mid_T$ and $\mid u_R \mid_B$ denoting the dimensionless magnitude of the particle slip velocity evaluated on the top and bottom of the particle, respectively, such that

$$\mid u_R \mid_T^2 = [u_p - u_f(y_p + \tfrac{1}{2})]^2 + v_p^2 \tag{3a}$$

$$\mid u_R \mid_B^2 = [u_p - u_f(y_p - \tfrac{1}{2})]^2 + v_p^2 \tag{3b}$$

Also, t denotes the dimensionless time; τ denotes a dummy variable for the integration of the Basset term; C_D, C_L and C_m denote the drag, lift, and added mass coefficients, respectively; and S denotes the dimensionless particle angular velocity, defined as

$$S = \Omega\, d_p/u_* \tag{4}$$

with Ω denoting the particle angular velocity in Hz.

The dimensionless parameters appearing in Equations (1a) and (1b) are defined as

$$\alpha = (1 + R + C_m)^{-1}, \quad R = (\rho_s/\rho - 1) \tag{5a, b}$$

$$R_p = \sqrt{Rgd_p^3}/\nu, \quad \tau_* = u_*^2/(gRd_p) \tag{5c, d}$$

where ρ_s and ρ denote particle and fluid densities, respectively, g denotes the gravitational acceleration and v denotes the fluid kinematic viscosity.

Equations (1a) and (1b) have been averaged over the turbulence and terms of order higher than $O(\langle u'_{ki}u'_{lj}\rangle / |u_R|^2)$ have been neglected (Niño, 1995). The primes indicate velocity fluctuations, the subindices k and l indicate whether the velocity fluctuation corresponds to the particle or to the flow, the subindices i and j indicate whether the velocity fluctuation corresponds to the streamwise or vertical component and the angular brackets represent ensemble average over many realizations of the saltation process.

Neglecting second-order terms in Equations (1a) and (1b) implies neglecting turbulence effects on the saltation process. Although this may not introduce large errors in the case of saltation of large particles (e.g. gravel), which, because of their large inertia, tend not to respond to high frequency fluctuations of the flow velocity, it seems that this is not true in the case of smaller, sand-size particles. Keeping higher order terms in Equations (1a) and (1b) requires introducing closures for the auto- and cross-correlations of particle and fluid velocity, which would over-complicate the present model. As is shown later, this does not appear to have a major effect on the simulations of particle saltation as compared with the experimental observations of Niño and García (1997). Nevertheless, the effect of turbulence on particle saltation, as predicted by a saltation model that accounts for flow turbulence through a random walk model for the flow velocity fluctuations (Niño, 1995) is discussed in the next section.

Consideration of non-linear drag in Equations (1a) and (1b) requires definition of the drag coefficient C_D. Unfortunately, the behavior of C_D for unsteady particle motion is not known (Yen, 1992), and the only alternative is to estimate it as that of a single, steady, free-falling sphere. Herein, the following approximation for C_D, proposed by Yen (1992), is used

$$C_D = \frac{24}{Re}(1 + 0{\cdot}15\sqrt{Re} + 0{\cdot}017Re) - \frac{0{\cdot}208}{1 + 10^4 Re^{-0{\cdot}5}} \tag{6}$$

where $Re = |u_R| d_p / v$ denotes the particle Reynolds number.

As the mean flow velocity gradient in the bed-normal direction can be large in the region close to the bed, a lift force acting on the saltating particle must be taken into account. In Equations (1a) and (1b), the lift force is estimated using the expression proposed by Wiberg and Smith (1985), requiring evaluation of the lift coefficient, C_L. Much less is known about C_L than about C_D. Based on the measurements of lift and drag forces on immobile hemispheres near a boundary made by Chepil (1958), Wiberg and Smith (1985) proposed a mean value $C_L = 0{\cdot}2$, which is the one adopted in the present analysis.

An alternative way of estimating the lift force term is that proposed by Patnaik *et al.* (1994), who measured the lift force acting on fixed spheres immersed in a wind tunnel flow. Their definition of the lift coefficient is slightly different than that used in Equations (1a) and (1b). Nevertheless, Patnaik *et al.* (1994) indicate that C_L decreases with increasing values of Re and decreasing values of the dimensionless coordinate y_p. Unfortunately, Patnaik *et al.* present only a few experimental points for y_p smaller than 2, which corresponds to the range of values of this parameter associated with particle saltation (Niño and García, 1997). Furthermore, those results show large scatter. Based on this, it seems that the expression for the lift force proposed by Wiberg and Smith (1985) gives a more reliable estimate of the lift force term than that proposed by Patnaik *et al.* (1994).

The Basset term in Equations (1a) and (1b) is a type of history term that indicates that the resistance to motion at any given time, caused by the unsteadiness of the flow, is, in part, a function of the resistance at a previous time (Brush *et al.*, 1964). As pointed out by Mei (1990), it is not known whether the Basset term in Equations (1a) and (1b) is valid in the non-linear drag range. Mei's numerical analysis seems to indicate that the memory in the classical expression for the Basset force is too long, and that the initial transient does not decay fast enough at large times. Mei concluded that the integrand in the Basset force term must decay much faster than $(t^* - \tau)^{-1/2}$ at large times, which was confirmed by Lawrence and Mei (1995), who found that a decay $(t^* - \tau)^{-2}$ would most likely prevail at large times.

Another source of uncertainty in Equations (1a) and (1b) is the added mass coefficient. Theoretically, a solid sphere immersed in a potential flow of an inviscid fluid has a value $C_m = 0{\cdot}5$. This value is commonly used to model the motion of particles in a viscous fluid (Rizk and Elghobashi, 1985; Elghobashi and Truesdell, 1992). However, according to Yen (1992), experimental evidence indicates that the value of C_m for a sphere is greater than 0·5. This appears to contradict the numerical analysis by Mei (1990), indicating that the force resulting from the added mass on a sphere at finite Reynolds number would be the same as that given by potential flow theory.

The rotation of the saltating particle can introduce another lift force, known as the Magnus force. In Equations (1a) and (1b), this force was estimated using the expression proposed by White and Schulz (1977), and derived originally by Rubinow and Keller (1961) for spheres at vanishingly small values of *Re*. The problem in estimating the Magnus term is that the particle angular velocity is an external variable that must be estimated *a priori*. Experimental observations indicate that the particle angular velocity is controlled mainly by the particle collision with the bed, which makes this variable highly random, and thus difficult to estimate. Niño and García (1997) proposed the following expression to estimate the mean value of S measured in their experiments with fine sand

$$S = 5{\cdot}11 - 1{\cdot}13\tau_*/\tau_{*c} \tag{7}$$

where τ_{*c} denotes the threshold value of the dimensionless bed shear stress for the initiation of motion, which was estimated by Niño and García (1997) using Shields' curve.

Equations (1a) and (1b) are complemented with the Lagrangian equations for the particle trajectory

$$u_p = \frac{dx_p}{dt}, \quad v_p = \frac{dy_p}{dt} \tag{8}$$

where x_p denotes the dimensionless streamwise coordinate of the particle centroid.

In Equations (1) and (8) the local mean flow velocity, u_f, must be specified. Herein, the semi-logarithmic law for the mean velocity profile in turbulent open channel flows is used, which can be written in dimensionless form as

$$u_f = \frac{1}{\kappa}\ln(y) + B_0 \tag{9}$$

where κ denotes von Karman's constant taken to be equal to 0·4; the roughness height, k_s, has been taken to be equal to the particle diameter; and the coefficient B_0 is equal to 8·5 for fully rough flows, and is a function of $k_{s+} = u_* k_s/\nu$ for transitionally rough flows (Schlichting, 1968). The origin for the y-coordinate in Equation (9) should be located some distance below the top of the bed grains. For simplicity, it is assumed that such a distance corresponds to the radius of the bed grains.

Given proper initial conditions, Equations (1) and (8) are easily solved numerically using a fourth-order Runge–Kutta scheme. However, some attention must be given to the treatment of the integral (Basset) term. Following Brush *et al.* (1964), the Basset term corresponding to the direction i can be written in general form as

$$BI_i(t) = K\int_0^t \frac{d}{d\tau}(u_{s_i})\frac{d\tau}{\sqrt{t-\tau}} \tag{10}$$

where $K = (9\alpha)/(\sqrt{R_p}\tau_*^{1/4}\sqrt{\pi})$, and u_{s_i} denotes the ith component of the particle slip velocity. However, Equation (10) can be rewritten as

$$BI_i(t) = K\left[\int_0^{t-\delta t} \frac{d}{d\tau}(u_{s_i})\frac{d\tau}{\sqrt{t-\tau}} + \int_{t-\delta t}^{t} \frac{d}{d\tau}(u_{s_i})\frac{d\tau}{\sqrt{t-\tau}}\right] \tag{11}$$

and assuming δt is very small so that the derivative $\mathrm{d}u_{s_i}/\mathrm{d}\tau$ can be considered approximately constant in the interval $t - \delta t < \tau < t$, then Equation (11) simplifies to

$$BI_i(t) = BI_i(t - \delta t) + 2K\sqrt{\delta t}\,\frac{\mathrm{d}}{\mathrm{d}t}(u_{s_i}) \tag{12}$$

Replacing Equation (12) in Equation (1a) and (1b) and using the following definition (Mei, 1990)

$$\frac{\mathrm{d}u_f}{\mathrm{d}t} = \frac{\partial u_f}{\partial t} + u_p\frac{\partial u_f}{\partial x} + v_p\frac{\partial u_f}{\partial y} \tag{13}$$

then the new system of governing equations for the saltating particle velocities are

$$\frac{\mathrm{d}u_p}{\mathrm{d}t}(1 + 2K\sqrt{\delta t}) = -\frac{3}{4}\alpha C_D\,|u_R|\,(u_p - u_f) + (\alpha C_m + 2K\sqrt{\delta t})\frac{\mathrm{d}u_f}{\mathrm{d}y}v_p + \frac{\alpha\sin\theta}{\tau_*} + BI_u(t - \delta t) \tag{14a}$$

$$\frac{\mathrm{d}v_p}{\mathrm{d}t}(1 + 2K\sqrt{\delta t}) = -\frac{3}{4}\alpha C_D\,|u_R|\,v_p + \frac{3}{4}\alpha C_L(|u_R|_T^2 - |u_R|_B^2) - \frac{\alpha\cos\theta}{\tau_*} + BI_v(t - \delta t) + \frac{3}{4}\alpha\,|u_R|\left(S - \frac{1}{2}\frac{\mathrm{d}u_f}{\mathrm{d}y}\right) \tag{14b}$$

where

$$BI_u(t - \delta t) = K\int_0^{t-\delta t}\left(\frac{\mathrm{d}u_f}{\mathrm{d}y}v_p - \frac{\mathrm{d}u_p}{\mathrm{d}\tau}\right)\frac{\mathrm{d}\tau}{\sqrt{t - \tau}} \tag{15a}$$

$$BI_v(t - \delta t) = -K\int_0^{t-\delta t}\frac{\mathrm{d}v_p}{\mathrm{d}\tau}\frac{\mathrm{d}\tau}{\sqrt{t - \tau}} \tag{15b}$$

The numerical evaluation of Equations (15a) and (15b) was made herein by using Simpson's quadrature, and a value of δt equal to the time interval used in the Runge–Kutta scheme.

ANALYSIS OF THE EQUATIONS FOR PARTICLE SALTATION

The importance of different terms in the equation of motion are analysed and the numerical results of the model for particle motion are compared with the experimental observations of saltation of fine sand reported by Niño and García (1997). This comparison is done by giving the model initial conditions consisting of values of normal and streamwise components of particle velocity at the beginning of a saltation estimated from measured saltation trajectories. Examples of the results obtained are presented in Figure 2, where the Magnus term has been neglected and the following values have been used in the computations: $C_m = 0{\cdot}5$, $C_L = 0{\cdot}2$ and C_D is estimated from Equation (6). A value $R = 1{\cdot}65$, valid for natural sediment, was used throughout.

As seen in Figure 2, the model performs generally very well, reproducing the shape as well as the height and length of the saltations satisfactorily. As in the case of gravel saltation (Niño and García, 1994), the lift force plays an essential role in the motion of sand particles. Also, the Basset term appears to have much more importance than in the case of the saltation of gravel, for which its effect was practically negligible (Niño and García, 1994). The Basset force tends to increase the length of the trajectories of sand particles significantly (Figure 3).

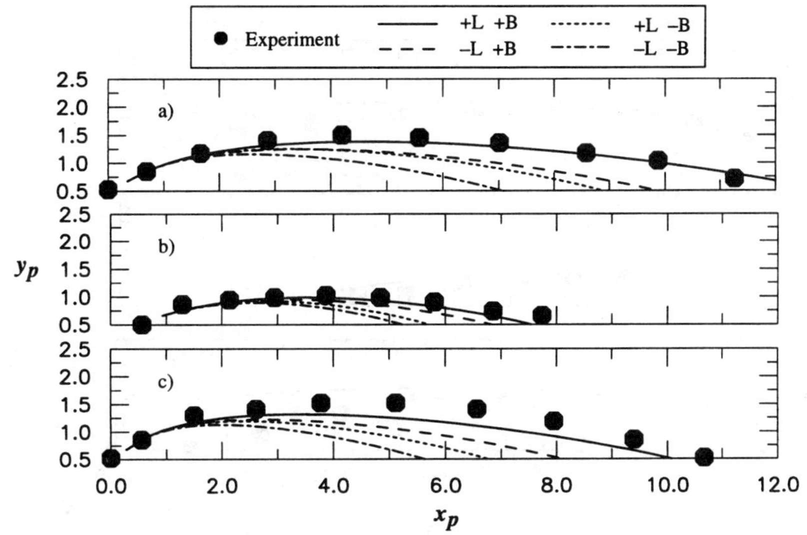

Figure 2. Comparison between simulated dimensionless saltation trajectories and experimental observations of Niño and García (1997). Experimental conditions correspond to: (a) $\tau_* = 0{\cdot}067$, $R_p = 53$; (b) $\tau_* = 0{\cdot}056$, $R_p = 88$; (c) $\tau_* = 0{\cdot}095$, $R_p = 70$. In the key, L denotes lift term, and B denotes Basset term, and the symbols + and − indicate that the respective term is included or neglected in the simulation, respectively

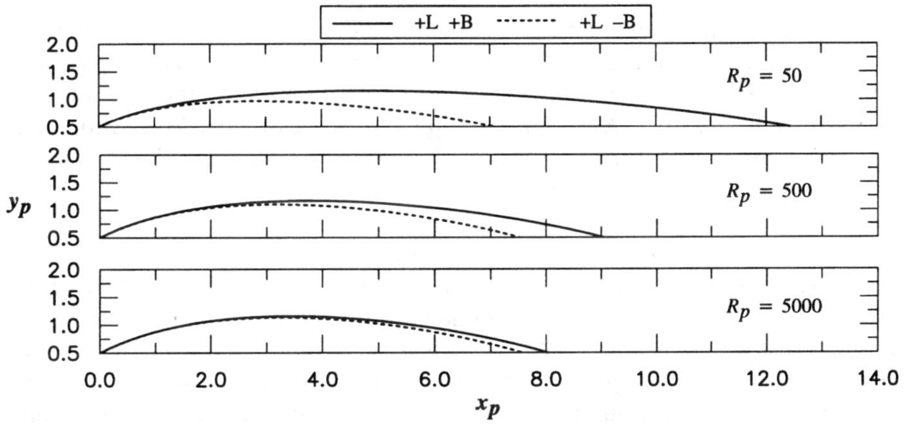

Figure 3. Effect of Basset term on particle saltation as a function of R_p, for the values of this parameter: 50, 500, 5000. All other control parameters are held constant with values: $\tau_* = 0{\cdot}10$, $R = 1{\cdot}65$, $\tan\theta = 0{\cdot}001$. Initial conditions are $x_p = 0$, $y_p = 0{\cdot}5$, $u_p = 3{\cdot}5$, $v_p = 1{\cdot}8$. Key as in Figure 2

Figure 3 shows the increasing importance of the Basset force on the length of the saltation trajectory as R_p decreases. It is apparent that when the Basset term is neglected, the length of the simulated saltation does not change much with R_p in the range 50–5000; however, when the Basset term is included in the simulation, the length of the saltation corresponding to $R_p = 50$ seems to be about 60% longer than that corresponding to $R_p = 5000$, keeping all other parameters constant. On the other hand, the height of the jumps appears to be independent of R_p when the Basset term is included in the simulation, and it seems that this term tends to increase somewhat the saltation height, although, again, its effect is higher for smaller values of R_p.

In order to analyse the effect of particle rotation on saltation trajectories, a few simulations of particle saltation were conducted including the Magnus force term in the equations for particle motion (Figure 4). The value of the dimensionless particle angular velocity, S, was estimated in each case using Equation (7).

The Magnus term increases the lift acting on the particle such that the saltation height and length tend to be somewhat larger than those obtained by neglecting the effect of particle rotation (Figure 4). In general,

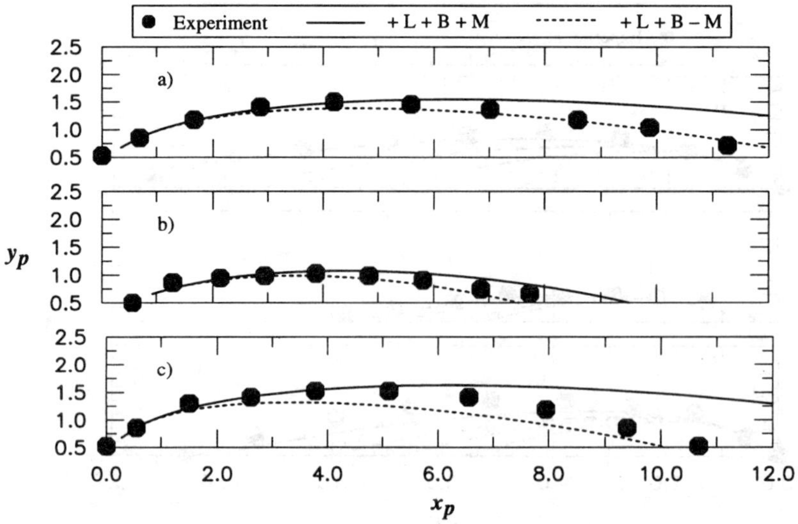

Figure 4. Effect of Magnus term on particle saltation, comparison with experimental observations of Niño and García (1997). Experimental conditions correspond to: (a) $\tau_* = 0{\cdot}067$, $R_p = 53$, $S = 2{\cdot}8$; (b) $\tau_* = 0{\cdot}056$, $R_p = 88$, $S = 3{\cdot}2$; (c) $\tau_* = 0{\cdot}095$, $R_p = 70$, $S = 1{\cdot}8$. In the key, L denotes lift term, B denotes Basset term and M denotes Magnus term, and the symbols + and − indicate that the respective term is included or neglected in the simulation, respectively

computations including the Magnus term tend to overestimate the observed saltation length, particularly in cases (a) and (c), by a rather large amount. This may imply that estimating the Magnus force term using Rubinow and Keller's (1961) expression, which was originally derived for the case of vanishingly small values of the particle Reynolds number, *Re*, is not accurate in the case of particle saltation at larger values of *Re*. Nevertheless, it may also be possible that there is another factor, not considered in the present model, that tends to cancel out the effect of particle rotation, such that simulations that neglect this effect predict observed trajectories reasonably well.

Niño (1995) analysed turbulence effects on particle saltation based on a random walk model for flow velocity fluctuations. An example of the results obtained is shown in Figure 5, where ensemble-averaged values of the dimensionless saltation length, λ_s, simulated with the random walk model are plotted as a function of R_p for the values $\tau_* = 0{\cdot}1$ and $0{\cdot}2$. The values of λ_s for each combination of R_p and τ_* were computed over 150 saltations, where initial conditions for all saltations were held constant with values $x_p = 0$, $y_p = 0{\cdot}5$, $u_p = 3{\cdot}5$, $v_p = 1{\cdot}8$, similar to the saltations shown in Figure 3. The variability obtained in the saltation characteristics in any set of simulations results only from turbulence effects. In Figure 5 the predictions of the present model without turbulence effects are also plotted for comparison. As R_p increases, the values of λ_s, computed with turbulence neglected, approach asymptotically the mean values of λ_s predicted by the random walk model. At values of $R_p = 50$, neglecting turbulence leads to an overestimate of the mean values of the dimensionless saltation length in about 40% for values of $\tau_* = 0{\cdot}1$, and about 60% for values of $\tau_* = 0{\cdot}2$. At values of $R_p = 200$ these values have been reduced to about 5 and 15%, respectively. At values of R_p larger than about 2000, the turbulence does not appear to influence the mean values of λ_s.

Similar results are obtained for the dimensionless saltation heights, although, in this case, neglecting turbulence leads to much less overestimation of the ensemble-averaged values of this variable (Niño, 1995). Results of the analysis indicate that, in general, turbulence effects can be neglected in modelling mean trajectories of saltations without incurring important errors, for values of R_p larger than, say, 500. In terms of the particle size, this limit corresponds to about 2·5 mm.

Since particle rotation tends to increase the length and height of the saltation trajectories and the turbulence of the flow seems to induce the opposite effect, this suggests that particle rotation and turbulence

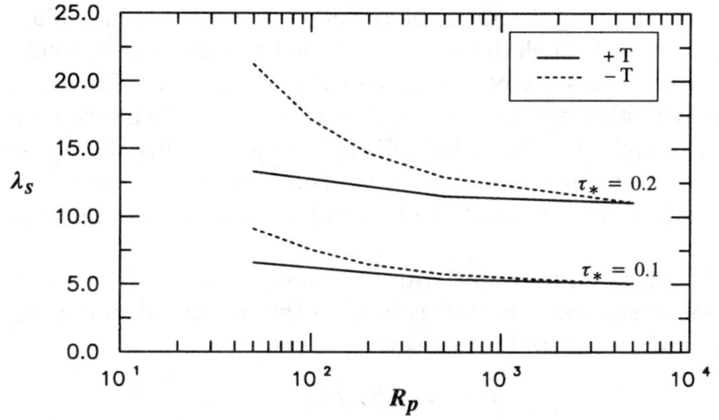

Figure 5. Effect of turbulence on mean saltation length as a function of R_p and τ_*. In the key, −T denotes computations that do not include turbulence, and +T denotes computations carried out with the random walk model of Niño (1995)

effects tend to cancel each other, such that neglecting both of them in the simulation of particle saltation results in a reasonable estimation of the mean particle trajectory, as shown in Figures 2 and 4.

A comprehensive comparison between experimental and theoretical trajectories of sand saltations, of which the results presented in Figure 2 are examples, demonstrates that, given appropriate initial conditions and neglecting the Magnus term, the equations for particle saltation proposed herein perform satisfactorily. It is acknowledged, however, that such equations are only approximate, in that they neglect rotation and turbulence effects, and also because of the uncertainties regarding the added mass, drag and lift coefficients and the Basset term.

STOCHASTIC SIMULATION OF PARTICLE SALTATION

In order to solve Equations (8) and (14), initial conditions are required for particle position and velocity. In the case of successive hops of the particle, those initial conditions are controlled mainly by the particle collision with the bed.

Particle velocities immediately after the collision are estimated from particle velocities immediately before the collision, by introducing coefficients of restitution, e, and friction, f, which reduce the incident particle velocity components in the normal and tangential directions, respectively, with respect to the surface of contact at the bed. For the situation shown in Figure 6, the angle of contact corresponds to $\theta_{\mathrm{in}} + \theta_{\mathrm{b}}$, where the bed angle, θ_{b}, is clearly a random variable, even in the idealized situation of a bed formed by uniformly packed spheres.

In order to make a stochastic evaluation of θ_{b}, the method proposed originally by García and Niño (1992), and akin to those proposed earlier by Wiberg and Smith (1985) and Tsujimoto and Nakagawa (1983), is followed, which assumes a bed formed by uniformly packed spheres. With reference to Figure 6, θ_{b} can take

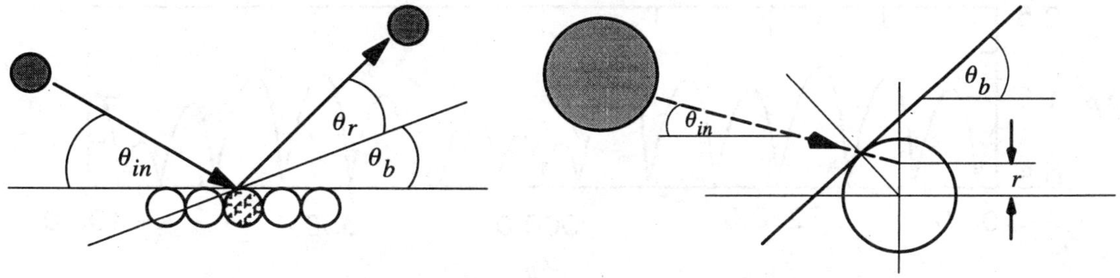

Figure 6. Definition diagram for particle collision with the bed

values in the range −30 to 30°. Mapping this range of angles on to a set of values of an auxiliary variable, r, defined along a vertical line passing through the centre of the bed particle, then geometrical transformations, allow determination of the values θ_b associated with given values of θ_{in} and r. The stochastic estimation of θ_b is obtained by using a random number generator to produce values of r with uniform probability density, for given values of the incidence angle θ_{in}. Prescribing a uniform probability density for r is equivalent to assuming that the bed particle has a uniform probability of being located anywhere in the bed. Values of θ_{in} can be obtained deterministically from the trajectory of the saltating particle prior to the current collision with the bed.

Based on the analysis of experimental results with fine sand presented by Niño and García (1997), a constant value of the friction coefficient $f = 0{\cdot}89$ is used in the numerical simulation, as well as a linear relationship for the variation of the restitution coefficient, e, with τ_*

$$e = A - B\tau_*/\tau_{*c} \tag{16}$$

where A and B take the values 0·75 and 0·25, respectively, and, as before, τ_{*c} was estimated using Shields. Niño and García's results correspond to values of τ_*/τ_{*c} lower than about 2·5. For larger values of this region, the experimental results suggest values of e from 0·0 to about 0·2. Herein, a minimum value of e of 0·2 in the range of $\tau_*/\tau_{*c} > 2{\cdot}2$ is adopted, because this gave results for the simulation of particle saltation in good agreement with the experimental observations.

Four hundred successive hops of a saltating sand particle were simulated for different values of governing dimensionless parameters with the aim of estimating the statistics of the saltation parameters. This was done by solving the equations for gravel saltation numerically, using a fourth-order Runge–Kutta scheme, coupled with the model of particle collision with the bed explained above. The particle is initially released at a distance of two particle diameters from the bed, with zero velocity. The first 20 hops are discarded when computing statistics of saltation, in order to get results that are independent of the arbitrary initial condition. A succession of simulated saltations of a sediment particle is shown in Figure 7 as an example of the results obtained.

STATISTICS OF SALTATION OF SAND

The length, duration, height and mean streamwise particle velocity of individual saltations, are denoted by λ_s^*, t_s^*, h_s^* and u_s^*, respectively, where u_s^* is computed as

$$u_s^* = \frac{1}{t_s^*}\int_0^{t_s^*} \bar{u}_p(t^*)\, dt^* \tag{17}$$

Mean values and standard deviations of the dimensionless variables λ_s, h_s and u_s, defined as

$$\lambda_s = \frac{\lambda_s^*}{d_p}, \quad h_s = \frac{h_s^*}{d_p}, \quad u_s = \frac{u_s^*}{u_*} \tag{18a–c}$$

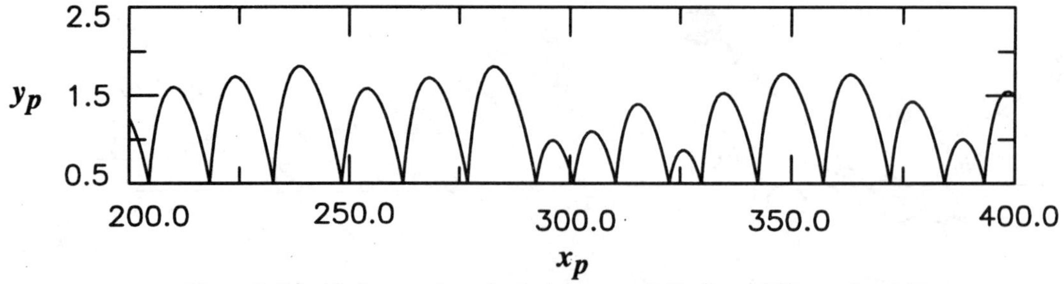

Figure 7. Simulated succession of saltations, $\tau_* = 0{\cdot}15$, $R_p = 7400$, $\tan\theta = 0{\cdot}05$

were computed by ensemble averaging over the four hundred simulated saltations. In addition, the dynamic friction coefficient, μ_d, estimated as (Abbott and Francis, 1977; Sekine and Kikkawa, 1992)

$$\mu_d = (R + 1)\tau_*(\Delta u_s u_s)/\lambda_s \tag{19}$$

where Δu_s denotes the change in streamwise particle velocity before and after the collision with the bed, made dimensionless with u_*, was computed from the results of the ensemble averaged characteristics of the simulated saltations. The variables λ_s, h_s, u_s and μ_d are functions of the dimensionless parameters R, τ_*, R_p and θ as suggested by the system of governing Equations (1).

Mean values and standard deviations of λ_s, h_s, u_s and μ_d computed from the numerical results for different values of the ratio τ_*/τ_{*c}, and for the value $R_p = 73$, representative of the range of values of this parameter in the experiments with sand of Niño and García (1995), are plotted in Figures 8 to 11. In these figures, the experimental results of Niño and García (1995), Francis (1973), Fernandez-Luque and van Beek (1976), Abbott and Francis (1977) and Lee and Hsu (1994) are also plotted for comparison. In the computations, the values $\tan\theta = 0{\cdot}001$, $R = 1{\cdot}65$, $C_m = 0{\cdot}5$ and $C_L = 0{\cdot}2$ were used, and the values of C_D were estimated from Equation (6). Using the ratio τ_*/τ_{*c} as an independent variable instead of τ_* allows a better comparison of data with extremely different values of R_p, such as those corresponding to the experiments with coarse particles of Abbott and Francis (1977) and Niño *et al.* (1994), and those with sand of Lee and Hsu (1994) and Niño and García (1997).

ANALYSIS OF NUMERICAL SIMULATION

As seen in Figure 8, very good agreement between the simulated mean values and standard deviations of h_s and those observed by Niño and García (1995) is obtained. Very good agreement is also apparent between the simulated mean values of h_s and corresponding observations by Abbott and Francis (1977). On the other hand, the model tends to underestimate the experimental values of h_s of Lee and Hsu (1994), which nevertheless appear to be high compared with all other experimental data available.

The numerical model reproduces mean values and standard deviations of λ_s observed in the experiments of Niño and García (1997) very well (Figure 9). However, the simulated mean values of λ_s seem to be larger than

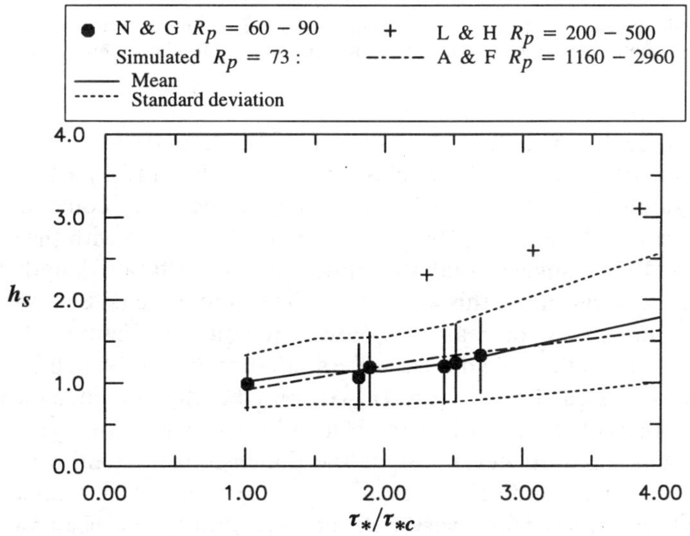

Figure 8. Comparison between simulated and observed dimensionless saltation height. Symbols represent mean experimental values and vertical lines represent a total length of two corresponding standard deviations. Solid lines represent simulated values and dashed lines correspond to mean values plus and minus one standard deviation. In the key, N & G denotes Niño and García (1997), L & H denotes Lee and Hsu (1994) and A & B denotes Abbott and Francis (1977)

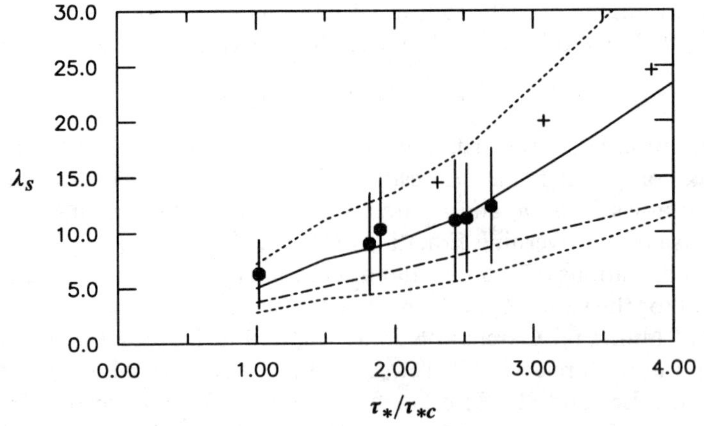

Figure 9. Comparison between simulated and observed dimensionless saltation length. Key as in Figure 8

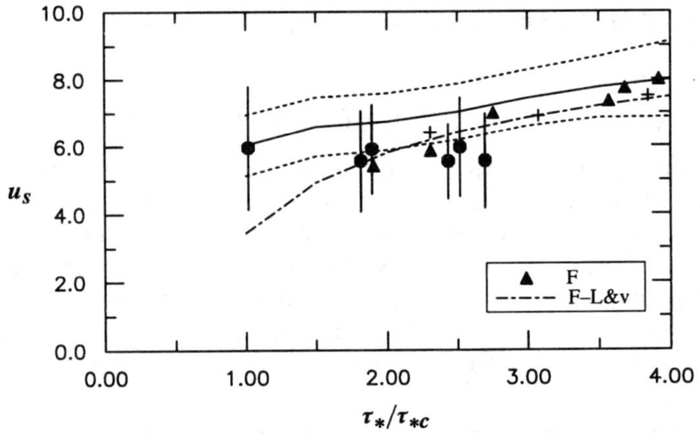

Figure 10. Comparison between simulated and observed dimensionless saltation streamwise velocity. In the key, F denotes Francis (1973) and F-L & v denotes Fernandez-Luque and van Beek (1976). All other symbols are as in Figure 8

corresponding values observed by Abbott and Francis (1977). Abbott and Francis' experiments were performed with rather coarse particles, which are closer in size to the gravel particles used by Niño *et al.* (1994). In fact, the dimensionless saltation lengths measured by the latter coincide very well with those measured by Abbott and Francis (Niño *et al.*, 1994). These results, together with the experimental observations of Niño and García (1997), suggest that the dimensionless saltation length would increase with decreasing values of R_p. The explanation for this would be in the fact that the Basset force tends to increase the dimensionless length of the jumps as R_p decreases, as shown previously in Figure 3. From the same figure it seems that this does not appear to affect the dimensionless saltation height, which is in agreement with the results shown in Figure 8, as discussed above. Figure 9 also shows that the simulated mean values of λ_s tend to get somewhat closer to the values observed by Lee and Hsu (1994) for values of τ_*/τ_{*c} larger than about 3·5.

Simulated mean values and standard deviations of the dimensionless streamwise velocity of saltating grains seem to overestimate corresponding values observed by Niño and García (1997) (Figure 10). Nonetheless, the numerical results reproduce very well the apparently high mean value of u_s observed by Niño and García (1997) at τ_*/τ_{*c} of about 1. As seen in Figure 10, the numerical results tend also to overestimate experimental observations by Francis (1973), Fernandez-Luque and van Beek (1976) and Lee and Hsu (1994) for values of τ_*/τ_{*c} less than about 2·5, but are in much better agreement with them for larger

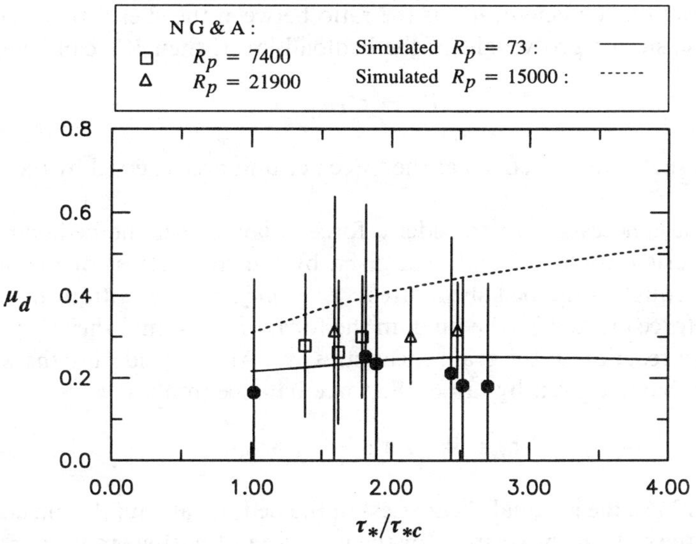

Figure 11. Comparison between simulated and observed values of the dynamic friction coefficient. In the legend N G & A denotes Niño *et al.* (1994). All other symbols are as in Figure 8

values of this parameter. A more thorough analysis of the saltation streamwise velocity predicted by the numerical model is presented in the next section.

The simulated values of μ_d shown in Figure 11 agree fairly well with the experimental estimations of this variable by Niño and García (1997). In Figure 11, the values of μ_d estimated by Niño *et al.* (1994) for the case of gravel saltation are also plotted, together with the numerical prediction of μ_d for $R_p = 15\,000$, corresponding to the typical value of this parameter in Niño *et al.*'s experiments.

The values of μ_d estimated from the numerical simulations are somewhat greater in the case of saltation of gravel than in the case of saltation of sand, which is again supported by experimental data (Figure 11). A dependence of μ_d on the particle Reynolds number has been observed experimentally in the case of rolling particles (Garde and Sethuraman, 1969). The present results suggest that a similar effect occurs in the case of saltation. The simulated values of μ_d in the case of saltation of sand are much closer to the corresponding observed values than in the case of saltation of gravel, in which the simulated values tend to overestimate the observed values of μ_d. In any case, simulated and observed values of μ_d seem to be, in general, much smaller than the value 0·63 predicted by Bagnold (1973). Further discussion of this particular result is presented in the next section.

APPLICATION TO BEDLOAD MODELLING

The results of the numerical simulations of particle saltation can be used to analyse bedload transport in open channel flows. The situation of equilibrium sediment transport is considered, and it is assumed that the bedload transport occurs in a thin, constant stress layer. Following the theoretical derivations of Bridge and Dominic (1984) (see also, Engelund and Fredsøe, 1976; Leeder, 1979b; Bridge, 1981; Bridge and Bennett, 1992; Niño and García, 1994), the bedload transport rate i_b (in terms of submerged weight per unit width) can be expressed as

$$i_b = W_b u_b^* \tag{20}$$

where W_b denotes the submerged weight of the bedload layer per unit bed area and u_b^* denotes the mean velocity of the bedload grains.

Defining the dynamic friction coefficient, μ_d, as the ratio between the shear stress exerted by the bedload grains over the bed and the submerged weight of the bedload layer, then W_b can be expressed as

$$W_b = \tau_g/\mu_d \tag{21}$$

where τ_g denotes the grain shear stress, defined as the force per unit area exerted by the moving grains on the bed.

In order to evaluate τ_g it is necessary to consider a force balance over the bedload layer. On top of the bedload layer, the fluid exerts a force per unit area given by the shear stress of the flow τ_0, which can be considered approximately equal to the bed shear stress: $\tau_0 = \rho u_*^2$. At the bottom of the bedload layer the total force is given by the force exerted by the fluid, in the form of a residual shear stress at the bed, τ_b, and that exerted by the bed as a reaction to the grain shear stress τ_g. Also considering the weight of the bedload layer and the slope of the channel given by tan θ, the force balance implies

$$\tau_0 + W_b \sin\theta = \tau_g + \tau_b \tag{22}$$

According to Bagnold (1956), the residual shear stress at the bed, τ_b, at equilibrium conditions for bedload transport, corresponds to the critical shear stress for the initiation of particle motion, τ_c. The argument is as follows. Within the bedload layer and because of the presence of moving particles, the fluid shear stress decreases towards the bed, from the value τ_0 on top of the bedload layer to the value τ_b at the bed. The amount of shear stress reduction increases with the concentration of particles within the bedload layer. Given a value of τ_b larger than τ_c, more and more particles are entrained into motion; however, this produces a drop in the value of τ_b. If equilibrium conditions for bedload transport prevail, then a maximum concentration of particles in the bedload layer is attained, such that the flow shear stress at the bed drops to τ_c and no more particles can be entrained into motion. In this situation: $\tau_b = \tau_c$. Hence, from Equation (22) and assuming that the slope of the channel is small

$$\tau_g = \tau_0 - \tau_c \tag{23}$$

Replacing Equations (23) and (21) in Equation (20), i_b can be written as

$$i_b = \frac{1}{\mu_d}(\tau_0 - \tau_c)u_b^* \tag{24}$$

and expressing i_b in terms of Einstein's form for the dimensionless volumetric bedload transport rate given by

$$q^* = i_b/\sqrt{\rho^2(Rgd_p)^3}$$

then Equation (24) can be rewritten as

$$q^* = \frac{1}{\mu_d}\tau_*^{1/2}(\tau_* - \tau_{*c})u_b \tag{25}$$

where u_b denotes the mean velocity of the bedload grains made dimensionless with u_*. By identifying the value of u_b with the mean dimensionless streamwise saltation velocity, u_s, and the dynamic friction coefficient, μ_d, defined as in Equation (21) with the value estimated as in Equation (19), then results of the numerical simulation of particle saltation can be used to evaluate Equation (25).

Simulated values of u_s for values of R_p in the range 75 to 200, representative of the experimental conditions for the bedload transport of sand in the experiments reported by Niño and García (1997) and by

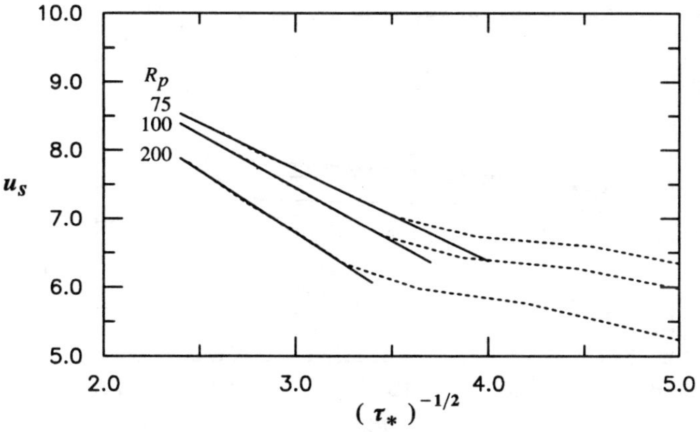

Figure 12. Simulated dimensionless streamwise saltation velocity. Dashed lines represent numerically simulated values, solid lines represent tendencies as τ_* increases

Fernandez-Luque and van Beek (1976), are plotted in Figure 12 as a function of $(\tau_*)^{-1/2}$. The numerical results tend to define straight lines of variable slope as τ_* increases. Based on this, the following general form for the dimensionless streamwise saltation velocity can be proposed

$$u_s = a - b(\tau_*)^{-\frac{1}{2}} \tag{26}$$

Table I shows values of a and b as functions of R_p fitted to the numerical results shown in Figure 12. The value of a is fairly constant in the range $75 < R_p < 200$, equal to about 12·0, with a variation of about 2%. The value of b seems to vary relatively more than a in the same range of values of R_p, with a mean value of about 1·6.

Equation (26) can be rewritten as

$$u_s = a[1 - b'(\tau_*/\tau_{*c})^{-\frac{1}{2}}] \tag{27}$$

where $b' = b/(a\sqrt{\tau_{*c}})$. Values of b' are also shown in Table I, where values of τ_{*c} were estimated using Shields' curve. As seen therein, values of b' in the range $75 < R_p < 200$ are, on average, about 0·7.

The form of Equation (27) is identical to that of relationships for u_s proposed analytically and experimentally by different researchers (e.g. Ashida and Michiue, 1972; Fernandez-Luque and van Beek, 1976; Bridge and Dominic, 1984; Bridge and Bennett, 1992). Most of the models for u_s consider $b' = 1$; however, Fernandez-Luque and van Beek (1976) found, experimentally, a value of $b' = 0·7$, which is similar to the mean value of b' in the range $75 < R_p < 200$ obtained from the numerical simulation of saltation of sand as shown in Table I. In general, according to Bridge and Dominic (1984), different researchers have proposed values of a in the range 8·5 to 14·3. Fernandez-Luque and van Beek (1976) propose a value of $a = 11·5$, which is also similar to the values of this parameter shown in Table I.

Simulated values of μ_d for the same values of R_p used in the analysis of $u_s (75 < R_p < 200)$ are plotted in Figure 13 as functions of R_p and τ_*/τ_{*c}. Values of μ_d tend to increase slightly with R_p in the range

Table I. Values of constants in equation for dimensionless streamwise saltation velocity

R_p	a	b	b'
75	11·8	1·4	0·6
100	12·1	1·6	0·7
200	12·3	1·8	0·8

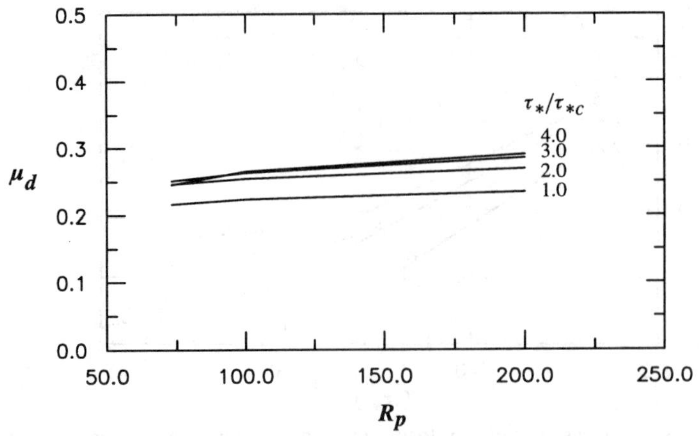

Figure 13. Simulated dynamic friction coefficient

considered. They also increase with τ_*/τ_{*c}, but their change is rather small for values of τ_*/τ_{*c} larger than about 2. Figure 13 indicates that a reasonable approximation is to use a constant value of μ_d of about 0·28, for values of R_p in the range 75–200 and for values of τ_*/τ_{*c} larger than about 2. This value of μ_d is in agreement with the experimental observations of Niño and García (1997), as has already been discussed.

Replacing Equation (27) in Equation (25), q^* can be expressed as

$$q^* = \frac{a}{\mu_d}(\tau_* - \tau_{*c})(\tau_*^{1/2} - b'\tau_{*c}^{1/2}) \tag{28}$$

where, according to the above discussion, a, b' and μ_d can be approximated as constants, independent of R_p. The form of Equation (28) is identical to that of Engelund and Fredsøe (1976). The asymptotic behaviour of Equation (28) at large values of τ_* is $q_* \propto \tau_*^{3/2}$, which is in agreement with most of the bedload formulae available in the literature (e.g. Meyer-Peter and Muller, 1948, and others). Replacing the values $a = 12$, $b' = 0{\cdot}7$ and $\mu_d = 0{\cdot}28$, the following equation is obtained

$$q^* = 43{\cdot}0(\tau_* - \tau_{*c})(\tau_*^{1/2} - 0{\cdot}7\tau_{*c}^{1/2}) \tag{29}$$

In Figure 14, Equation (29) is compared with Meyer-Peter and Muller's (1948) and Engelund and Fredsøe's (1976) formulae, which are given by

$$q^* = 8(\tau_* - \tau_{*c})^{3/2} \tag{30}$$

$$q^* = 18{\cdot}74(\tau_* - \tau_{*c})(\tau_*^{1/2} - \tau_{*c}^{1/2}) \tag{31}$$

respectively, where τ_{*c} is approximated as 0·047 in Equation (30) and as 0·05 in Equation (31). The values of q^* are plotted as a function of $\tau_* - \tau_{*c}$ in Figure 14. Since Equation (29) depends on τ_{*c}, parametric curves with values of this parameter in the range 0·02 to 0·05, typical for sediment particles in the sand size range, are plotted. The values of q^* given by Equation (29) are not too sensitive to parametric variations of τ_{*c} in the range considered, and depend mainly on $\tau_* - \tau_{*c}$. As $\tau_* - \tau_{*c}$ increases, the curves for q^* tend to collapse into one, which has a constant slope of $\frac{3}{2}$ in the log–log plot, and is parallel to those given by Equations (30) and (31). Nevertheless, it is also apparent that Equation (29) significantly overestimate the values of q^* predicted by those equations. In fact, the coefficient $a/\mu_d = 43$ in Equation (29) is about 2·3 times larger than the corresponding coefficient appearing in Engelund and Fredsøe, Equation (31).

Since the expression in Equation (27) for u_s with the values of $a = 12$ and $b' = 0{\cdot}7$ is very close to that of Fernandez-Luque and van Beek (1976), and in general agrees satisfactorily with other experimental observations, including those reported by Niño and García (1997), it must be concluded that the dynamic

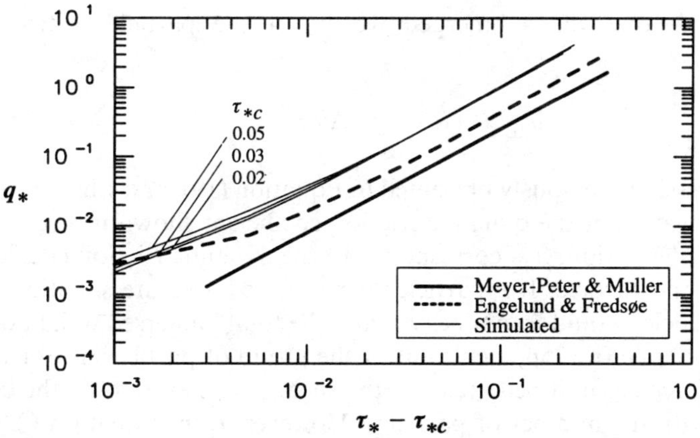

Figure 14. Simulated dimensionless bedload transport rates. Comparison with formulae by Meyer-Peter and Muller (1948) and Engelund and Fredsøe (1976)

friction coefficient, μ_d, estimated from the numerical simulation of particle saltation using Equation (19) seems to be about 2·3 times smaller than expected. None the less, the simulated values of μ_d reproduce experimental observations of this parameter well, as discussed previously (Figure 11). To elucidate this paradox it is necessary to pay some attention to the different definitions of the dynamic friction coefficient used so far, and the method of estimating this variable from the experimental observations.

As indicated in Equation (21), the definition of μ_d used in the derivation of the bedload equation, (25), considers the dynamic friction coefficient as the ratio between the shear stress exerted by the bedload grains over the bed and the submerged weight of the bedload layer per unit area. Furthermore, the grain shear stress is assumed to equal, by force balance considerations, the reduction of the shear stress of the fluid at the bed owing to the presence of the saltating grains. On the other hand, Equation (19) corresponds to a Lagrangian estimation of μ_d, given by the ratio between the local streamwise force exerted over the bed by a saltating particle through successive collisions, and the submerged weight of the particle. The derivation of Equation (19) is as follows (Abbott and Francis, 1977).

Considering a particle in continuous saltation, its streamwise velocity just before an impact with the bed, $u_p^*|_{in}$, is larger than that just after the impact, $u_p^*|_{out}$, implying the existence of a retarding force exerted by the bed over the saltating grain during the collision. In a steady saltation, the mean retarding force exerted by the bed must equal the mean forward thrust exerted by the fluid during the saltating trajectory. The mean forward thrust can be estimated from the streamwise acceleration of the grain from a streamwise velocity $u_p^*|_{out}$ just after an impact with the bed, to the streamwise velocity $u_p^*|_{in}$ immediately before the next collision with the bed. The time during which this acceleration takes place corresponds to the duration of the saltation, given by

$$t_s^* = \lambda_s^*/u_s^* \tag{32}$$

The mean streamwise momentum change of the particle per unit time during the saltation event, which is equivalent to the fluid thrust, is given by

$$F_f = \rho_s V_p(u_p^*|_{in} - u_p^*|_{out})u_s^*/\lambda_s^* \tag{33}$$

where V_p denotes the particle volume. Calling $(u_p^*|_{in} - u_p^*|_{out}) = \Delta u_s^*$, then the Lagrangian dynamic friction coefficient, μ_{dL}, can be written as

$$\mu_{dL} = F_f/W_g = (\rho_s \Delta u_s^* u_s^*)/((\rho_s - \rho)g\lambda_s^*) \tag{34}$$

where $W_g = (\rho_s - \rho)gV_p$, denotes the submerged weight of the particle. Introducing dimensionless parameters already defined

$$\mu_{dL} = (R + 1)\tau_* \Delta u_s u_s / \lambda_s \tag{35}$$

which is identical to the expression previously presented in Equation (19), and which was used to estimate the experimental and numerical values of the dynamic friction coefficient shown in Figure 11.

According to the previous derivation, μ_{dL} corresponds to the dynamic friction coefficient associated with the continuous saltation of one particle. If n particles per unit bed area are saltating, then the total force exerted by the bed on the particles would be n times F_f, and the total immersed weight of the particles would be n times W_g. Thus, the dynamic friction coefficient of the group of particles is exactly equal to μ_{dL}. The latter implies that the total force per unit bed area, i.e. the shear stress, exerted by the bed over the saltating particles increases linearly with the number of particles. However, from Equation (22), the effective shear stress exerted by the bed over the saltating grains is given by the reduction of the fluid shear stress owing to the presence of particles, which, most likely, is a non-linear function of the number of particles in the bedload layer, since it involves non-linear interactions between fluid and solid phases, including turbulence modification by the presence of the sediment. This indicates that the two concepts, the Lagrangian dynamic friction coefficient, μ_{dL}, and the dynamic friction coefficient, μ_d, used in the bedload equation, (25), do not necessarily coincide. In fact, the latter involves a continuum assumption for the bedload layer, different from the Lagrangian approach of the former, which may be valid only under conditions of relatively high concentration of particles in the bedload layer.

The previous analysis seems to justify the observed differences between the Lagrangian estimation of the dynamic friction coefficient by Equations (19) or (35) and the value of this coefficient required to fit typical bedload formulae, which according to Bagnold (1973), should be close to the value of the static friction coefficient, tan ϕ, which for natural sediment is about 0·63. Interestingly, the ratio between tan $\phi = 0{\cdot}63$ and the numerically estimated value of $\mu_{dL} = 0{\cdot}28$ gives a value of 2·3, which is equal to the factor by which the numerically derived bedload equation, (29), overestimates the values of q^* with respect to those given by Engelund and Fredsøe, Equation (31).

The above analysis would also explain why experimental estimations of μ_{dL} have been found to be lower than the value of μ_d predicted by Bagnold. In fact, Francis' (1973) and Abbott and Francis' (1977) experiments involved the saltation of single particles, while those of Niño *et al.* (1994) and Niño and García (1997) were conducted under low bedload transport rates. In all these cases the Lagrangian approach for estimating the dynamic friction coefficient seems more appropriate. Accordingly, the observed differences between μ_{dL} and the value of μ_d predicted by Bagnold would not be a result of experimental inaccuracies in the temporal resolution of the moment of collision, as has been argued in the past (e.g. Abbott and Francis, 1977). Indeed, Niño and García's (1997) experimental results showed similar such differences, although a high-speed video system was used, thus improving the temporal resolution of the moment of collision substantially with respect to previous experimental work.

The above results regarding the modelling of bedload transport of sand, and those discussed by Niño and García (1994) for the case of gravel, seem to indicate a few problems with the Bagnoldean formulation. In the first place, it assumes that the bedload layer is a continuum, such that a reduction in the fluid shear stress takes place within such a layer as a result of the presence of the particles. This is likely to be valid only under conditions of relatively high concentrations of particles, and relatively small particle size. Secondly, it assumes that the net entrainment of particles into saltation is somehow proportional to the excess bed shear stress over the critical value τ_{*c}. Since, in equilibrium bedload transport, a constant number of particles forming part of the bedload layer are kept in motion, the latter assumption implies that in this case a reduction of fluid shear stress at the bed to the critical value τ_{*c} must take place owing to the presence of the bedload layer. This is very unlikely to occur in the case of low bedload transport rates such as those used in the experiments of Niño *et al.* (1994) and Niño and García (1997), where only a few particles were moving over the bed at a given time. Thirdly, the resulting model for bedload transport appears to be very dependent on the value of τ_{*c}, at least for relatively low bedload transport rates. This parameter, as used in the Bagnoldean formulation, does not necessarily correspond to the value associated with the threshold conditions for particle motion in the absence of bedload transport, and thus is difficult to estimate. In fact, Lavelle and Mofjeld (1987) have criticized the existence of the concept of threshold of particle motion and proposed the formulation of transport models that do not rely on such a concept. An alternative analysis to estimate bedload transport rates by saltating particles would be to follow an Einstenian formulation (e.g. Einstein, 1950), in which saltation characteristics are used but there is no need to assess the shear stress exerted on the bed by the particles (Parker, 1991).

CONCLUSIONS

The model for saltation of sand developed herein performs satisfactorily when compared with experimental measurements of single particle trajectories by Niño and García (1997). As in the case of the saltation of gravel, lift forces must be included for the adequate modelling of observed trajectories of sand particles, indicating that upward forces arising from the particle collisions with the bed are insufficient to maintain such trajectories. The Basset force plays a much more important role in the saltation of sand than in the saltation of gravel. The dimensionless saltation length increases substantially as R_p decreases, which is almost solely a result of the effect of the Basset force. Simulations of particle saltation with the random walk model of Niño (1995) suggest that the effects of turbulence of the flow and particle rotation on the mean saltation trajectories tend to cancel each other, such that the proposed model of particle saltation, which neglects both effects, performs satisfactorily when compared with the experimental results reported by Niño and García (1997).

For the case of successive saltations, simulated mean values and standard deviations of h_s and λ_s agree very well with corresponding observations of saltation of sand by Niño and García (1995). Simulated mean values of h_s also agree well with the experimental results of Abbott and Francis (1977) and Niño *et al.* (1994), corresponding to larger particles, which implies that the dimensionless saltation height appears to be independent of particle size. Simulated mean values of λ_s for the saltation of sand tend to be larger than those corresponding to the saltation of gravel, and this is supported by experimental observations reported by Niño and García (1997) and also by Abbott and Francis (1977), which implies that the dimensionless saltation length appears to decrease as the particle size increases.

Simulated mean values and standard deviations of u_s tend to overestimate the corresponding values observed by Niño and García (1997). However, simulated mean values of u_s at values of τ_*/τ_{*c} close to unity reproduce fairly well the rather high observed values of this variable in the experiments of Niño and García, in that range of values of τ_*/τ_{*c}, when compared with Fernandez-Luque and van Beek's (1976) results. In the range τ_*/τ_{*c} larger than about 2·5, on the other hand, simulated values of u_s are in much better agreement with the results of these researchers and as well as with observations of Francis (1973) and Lee and Hsu (1994).

The values of the dynamic friction coefficient predicted by the model are about 0·25 to 0·30, which is in agreement with the experimental observations of Niño and García (1997). These values are slightly lower than the values simulated for the saltation of gravel, which are also supported by the experimental observations. These values are much lower than the value 0·63 proposed by Bagnold (1973) for μ_d, and would indicate that the experimental evaluations of this coefficient are not prone to underestimation as suggested previously by Abbott and Francis (1977).

The application of the bedload model based on a Bagnoldean formulation, using the numerically simulated characteristics of particle saltation, yields an equation that is identical in structure to that of typical bedload formulae, in particular to that of Engelund and Fredsøe (1976). Nevertheless, it was found that the bedload rates predicted by the proposed formula are about 2·3 times larger than those given by Engelund and Fredsøe, which appears to be a consequence of the underestimation of the dynamic friction coefficient, μ_d, by the same amount.

The analysis of the Lagrangian dynamic friction coefficient, μ_{dL}, which is the value estimated from the saltation model, indicates that this variable differs from the concept of dynamic friction coefficient used in the derivation of the bedload model. The latter seems to be given by $\tan \phi = 0{\cdot}63$, in agreement with Bagnold (1973), whereas the values of μ_{dL} would be in the range 0·25–0·4, in agreement with the values observed in the experiments of Niño and García (1997) and also with the numerical simulations of particle saltation presented herein.

Results regarding the modelling of bedload transport seem to indicate that a major problem with the Bagnoldean formulation is related to the continuum assumption for the bedload layer, which would be valid only for very high particle concentrations and small particle diameters. Also, this formulation appears to be very dependent on the value of τ_{*c}, which in general is difficult to estimate because it does not necessarily correspond to the value associated with the threshold of particle motion in the absence of bedload transport.

ACKNOWLEDGEMENTS

We thank the National Science Foundation (grant CTS-9210211), and the donors of the Petroleum Research Fund, administered by the American Chemical Society (grant PRF 24328-G2) for supporting this research.

NOTATION

The following symbols are used in this paper.

A = coefficient in the expression for e
a = coefficient in the equation for particle saltation velocity
B = coefficient in the expression for e
BI_i = Basset term in the ith direction
BI_u = Basset term in the streamwise direction
BI_v = Basset term in the bed-normal direction
B_0 = coefficient in the expression for the mean flow velocity
b = coefficient in the equation for particle saltation velocity
b' = coefficient in the equation for particle saltation velocity
C_D = drag coefficient
C_L = lift coefficient
C_m = added mass coefficient
d_p = particle diameter
e = restitution coefficient for particle collision with the bed
F_f = fluid thrust during saltation
f = friction coefficient for particle collision with the bed
g = gravitational acceleration
h_s = dimensionless saltation height
h_s^* = saltation height
i_b = bedload transport rate in terms of submerged weight per unit width
K = dimensionless parameter in Basset term
k_s = roughness height
k_{s+} = dimensionless roughness height
q_* = dimensionless volumetric bedload transport rate
R = particle submerged specific density
Re = particle Reynolds number

R_p = dimensionless parameter involving the particle diameter
r = random variable with uniform probability density
S = dimensionless particle angular velocity
t = dimensionless time
$\tan\phi$ = static friction coefficient
t_s^* = duration of individual saltation
u_b = dimensionless mean velocity of bedload grains
u_b^* = mean velocity of bedload grains
u_f = dimensionless local mean flow velocity
u_p = dimensionless streamwise particle velocity
$u_p^*|_{in}$ = streamwise component of particle velocity just before collision
$u_p^*|_{out}$ = streamwise component of particle velocity just after collision
$|u_R|$ = dimensionless magnitude of particle slip velocity
$|u_R|_B$ = dimensionless magnitude of particle slip velocity evaluated on particle bottom
$|u_R|_T$ = dimensionless magnitude of particle slip velocity evaluated on particle top
u_s = dimensionless particle saltation velocity
u_{s_i} = dimensionless particle slip velocity in the ith direction
u_s^* = mean streamwise particle velocity
u_* = flow shear velocity
V_p = particle volume
v_p = dimensionless bed normal particle velocity
W_b = submerged weight of bedload layer per unit bed area
W_g = particle submerged weight
x_p = dimensionless streamwise coordinate of particle centroid
x^* = streamwise coordinate
y = dimensionless bed-normal coordinate
y_p = dimensionless bed-normal coordinate of particle centroid
y^* = bed-normal coordinate
α = dimensionless parameter involving R and C_m
Δu_s = dimensionless streamwise particle velocity change at collision
Δu_s^* = streamwise particle velocity change at collision
δt = dimensionless time interval in Basset term
θ = angle of inclination of the bed with respect to the horizontal
θ_b = angle of surface of collision measured with respect to the streamwise direction
θ_{in} = angle of incidence at collision
κ = von Karman's constant
λ_s = dimensionless saltation length
λ_s^* = saltation length
μ_d = dynamic friction coefficient
μ_{dL} = Lagrangian dynamic friction coefficient
ν = kinematic viscosity of the fluid
ρ = fluid density
ρ_s = particle density
τ_c = critical bed shear stress for initiation of sediment motion
τ_g = grain shear stress
τ_0 = bed shear stress
τ_* = dimensionless bed shear stress
τ_{*c} = dimensionless critical bed shear stress for initiation of sediment motion
Ω = particle angular velocity

REFERENCES

Abbott, J. E. and Francis, J. R. D. 1977. 'Saltation and suspension trajectories of solid grains in a water stream', *Phil. Trans. R. Soc. London, Ser. A*, **284**, 225–254.

Ashida, K. and Michiue, M. 1972. 'Study on hydraulic resistance and bedload transport rate in alluvial streams', *Proc. Jpn Soc. Civ. Engng.*, **206**, 59–69.

Bagnold, R. A. 1956. 'The flow of cohesionless grains in fluids', *Phil. Trans. R. Soc. London, Ser. A*, **249**, 235–297.

Bagnold, R. A. 1973. 'The nature of saltation and of 'bed-load' transport in water', *Proc. R. Soc. London, Ser. A*, **332**, 473–504.

Bridge, J. S. 1981. 'A discussion of Bagnold's (1956) bedload transport theory in relation to recent developments in bedload modelling', *Earth Surf. Process. Landf.*, **6**, 187–190.

Bridge, J. S. and Bennett, S. J. 1992. 'A model for the entrainment and transport of sediment grains of mixed sizes, shapes, and densities', *Wat. Resour. Res.*, **28**, 337–363.

Bridge, J. S. and Dominic, D. F. 1984. 'Bed load grain velocities and sediment transport rates', *Wat. Resour. Res.*, **20**, 476–490.

Brush, L. M., Ho, H. W., and Yen, B. C. 1964. 'Accelerated motion of a sphere in a viscous fluid', *ASCE J. Hydraul. Div.*, **90**, 149–160.

Chepil, W. S. 1958. 'Use of evenly spaced hemispheres to evaluate aerodynamic forces on soil surfaces', *Eos. Trans. AGU*, **39**, 397–404.

Einstein, H. A. 1950. 'The bedload function for sediment transportation in open channel flow', *US Dept Agric., Tech. Bull. No 1026*. US Dept. Agriculture, Washington, D.C.

Elghobashi, S. and Truesdell, G. C. 1992. 'Direct simulation of particle dispersion in a decaying isotropic turbulence', *J. Fluid Mech.*, **242**, 665–700.

Engelund, F. and Fredsøe, J. 1976. 'A sediment transport model for straight alluvial channels', *Nord. Hydrol.*, **7**, 293–306.

Fernandez-Luque, R. and Van Beek, R. 1976. 'Erosion and transport of bed sediment', *J. Hydrol. Res.*, **14**, 127–144.

Francis, J. R. D. 1973. 'Experiments on the motion of solitary grains along the bed of a water stream', *Proc. R. Soc. London, A*, **332**, 443–471.

García, M. and Niño, Y. 1992. 'Lagrangian description of bedload transport by saltating particles', *Proc. VI IAHR Int. Symp. on Stochastic Hydraulics, Taipei, Taiwan*. pp. 259–266.

Garde, R. J. and Sethuraman, S. 1969. 'Variation of the drag coefficient of a sphere rolling along a boundary', *La Houille Blanche*, **7**, 727–732.

Hayashi, T. and Ozaki, S. 1980. 'On the unit step length of saltation of sediment particles in the bed-load layer'. *Third Int. Symp. on Stochastic Hydraulics, Tokyo, Japan*. pp. 337–346.

Lavelle, J. W. and Mofjeld, H. O. 1987. 'Do critical stresses for incipient motion and erosion really exist?' *ASCE J. Hydraul. Engng*, **113**, 370–385.

Lawrence, C. J. and Mei, R. 1995. 'Long-time behaviour of the drag on a body in impulsive motion', *J. Fluid Mech.*, **283**, 307–327.

Lee, H. and Hsu, I. 1994. 'Investigation of saltating particle motions', *ASCE J. Hydraul. Engng*, **120**, 831–845.

Leeder, M. R. 1979a. 'Bedload dynamics: grain impacts, momentum transfer and derivation of a grain Froude number', *Earth Surf. Process. Landf.*, **4**, 291–295.

Leeder, M. R. 1979b. 'Bedload dynamics: grain–grain interactions in water flows', *Earth Surf. Process. Landf.*, **4**, 229–240.

Maxey, M. R. and Riley, J. J. 1983. 'Equation of motion for a small rigid sphere in a nonuniform flow', *Phys. Fluids*, **26**, 883–889.

Mei, R. 1990. 'Particle dispersion in isotropic turbulence and unsteady particle dynamics at finite Reynolds number', *PhD Thesis*, University of Illinois at Urbana-Champaign, Urbana, Illinois.

Mei, R., Adrian, R. J., and Hanratty, T. J. 1991. 'Particle dispersion in isotropic turbulence under Stokes drag and Basset force with gravitational settling', *J. Fluid Mech.*, **225**, 481–495.

Meyer-Peter, E. and Muller, R. 1948. 'Formulas for bedload transport', *Proc. 2nd Congress IAHR, Stockholm*.

Murphy, P. J. and Hooshiari, H. 1982. 'Saltation in water dynamics', *ASCE J. Hydraul. Div.*, **108**, 1251–1267.

Niño, Y. 1995. 'Particle motion in the near wall region of a turbulent open channel flow: implications for bedload transport by saltation and sediment entrainment into suspension', *PhD Thesis*, University of Illinois at Urbana-Champaign, Urbana, Illinois.

Niño, Y. and García, M. 1994. 'Gravel saltation 2. Modelling', *Wat. Resour. Res.*, **30**, 1915–1924.

Niño, Y. and García, M. 1997. 'Experiments on saltation of fine sand in water', *ASCE J. Hydraul. Engng*, (accepted for publication).

Niño, Y., García, M., and Ayala, L. 1994. 'Gravel saltation 1. Experiments', *Wat. Resour. Res.*, **30**, 1907–1914.

Parker, G. 1991. 'Some random notes on grain sorting', *Proc. Int. Grain Sorting Seminar, Ascona, Switzerland*. pp. 19–76.

Patnaik, P. C., Vittal, N., and Pande, P. K. 1994. 'Lift coefficient of a stationary sphere in gradient flow', *J. Hydraul. Res.*, **32**, 471–480.

Reizes, J. A. 1978. 'Numerical study of continuous saltation', *ASCE J. Hydraul. Div.*, **104**, 1303–1321.

Rizk, M. A. and Elghobashi, S. E. 1985. 'The motion of a spherical particle suspended in a turbulent flow near a plane wall', *Phys. Fluids*, **28**, 806–817.

Rubinow, S. and Keller, J. 1961. 'The transverse force on a spinning sphere moving in a viscous fluid', *J. Fluid Mech.*, **11**, 447–459.

Schlichting, H. 1968. *Boundary-Layer Theory*. McGraw-Hill Book Company, New York.

Sekine, M. and Kikkawa, H. 1992. 'Mechanics of saltating grains. II', *ASCE J. Hydraul. Engng*, **118**, 536–558.

Tsuchiya, Y. 1969. 'On the mechanics of saltation of a spherical sand particle in a turbulent stream', *Proc. 13th Congress IAHR*, **2**, 191–198.

Tsujimoto, T. and Nakagawa, H. 1983. 'Stochastic study on successive saltation by flowing water', *Proc. 2nd Int. Symp. on River Sedimentation, Beijing, China*. pp. 187–201.

Van Rijn, L. C. 1984. 'Sediment transport. Part I: bed load transport', *ASCE J. Hydraul. Engng*, **110**, 1431–1456.

White, B. R. and Schulz, J. C. 1977. 'Magnus effect in saltation', *J. Fluid Mech.*, **81**, 497–512.

Wiberg, P. L. and Smith, J. D. 1985. 'A theoretical model for saltating grains in water', *J. Geophys. Res.*, **90**, 7341–7354.

Yen, B. C. 1992. 'Sediment fall velocity in oscillating flow', *Wat. Resour. Environ. Engng Res.*, Report No. 11. Dept. of Civil Engng, University of Virginia.

SECTION II

HYDRAULIC MODELLING

5

HIGH RESOLUTION, TWO-DIMENSIONAL SPATIAL MODELLING OF FLOW PROCESSES IN A MULTI-THREAD CHANNEL

STUART N. LANE* AND KEITH S. RICHARDS

Department of Geography, University of Cambridge, Downing Place, Cambridge CB2 3EN, UK

ABSTRACT

This paper describes application and testing of a two-dimensional numerical flow model in a multi-thread reach of a proglacial stream. The model solves the depth-averaged form of the Navier–Stokes equations for open channel flow, incorporating a two-equation turbulence closure, an analytical correction for the effects of secondary circulation and a rigid lid approximation. The model requires as input the channel bed topography, the water surface, bed roughness and inflow discharge information, and predicts the spatial distribution of depth-averaged velocity and eddy viscosity. The results have been subjected to intensive testing using simple assessment of numerical performance (e.g. conservation of mass and momentum, numerical convergence), distributed sensitivity analysis and comparison of model predictions with field measurements of velocity. The results are encouraging, particularly given some of the difficulties in obtaining accurate, distributed cross-stream and downstream velocities. Distributed sensitivity analysis allowed more detailed consideration of the necessary development of the model. This suggested that significant errors in the velocity predictions were largely a result of uncertainty in the specification of both the magnitude and the spatial variation of bed roughness. Secondly, the two-equation turbulence closure was observed to have little effect upon model predictions, except in the vicinity of the side walls. In applications of models of this type to irregular, coarse-bedded channels, improvements in the specification of the topographic boundary condition specification and bed roughness are likely to be more important than a sophisticated turbulence closure scheme. Thirdly, although the secondary circulation correction was observed to reproduce some of the expected streamwise transfer of momentum, the effects were seen to be relatively small. Given the intensity of secondary circulation defined in field contexts, the inability of the model to correct effectively for the momentum transfer associated with secondary circulation processes related to topographic discordance and shear-generated turbulence suggests that further work is required in this respect.

KEY WORDS braided rivers; numerical modelling; secondary circulation; turbulence closure

INTRODUCTION

The nature of form–process interrelationships in dynamic braided channels has been investigated through intensive field study (e.g. Ashworth and Ferguson, 1986; Ferguson and Ashworth, 1992). This has resulted in substantial increases in our understanding of the interrelationships between river channel topography, the spatially distributed flow field and sediment entrainment and transport, as the river channel is subjected to discharge and sediment supply events of various magnitudes. However, process-based studies have only just begun to attempt to simulate *numerically* the nature of these interactions in irregular natural channels. This

* Correspondence to: Stuart N. Lane, Department of Geography, University of Cambridge, Downing Place, Cambridge, CB2 3EN, UK.

Contract grant sponsor: NERC.
Contract grant number: GR3/9715.

involves coupling river channel topography to two- and three-dimensional models of river channel flow and sediment transport processes (Ferguson, 1993). Such a methodological development is important because it may provide: (i) an alternative methodology for understanding the dynamics of the braiding process, where the effects of combinations of different boundary conditions on river channel development may be explored through numerical simulation; and (ii) the means of predicting short and medium time-scale river channel dynamics in response to externally imposed boundary conditions.

Several authors have indicated the potential of this approach (e.g. Ferguson and Ashworth, 1992; Ferguson *et al.*, 1992), and a few examples exist of the application of such distributed coupled numerical models to alluvial problems. Nelson and Smith (1989) developed a two-dimensional coupled flow and sediment transport model that was applied to laboratory-formed channels and, specifically, to the development of point bars in a sinuous channel and alternate bars in a straight channel. This method was extended by Andrews and Nelson (1989) for the evolution of a single bar in a natural channel in response to time varying discharge. However, distributed numerical models have yet to be developed for the more complicated scenario of multi-thread channels. This is an ambitious project, but one that has the potential to improve our understanding of the way in which braided channels behave.

If some confidence can be placed in model performance through application and testing in specific field investigations, models may be used for prediction and simulation purposes. This paper describes the first stage of development of such a model, in which spatially distributed topographic information provides the boundary condition for a two-dimensional, depth-averaged numerical model of turbulent open channel flow. The model was applied to a multi-thread reach of proglacial stream. A detailed assessment of model performance is undertaken, using field observations of flow velocity and distributed sensitivity analysis. The results from this are used as a basis for identifying those areas of model application that require further research.

THE NUMERICAL MODEL

The model used in this study is that developed by the STREMR Group at the US Army Corps of Engineers Waterways Experimental Station, Vicksburg, Mississippi (Bernard, 1992). The model solves the Reynolds-averaged and depth-averaged mass conservation and Navier–Stokes momentum equations for an incompressible fluid (see Lane, this volume). It makes a rigid lid approximation, and therefore requires specification of the water surface elevation, but accounts for the potential effects of a free surface through a correction to the pressure equation. The turbulence closure uses a two-equation model of the effects of turbulence on the mean flow properties, following Prandtl (1945); the eddy viscosity is calculated from two terms, k (the turbulent energy per unit mass) and l (a mixing length), where both are determined from additional equations that model their rates of change as controlled by processes of convective and diffusive transport and viscous dissipation. Given that the dispersion terms introduced during the depth-averaging process may be important in natural channels with significant streamline curvature, STREMR attempts to model their effects on the depth-averaged velocities (see Lane, 1998, this issue) such that the magnitudes of the dispersion term effects are directly dependent upon the magnitude of the bottom friction and the depth-averaged velocity, and inversely proportional to the water depth and the radius of curvature of the depth-averaged streamlines. The model imposes a rigid lid solution (which therefore requires water surface specification), but with correction for the effects of a free surface on the momentum equation, such that it is equivalent to the displacement of a free surface (see Lane, this volume).

This model is sufficient for calculating the depth-averaged mean flow, provided suitable boundary conditions and parameters are provided (see below), and the equations are modified in the vicinity of channel side walls. Boundary layer calculations require three-dimensional treatment and grid and time resolutions on an extremely small scale (the grid must be fine enough to resolve the flow in the boundary layer before and after separation) and boundary condition information at a spatial resolution that reflects these resolutions. It is therefore inappropriate to include a detailed treatment of the boundary layers, and STREMR approximates their influence well enough to predict the flow at larger scales (see below).

The model is based upon a finite difference (volume) solution with a computational grid established over the section of river channel being studied. The depth-averaged flow equations are transformed from Cartesian coordinates, with arbitrary spacing (Δx, Δy), to curvilinear (i, j) co ordinates with unit spacing (Δi, Δj) such that $\Delta i = \Delta j = 1$ between the grid lines. The grid cells are regular in the computational plane but may be irregular in the Cartesian plane. Although this complicates the equations, it makes their solution more efficient. The location of grid nodes is user-determined but follows certain well-established principles concerning grid characteristics in the Cartesian plane. Most important is the need to avoid sharply skewed grid lines and abrupt changes in grid spacing, since these degrade accuracy and promote numerical instability. Bernard (1992) provides some basic rules of thumb in this respect: (i) the angle between intersecting grid lines should be between 45 and 135°; (ii) the local change in grid spacing should be no more than 30% from one cell to the next; (iii) cell aspect ratios (length/width) should be no greater than 10 wherever sharp gradients may occur in the *lengthwise* direction; and (iv) grid cells should be finest in the direction of the strongest depth or velocity gradient. This last condition implies that the grid spacing must be finest closest to channel edges as it is in this zone where velocity and depth gradients tend to be greatest.

Grid generation is a particularly difficult issue in application to arbitrarily and irregularly shaped channels of the kind to which this model is being applied here. The river channel represents a continuous feature which, when discretized, becomes discontinuous and irregular, with the nature of this irregularity reflecting the scale of the discretization. This emphasizes the point that the scale of discretization critically controls the nature of model predictions and the scale at which they are interpretable. At coarse grid spacings, the model may be unable to resolve the flow properly. One solution is to reduce progressively the grid spacing (e.g. Bernard, 1992) until flow model predictions become independent of grid spacing, but this results in two problems. First, it may result in unacceptable computational demands (cf. Emmons, 1970). Secondly, the grid spacing must be of the same order as the resolution of the topographic (boundary condition) representation. If the spacing is finer than the topographic resolution, the flow model predictions will, in part, be a product of the nature of the topographic interpolation. If coarser than this, the flow model predictions would be a product of the discretization process and not the input topographic information. However, when dealing with the continuous surface of a natural, irregular channel, this is problematic. The topographic information will have been sampled from the surface, perhaps using structured sampling to include key topographic features such as breaks of slope (e.g. Lane *et al.*, 1994a). However, there will be scales of topography (e.g. the bedform scale or the grain scale) that will not have been included in full (as the topographic sampling density will not be sufficient to map either every bedform or every grain). Making the grid resolution progressively finer should make model solution independent of coarser scales of topographic variation, but if this is undertaken until the solution is independent of even the small scale, the model will be resolving flow around topography that is a product of the sampling method used to represent the surface rather than any real topographic variation. In natural, irregular channels, a trade-off will be required between the need for the model solution to be independent of coarser scales of topography (that are well represented in the boundary condition information) and the need to ensure that model predictions are not resolving flow around artificially constructed components of bed topography. This trade-off will also be important in controlling the subgrid scale parameterization of smaller than grid dimension topographic effects through the use of roughness parameters (see below).

Grid spacing should be especially small in the vicinity of the channel side walls, but computational limitations tend to prevent it from being small enough. Hence, the governing equations require further modification in the vicinity of the side walls, although there is no physical theory on which this can be based. The model therefore assumes a functional form of the velocity near the wall. This is a power law, because logarithmic distributions are only valid where the pressure gradients are weak and with the fine grids that are able to resolve the boundary layer (Bernard, 1992), and which this type of correction is designed to avoid. The model calculates a tangential shear stress on the sides of cells adjacent to the side wall, assuming a power law distribution, and including a parameter to increaase the magnitude of the side wall resistance if necessary. The turbulence model is also corrected in the vicinity of side walls.

Each grid cell [which is square in the computational (i, j) plane] is a miniature control volume for which STREMR calculates two-dimensional face-centred fluxes, cell-centred velocities and cell-centred pressures. The numerical solution is based upon using the inflow information to calculate volumetric fluxes through the boundary faces of all flux and open cells and these are converted to mass-conserving fluxes that define the initial flow. The code then marches forward in time, using a MacCormack predictor–corrector scheme (MacCormack, 1969; Bernard, 1992) for the momentum equations, and an (Euler) upwind scheme (Bernard, 1992; Anderson *et al.*, 1984) for the turbulence and secondary flow models. The time step employed can be varied. This is important as both the MacCormack and upwind schemes are explicit in that they use only the existing values to calculate new values. This can result in problems if the time increments are too large. Thus the code uses numerical stability criteria to constrain the maximum allowable time step, such that as small a time step as possible is used and the solution reaches a steady state much more rapidly than would otherwise be the case.

Model application requires specification of both parameters and boundary conditions. The former are specific to the model and its particular formulation and are uniformly imposed (Table I). The latter are specific to the particular model application and may be spatially distributed (Table II).

The two parameters 'Turbin' and 'Peclet' defined in Table I must be specified as starting values for the initial model solution. They are applied to a single reference cell, normally located in a non-critical area of the flow, but their effects should be limited to grid cells in the immediate vicinity. The parameter 'Iters' is important since increasing it generally improves model convergence, but should have little effect on the final model predictions. 'Drag' and 'Avis' both give the user additional empirical control over the model solution. 'Drag' is important for solutions where the grid is coarse in proximity to the side walls. At side walls, gradients in turbulence and the mean flow tend to be large. It is possible, by having a fine enough grid, to model these effects directly but this results in solutions that are computationally demanding. To limit the need to do this STREMR approximates side wall influence on flow at larger scales through a correction to

Table I. Model parameters used in STREMR

Parameter	Description	Determination
Turbin	Reference value for ratio of turbulence energy to velocity squared for a single reference cell	Default = 0·03 — should have little effect on model solution outside reference cell location
Peclet	Reference value for initial Peclet number (a function of initial viscosity) for a single reference cell	Default = 50 — should have little effect on model solution outside reference cell location
Iters	Number of iterations used in solution of the Poisson pressure equation	Default = 3 — but can be increased to increase convergence of the numerical solution
Drag	Coefficient for increasing or decreasing side wall shear stress with turbulence model	Default = 1·0 — but can be increased if grid is excessively coarse in the vicinity of side walls
Avis	Coefficient for uniformly increasing or decreasing viscosity whether input by user (zero-equation model) or computed by turbulence (two-equation model)	Default = 1·0
Recap	Turbulence adjusted parameter to reduce the eddy viscosity in regions of low velocity where turbulence over-prediction is likely to occur	Default = 2·5 — experimental research suggests a default value of 1·5–2·0 but the STREMR default is set higher than this to account for potential grid error with coarse grids. Once grid independence is achieved Recap should be reduced to 1·75. The standard k–ε model can be imposed by setting Recap = 1·0

Table II. The required input boundary conditions

Boundary condition	Description
Input velocity	This is used to specify the velocity distribution in a cross-section at the top or bottom of the reach. It is used both to calculate the total discharge and to determine the distribution of that discharge within that particular cross-section. It is specified in terms of both x and y directions or both i and j directions.
Manning's n	This controls (along with the calculated water depth from water surface and bottom elevations) the bottom frictional resistance.
Water suface elevation	This is used to 'fix' the lid. Random water surface data are transformed into a water surface elevation for each grid cell through a Laplacian interpolation. Values at specified locations are treated as though they were fixed boundary conditions and a Laplace equation is solved for the remaining (unspecified) locations.
Bottom elevation	This is used to 'fix' the bed. Random bottom elevation data are transformed into bottom elevations for each grid cell theough a Laplacian interpolatation. Values at specified locations are treated as through they were fixed boundary conditions and a Laplace equation is solved for the remaining (unspecified) locations. Bottom elevation is subtracted from water surface elevation for each grid cell to determine the water depth.
Eddy viscosity	This is specified when the two-equation turbulence model is switched off and the flow model uses a zero-equation turbulence model with a uniform eddy viscosity value. Spatially distributed viscosity values may be specified but this is unusual because of difficulties in field determination.

the turbulence model using the parameter 'Drag'. 'Avis' is used to increase the eddy viscosity uniformly and is useful with coarse grids in general, where the flow may be insufficiently resolved to model, in full, the effects of turbulence on the transport of momentum. 'Recap' is used to correct the standard two-equation turbulence model for over-prediction of turbulence effects in zones of low velocity, particularly important where there is grid error. Experimental results suggest an optimal value of 1·75, but only after grid independence has been achieved. In theory, only 'Drag', 'Avis' and 'Recap' should have any effect on the final model predictions. The extent to which this is the case, and the magnitude of the effects of all parameters must be investigated in the context of each specific application.

All model boundary conditions (Table II) can be spatially variable should the user wish, but this depends upon the availability of spatially distributed information. If available, a Laplace interpolation procedure is used to assign a value for every boundary condition to every grid cell, treating the specified values as fixed.

It is possible to specify a total discharge, which is then automatically distributed across the inflow cells such that they each have the same inflow velocity. However, distributed velocity information across the upstream river cross-section can be used to calculate the inflow discharge distribution, and this has an important effect on the subsequent downstream velocity predictions for irregular natural channels. During solution, STREMR does not alter either the total discharge or its distribution across the input section. Distributed velocity information is also used to provide model starting values through uniform specification of an initial starting velocity, since this can increase the rate of convergence. Detailed testing suggested that this has no effect on final model predictions.

Information on Manning's n is required since this controls the magnitude of the bottom friction and secondary circulation correction. This is given by

$$C_{\mathrm{f}} = gn^2h^{-\frac{1}{3}} \tag{1}$$

and is the Darcy–Weisbach friction factor specified in terms of Manning's n, where h is water depth. Manning's n can be uniformly specified, or spatially distributed. Water surface values are required to 'fix the lid'. In combination with bottom elevation values they define the water depth in every grid cell. The bottom elevation input is designed so as to allow irregular topography. A different bottom elevation and water surface is allowed in every grid cell. This produces a stair-stepped depth discretization in the computational space, which is interpolated from distributed input data and implies that grid node locations must be spaced to reflect the local rate of change of topography. The eddy viscosity only needs to be specified when the second-order turbulence model is replaced with a zero-equation model. There is no simple means of determining this value in the field, and thus the zero-equation model requires substantial guesswork.

MODEL APPLICATION AND ASSESSMENT

The model was applied to a multi-thread reach of river channel in the proglacial area of the Haut Glacier d'Arolla, Valais, Switzerland, using data collected during the summer of 1992. A full summary of the details of model application is given in Lane *et al.* (1995). Accurate and dense digital terrain models (DTMs) of the stream bed were acquired using combined analytical photogrammetry and digital tacheometric survey (see Lane *et al.*, 1994a). The photogrammetry also provided detailed information on the position and elevation of the water surface at the channel margins. The average density of bed elevations was 3·2 points per square metre, approximating to a point spacing of 0·5 m. This was used to define both an appropriate grid resolution and the form of the subgrid-scale parameterization. Experiments with different grid densities suggested that a grid spacing of approximately 0·05 m would produce a grid-independent solution. To avoid modelling an artificially constructed surface, resulting from the effects of bed sampling (see above), it was decided to coarsen this grid spacing to 0·10 m, but with reduced spacing in critical areas (e.g. zones of flow recirculation). This had the advantage of increasing computational efficiency, but the nature of the interaction between grid spacing, topography and grid independence is an issue that needs further and more rigorous research. The parameterization of surface topographic information that was of a finer scale than that recorded in the DTMs was done through Manning's n. The latter was determined from the Strickler (1923) equation, based upon the median particle diameters obtained from Wolman grid by number sedimentological sampling, but only of *exposed* bar surfaces. This produced a value of n for each sediment sampling location and these were averaged to obtain a reach-scale estimate. On the basis that the topographic survey had an average spacing of 0·5 m, the largest scale of topography that could be detected was 1·0 m and observations suggested that this was coarser than the average micro-scale bedform dimensions. Hence, n had to be increased to represent the micro-scale bedform effect, using the conventional multiplier of roughness length, in which the roughness height is scaled by $6{\cdot}8D_{50}$ (Bray, 1982). The corrected value of Manning's n used was 0·035. This approach suffered from three problems. First, because of the necessary restriction of sampling to bar surfaces, no information was available for subaqueous zones, which implies that the estimate of roughness will be biased. Secondly, it was not considered possible to allow spatial variation in the magnitude of n, because interpolation from the exposed surface data to channel bed locations would be spurious. Thirdly, the nature of the micro-scale bedform correction, in terms of its relationship to spatial variation in bedform size and structure, remains a major area of uncertainty (see below).

Distributed discharge information across the upstream entry to the reach was obtained from regularly and closely spaced depth-averaged velocity measurements using an impeller-type current meter, obtained such that not more than 10% of the total discharge passed between adjacent velocity measurements. The results in this paper are based upon model application to a steady state problem, using data collected on the 5 July 1992, using the 10:00 a.m. discharge. Figure 1 shows the DTM and Figure 2 the distributed velocity information across the entry to the reach.

Model assessment involved the following: (i) assessment of numerical stability, convergence and mass conservation; (ii) comparison of predicted velocity patterns with those measured in the field; and (iii) distributed sensitivity analysis of one region to which the model was applied. This paper will not present the results from (i) but will focus on (ii) and (iii). Results from (i) are provided in full in Lane (1994). For the purpose of (ii), spatially distributed flow velocity information was obtained. The methodology behind this is described in Lane *et al.* (1995) and, in summary, involved distributed measurement of the two planform velocity components using an electromagnetic current meter. The current meter was positioned and oriented

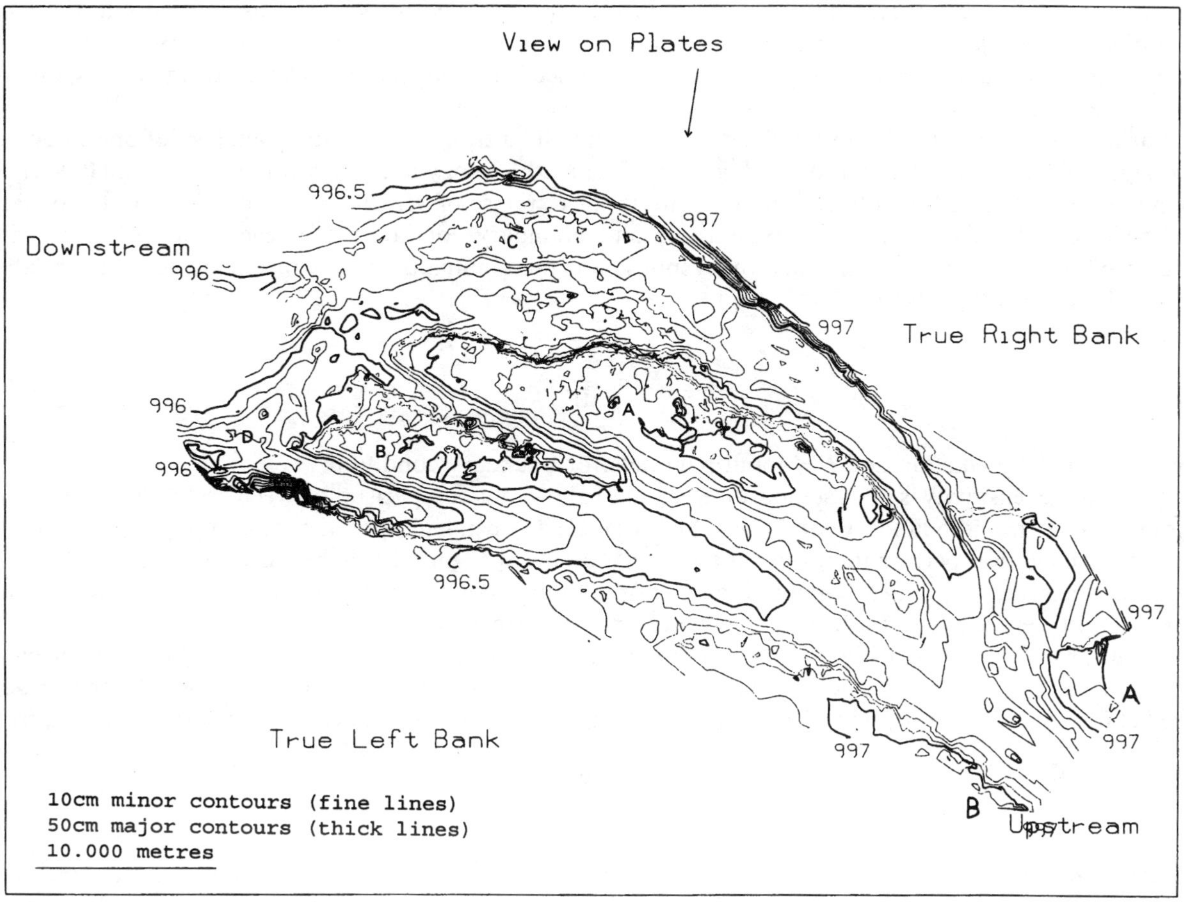

Figure 1. The DTM acquired on 5 July using combined terrestrial oblique photogrammetry and bed survey and used as input to the flow model

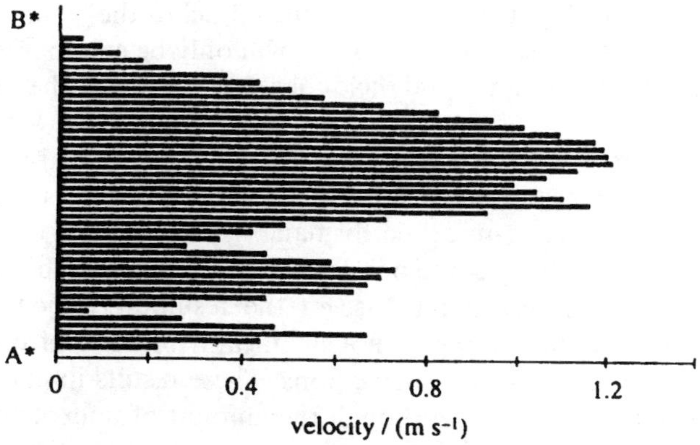

Figure 2. The 10:00 a.m. distributed, depth-average velocity information supplied to the model at the upstream entry to the reach (total discharge = $1{\cdot}18\ m^3\ s^{-1}$)

into the tacheometric survey. This allowed observed flow model velocities to be transformed into the same coordinate system as $\bar{U}$- (downstream) and $\bar{V}$- (cross-stream) velocities predicted by the model. A total of 44 velocity observations of known position were obtained and matched by location to the corresponding flow model predictions.

For the purpose of (iii), the authors undertook a spatially distributed sensitivity analysis of one section of the reach (the upstream bifurcation in Figure 1). The methodology is described in Lane *et al.* (1994b). In summary, this involved quantification of sensitivity of response of velocity and eddy viscosity to spatially uniform perturbation of the input parameters and boundary conditions, for each grid cell. Following McCuen (1973), in situations in which the response was predominantly linear, this made use of the relative sensitivity factor (R_s), where

$$R_s = \frac{dF_o}{dF_i}\frac{\bar{F}_i}{\bar{F}_o} \tag{2}$$

and F_o = model prediction and F_i = input parameter or boundary condition.

The relative sensitivity is in its standardized form to allow intercomparison between grid cells and different input parameters and boundary conditions. Equation (2) is dependent upon the range of input parameter or boundary condition perturbation, and the research design recognized this by only perturbing within the physically realistic range of each input parameter or boundary condition. With a relative sensitivity obtained for each grid cell, it was possible: (i) to calculate an average relative sensitivity; and (ii) to assess the spatial distributions of relative sensitivity, for each model output in response to perturbation of each input parameter. It was also possible to calculate the prediction confidence (C), an estimate of the degradation in model predictions owing to error (ε) in specification of a particular input parameter or boundary condition, from

$$C = \frac{dF_o}{dF_i}\varepsilon \tag{3}$$

In addition to determining R_s and C values for the responses of model predictions to the input parameters and boundary conditions given in Tables I and II, a simpler form of sensitivity analysis was undertaken to assess the effects of the turbulence model and secondary circulation correction upon model predictions, by comparing predictions with and without the corrections.

RESULTS

Comparison of model predictions with EMCM observations

Figure 3 shows model predictions from application of the model to the 5 July dataset. These results have been interpreted in greater detail in Lane *et al.* (1995) and will only be summarized here. There was a good correspondence between model predictions and field observations and the results are summarized in Table III.

This comparison suggested a reasonable correlation between observed and predicted resultant velocities and $\bar{U}$-velocities (0·783 and 0·717, respectively), but a poorer correlation between observed and predicted $\bar{V}$-velocities (0·566). Further confidence in model performance is suggested by consideration of the slope and intercept of the regression lines. For perfect correspondence, the slope should be unity and the intercept zero. Applying *t*-tests (95% significance level) in the case of the resultant velocity, the slope (0·913) is not significantly different from one and the intercept is not significantly different from zero. This is not the case for the $\bar{U}$- and $\bar{V}$-velocities, despite significant correlations. These results imply that model predictions are sensible in the case of the resultant velocities, although the amount of unexplained variance remains high. This may be explained by the problem of using sampled instantaneous velocities, which are then averaged, to assess time-averaged predictions. Long-term monitoring of flow structure in this river (Clifford *et al.*, 1995)

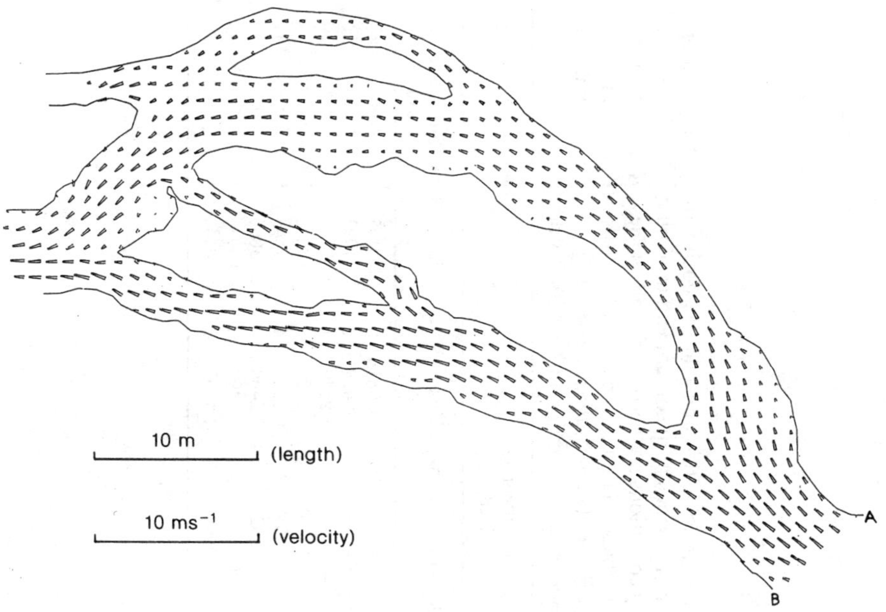

Figure 3. Model predictions for 5 July DTM and 10:00 a.m. discharge

Table III. Results from comparison of model predictions with distributed velocity information

	Correlation with flow model velocities	Regression analysis				
		Slope	Standard error	Intercept	Standard error	F ratio
$\bar{U}$-velocity	0·717	0·799	0·121	−0·139	0·0589	43·3
$\bar{V}$-velocity	0·566	0·404	0·0919	0·0551	0·0248	19·4
Resultant velocity	0·783	0·913	0·113	0·0555	0·0628	65·1

has suggested medium-term flow periodicities, with time-scales of variation longer than those used to determine the time-averaged velocities, which could contribute to the relatively high level of unexplained variance. Other explanations are dealt with below. The time-averaging effect will also affect the $\bar{U}$-velocity and $\bar{V}$-velocity observations, but these will also be particularly sensitive to the orientation of the sensor head, which in this situation is difficult to guarantee to better than $\pm 20°$ of the correct orientation (Lane *et al.*, 1995). This orientation problem does not affect resultant velocities, which are independent of sensor head orientation, but does result in error in the individual component velocities. Orientation problems suggest that model predictions may even be of a higher quality than the field observations.

Distributed sensitivity analysis

Table IV provides summary R_s statistics for all of the input parameters and boundary conditions subjected to sensitivity testing (the average R_s from the sensitivity surface, and the R_s at the locations of the maximum and minimum velocity* and eddy viscosity). Figure 4 provides R_s surfaces for the response of resultant

* Maximum and minimum R_s data for resultant velocity are not provided since resultant velocity is not calculated by the model but from (U, V) velocity pairs. This meant that the maximum U-velocity might not coincide with the maximum V-velocity and that a resultant velocity calculated from these would be meaningless.

Table IV. Results from distributed sensitivity analysis for the response of $\bar{U}$-, $\bar{V}$- and resultant velocities and eddy viscocity to spatially uniform perturbation of input parameters and boundary conditions. Note that Max. and Min. refer to the locations of maximum and minimum velocity predicted by STREMR, which do not necessarily occur in the same location as parameters and boundary conditions are perturbed, and so these R_s values must be treated with some caution. The relative sensitivity of resultant velocity to input parameter and boundary condition perturbation was determined for each grid cell and was used to determine a surface average. As the maximum value of $\bar{U}$-velocity was generally not in the same location as the maximum value $\bar{V}$-velocity, it was not possible to calculate a value of R_s for the response of maximum and minimum resultant velocity

Parameter/ boundary condition	$\bar{U}$-velocity						$\bar{V}$-velocity						Resultant velocity		Eddy viscosity					
	Max.	Rank	Min.	Rank	Surface mean	Rank	Max.	Rank	Min.	Rank	Surface mean	Rank	Surface mean	Rank	Max.	Rank	Min.	Rank	Surface mean	Rank
Turbin	−0·0112	8	−0·0055	6	0·0012	7	0·0000	8	−0·3790	4	0·0015	8	0·0009	7	−0·1961	8	0·0198	8	−0·0021	8
Peclet	0·5683	3	0·0437	4	−0·0032	6	0·0115	5	0·1951	5	−0·0054	6	−0·0033	6	−0·9459	3	0·1993	6	−0·3073	3
Avis	−0·0958	4	−0·0049	7	−0·0510	3	−0·0530	4	−0·4425	3	−0·0209	4	−0·0151	4	1·9849	1	−13·050	1	0·9659	2
Drag	0·0473	6	−0·0123	5	−0·0209	5	0·0020	6	−0·0173	8	−0·0208	5	−0·0193	3	−0·3501	5	0·8286	3	−0·0307	7
Recap	0·0390	7	0·0000	8	0·0011	8	0·0001	7	0·1923	6	0·0019	7	0·0001	8	−0·3542	4	−0·2543	5	−0·0386	6
n	−0·7545	2	−0·2518	2	−0·0364	4	−0·3625	2	0·6626	2	−0·0561	3	−0·0132	5	0·2460	7	0·3006	4	0·0589	5
Depth	0·0572	5	−0·1005	3	−0·1093	2	−0·1409	3	−0·1590	7	−0·2016	2	−0·1715	2	−0·3493	6	−0·0936	7	−0·1320	4
Discharge	1·0537	1	1·0685	1	0·9999	1	1·0591	1	1·1658	1	0·9996	1	0·9998	1	1·1292	2	1·0444	2	1·0001	1

velocity and eddy viscosity. Figure 5 gives the differences in model predictions with and without the turbulence model, and Figure 6 gives the differences with and without the secondary circulation correction for both the confluence and a diffluence.

The input parameters 'Turbin' and 'Peclet' should have little effect on model predictions except close to the reference cell location (A on Figure 3). This is supported by the velocity sensitivity surfaces for both input parameters, although velocity predictions are slightly more sensitive with respect to 'Peclet' (Figure 4c). The response of eddy viscosity to variation in both 'Turbin' and 'Peclet' was much stronger (Figure 4b and d), as would be expected given that both parameters are related directly to the magnitude of the eddy viscosity (Table I).

'Avis' was used to increase (or decrease) the magnitude of the calculated eddy viscosity uniformly. The effects of this on the calculated eddy viscosity are as expected, and reflected in Table IV and Figure 4f. The R_s values are distributed around unity on the sensitivity surface, although there is some spatial variation in the sensitivity, with values greater than unity concentrated on the barhead. This reflects the greater production and dissipation of turbulent energy in this shallower zone. 'Avis' has a much smaller effect on velocity predictions (Table IV, Figure 4e) and these are also concentrated in the zone where turbulent energy production and dissipation would be expected to be highest (in the vicinity of channel side walls). These would be associated in part with the changes in the eddy viscosity illustrated in Figure 4f. In these zones the sign of R_s is sensible, suggesting that as 'Avis' is increased, and there is an increase in the transport of momentum by turbulence with respect to the primary flow, so the primary flow velocity (as represented by the resultant velocity) decreases.

'Drag' was the parameter used to increase or decrease the magnitude of the side wall shear stress correction, and, along with the parameter 'Recap', is important when grids are coarse. The response of the resultant velocity to variation in 'Drag' is limited to the side walls (Figure 4g) where values of R_s can be quite high (up to $-0{\cdot}6$). This decrease in velocity in the vicinity of the side wall might be expected to result in a slight increase in velocity just into the flow as a means of conserving mass. Although there were such velocity increases, they were extremely small, no doubt reflecting the small discharge through the zones of small depth and low velocity adjacent to the side walls. The sensitivity of side wall velocities to 'Drag' will be considered further in the Discussion. 'Recap' is an adjustment parameter used by the turbulence model to correct for locations of eddy viscosity over-prediction in zones of low velocity, when grids are coarse. Figure 4j shows the importance of this parameter in controlling the eddy viscosity in zones of low velocity, particularly around the channel side walls and the sign of the response is logical with respect to the effects of increasing 'Recap'. As with 'Avis', however, there is little effect on velocity predictions (Figure 4i).

Model sensitivity to uniform variation in the roughness boundary condition produces interesting results, and these have been detailed in Lane *et al.* (1994b). A uniform increase in n in general results in a decrease in the velocity magnitudes (Figure 4k). This is as expected given that increasing n would increase the magnitude of the bottom friction and hence decrease flow velocities. However, the minimum location $\bar{V}$-velocity has an extremely high and *positive* relative sensitivity (0·6626; Table IV). This is clearly a counterintuitive tendency. This is not an exception to the rule and the associated sensitivity surface (Figure 4k) shows not only marked spatial variability in the relative sensitivity, but also quite substantial zones of positive sensitivity (30% of all cells). Around the bar head, and extending for a small distance upstream, the relative sensitivity is essentially negative, reaching a minimum in close proximity to the bar head. Either side of this area of negative values, the relative sensitivity becomes positive.

The distributed nature of the analysis, however, allows a logical interpretation of these results, and results in the conclusion that the model is performing effectively, even though there are counterintuitive sensitivities of velocity in certain zones. It is noticeable (Figure 4k) that the effect of n on flow retardation was maximized around and upstream of the bar head, coincident with the most shallow zones of flow. In the zones either side of this area the positive relative sensitivity was most marked. This can be explained by the spatial feedback effect associated with a distributed model. The model assumes that discharge was fixed, whilst specification of depth was through bed and water surface boundary conditions and was not altered during model calcula-

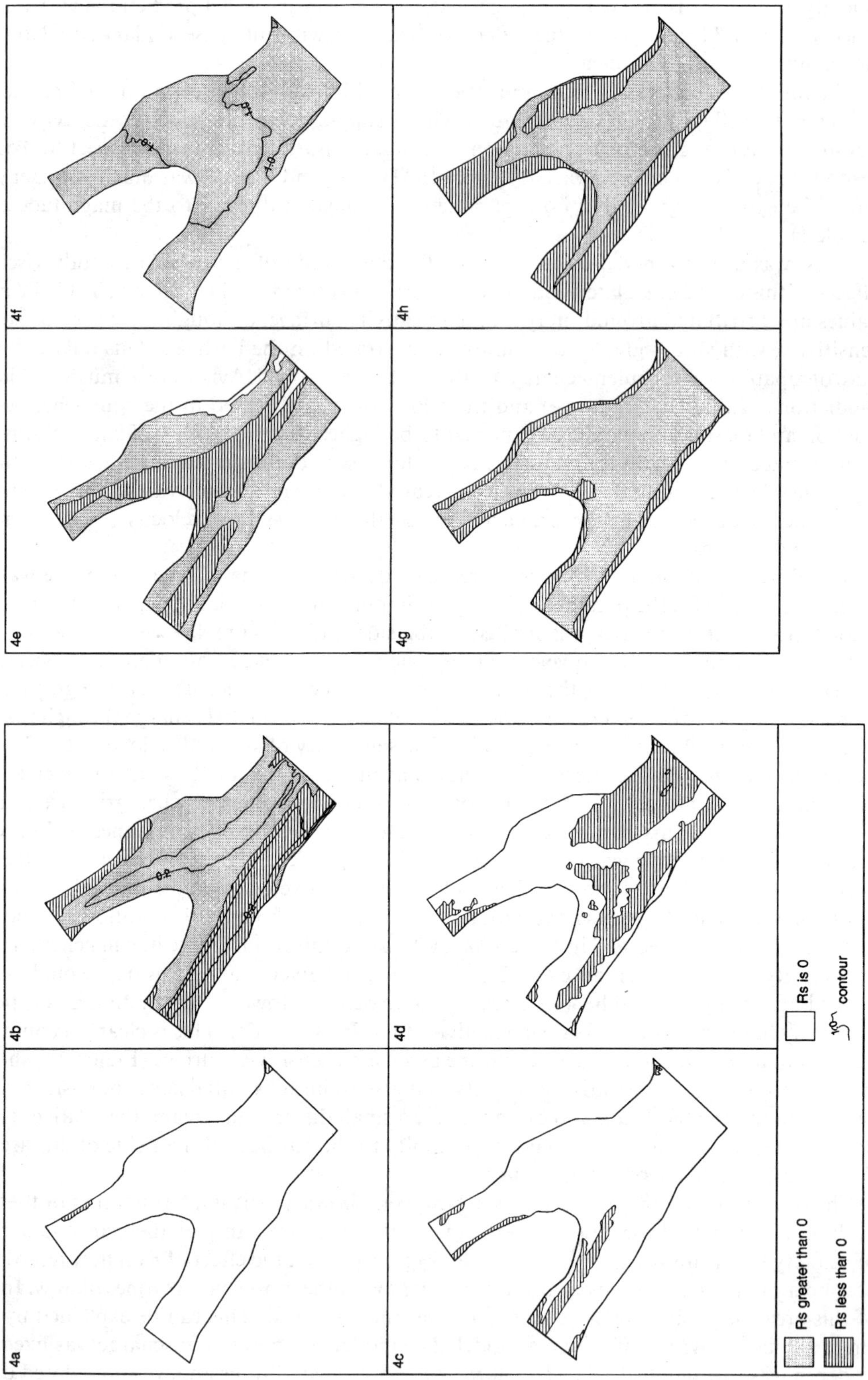
4a
4b
4c
4d
4e
4f
4g
4h
Rs greater than 0
Rs less than 0
Rs is 0
contour

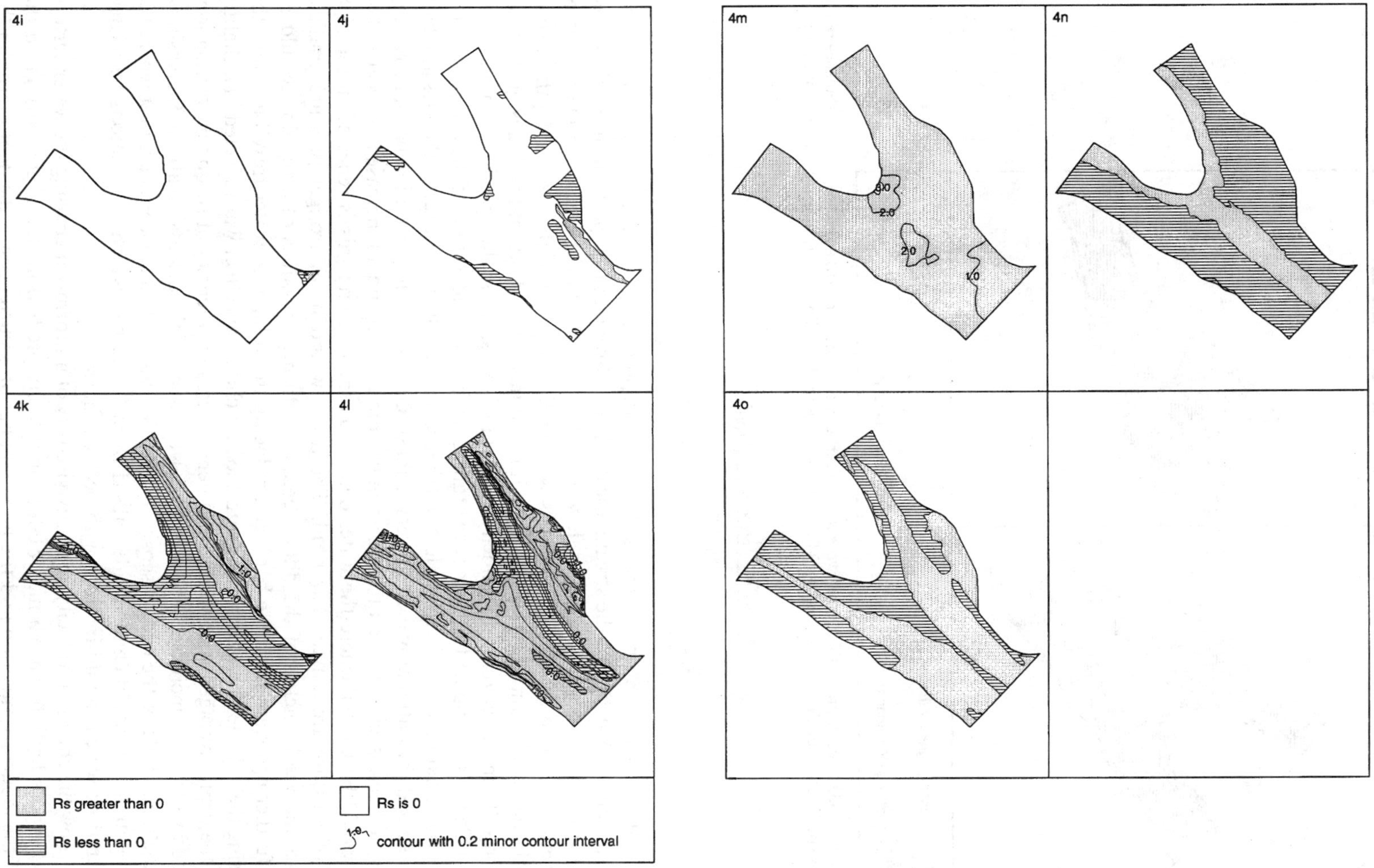

Figure 4. Relative sensitivity surfaces for the response of: (i) resultant velocity to spacially uniform perturbation of 'Turbin' (a) 'Peclet' (c), 'Avis' (e), 'Drag' (g), 'Recap' (i), *n* (k) and depth (n); (ii) eddy viscosity to spatially uniform pertubation of 'Turbin' (b), 'Peclet' (d), 'Avis' (f), 'Drag' (h), 'Recap' (j), *n* (l) and depth (o); and (iii) pressure to spatially uniform perturbation of *n* (m). Contours are only included where the R_s values exceed 0·1

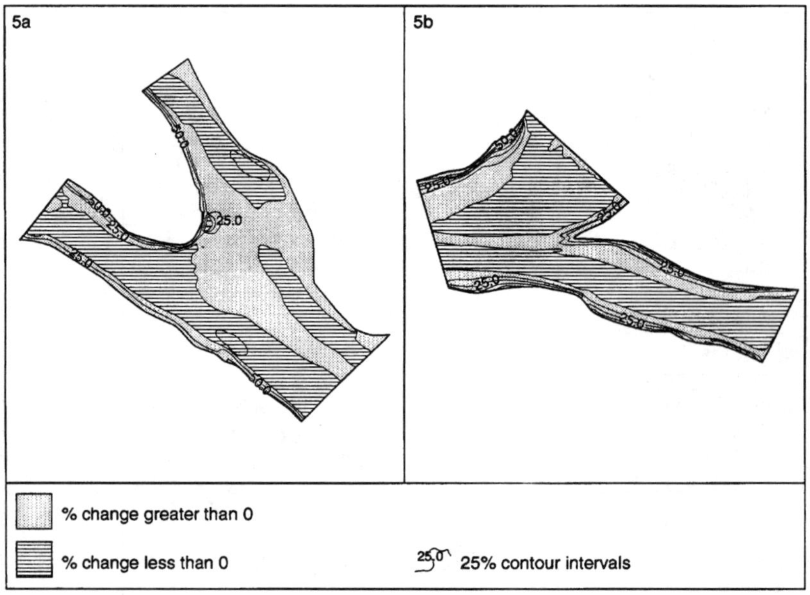

Figure 5. Comparison of flow model predictions with: (a) the two-equation turbulence model; and (b) a zero-equation turbulence model in which there is uniform specification of the eddy viscosity (0·01 $m^2 s^{-1}$) for both the diffluence (a) and the major confluence (b). The percentages are defined by;

$$\frac{\bar{V}_{\nu\text{uniform}} - \bar{V}_{\nu\text{twoequation}}}{\bar{V}_{\nu\text{twoequation}}} \times 100$$

tion. Thus, if the velocity decreased in one region, and because depths are fixed, in order to maintain the constant flow rate across any one section of the stream, the discharge through other parts of the section must increase. The result is that more water is forced to flow in areas of the diffluence where the magnitude of change in bottom friction owing to change in roughness was smaller, because of the greater depth, than in zones where the depth was smaller and the absolute magnitude of increase in bottom friction is greater [see Equation (1)]. This results in an *increase* in velocity in certain zones, even though the roughness was increased. It was possible to find support for the above arguments from consideration of the response of the pressure field to uniform variation in the magnitude of n (Figure 4m). In the first instance the pressure field would be determined so that the initial flow field conserved mass. On a location by location basis, increasing the magnitude of n would be associated with an increase in bottom friction and an increase in pressure. The increase in pressure would be greater where the water depth was smaller, and this is reflected in the high relative sensitivities of pressure around the bar head. The result would be a stronger pressure gradient between the bar head and the branches of the diffluence, and with higher values of n the flow would be observed to diverge further upstream from the bar head, reducing the volume of water travelling on to the bar head and increasing the volume of water flowing into the branches. This effect was reflected in a slightly greater curvature in the depth-averaged streamlines, and hence a greater intensity of secondary circulation.

The response of eddy viscosity to increased n has parallels with much of the discussion above. Increasing n generally increased the magnitude of the eddy viscosity (Table IV), but this increase was greatest where the flow was shallower and in the vicinity of channel side walls (Figure 4l), both of which are locations associated with turbulence generation, which should increase with bed roughness.

Figures 4n and 4o show the effects on velocity and eddy viscosity of uniformly increasing the water surface elevation. One would expect depth to have a major effect on velocity predictions (cf. Figure 4n), given the distributed nature of the model formulation and the large number of occurrences of the depth term (in the mass and momentum equations, and the secondary circulation model; see Lane, this volume). The resultant velocity sensitivity surface (Figure 4n) suggests a complicated response, with a general velocity decrease in the channels on either side of the bar head, but a velocity increase on to the bar head.

Three effects need consideration in accounting for the spatial patterns of velocity response. First, as the water depth is progressively increased, the magnitude of the pressure divergence term is decreased owing to reduction in the *relative* pressure differences. With a sufficient increase in water depth, the pressure divergence terms would become extremely small. The magnitude of this effect is greatest in shallow zones and this reduces the magnitude of the pressure gradient force in driving flow out of these zones. This would result in velocity decreases in the deeper locations and velocity increases in shallower locations. Secondly, as the water depth is increased, the magnitude of the bottom friction term [C_f in Equation (1)] is decreased. This would result in a general increase in velocity, whose effect would be felt most in shallow zones, which would be able to transport more of the total flow. This could result in velocity decreases in deeper zones for exactly the same reasons described with respect to roughness above. Thirdly, the equation of mass conservation implies a general velocity decrease in response to increasing depth owing to conservation of mass. Thus, one would expect a general velocity decrease, most marked in deeper zones but less in the shallower zones, and, given the reduction in the pressure gradient force and friction in the shallower zones as depth, increases, potentially an increase in velocity as these effects combine. These observations, in combination, explain the spatial patterns of sensitivity observed in Figure 4n. The response of eddy viscosity to increases in depth (Figure 4o) are also as expected. Increasing water depth reduces the eddy viscosity, as the frictional effects that are central to turbulent energy production and dissipation are drowned out. This is reflected in the focus of eddy viscosity decreases on the bar head and around the side walls.

The response of velocity to unit increases in discharge (Table IV) is as expected. In a fixed lid, steady-state flow model, increases in discharge can only be manifest by increases in velocity, such that a unit relative increase in discharge would result in a unit relative increase in velocity.

Figure 5 shows the comparison between flow model predictions with: (a) the two-equation turbulence model; and (b) a zero-equation turbulence model in which there is uniform specification of the eddy viscosity ($0{\cdot}01\ m^2\ s^{-1}$) for both the diffluence (Figure 5a) and the major confluence (Figure 5b). The values given are the percentage change in resultant velocity prediction owing to using the less sophisticated (uniform) eddy viscosity specification. In both the diffluence and the confluence, the effects of the more sophisticated eddy viscosity calculations are limited to the vicinity of the channel side walls. There are more minor and slightly positive changes in velocity in other regions. These observations reflect the importance of channel side wall turbulence generation. The more sophisticated turbulence approximation results in reduced primary velocities as more momentum transfer is achieved through the turbulence production, transport and dissipation processes. Away from the side wall, and to maintain mass conservation, the more sophisticated turbulence model results in slightly higher velocities. However, these observations aside, the effects of the more sophisticated turbulence model are not particularly great. At least some of the large percentage changes close to the side wall can be attributed to the small magnitude of velocity predictions in this zone (which amplifies the percentage difference). The one exception to this is in the true right branch of the diffluence, where the changes in velocity predictions are greater (Figure 5a).

Figure 6 shows the effects of removing the secondary circulation correction on model predictions for the diffluence (Figure 6a) and the major confluence (Figure 6b). The results again suggest that the secondary circulation correction has little effect upon velocity predictions. Percentage differences are high in the vicinity of the side walls, which may be a result of low velocities in this region. However, the signs of the differences vary spatially (positive indicates that the velocity predictions are greater with the secondary circulation than without it). Secondary circulation is thought to occur wherever the streamlines are curved and to result in a net transfer of momentum at right angles to the direction of the primary flow, away from the local centre of the radius of streamline curvature. When the secondary circulation correction is employed, there are higher resultant velocities on to the bar head in the diffluence and above the confluence scour hole, which seem to reflect this streamwise momentum transfer.

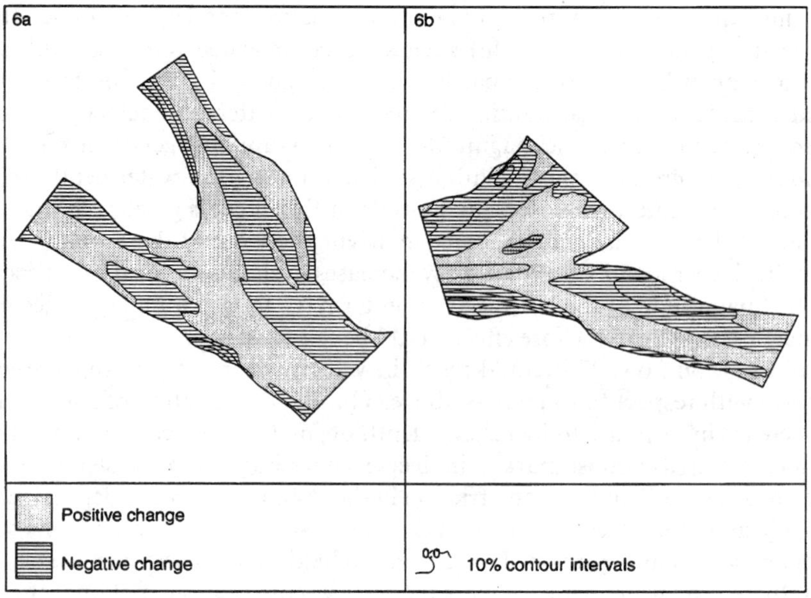

Figure 6. Plots to illustrate the effects of removing the secondary circulation correction upon model predictions for (a) the diffluence and (b) the major confluence. The percentages are defined by

$$\frac{\bar{V}_{\text{with}} - \bar{V}_{\text{without}}}{\bar{V}_{\text{with}}} \times 100$$

DISCUSSION

The results allow a number of important points to be made, some of a methodological nature, some of a more substantive nature. These results illustrate the utility of sensitivity analysis as a means of assessing model performance, allowing interpretation of the results in the context of model formulation. As such, this becomes a basis of identifying those assumptions involved in model formulation that need further consideration and development. Comparison of the results in Table IV with those in Figure 4g illustrates the need for the sensitivity analysis to be distributed in models of this kind. Neither the R_s values at specific locations within the model (in this case the locations of velocity or eddy viscosity maxima or minima), nor the surface mean value of R_s, provide a reliable indication of model sensitivity. To understand the real nature of model sensitivity it is necessary to refer to the total sensitivity surface. The ranks in Table IV illustrate this. For instance, with respect to the parameter 'Drag', the sensitivity of the model output varies according to whether or not a specific location or a surface average is chosen and, further, which location is chosen. This point is further amplified by consideration of Figure 4, in which the only zones that are sensitive to the parameter 'Drag', as one would expect, are in the vicinity of the model side walls.

It is apparent from the results that the effects of the turbulence model upon model predictions are perhaps less than might be expected (Figure 5a and b). Indeed, investigation of the parameters that control the magnitude of the eddy viscosity ('Turbin', 'Peclet', 'Avis' and 'Recap') through sensitivity analysis, indicates that, although they have some effects on the magnitude of the eddy viscosity, they have few effects on the magnitude of the resultant velocity. This might imply that the effects of turbulence are unimportant in this type of river channel. This is unlikely to be the case (cf. Kirkbride, 1993; Clifford *et al.*, 1995), and it is perhaps more likely that the apparent lack of importance of the turbulence model reflects the nature of the model itself rather than the importance of turbulence in the momentum transfer process. It is possible that neither the Boussinesq approximation on which the model is based, nor the way the eddy viscosity is calculated in the approximation, are appropriate to this problem, and there has been some criticism of this type of model in the engineering literature (see Lane, 1988, this issue, for review). The fact that most turbulence is generated in the vicinity of side walls reflects the depth-averaging method used in this model, and would be expected to constrain turbulence generation to such zones. However, although this might be acceptable for regular trapezoidal channels, where the ratio between side wall length and channel width is high, this is not always the case in natural channels. In this situation, water depths in the vicinity of the side walls were small (0·02–0·04 m), and in most cases the cross-stream bed slope was small. This would result in some turbulence generation in the vicinity of the wall, but its rapid dissipation before it could penetrate sufficiently far into the flow to effect momentum transfer, and hence primary velocities, further away from the sidewall. In natural channels, with complex geometries, depth-averaged turbulence modelling may therefore be unimportant relative to accurate specification of boundary conditions. This point is implicit in a review of the use of turbulent models in open channel flow problems by the ASCE (1988), where it is argued that a continuing increase in the sophistication of turbulence models had paid scant attention to the extent to which this was necessary. In this application, the uncertainty in model predictions associated with inaccurate boundary condition specification may be greater than the improvement in model predictions that comes from a more sophisticated turbulence treatment. However, research presented elsewhere in this issue (Bradbrook *et al.*, this volume; Hodskinson and Ferguson, this volume) suggests that the criticisms of turbulence modelling made above do not apply to three-dimensional solutions of the Navier–Stokes equations, particularly in the presence of separation zones.

The need for close attention to be given to the determination of boundary conditions is reflected in the significant degradation of prediction confidence owing to possible errors in roughness specification. Figure 7 shows the prediction confidence surface, with the error term in Equation (3) estimated from the standard deviation of n acquired from the distributed Wolman sampling. The pattern of variation in prediction confidence reflects the findings of the sensitivity analysis (Figure 4k). The greatest errors are in the vicinity of the bar head and in the backwater zone on the right. The signs of the errors suggest that under-estimation of n results in over-prediction of velocity in the bar head zone and under-prediction in the backwater zone as implied by the sensitivity analysis. These implications are critical given qualitative observations of local variation in sediment sorting within this diffluence (e.g. coarsening on to the bar head), which mean that there could be substantial spatial variation in n. The errors could be even greater than Figure 7 suggests since bar head coarsening would increase n, while the deposition of fines in the backwater zone would decrease n. Frictional effects on the bar head would be increased and those in the deeper zone decreased, which, through

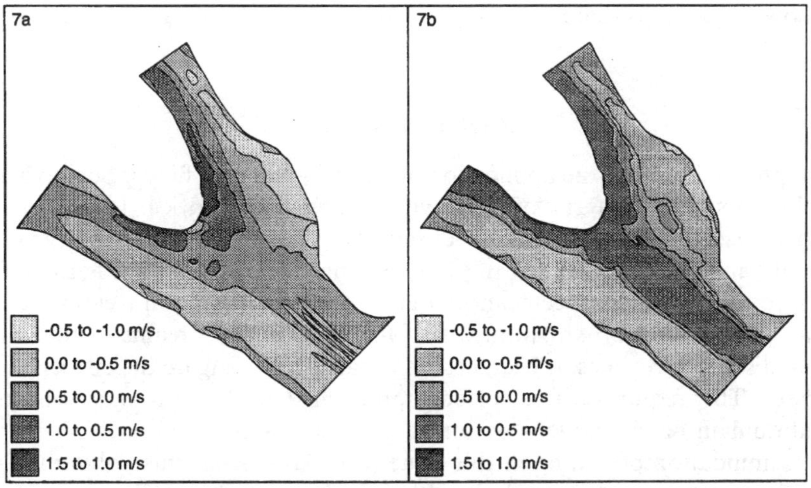

Figure 7. Prediction confidence surface obtained from Equation (3) for the velocity predictions. The error term was taken as the standard deviation of spatially distributed values of n

mass conservation and effects on the pressure field, would result in an amplified spatial feedback process. In the case of natural river channels with complex geometries, improved descriptions of topographical boundary conditions, of which roughness can be viewed as one element, may be critical and perhaps more important than further refinement of turbulence models. This is particularly the case given that turbulent processes will be a result of a range of different scales of river bed topography, and not just those associated with general planform morphologies (e.g. Nezu *et al.*, 1993; Colombini and Parker, 1995). Recognizing the need to incorporate better representation of bed roughness information raises difficult questions regarding the modelling of flow processes, particularly in gravel-bed rivers where spatial variability in grain size combines with the organization of individual grains into bedforms, and where accurate information on the spatiotemporal variability of roughness will be difficult to obtain. It requires that bed roughness be explicitly regarded as one part of a topographical continuum from large- to small-scales, and much more research into the way in which roughness information is incorporated as a subgrid-scale process, the scale being defined by both the typical topographic characteristics of the reach chosen for study and the grid or element size being used for model application.

As with the turbulence model, the secondary circulation correction affected primary velocities in a sensible manner, but the effect of the correction was small. In part, this must reflect the fact that at least two secondary circulation correction mechanisms were not included in the model; that associated with vertical variation in topography in the downstream direction, and that associated with shear-generated turbulence (see Lane, this volume). Ashmore *et al.* (1992) and Biron *et al.* (1993) have found these secondary circulation generating mechanisms to be important, particularly in confluences. At least some of these problems could be addressed by using a three-dimensional flow model (to model the effects of shear stress generated by bed discordance) that incorporates an algebraic or Reynolds stress turbulence model (to model the effects of shear-generated secondary circulation). Such an approach would allow some estimate of the relative importance of secondary circulation with respect to other areas of model uncertainty and, if found to be necessary, it might be possible to begin to explore different means of parameterizing secondary circulation generating mechanisms. A possible result might be a set of parameterizations for depth-averaged modelling of confluences and diffluences, whose relative importance would be morphologically controlled.

The final major area of uncertainty in model performance is with respect to the 'Drag' parameter, which emphasizes an important general problem in modelling flows of this type. If planform migration dominates river channel dynamics, accurate velocity and shear stress prediction in the side wall zone is critical. However, side wall velocity (and, by implication, shear stress) predictions are critically dependent upon an empirical side wall correction that requires empirical coefficients that have little physical basis (e.g. 'Drag'). Understanding side wall flow processes is therefore a key, underdetermined aspect of numerical modelling studies of this type.

CONCLUSION

This paper has attempted to illustrate the application of a depth-averaged flow model to a multi-thread reach of river. High resolution spatial patterns of sensitivity analysis were critical to understanding the model validity, illustrating how comparison with field observations alone does not uniquely determine model success. The sensitivity analysis has outlined the way in which the research needs to be furthered through consideration of improved roughness specification and secondary circulation treatments in particular. The model used made a steady-flow approximation with a fixed lid. In the streams of the kind described here, with strongly diurnal discharge regimes, it becomes important to investigate the effects of discharge fluctuation on flow processes. This requires a free surface approach that allows inundation extent to alter, and therefore the spatial domain of the model solution to alter, as discharge changes. This provides a link to studies of floodplain inundation presented in this issue (e.g. Bates *et al.* this volume). Current research is focusing on the development of an improved algorithm for handling unsteady free surface flows in which wetting and drying of topography occurs. This is being extended to the incorporation of a bed adjustment model into the flow model, incorporating an entrainment function and a two-dimensional form of the sediment continuity fraction for individual grain sizes. These types of models have been developed for relatively simple straight and curved laboratory-generated channels (e.g. Shimizu and Itakura, 1989), but have yet to be applied to either natural or more complicated multi-thread reaches, at least in part because of problems of boundary condition specification.

ACKNOWLEDGEMENTS

Paul Bates provided critical and constructive comments on an earlier draft of this paper. This research is supported by NERC Grant GR3/9715 (to S.N.L., K.S.R. and Dr J. H. Chandler). Owen Tucker prepared the diagrams.

REFERENCES

Anderson, D. A., Tannehill, J. C., and Pletcher, R. H. 1984, *Computational Fluid Mechanics and Heat Transfer*. Hemisphere Publishing, McGraw-Hill, New York. 599 pp.

Andrews, E. D. and Nelson, J. M. 1989. 'Topographic response of a bar in the Green River, Utah, to variation in discharge', in Ikeda, S. and Parker, G. (Eds), *River Meandering*, American Geophysical Union Monograph, 12. American Geophysical Union, Washington, D.C. pp. 463–485.

ASCE, 1988. 'Turbulence modelling of surface water flow and transport: Part V', *ASCE J. Hydraul. Engng*, **114**, 1052–1073.

Ashmore, P. E., Ferguson, R. I., Prestegaard, K. L., Ashworth, P. J., and Paola, C. 1992. 'Secondary flow in anabranch confluences of a braided gravel-bed stream', *Earth Surf. Process. Landf.*, **17**, 299–311.

Ashworth, P. J. and Ferguson, R. I. 1986. 'Interrelationships of channel processes, changes and sediments in a proglacial braided river', *Geograf. Annal.*, **68A**, 361–371.

Bates, P. D. *et al.*, (this volume).

Bernard, R. S. 1992. 'STREMR: numerical model for depth-averaged incompressible flow', *Technical Report HY-105*. US Army Corps Engineers, Waterways Experiment Research Station, Vicksburg, Mississippi.

Biron, P., De Serres, B., Roy, A. G., and Best, J. L. 1993. 'Shear layer turbulence at an unequal depth confluence', in Clifford, N. J., French, J. R., and Hardisty, J. (Eds), *Turbulence: Perspectives on Flow and Sediment Transport*. Wiley, Chichester. pp. 197–214.

Bradbrook, K. F., Biron, P. M., Lare, S. N., Richards, K. S., and Roy, A. G. 1998. 'Investigation of controls on secondary circulation in a simple confluence geometry using a three-dimensional numerical model', (this volume).

Bray, D. I. 1982. 'Flow resistance in gravel-bed rivers', in Hey, R. D., Bathurst, J. C., and Thorne, C. R. (Eds), *Gravel-bed Rivers*. Wiley, Chichester. pp. 109–133.

Clifford, N. J., Richards, K. S., Brown, R. A., and Lane, S. N. 1995. 'Scales of variation of suspended sediment concentration and turbidity in a glacial meltwater stream', *Geograf. Annal.* **77A**, 45–65.

Colombini, M. and Parker, G. 1995. 'Longitudinal streaks', *J. Fluid Mech.*, **304**, 161–183.

Emmons, H. W. 1970. *Ann. Rev. Fluid Mech.*, **2**, 15–36.

Ferguson, R. I. 1993. 'Understanding braiding processes in gravel-bed rivers: progress and unresolved problems', in Best, J. L. and Bristow, C. S. (Eds), *Braided Rivers*, Geological Society Special Publication, 75. Geological Society, London. pp. 13–71.

Ferguson, R. I. and Ashworth, P. J. 1992. 'Spatial patterns of bedload transport and channel change in braided and near-braided rivers', in Billi, P., Hey, R. D., Thorne, C. R., and Tacconi, P. (Eds), *Dynamics of Gravel-bed Rivers*. Wiley, Chichester. pp. 477–492.

Ferguson, R. I., Ashmore, P. E., Ashworth, P. J., Paola, C., and Prestegaard, K. L. 1992. 'Measurements in a braided river chute and lobe: I. Flow pattern, sediment transport and channel change', *Wat. Resour. Res.*, **28**, 1877–1886.

Hodskinson, A. and Ferguson, R. I. 1998. 'Numerical modelling of separated flow in river bends: model testing and experimental investigation of geometric controls on the extent of flow separation at the concave bank', (this volume).

Kirkbride, A. D. 1993. 'Turbulence structure in straight gravel-bed rivers', *PhD Thesis*, University of Sheffield.

Lane, S. N. 1994. 'Monitoring and modelling morphology, flow and sediment transport in a gravel-bed streasm', *PhD Thesis*, University of Cambridge. 372 pp.

Lane, S. N. 1998. 'Hydraulic modelling in hydrology and geomorphology: a review of high resolution approaches', (this volume).

Lane, S. N., Chandler, J. H., and Richards, K. S. 1994a. 'Developments in monitoring and terrain modelling small-scale river-bed topography', *Earth Surf. Process. Landf.*, **19**, 349–368.

Lane, S. N., Richards, K. S., and Chandler, J. H. 1994b. 'Distributed sensitivity analysis in modelling environmental systems', *Proc. R. Soc. London. Ser. A*, **447**, 49–63.

Lane, S. N., Richards, K. S., and Chandler, J. H. 1995. 'Within reach spatial patterns of process and channel adjustment', in Hickin, E. J. (Ed.), *River*. Geomorphology, Wiley, Chichester. pp. 105–130.

Livesey, J. R. and McLelland, S. J. 1995. 'Streamwise vorticity and sediment sorting in a narrow, open channel', Presented at the *Conference on Coherent Flow Structures in Open Channels: Origins, Scales and Interactions with Sediment Transport and Morphology, Leeds, 10–12 April, 1995.*

MacCormack, R. W. 1969. 'The effect of viscosity in hypervelocity impact cratering', *AIAA Paper 69–354*. American Institute of Aeronautics and Astronautics, Cincinatti.

McCuen, R. H. 1973. 'The role of sensitivity analysis in hydrologic modelling', *J. Hydrol.*, **18**, 37–53.

Nelson, J. M. and Smith, J. D., 1989. 'Flow in meandering channel with natural topography', in Parker, G. (Ed.), *River Meandering*, American Geophysical Union Mongraph, 12. American Geophysical Union. Washington, D.C. pp. 69–102.

Nezu, I., Tominaga, A. and Nakagawa, H., 1993. 'Field measurements of secondary currrents in straight rivers', *ASCE J. Hydraul. Engng*, **119**, 598–614.

Prandtl, L. 1945. 'Uber ein neues formelsystem für die ausgebildete turbulenz', Translated as *Jet Propulsion Lab. Publ.*, **13**, 1952.

Shimizu, Y. and Itakura, T. 1989. 'Calculation of bed variation in alluvial channels', *ASCE J. Hydraul. Engng*, **115**, 367–384.

Strickler, A. 1923. 'Beitrage zur Frage der Geschwindigheitsformel und der Rauhigkeitszahlen fur Strome, Kanale und Geschlossene Leitungen', *Mitteilungen des Eidgenossischer Amtes fur Wasserwirtschaft*. Bern, Switzerland. 16 pp.

6

NUMERICAL MODELLING OF SEPARATED FLOW IN RIVER BENDS: MODEL TESTING AND EXPERIMENTAL INVESTIGATION OF GEOMETRIC CONTROLS ON THE EXTENT OF FLOW SEPARATION AT THE CONCAVE BANK

A. HODSKINSON* AND R. I. FERGUSON

Department of Geography, University of Sheffield, Winter Street, Sheffield S10 2TN, UK

ABSTRACT

The occurrence of separated flow in sharply curving river bends, and the associated deposition of fine-grained sediment, is widely recognized, but the factors controlling the existence and extent of separated flow have not been clearly identified. A three-dimensional, fixed lid computational fluid dynamics program was used to simulate time-averaged flow in idealized channel bends in order to investigate geometric controls on flow separation. The code was applied first to a field site; it successfully modelled the qualitative features of the flow, including a separation zone, and was insensitive to choice of turbulence closure or exclusion of lateral variation in water surface topography. The simulations of idealized bends demonstrate that the existence and extent of concave bank flow separation can be significantly influenced by changes in bend planform, point bar topography and upstream planform.

KEY WORDS numerical modelling; flow separation; meander bends

BACKGROUND AND AIMS

In sharply curving river bends the main flow can become separated from the channel boundary, either just upstream of the bend apex at the concave (outer) bank, or just downstream of the apex at the convex (inner) bank. This has long been recognized as important in terms of channel and floodplain deposition, since the low velocities in these regions allow the progressive deposition of fine-grained sediments and organic matter advected into them, either in the form of a concave bank bench if the separation occurs at the outer bank, or on the downstream part of the point bar if it occurs at the inner bank (Carey, 1969; Taylor *et al.*, 1971; Leeder and Bridges, 1975; Woodyer, 1975; Woodyer *et al.*, 1979; Hickin, 1977, 1986; Page and Nanson, 1982; Reid, 1984; Hattingh and Rust, 1993). Separation and associated deposition also occur downstream of channel constrictions in straight channels (e.g. Rubin *et al.*, 1990). Concave bank bench deposits can contribute significantly to floodplain sedimentation in situations where the migration of a bend is confined and the sharply curving confined bend migrates down-valley along the valley side, maintaining its form and the associated flow separation and deposition (Hickin, 1986).

Several field workers have proposed hypotheses concerning the origin of concave bank flow separation based on sequential changes in individual bends or comparisons of bends at different stages of development.

---

* Correspondence to: Dr A. Hodskinson, Manchester Grammar School, Manchester, M13 0XT, UK.

Contract grant sponsors: NERC and University of Sheffield.
Contract grant number: GT4/93/27P.

Concave bank flow separation is known to occur at the apex of sharply curving bends; Hickin (1977) suggested a curvature of $r_c/w \leqslant 2$ is required, where r_c is the centreline radius and w the channel width. If high curvature is reduced through continued bend migration, flow separation and associated deposition will cease (e.g. Reid, 1984). Sharpening of bends to the critical curvature has been attributed to confinement of a migrating channel by erosion-resistant features, such as cohesive floodplain sediment, bedrock obstructions or valley sides. Hickin (1977) and Page and Nanson (1982) proposed that channel widening at the apex of a sharply curving bend facilitates flow expansion and allows separation to develop at the concave bank. The work of Page and Nanson (1982), Lewin (1978) and Andrle (1994) demonstrated that it is erosion at the inner bank of the sharp bend that leads to widening. As these studies were based on observations of bend planform, the influence on flow separation of the three-dimensional bed topography in the bends is unknown. Also, whilst migration of the upstream channel reach has been shown to influence the development of sharp curvature in individual bends (Reid, 1984), the diversity in the planforms of many meandering rivers and individual bends (e.g. Brice, 1974; Lewin and Brindle, 1977; Hooke and Harvey, 1983) means that the configuration of the channel upstream of individual bends may vary considerably. The effects of such variation on the flow distribution entering a bend and on the development of flow separation in that bend are also unknown.

The exact controls on the development of concave bank flow separation in river bends are therefore unclear. Knowledge of these controls is necessary in order to understand the development and persistence of concave bank flow separation and the associated deposition in individual bends, and to predict effects of channel change on the onset and extent of flow separation. Computational fluid dynamics (CFD) software offers the potential to construct model channel geometries and simulate the three-dimensional flow patterns in them. Flow patterns have been successfully reproduced using such code applied to a small laboratory channel with realistic bed topography (Demuren, 1993), although in this case no flow separation was evident, and a two-dimensional code has been applied to a divided, gravel-bed stream with encouraging results (Lane *et al.*, 1995). However, the potential for applying CFD codes to natural gravel- or sand-bed river bends that exhibit separated flow has yet to be explored. By setting up an idealized version of the geometry of a river bend known to exhibit separated flow, and then systematically altering various aspects of the channel geometry and observing the effects on simulated flow patterns, it should be possible to identify the geometric properties of the channel that significantly influence the development of flow separation and the effects of changing them. This is the principal aim of the work presented in this paper.

Model predictions for idealized channels must be reliable. To test the capabilities of the CFD code used in this work, it was applied to a real river bend with flow separation, where the channel flow geometry and boundary conditions could be quantified and the flow predictions tested against field measurements. This experiment also gave the opportunity to examine the sensitivity of the model predictions to the water surface specification used and the choice of turbulence closure in the flow calculations. The dependence of flow predictions for the idealized channel on the choice of computational grid resolution was also assessed. These checks on the robustness of the predictions to various technical choices in the model set-up allow confidence to be placed in the flow predictions and results of the idealized bend experiments.

USE OF THE FLUENT CFD CODE

The CFD code used in this work was Version 4.2 of Fluent (Fluent Inc. 1993a,b). This software allows the construction of model flow geometries, interpolation of a computational grid, specification of model boundary conditions and solution of the three-dimensional Navier–Stokes flow equations in order to calculate the flow field. Flow geometries are constructed by defining points in three-dimensional space that lie along the edges of the geometry, joining these points to create edge curves and then joining these curves to create the boundary surfaces. For open channels the geometry construction must include both the channel geometry and the position of the water surface, as Fluent 4.2 approximates a free surface as a frictionless, rigid lid. Models of real river bends therefore require a topographic survey of the channel and water surface, whilst for idealized bends this information must be assumed (see below). A computational grid is interpolated within the fluid geometry by the modelling programme, with an additional layer of cells adjacent to the boundary to store boundary information. In this work, three-dimensional grids were used to allow calculation of the three-dimensional flow structure. The grid resolutions are described with the results.

Although the aim was to simulate the time-averaged flow in bends, the turbulent nature of river flow necessitates Reynolds averaging of the Navier–Stokes flow equations and this in turn introduces terms that require semi-empirical models to be quantified (see Lane, 1998, this issue, for a discussion). Most applications of CFD codes to environmental hydraulics (e.g. ASCE, 1988; Demuren, 1993; Lane *et al.*, 1995) have used the two-equation k–ε turbulence closure model based on the Boussinesq hypothesis and empirically calibrated equations for turbulent kinetic energy, k, and energy dissipation, ε (Nezu and Nakegawa, 1993; Rodi, 1993). Fluent offers this standard k–ε model as a default, but also provides an additional k–ε model based on renormalization group theory (the RNG k–ε model) which is recommended for simulating separated flows and flows strained by stagnation and strong curvature (Fluent Inc., 1993a). Bradbrook *et al.* (1998, this issue) discuss the physical basis of the RNG model. Zhou *et al.* (1994) demonstrated that an RNG-based turbulence model gave improved predictions in a high resolution simulation of laboratory measurements of flow over a backward-facing step, and Bradbrook *et al.* (1998, this issue) found that it gave better predictions than the standard k–ε model in simulating laboratory data for sheared flows in a confluence of parallel channels. Such studies suggest that an RNG-based turbulence model is the most appropriate for our simulations, given that the flow in sharply curved river bends is characterized by high curvature and often, also, stagnation and massive flow separation. Fluent's RNG k–ε model was therefore used in this work, although, as discussed below, the choice between it and the standard k–ε model makes little quantitative difference to flow predictions for the test bend. Fluid density, ρ, and viscosity, μ, were set at 1000 kg m^{-3} and 0·0013 kg m^{-1} s^{-1} respectively.

In addition to the channel and water surface geometry the boundary conditions that must be specified are the bed and bank roughness and the flow distributions entering the channel from upstream. Fluent allows the incorporation of rough boundaries using a parameter E' which is related to the log-law via

$$E' = \frac{E}{1 + 0{\cdot}3k_s^+} \tag{1}$$

where E is the value of E' for smooth walls (9·8; Fluent Inc., 1993a) and k_s^+ is defined as

$$k_s^+ = \frac{\rho k_s u^*}{\mu} \tag{2}$$

Quantification of the shear velocity, u^*, and roughness height, k_s, for the boundaries of the model river channels allows E' to be calculated and a realistic boundary roughness to be incorporated in the model. The shear velocity, u^*, was estimated as $u^* = \sqrt{gRS}$, where R is the hydraulic radius of the channel, S is the slope and g is the gravitational acceleration. The roughness height, k_s, was quantified from surface counts of the gravel beds of the field site and the channel upon which the idealized bends were based (see below) via the relationship $k_s = 3{\cdot}5D_{84}$, as recommended by Bray (1982) for gravel-bed rivers. Values of E' for the channel banks were set an order of magnitude higher than those for the bed, corresponding to a lower roughness.

The inlet flow field in the Fluent model of the test bend was based on field measurements as described below. For simulations in idealized bends the mean channel velocity $\bar{U}$ at the inflow cross-section was estimated using the Darcy–Weisbach equation

$$\bar{U} = \sqrt{\frac{8gRS}{f}} \tag{3}$$

with the friction factor, f, estimated from the Keulegan-type relationship

$$\sqrt{\frac{1}{f}} = a \ln\left(b\,\frac{R}{k_s}\right) \tag{4}$$

with $k_s = 3{\cdot}5D_{84}$, as above, and $a = 0{\cdot}82$, $b = 15{\cdot}05$ following Knighton (1984); this method of calculating $\bar{U}$ is compatible with the use of the log-law for boundary roughness. The derived mean velocity was specified at all grid cells across the inlet cross-section. The inlet was separated from the experimental bend by a long, straight or curved entrance channel, which allowed lateral and vertical velocity gradients to develop before the experimental bend. Turbulence intensity at the inlet was set at 10% and the characteristic length of the turbulence was set as the hydraulic radius of the channel. These are standard assumptions in such modelling, with some support from field measurements (e.g. Clifford, 1993).

FLOW MODEL TESTING AT A FIELD SITE

Model set-up

The Fluent code was used to attempt to predict the observed flow patterns in a natural river bend with flow separation in order to test the model's capabilities and the sensitivity of the predictions to aspects of the model set-up. The bend chosen to conduct these tests on was a sharply curving right-hand bend of the River Dean, Cheshire, UK, in which the flow converged strongly towards the outer bank whilst a pronounced zone of separated flow existed over the point bar at the inner bank. The bend is illustrated in Figure 1. The mean reach slope is 0·0027, the channel width in the bend varies from 5 to 6 m and the bend exhibits a pronounced pool/point bar bed topography. The flow at which measurements were made was fairly low, with a discharge of about 0·5 $m^3\ s^{-1}$ and depth of only 0·2 m in the entrance riffle and 0·1 m over the point bar, although far deeper (up to 1·5 m) in the pool.

Twelve cross-sections around the bend were surveyed using an EDM. The sections were approximately perpendicular to the channel (Figure 1). They extended upstream of the bend to allow the approach flow to be modelled but were more closely spaced in the bend itself, where the bed topography changed rapidly.

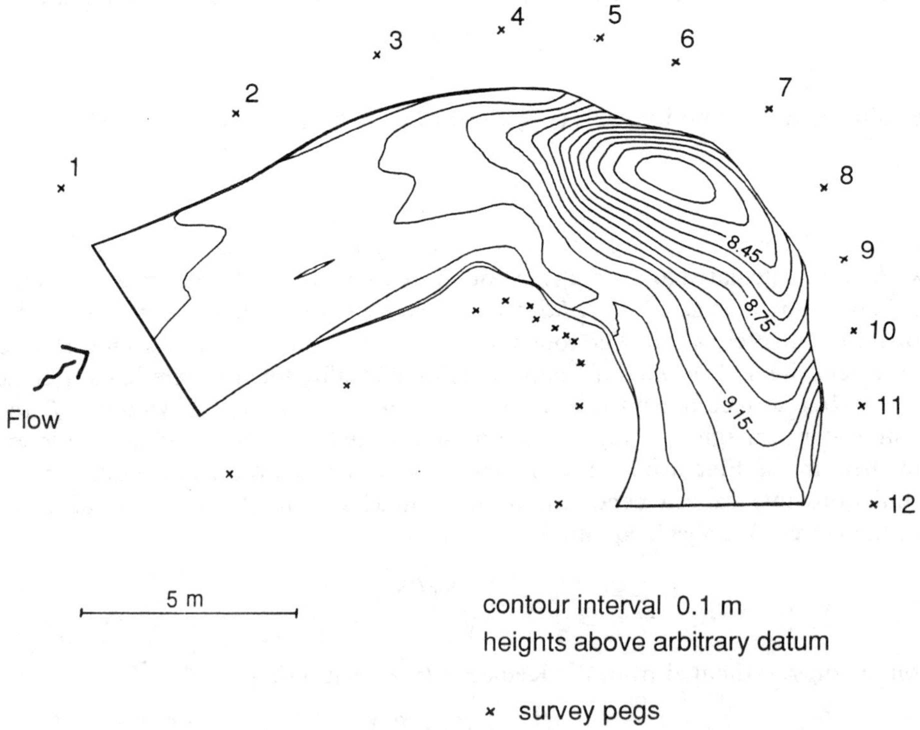

Figure 1. Plan view of the field site on the River Dean used to test the model, showing end-points of surveyed cross-sections. Inflow velocities were measured at cross-section 1, and near-surface and near-bed predictions were tested at cross-sections 4 and 11

Water surface elevation was surveyed at left and right edges of each section to quantify the water surface topography. The 12 surveyed cross-sections were then joined together along the channel to create four surfaces representing the bed, left and right banks, and water surface. The cross-sections at the upstream and downstream ends of the reach (1 and 12 in Figure 1) were defined as inlet and outlet planes in the model.

A computational grid of 18 000 cells was used in the model: 75 uniformly along the channel, 30 non-uniformly across the width, and 8 uniformly through the flow depth. Because the fastest current was close to the outer bank, giving high velocity gradients there, the cell width was arranged to increase linearly across the channel from 8–10 cm (according to local channel width) at the outer bank to 33–40 cm at the inner bank. This non-uniform spacing of the grid cells can be seen in the plotting position of the vector output in Figure 4 (see later). Cell height varies with flow depth, increasing from 3 to 19 cm as the channel deepens from riffle and point bar to pool. The curvature of the channel means that the cell length is greater at the outer bank (33 cm) than the inner bank (21 cm). The adequacy of this 75 × 30 × 8 grid is discussed below.

The flow field at the entrance to the bend (cross-section 1) was measured using an electromagnetic current meter (ECM). Velocities at five heights in each of seven verticals were well described by fitting the law of the wall, which was then used to interpolate laterally and vertically to estimate velocity in each of the 240 cells making up the inlet cross-section.

The estimation of bed roughness was based on a random sample of 500 stones from the surface of the shallow reach at the upstream end of the bend. This gave $D_{84} = 0{\cdot}066$ m, which, together with a surveyed water surface slope of 0·010 and mean depth of 0·20 m for the entrance channel, gave $E' = 0{\cdot}00131$ using Equations (1) and (2). However, when this value of E' was specified for the entire channel bed throughout the bend, and using the computational grid just described, the model solution diverged repeatedly, despite considerable reduction of all under-relaxation parameters. The smallest value of E' for which the model would run and reach a converged solution was 0·025, specified uniformly across bed and banks. This corresponds to $D = 3{\cdot}5$ mm, which is considerably less than the actual measured value on the riffle but higher than that corresponding to the sandy point bar top. A series of two-dimensional flow simulations for 10, 20 and 50 cm deep vertical profiles of slope 0·001, with various distributions of grid cells and values of bed roughness, showed that for a given value of bed roughness and channel depth there is a minimum grid cell size that may be used near the bed. The relatively small cells (in relation to those in the pool) near the bed in the shallow, 20 cm deep riffle were therefore too small to accommodate the coarse gravel roughness present. Fewer cells in the vertical dimension could be used in order to allow the use of a realistic roughness but this would reduce the resolution of flow patterns in the deeper parts of the river where cell dimensions increased. Our choices of vertical grid resolution and roughness value are a compromise forced by the extremely high relative roughness of the bend entrance in the low flow that was measured ($R/D_{84} \approx 3$). In other situations (this bend at higher discharge, or a bigger river even at low flow) it should be possible to use a locally accurate roughness height as well as good vertical grid resolution. However, specification of realistic bed roughness in shallow flow over coarse bed material could be problematic in applications to steep streams and warrants further attention.

Results of the field test

The model predictions were compared with two sets of field data. The first was a series of ECM measurements of downstream and cross-stream velocity made on the same day as the inflow and water surface measurements. This data set comprised near-surface and near-bed measurements at cross-sections 4 and 11 (see Figure 1) with additional near-surface measurements along the outer bank at cross-sections 5–10 and near-surface and near-bed measurements in the separation zone at cross-section 8; near-bed measurements could not be obtained in the 1·5 m deep pool because the ECM wading rod was too short. The second data set comprised near-surface ECM measurements throughout the bend. These measurements were made on a different day but the discharge was only slightly lower, and no noticeable differences in flow structure were apparent, so they serve for qualitative comparison of the observed and predicted velocity pattern. Figures 2–4 illustrate the comparisons made between observed and predicted velocity patterns at the study bend.

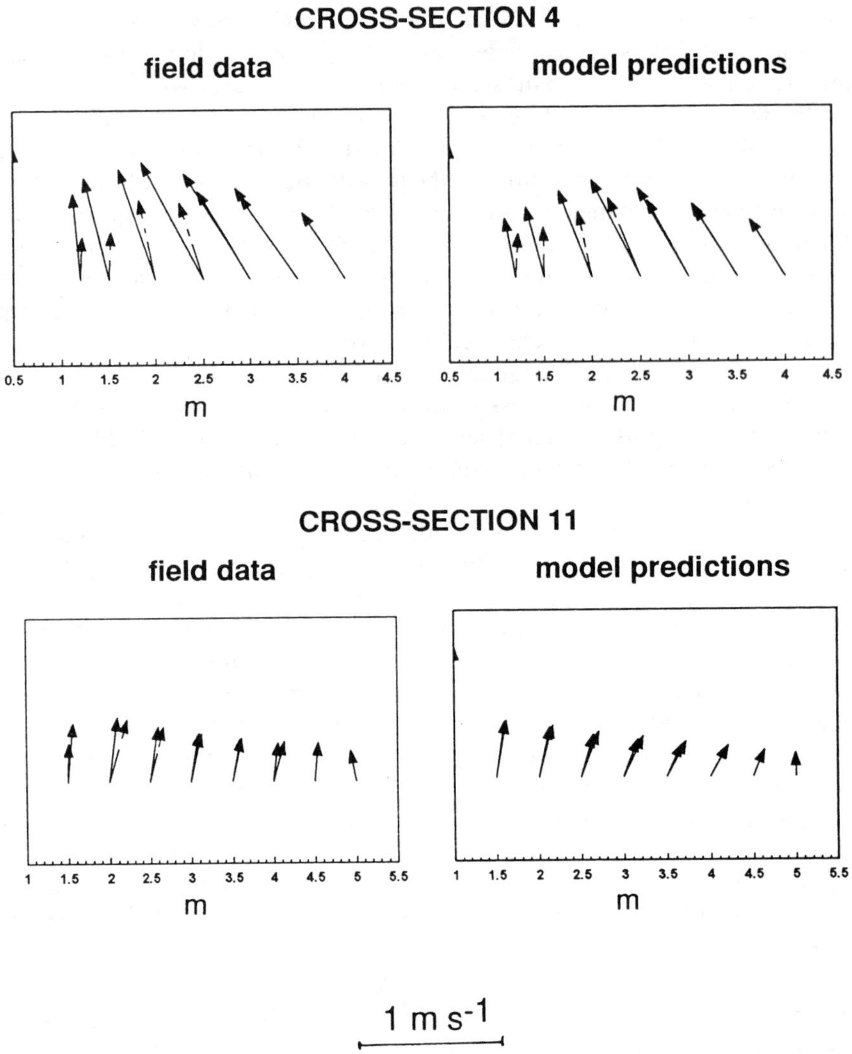

Figure 2. Comparison between observed and predicted near-surface (solid) and near-bed (dashed) velocity vectors at cross-sections 4 and 11 of the River Dean test bend

Figure 2 shows the observed and predicted near-surface and near-bed velocity vectors (cross-stream and downstream components only) for cross-sections 4 and 11 (see Figure 1 for the location of these sections). The comparison at cross-section 4 is qualitatively very good. The cross-channel flow distribution, with the maximum velocity in the channel centre and the vectors directed towards the left bank, is successfully reproduced by the model. The developing helical flow towards the left side of the channel is also reproduced very well, as are the stronger velocity gradients on the left side of the channel. At cross-section 11, the cross-channel flow asymmetry is reproduced, as are the lesser velocity gradients and deviations between surface and bed vectors compared with cross-section 4. The turning of the vectors near the right bank is also reproduced. However, the model predicts flow vectors at cross-section 11 to be directed more strongly towards the right bank than was observed in the field, and, in general, at both cross-sections, velocity magnitudes appear to be underestimated by the model.

Figure 3 illustrates the quantitative comparisons between predicted and observed velocity magnitude, along stream velocity (U) and cross-stream velocity (W) at a sample of points within the bend. The correlation is

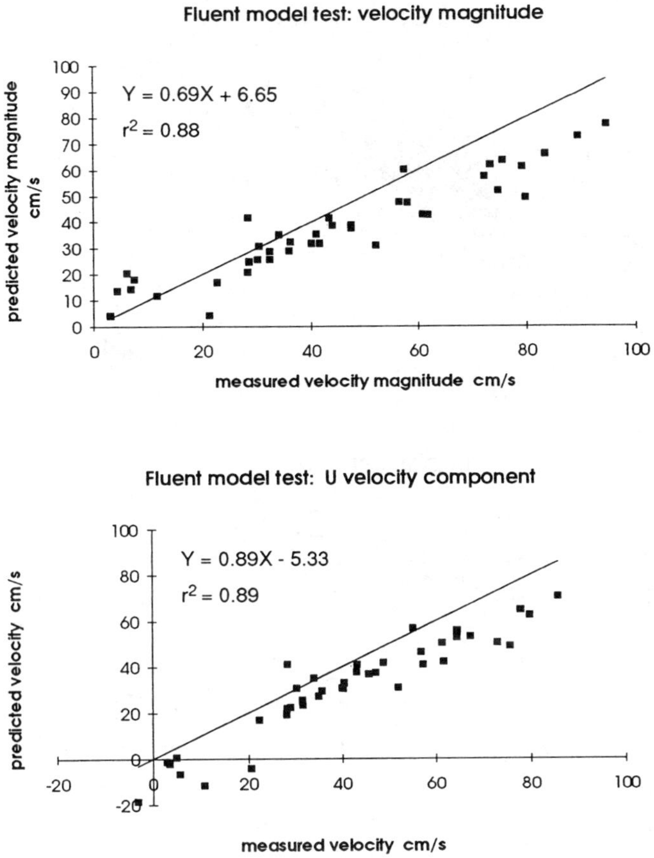

Figure 3. Comparison between observed and predicted values of velocity magnitude, streamwise (U) component and transverse (W) component at 40 points in the River Dean test site; 1 : 1 relationships shown as solid lines

high ($r = 0{\cdot}94$–$0{\cdot}96$) but, as suggested by Figure 2, the velocity tends to be underpredicted, particularly the higher values at near-surface points.

Figure 4 illustrates the qualitative comparison between near-surface flow patterns in the bend as predicted by the model and as observed in the field. A perfect match is not expected because the field measurements were at a slightly lower discharge than that modelled, and because the simulated velocities are for cell centres at different depths below the water surface according to the local flow depth, whereas the field measurements were all at 5 cm below the surface. Nevertheless, the agreement is very good. Both vector maps show convergence of the main flow towards the outer bank at the entrance to the sharply curved reach of the channel (cross-sections 4 and 5), and deceleration of a narrow stream of fast flow along the outer bank before the flow expands across the entire channel section and accelerates to the exit of the bend (cross-section 10 and 11). Both maps also show a large zone of gently circulating flow at the inner bank and divergence of the main flow near the inner bank (around cross-section 10), with part of the flow entering the separation zone at its downstream end.

It can be seen, therefore, that the general qualitative flow patterns present in the study bend are reproduced correctly, in particular the position and relative scale of the region of separated flow. The deviation of velocity vectors between the surface and bed is also reproduced for the two cross-sections where data were collected. This implies that the three-dimensional structure of the flow is also reproduced. Inspection of the vertical velocity components predicted by the model revealed downwelling near the outer bank at the approach to the most sharply curved part of the bend (cross-sections 4–6), as expected from the near-surface

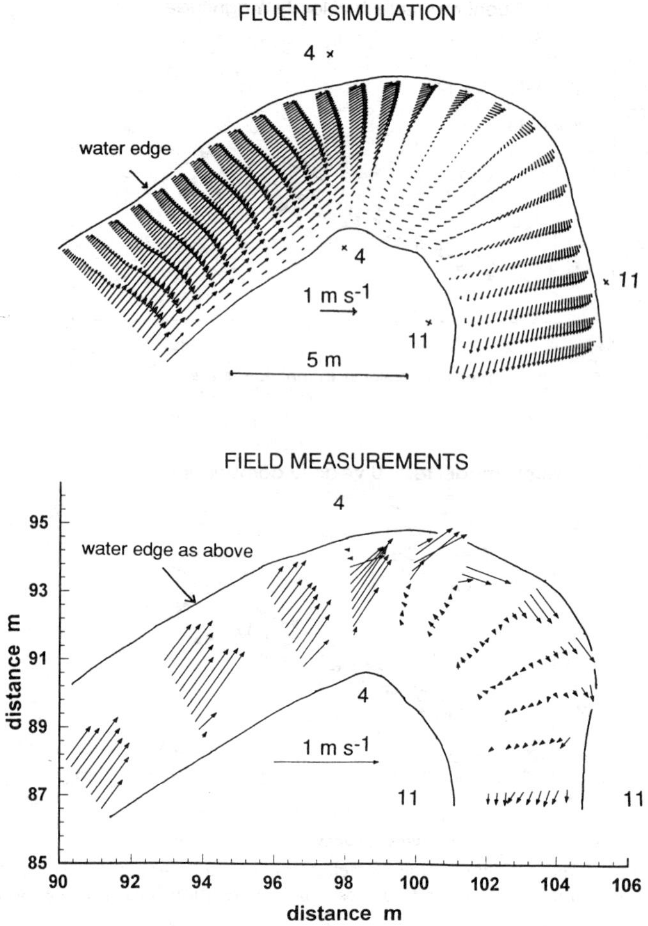

Figure 4. Simulated and observed (slightly lower discharge) near-surface velocity patterns at the River Dean test site, showing good qualitative agreement. Cross-sections 4 and 11 are labelled. The vector scale for simulated velocities is reduced for clarity

and near-bed velocities at cross-section 4. As the ECM could only measure two components no direct test of the vertical structure was possible and so this information is not presented.

Sensitivity tests

In order to assess the sensitivity of the model predictions to the choice of turbulence closure used in the calculations, the Fluent model of the field site was rerun using the standard k–ε model and the predictions were compared with those made using the RNG k–ε model, as in the evaluation of turbulence closure models presented by Basara and Younis (1995). Velocity magnitude, U and W were compared for the same sample points, as in Figure 3. Regression relationships between standard k–ε and RNG-based predictions were not significantly different from 1 : 1 relationships and the scatter around these relationships was very small (r from 0·98 to 1·00). This indicates that the choice of turbulence closure makes little difference to the model predictions, at least at the fairly coarse grid resolution employed in this study.

In order to construct model flow geometries for idealized experimental bends, assumptions must be made regarding the position of the water surface. Field evidence suggests that at bank-full discharge the downstream water surface slope in meandering channels becomes uniform (Leopold *et al.*, 1964; Carling, 1991). However, it is well known that the water surface elevation is not the same at either end of a cross-section but shows superelevation of the water surface against the outer bank, generating a cross-stream pressure gradient, which, in conjunction with curvature-induced centrifugal forces, drives the helical circulations observed in bends. The model of the River Dean field site was used to test the sensitivity of the predictions to the incorporation, or omission, of this lateral water surface slope. The water surface at the field site exhibited the expected elevation against the outer bank, with a maximum lateral height difference of 0·029 m at cross-section 9. When the model was rerun with the separate left- and right-bank water surface elevations replaced by their mean value the results were almost indistinguishable; for the sample points of Figure 3 there was never more than a 1% difference in simulated velocity magnitude, U or W. This suggests that specification of the correct lateral water surface topography is not strictly necessary when using this code, and the water surface profile in idealized bends can be approximated as a uniform downstream slope with no lateral variation.

GEOMETRICAL CONTROLS ON THE EXTENT OF CONCAVE BANK FLOW SEPARATION

Factors controlling the existence and extent of flow separation at the outer bank of a bend were investigated by constructing idealized bend geometries, systematically altering one aspect of the planform and bed geometry at a time and using Fluent to simulate the flow pattern for each configuration. The initial channel configuration for these experiments is loosely based on a prototype bend of Allt Dubhaig, Scotland (close to reach F of Ferguson and Ashworth, 1991) and is illustrated in Figure 5. The channel is 8 m wide, 0·6 m deep in straight sections and turns rightwards through 90° with a centreline curvature $r_c/w = 2$, the value that Hickin (1977) suggested was the threshold for flow separation at the outer bank. The water surface slope is assumed to be 0·001, with lateral variation neglected following the sensitivity test described above. The geometry was constructed by defining curves along the edges of the water surface and base of each bank and

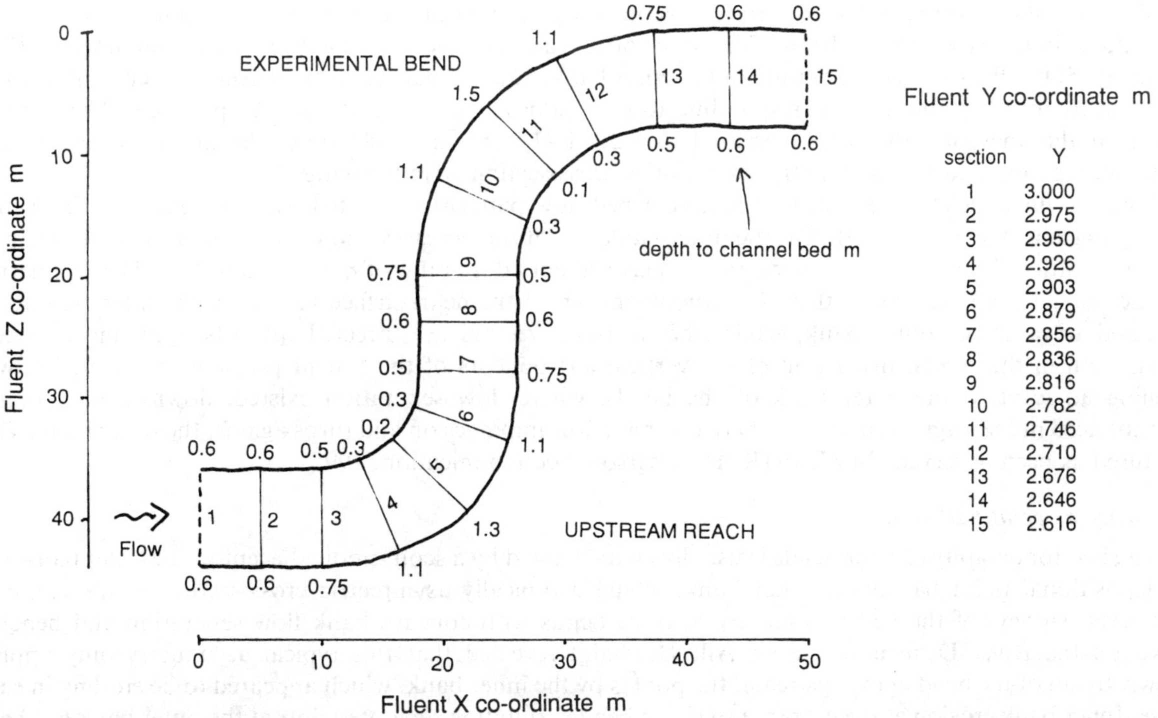

Figure 5. Plan view and bed elevations along the base of each bank of the basic geometry used in experimental investigations of bends with idealized geometry

joining these across the bed and water surface and down each bank. Varying the depth of the bed below the water surface allows a spatially variable bed topography to be incorporated. In the standard bend the bed deepened to 1·5 m at the outer bank and shallowed to 0·1 m at the inner bank to create a typical meander cross-section with maximum cross-sectional asymmetry occurring halfway round the bend, i.e. after a turn of 45° (Figure 5). As mentioned earlier a length of channel was added upstream of the experimental bend to allow the inlet flow to adjust to the boundary conditions. The prototype bend is known from field observations to exhibit flow separation at the outer bank and has a left-hand curving reach immediately upstream. The approach to the model bend was therefore set initially as a 90° left-hand bend, again with asymmetric cross-sections.

The prototype bend has a gravel bed, except in the separation zone where the bed is sandy, and grain size data from the river were used to calculate plausible wall functions and mean inflow velocity. E' was set at a uniform value of 0·00424 for the bed (corresponding to $k_s = 0{\cdot}14$ m, $D_{84} = 0{\cdot}040$ m) with the lower roughness of $E' = 0{\cdot}01$ for the banks. Using equations (3) and (4), the mean inflow velocity was calculated as $\bar{U} = 0{\cdot}67$ m s^{-1}. The computational grid consisted of 105 cells along the reach, 30 cells across the channel width and 5 cells through the flow depth. The dependence of the flow simulations on the choice of computational grid resolution is considered below. As in the field test previously described, the size of individual grid cells varies as the channel deepens and shallows; cell length was 0·7 m along the channel centreline, cell height ranged from 0·12 to 0·3 m and cell width ranged from 0·27 to 0·45 m.

The influence of three aspects of planform and bed geometry was investigated: the planform of the experimental bend itself, the form of the point bar in it and the planform of the bend in the approach channel upstream of the experimental bend.

Influence of outer bank curvature

In order to test Hickin's (1977) suggestion that concave bank flow separation may require channel expansion in the bend afforded by an increase in curvature of the outer bank compared with the inner, four model simulations were conducted using the geometry presented in Figure 5, but with differing outer bank curvatures in the experimental bend. The radius of curvature of the inner bank, r_i, was maintained at 12 m ($r_i/w = 1{\cdot}5$) but the radius of curvature of the outer bank, r_o, was reduced progressively from its initial value of 20 m to 16, 12, 8 and 4 m, corresponding to r_o/w values of 2, 1·5, 1 and 0·5, respectively. The channel width at the apex, w_a, therefore increased to 1·21, 1·41, 1·62 and 1·84 times the inflow width, w. The maximum depth, and the asymmetry, of the apex cross-section were unchanged.

Figure 6 illustrates the near-surface and near-bed flow patterns simulated using Fluent with these four configurations. It can be seen that as the outer bank curvature increases and the channel widens at the apex, flow separation develops at the outer bank at a critical curvature ratio of $r_o/w = 1{\cdot}5$ ($r_o/r_i = 1$) and enlarges as the outer bank sharpens further. All experiments show the near-surface vectors in the main flow to be directed towards the outer bank, whilst the near-bed vectors are directed inwards, implying a helical circulation in the bends. Inspection of the vertical components of the Fluent predictions revealed down-welling adjacent to the outer bank of the bends; where flow separation existed, downwelling became pronounced in the stagnation region where the main flow impinges on and turns against the outer bank. This feature has been observed elsewhere (R. Hey, personal communication, 1995).

Influence of point bar form

The bed topography of river bends is usually characterized by a scour pool adjacent to the outer bank and a depositional point bar at the inner bank, giving a typically asymmetric cross-section to the channel. However, surveys of the bed topography of three bends with concave bank flow separation and benches (two on the River Dean and one on Allt Dubhaig) revealed that this typical asymmetry only applies downstream of the bend apex; upstream, the pool is by the inner bank, which appeared to be eroding in each case. Inner bank erosion at the entrance to sharp bends exhibiting separated flow at the outer bank has been observed in other studies (Page and Nanson, 1982; Andrle, 1994). The influence of the form of the point bar

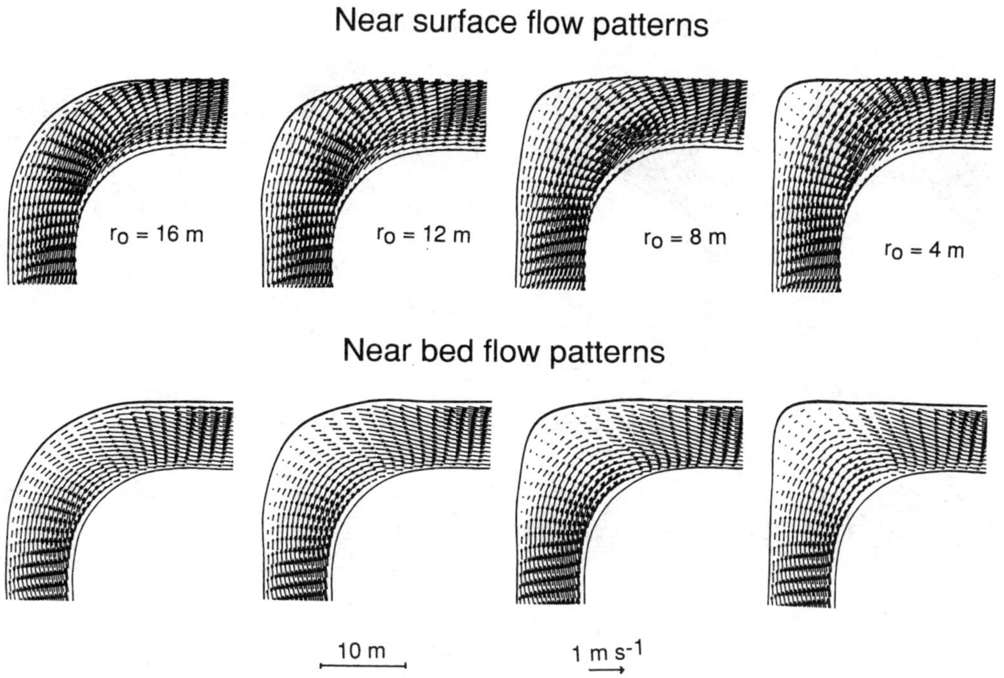

Figure 6. Simulated near-surface and near-bed flow patterns in idealized channels with varying outer bank curvature and apex width; see Figure 5 for upstream planform

topography on the development of flow separation at the concave bank was therefore investigated in a second series of simulations. This was done using the same range of planform configurations as just described, but with the elevation of the base of the inner bank in the upstream part of the experimental bend lowered to 0·6 m below the water surface, instead of 0·5–0·1 m as in Figure 5. Figure 7 illustrates the new bed topography for one planform configuration, and the simulated near-surface and near-bed flow patterns for each; these patterns should be compared with those in Figure 6. It can be seen that lowering the point bar head allows the main current to move towards the inner bank; the separation zone at the outer bank then grows in extent (if present in Figure 6), or develops in the bend in cases where it was not previously present. As in Figure 6 the extent of the separation zone is larger in the wider bends with sharpest outer bank curvature.

Influence of upstream channel configuration

The effect on the development of flow separation of the configuration of the upstream channel was also investigated. Four simulations were conducted. The geometry of the 90° experimental bend was kept constant in each of these experiments (as in the fourth configuration of Figure 6, with the largest region of flow separation; $r_o/w = 0{\cdot}5$, $r_i/w = 1{\cdot}5$). The 90° left-hand bend upstream of the right-hand experimental bend was replaced by a 45° left-hand bend with maximum and minimum pool and pointbar water depths of 1·1 and 0·3 m, respectively. This bend was separated from the experimental bend by a straight reach whose length was 2·25w, 1·63w and w in successive runs. A final run was done with inflow through a 45° right-hand bend followed by a straight reach of length w. Figure 8 illustrates these planform configurations and the near-surface and near-bed flow patterns simulated using Fluent. It can be seen that as the approach bend is moved closer to the experimental bend the zone of flow separation at the concave bank grows in extent, both at the surface and at the bed, although the increase is more pronounced near the surface. In the final simulation, where the approach bend has the same sense of curvature as the experimental bend, all evidence of flow

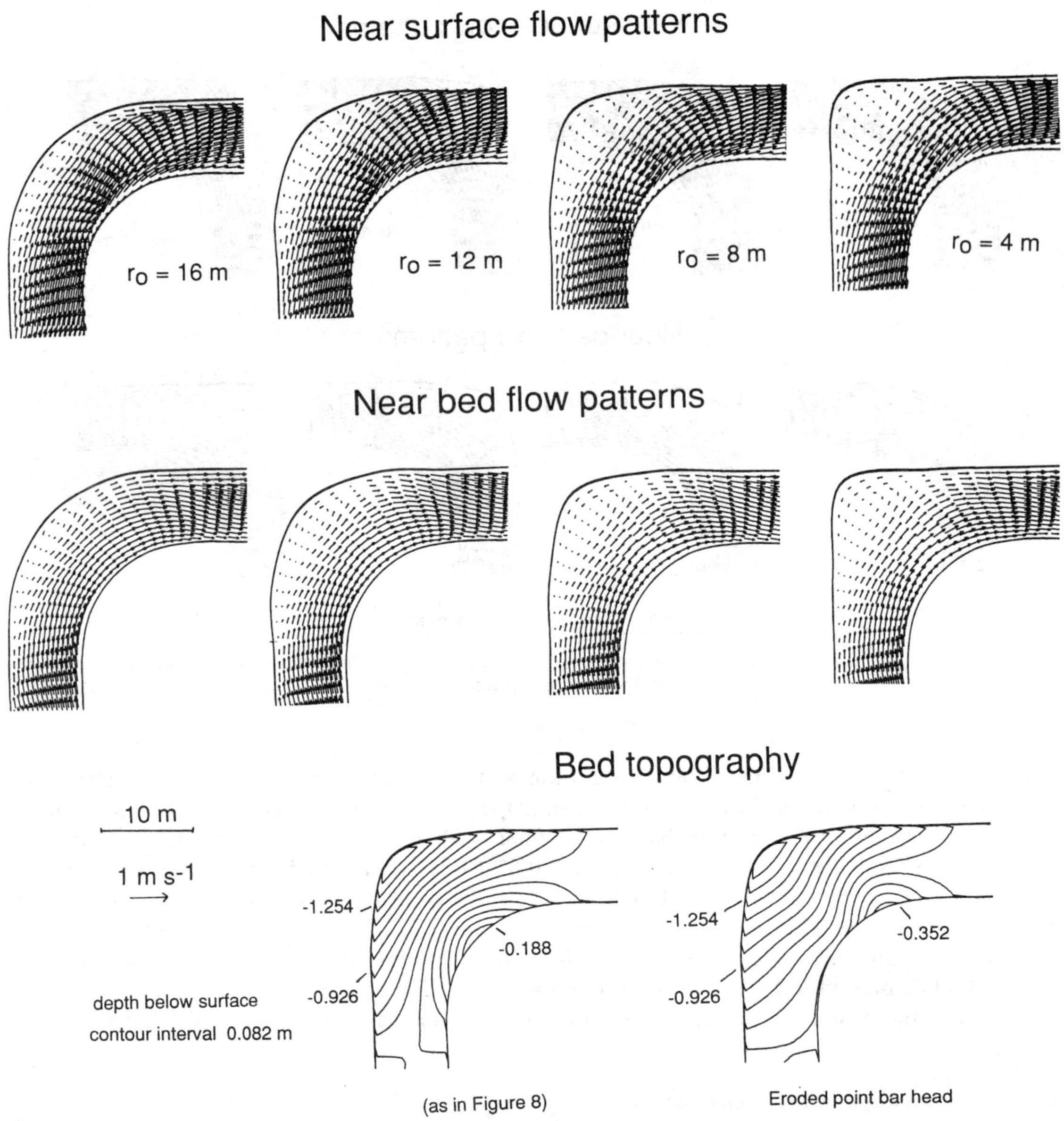

Figure 7. Simulated near-surface and near-bed flow patterns in the idealized channels with bend and approach planform as in Figure 6 but with eroded point bar head; the change in bed topography is illustrated for one of the simulations

separation at the concave bank disappears. These changes are associated with differences in the lateral velocity distribution at the entrance to the right-hand bend, as a consequence of the different secondary circulations developed in the upstream bend according to its geometry.

Grid dependence tests

It has already been shown that the model simulation of a natural bend with flow separation matched field measurements of velocity quite well and was insensitive to the choice of turbulence closure and lateral water surface topography; this insensitivity was assumed to carry over to the simulations of hypothetical bends. One other aspect of model set-up was, however, tested: the possible dependence of flow predictions on the

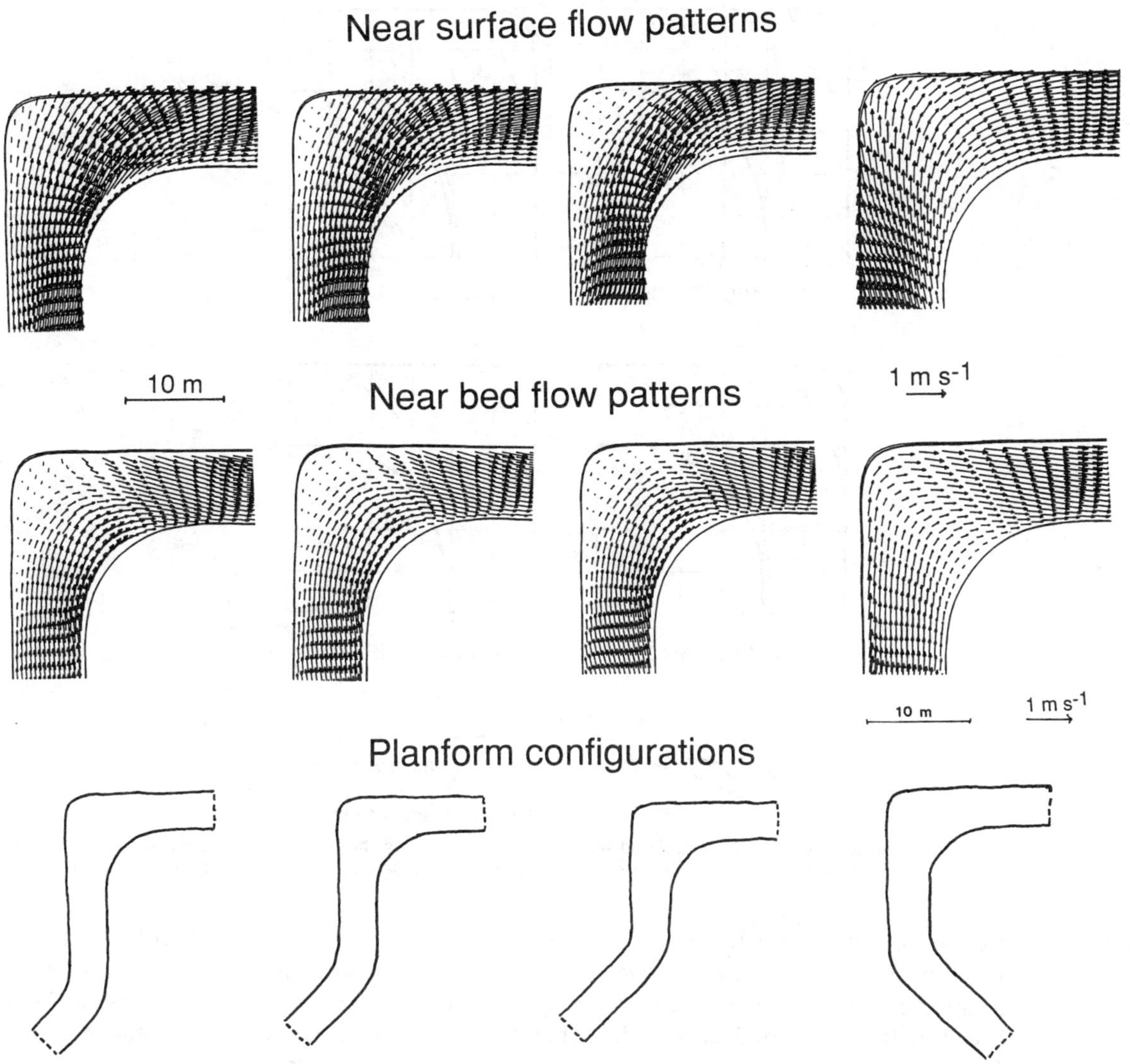

Figure 8. Simulated near-surface and near-bed flow patterns in idealized channels with differing configurations of the upstream channel as shown

resolution of the computational grid used. Using the geometry of run 4 of Figure 6, which exhibited a pronounced region of flow separation, three further simulations were conducted with the number of grid cells increased in one dimension at a time. The number of cells was increased from 105 to 205 along the channel, from 30 to 50 across the width and from 5 to 8 through the flow depth. Eight cells vertically was the maximum number before the model solution began to diverge owing to the problem of roughness-limited grid resolution near the bed (as described for the field test); this maximum vertical resolution is the same as in the successful simulation of the real bend.

Figure 9 shows, in map form, the size, shape and position of the separated region at the outer bank of the bend, with different modifications of the grid resolution; the position of the maximum velocity filament is also shown. This is not a full test of grid independence and may not reveal systematic effects on velocities in other parts of the bend. However, Figure 9 does show that the simulation of the phenomenon of interest is only slightly affected by grid resolution. Small differences in separation zone and maximum velocity filament

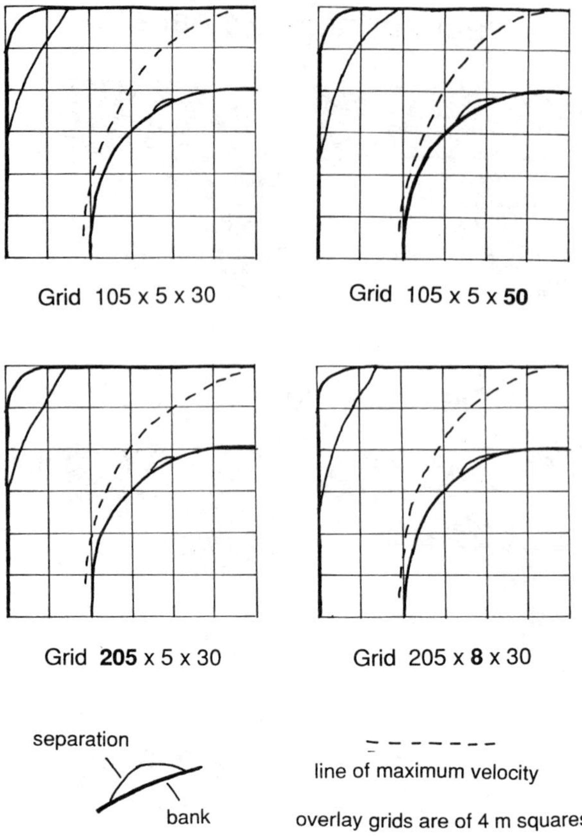

Figure 9. Summary diagrams of near-surface flow simulated with different grids using the bend and approach geometry of Figure 6, $r_o = 4$ m. Diagrams show line of maximum velocity and extent of flow separation in the experimental bend

can be seen on close inspection, but these grid-induced artefacts are minor compared with the differences induced by changes in channel geometry as illustrated in Figures 6–8. Basara and Younis (1995) showed that quantitative flow predictions in open channels may be significantly affected by the use of coarse grids. Their 'coarse' grid (22 cells in 0·2–0·4 m of flow) was finer than those used in this work, and it is shown above that high bed roughness constrains grid resolution, so the quantitative predictions of velocity in our idealized channels are likely to be in error. However, as the main interest is in the extent of the separation zone, which has been shown to be fairly robust to grid specification, the constraints on grid resolution imposed by the treatment of roughness in the Fluent code do not affect the qualitative conclusions of this work.

DISCUSSION

In order to conduct experiments using idealized channel geometries to investigate the controls on the extent of concave bank flow separation it is important to demonstrate that the predictions of the model are reliable. Use of the Fluent code to attempt to predict the flow patterns in a natural river bend with flow separation showed that the qualitative flow patterns can be reproduced very well, although the quantitative predictions were less satisfactory. In addition, it was shown that the choice of turbulence closure in the calculations, and the omission of the surveyed lateral water surface slope, had little effect on the predictions at this site. This means that the assumed water surface topography in the idealized channels consisting of a uniform downstream slope with no cross-channel variation should not invalidate the predictions; reproduction of helical circulations in the idealized channels in the absence of superelevation would appear to support this assertion. It was also shown that within the constraints of the Fluent code the qualitative predictions in an idealized channel with flow separation were not dependent upon the resolution of the computational grid used in the experiments. The results of the idealized experiments are therefore considered reliable.

The idealized channel experiments demonstrate that the extent of concave bank flow separation in a 90° bend is significantly influenced not just by the curvature of this bend but also by the form of the point bar within it and the planform of the upstream channel. The first finding, that flow separation is more extensive in bends that are wide at the bend apex, owing to a very sharply curving outer bank, supports the suggestions of Hickin (1977) and Page and Nanson (1982) that channel expansion is required for flow separation to develop.

In addition to this influence, the second series of idealized channel simulations demonstrates that erosion of the upstream part of the point bar leads to a redistribution of the flow in the bend and a considerable enlargement of the separation zone at the outer bank. Inner bank retreat has been noted in other studies (Lewin, 1978; Page and Nanson, 1982; Andrle, 1994), although these studies do not document changes in the bed topography. From the evidence presented here, it appears to be the modification of the bed topography in this region that allows *extensive* flow separation to develop at the outer bank, whilst the retreat of the bank leads to widening and flow expansion.

Finally, the planform of the channel upstream of the experimental bend was also shown to influence the development and extent of flow separation in the bend of interest, by altering the lateral velocity distribution entering it. In view of the diversity of channel planform observed in meandering streams (e.g. Lewin and Brindle, 1977; Hooke and Harvey, 1983), for any given *bend* planform considerable variability in the development of any flow separation may be expected.

This work has shown the potential for applying a three-dimensional CFD code to natural gravel-bed river bends; previously the only application of a CFD code to a gravel-bed river of this scale was of a two-dimensional model (Lane *et al.*, 1995). There is considerable scope for refinement and development of this first exploratory study. For example, the sources of error in the predictions at the field site should be identified, a starting point being the problem of roughness specification and grid resolution identified earlier where relative roughness is high. Grid dependence should be tested more thoroughly in a simulation of a bend with much lower relative roughness. The results of the CFD experiments presented here for hypothetical bends could be compared with field observations of the extent of flow separation in bends with different geometrical characteristics, or tested using physical models in a laboratory. The experiments could also be tailored to actual field sites to allow numerical comparison with flow measurements and investigation of the implications of envisaged channel change for flow structures at a later stage of bend evolution.

ACKNOWLEDGEMENTS

This work was undertaken whilst A.H. was in receipt of NERC studentship GT4/93/27P with additional financial support from the University of Sheffield. We thank Stuart Lane and three anonymous referees for helpful and constructive comments on the presentation of the results in earlier versions of this paper.

REFERENCES

Andrle, R. 1994. 'Flow structure and development of circular meander pools', *Geomorphology*, **9**, 261–270.

ASCE, 1988. 'Turbulence modeling of surface water flow and transport: Part I–IV', *J. Hydraul. Engng*, **114**, 970–1073.

Basara, B. and Younis, B. A. 1995. 'Prediction of turbulent flows in dredged trenches', *J. Hydraul. Res.*, **33**, 813–824.

Bradbrook, K. F., Biron, P. M., Lane, S. N., Richards, K. S., and Roy, A. G. 1998. 'Investigation of controls on secondary circulation in a simple confluence geometry using a three-dimensional numerical model', *Hydrol. Process.*, **12**, 1371–1396 (this issue).

Bray, D. I. 1982. 'Flow resistance in gravel bed rivers', in Hey, R. D., Bathurst, J. C., and Thorne, C. R. (Eds), *Gravel-bed Rivers. Fluvial Processes, Engineering and Management*. Wiley, Chichester. pp. 109–138.

Brice, J. C. 1974. 'Evolution of meander loops', *Geol. Soc. Am. Bull.*, **85**, 581–586.

Carey, W. C. 1969. 'Formation of flood plain lands', *ASCE J. Hydraul. Div.*, **95**, 981–994.

Carling, P. A. 1991. 'An appraisal of the velocity-reversal hypothesis for stable pool–riffle sequences in the River Severn, England', *Earth Surf. Process. Landf.*, **16**, 19–31.

Clifford, N. J. 1993. 'Formation of riffle–pool sequences: field evidence for an autogenic process', *Sediment. Geol.*, **85**, 39–51.

Demuren, A. D., 1993. 'A numerical model for flow in meandering channels with natural bed topography', *Wat. Resour. Res.*, **29**, 1269–1277.

Ferguson, R. I. and Ashworth, P. J. 1991. 'Slope-induced changes in channel character along a gravel-bed stream: the Allt Dubhaig, Scotland', *Earth Surf. Process. Landf.*, **16**, 65–82.

Fluent Inc., 1993a. *FLUENT User's Guide*. Fluent, Lebanon, New Hampshire. Distributor: Fluent Europe.

Fluent Inc., 1993b. *PreBFC User's Guide*. Fluent, Lebanon, New Hampshire. Distributor: Fluent Europe.

Hattingh, J. and Rust, I. C. 1993. 'Flood transport and deposition of tracer heavy minerals in a gravel-bed meander bend channel', *J. Sediment. Petrol.*, **63**, 828–834.

Hickin, E. J. 1977. 'Hydraulic factors controlling channel migration', in Davidson-Arnott, R. and Nickling, W. (Eds), *Research in Fluvial Systems, Proceedings of the 5th Guelph Geomorphology Symposium*. Geobooks, Norwich. pp. 59–72.

Hickin, E. J. 1979. 'Concave-bank benches on the Squamish River, British Columbia', *Can. J. Earth Sci.*, **16**, 200–203.

Hickin, E. J. 1986. 'Concave-bank benches in the flood plains of Muskwa and Fort Nelson Rivers, British Columbia', *Can. Geogr.*, **30**, 111–122.

Hooke, J. M. and Harvey, A. M. 1983. 'Meander changes in relation to bend morphology and secondary flows', in Collinson, J. D. and Lewin, J. (Eds), *Modern and Ancient Fluvial Systems*, International Association of Sedimentologists Spec. Publ. 6. pp. 121–132.

Knighton, A. D. 1984. *Fluvial Forms and Processes*. Edward Arnold, London. 218 pp.

Lane, S. N. 1998. 'Hydraulic modelling in hydrology and geomorphology: a review of high resolution approaches', *Hydrol. Process.*, **12**, 1279–1298 (this issue).

Lane, S. N., Richards, K. S., and Chandler, J. H. 1995. 'Within-reach spatial patterns of process and channel adjustment', in Hickin, E. J. (Ed.), *River Geomorphology*. Wiley, Chichester. pp. 105–130.

Leeder, M. R. and Bridges, P. H. 1975. 'Flow separation in meander bends', *Nature*, **253**, 338–339.

Leopold, L. B., Wolman, M. G., and Miller, J. P. 1964. *Fluvial Processes in Geomorphology*. Freeman, San Francisco. 522 pp.

Lewin, J. 1978. 'Meander development and floodplain sedimentation: a case study from mid-Wales', *Geol. J.*, **13**, 25–36.

Lewin, J. and Brindle, B. J. 1977. 'Confined meanders', in Gregory, K. J. (Ed.), *River Channel Changes*. Wiley, Chichester. pp. 221–233.

Nezu, I. and Nakagawa, H. 1993. *Turbulence in Open-Channel Flows*, IAHR Monograph Series. Balkema, Rotterdam. 281 pp.

Page, K. and Nanson, G. 1982. 'Concave-bank benches and associated floodplain formation', *Earth Surf. Process. Landf.*, **7**, 529–543.

Reid, J. B. 1984. 'Artificially induced concave bank deposition as a means of floodplain erosion control', in Elliot, M. (Ed.), *River Meandering. Proceedings of the Conference Rivers '83*, ASCE. pp. 295–304.

Rodi, W. 1993. *Turbulence Models and their Application in Hydraulics. A State-of-the Art Review*, 3rd edn, IAHR Monograph Series. Balkema, Rotterdam. 104 pp.

Rubin, D. M., Schmidt, J. C., and Moore, J. N. 1990. 'Origin, structure and evolution of a reattachment bar, Colorado River, Grand Canyon, Arizona', *J. Sediment. Petrol.*, **60**, 982–991.

Schmidt, J. C. 1990. 'Recirculating flow and sedimentation in the Colorado River in Grand Canyon, Arizona', *J. Geol.*, **98**, 709–724.

Taylor, G., Crook, K. A. W., and Woodyer, K. D. 1971. 'Upstream-dipping foreset cross stratification: origin implications for paleoslope analysis', *J. Sediment. Petrol.*, **41**, 578–581.

Woodyer, K. D. 1975. 'Concave bank benches on the Barwon River, N.S.W.', *Austr. Geogr.*, **13**, 36–40.

Woodyer, K. D., Taylor, G., and Crook, K. A. W. 1979. 'Depositional processes along a very low-gradient, suspended-load stream: the Barwon River, New South Wales', *Sediment. Geol.*, **22**, 97–120.

Zhou, Y., Vahala, G., and Thangam, S. 1994. 'Development of a turbulence model based on recursion renormalization group theory', *Phys. Rev. E*, **49**, 5195–5206.

7

INVESTIGATION OF CONTROLS ON SECONDARY CIRCULATION IN A SIMPLE CONFLUENCE GEOMETRY USING A THREE-DIMENSIONAL NUMERICAL MODEL

K. F. BRADBROOK,[1]* P. M. BIRON,[2] S. N. LANE,[1] K. S. RICHARDS[1] AND A. G. ROY[2]

[1]*Department of Geography, University of Cambridge, Downing Place, Cambridge CB2 3EN, UK*
[2]*Département de Géographie, Université de Montréal, CP 6128, succ. Centre-Ville, Montréal, Québec H3C 3J7, Canada*

ABSTRACT

Recent research into river channel confluences has identified confluence geometry, and particularly bed discordance, as a control on confluence flow structures and mixing processes, and this has been illustrated using both field measurements in natural confluences and laboratory measurements of simplified confluences. Generalization of the results obtained from these experiments is limited by the number of confluence geometries that can be examined in a reasonable amount of time. This limitation may be overcome by numerical models, in which confluence geometry is more readily varied, and data acquired more rapidly. This paper aims to: (i) validate the application of a three-dimensional numerical model to a simple confluence geometry; (ii) simulate the effects of different boundary condition values upon flow structures; and (iii) interpret the implications of these simulations for river channel confluence dynamics. The model used in this research solves the three-dimensional form of the Navier–Stokes equations and is used to simulate the flow in a parallel confluence of unequal depth channels and to investigate the effect of different combinations of velocity and depth ratio between the two tributaries. The results generally agree with empirical evidence that secondary circulation is generated in the absence of streamline curvature, but only for specific combinations of depth and velocity ratio. This research shows how understanding of the interaction of these controls is enhanced if pressure gradients are considered. The velocity ratio is the prime determinant of the cross-stream pressure gradient that initiates cross-stream velocities. However, for significant secondary circulation to form, cross-stream velocities must lead to significant transfer of fluid in the cross-stream direction. This depends on the vertical extent of the cross-stream pressure gradient which is controlled by the depth ratio. In this study, strong secondary circulation occurred for a depth differential of 25% or more, as long as the velocity in the shallower tributary was at least as great as that in the deeper channel. This provides an important context for interpretation of previous work and for the design of new experiments in both the field and the laboratory.

KEY WORDS river channel confluence; flow structure; secondary circulation; three-dimensional numerical modelling; bed discordance

---

* Correspondence to: K. F. Bradbrook, Department of Geography, University of Cambridge, Downing Place, Cambridge CB2 3EN, UK

Contract grant sponsor: NERC.
Contract grant numbers: GR3/09715, GT4/94/314/a.

INTRODUCTION

River confluences are important nodes in both dendritic (Richards, 1980; Best, 1987, 1988; Rhoads, 1996) and braided (Ashmore, 1993; Bridge, 1993) river systems. The mixing processes and flow structures associated with such confluences have important implications for sediment transport (Best, 1987, 1988; Rhoads and Kenworthy, 1995; Rhoads, 1996) and subsequent morphological change (Biron *et al.*, 1993a), both in the vicinity of the confluence and further downstream, as well as for solute and pollutant mixing processes (Gaudet and Roy, 1995).

Previous field and laboratory investigations have concentrated on examining flow structures consequent from two main controls: (i) planform streamline curvature, which depends on junction angle and discharge ratio and tends to promote 'back to back' helical cells that maintain segregated sediment transport patterns (e.g. Mosley, 1976; Best, 1987, 1988; Ashmore *et al.*, 1992; Rhoads and Kenworthy, 1995; Rhoads, 1996); and (ii) vertical separation owing to bed discordance at the junction, which can lead to shear layer distortion and thus enhanced mixing (e.g. Best and Roy, 1991; Gaudet and Roy, 1995; Biron *et al.*, 1996a,b). There is a wide range of variables associated with these geometries and flow structures, and laboratory investigations have been used to provide better experimental design to assess their effects than is possible in the field. However, even though laboratory experiments allow researchers to test systematically the effects of the main controls on flow structure, this is time-consuming and there remain many possible combinations of boundary condition values that need consideration (Table I). The limited number of experimental set-ups that have been examined leaves large gaps in our knowledge and impinges on the intercomparison of different types of experiment. For instance, Mosley (1976) and Best (1988) investigated systematically the effect of variation of junction angle and discharge ratio on a single morphological variable, confluence scour hole depth, and inferred the flow structures using dye tracing. The experiments conducted by Best and Roy (1991), Biron *et al.* (1996a,b) and McLelland *et al.* (1996) obtained detailed velocity measurements, but the time required to obtain data of sufficient density and quality in the laboratory flume is such that only a small number of boundary condition values could be considered. Best and Roy (1991) considered a single depth ratio with discordance and a single momentum ratio. Biron *et al.* (1996a,b) compared results from an experiment with a discordant depth ratio with those from experiments with a concordant depth ratio for a given momentum ratio and confluence angle. McLelland *et al.* (1996) considered the confluence of parallel channels with angled, concordant, steps for a single momentum ratio, analogous to a particular field situation. Although these experiments allow useful assessment of the flow structures that develop under particular combinations of boundary condition values, they do not allow assessment of the relative effects of different controls upon the development of flow structures (e.g. strength of upwelling, intensity of secondary circulation). Given the importance of flow structures for flow entrainment, and hence mixing processes, erosion and deposition (Biron *et al.*, 1993a), systematic investigation of the effects and interactions of bed discordance, depth ratio, momentum ratio and confluence angle is required, but is precluded by the time required to undertake flume investigations.

This paper shows that numerical modelling provides an alternative method with which to investigate the relative importance of these key controls. The possibility of applying numerical models to river channel flow problems has been recognized, both for investigating flow patterns (e.g. Bridge and Gabel, 1992; Lane *et al.*, 1995) and for simulating the evolution of channel topography once the model is coupled with sediment transport equations (Andrews and Nelson, 1989; Shimizu and Itakura, 1989). The development and complexity of such models has improved in recent years, especially in the area of turbulence modelling (see Lane, 1998, this issue), but most applications remain two-dimensional. This may be appropriate for reach-scale problems, but such models are unable to represent accurately the complex secondary flow structures associated with river channel confluences (Lane and Richards, 1998, this issue). Any detailed investigation of the form of secondary flow structures produced at confluences requires the use of fully three-dimensional numerical models, such as that used by Weerakoon *et al.* (1991), if only to suggest where two-dimensional models are most inappropriate and hence where most model development is needed. The study by

Table I. Summary of previous work on laboratory-style confluences. (Depth and velocity ratios are defined by the value in the shallower tributary divided by the value in the deeper tributary unless otherwise stated)

Author(s)	Geometry	Method	Specific investigation	Results	Key flow characteristics identified
Mosley (1976)	Symmetrical, 'Y'-shaped	Laboratory: self-formed in sand	Various junction angles and discharge ratios	Scour hold depth increased with both variables	Dye showed two helical cells
Best (1988)	Asymmetrical, '⊥'-shaped	Laboratory: self-formed in sand	Various junction angles and discharge ratios	Scour hole depth increased with angle and decreased with discharge ratio	Flow separation over avalanche faces
Best and Reid (1984)	Asymmetrical, '⊥'-shaped	Laboratory: rectangular cross-section; fixed bed	Various junction angles and discharge ratios	Size of lateral recirculation zone increased with both variables	The recirculation zone decreased effective cross-sectional area and led to flow acceleration
Weerakoon and Tamai (1989)	Asymmetrical, 60° angle	Numerical simulation: rectangular cross-section; fixed bed	Effect of 'training wall' delaying mixing of the two flows	Comparison with experimental results is satisfactory	Strong helical cell in the tributary, weak cell in the main channel
Weerakoon *et al.* (1991)	Asymmetrical, 60° angle	Numerical simulation: rectangular cross-section; fixed bed	'Training wall' removed	Experimental comparison good, but size of lateral separation zone underpredicted	Tributary cell rapidly dominates over cell in main channel and spreads across channel
Best and Roy (1991)	Parallel channels of unequal depths	Laboratory: rectangular cross-section; fixed bed	Effect of bed discordance; depth ratio of 0·5, velocity ratio of 1·0	Shear layer distortion towards the shallower channel leads to enhanced mixing and cross-stream entrainment of fluid	Kelvin–Helmholtz instabilities in the mixing layer are distorted at the base leading to subsequent break-up and upwelling
Biron *et al.* (1996a,b)	Asymmetrical, 30° angle	Laboratory: rectangular cross-section; fixed bed	Effect of bed discordance; depth ratio of 0·81; velocity ratio of 1·6	Bed discordance led to intensification of secondary circulation	Vertical and lateral separation zones formed; upwelling also occurred along the wall of the shallower channel
McLelland *et al.* (1996)	Parallel channels with avalanche faces at 45° to flow direction	Laboratory: rectangular cross-section; fixed bed	To simulate effect of avalanche faces into a scour hole; depth ratio of 0·5 (tributary to post-confluence depth); tributary velocity ratio of 1·9	Two counter-rotating cells form, in the absence of channel curvature	Flow separation over the avalanche faces is the primary control in this case

Weerakoon *et al.* (1991) showed the ability of such models to reproduce effectively the secondary circulation patterns associated with planform curvature.

However, research has yet to use numerical models to investigate systematically the effects of different values of confluence boundary conditions upon flow structures and so address uncertainties consequent from the finite number of boundary condition values assessed in laboratory situations. The present paper aims to validate the use of a numerical model, similar to that used by Weerakoon *et al.* (1991), to simulate the effect of vertical separation and mixing layer distortion at a confluence with bed discordance, by comparing the numerical results with a laboratory experiment. It then aims to move beyond using a numerical model to replicate field or laboratory observations by demonstrating the potential of such models for exploratory analysis to investigate the effect of different combinations of boundary condition values on flow structures. The experiments presented in this paper concentrate on examining the relative effects of changing depth and velocity ratios between the two tributaries at the confluence of unequal depth parallel streams. Although using a simple geometry, it represents a first attempt to quantify the relative importance of different controls on secondary circulation and mixing processes at river confluences, and the understanding gained from such investigations can aid interpretation of flow structures at more complex geometries.

THE NUMERICAL MODEL

Basic equations and solution. The model used solves the three-dimensional elliptic form of the Reynolds-averaged Navier–Stokes equations for the case of a steady-state flow. In Cartesian coordinates these involve the continuity equation

$$\frac{\partial u}{\partial x}+\frac{\partial v}{\partial y}+\frac{\partial w}{\partial z}=0, \tag{1}$$

and the momentum equations

$$\begin{aligned}
&x\text{-direction}: u\frac{\partial u}{\partial x}+v\frac{\partial u}{\partial y}+w\frac{\partial u}{\partial z}=-\frac{1}{\rho}\frac{\partial p}{\partial x}+\frac{1}{\rho}\left(\frac{\partial \tau_{ii}}{\partial x}+\frac{\partial \tau_{ij}}{\partial y}+\frac{\partial \tau_{ik}}{\partial z}\right)\\
&y\text{-direction}: u\frac{\partial v}{\partial x}+v\frac{\partial v}{\partial y}+w\frac{\partial v}{\partial z}=-\frac{1}{\rho}\frac{\partial p}{\partial y}+\frac{1}{\rho}\left(\frac{\partial \tau_{ij}}{\partial x}+\frac{\partial \tau_{jj}}{\partial y}+\frac{\partial \tau_{jk}}{\partial z}\right)\\
&z\text{-direction}: u\frac{\partial w}{\partial x}+x\frac{\partial w}{\partial y}+w\frac{\partial w}{\partial z}=-\frac{1}{\rho}\frac{\partial p}{\partial z}+\frac{1}{\rho}\left(\frac{\partial \tau_{ik}}{\partial x}+\frac{\partial \tau_{jk}}{\partial y}+\frac{\partial \tau_{kk}}{\partial z}\right)
\end{aligned} \tag{2}$$

where u, v and w are the time-averaged Cartesian velocity components in the x, y and z directions, respectively, ρ is the density, p is the pressure,

$$\tau_{ij}=\nu\left(\frac{\partial u_j}{\partial x_i}+\frac{\partial u_i}{\partial x_j}\right)-\overline{u_i'u_j'},$$

ν is the kinematic viscosity and $\overline{u_i'u_j'}$ are the time-averaged turbulent Reynolds shear stresses (see Lane, 1998, this issue) where i and j are standard tensor notation indicating two out of the three orthogonal coordinate directions.

A finite volume method is used to discretize the equations, with a cuboid grid in the Cartesian frame. The interpolation scheme used is hybrid upwind, where upwind differences are used in high convection areas (Peclet number >2) and central differences where diffusion dominates (Peclet number <2). This scheme is very stable although it can suffer from numerical diffusion. However, a comparison with other schemes for simulation of flow over a backward-facing step suggests that the errors owing to the interpolation scheme

Table II. Comparison of different interpolation schemes for a backward-facing step. (H is the step height, X_1 the length of the primary recirculation zone, and X_2 and X_3 the upper and lower bounds of a secondary separation cell on the upper wall of the channel. The same basic CFD code was used with a grid mesh of 32 × 200 cells, but Waterson performed no grid refinement studies) (from Waterson, unpublished)

$Re = 150$	Data	HDS	Cubic-upwind	van Albada non-linear
X_1/H	4·2	4·17	4·24	4·25
$Re = 450$				
X_1/H	9·5	8·79	9·13	9·06
X_2/H	7·6	7·66	8·26	8·09
X_3/H	11·3	11·14	11·39	11·52
DX/H	3·7	3·48	3·13	3·43

used are not likely to be significant (Table II). Coupling of the pressure and momentum equations is achieved using SIMPLEST, a variation on the SIMPLE algorithm of Pantankar and Spalding (1972).

The turbulence model. The turbulent Reynolds stress terms in Equation (2) cannot be calculated directly and require an additional model. This approach takes the two-equation k–ε model (see Lane, 1998, this issue) modified using renormalization group theory (Yakhot *et al.*, 1992) as this has been shown to perform better than the standard k–ε model for flows with zones of separation and reattachment (Yakhot *et al.*, 1992; Lien and Leschziner, 1994). In the standard k–ε model, the eddy viscosity [Equation (11) in Lane, 1998, this issue] is related to the turbulent kinetic energy, k, and the turbulent dissipation, ε, via the Kolomogorov–Prandtl expression [Equation (12) in Lane, 1998, this issue, but with the length scale here expressed as: $k^{3/2}/\varepsilon$]

$$\nu_t = c'_\mu \frac{k^2}{\varepsilon} \tag{3}$$

Transport equations can then be derived for turbulent kinetic energy and turbulent dissipation (Launder and Spalding, 1974)

$$u_i \frac{\partial k}{\partial x_i} = \frac{\partial}{\partial x_i}\left(\frac{\nu_t}{\sigma_k}\frac{\partial k}{\partial x_i}\right) - \nu_t\left(\frac{\partial u_i}{\partial x_j} + \frac{\partial u_j}{\partial x_i}\right)\frac{\partial u_i}{\partial x_j} - \varepsilon$$

convection owing to transport by mean flow; diffusion into mean flow; production through shear ($=P$); viscous dissipation rate (4a)

$$u_i \frac{\partial \varepsilon}{\partial x_i} = \frac{\partial}{\partial x_i}\left(\frac{\nu_t}{\sigma_e}\frac{\partial \varepsilon}{\partial x_i}\right) + c_{1\varepsilon}\frac{\varepsilon}{k}P - c_{2\varepsilon}\frac{\varepsilon^2}{k}$$

convection; diffusion; generation–destruction (P = production from above) (4b)

The resulting two-equation k–ε model includes five empirical constants: c'_μ, $c_{1\varepsilon}$, $c_{2\varepsilon}$, σ_k and σ_ε. Values for these constants, which have been determined experimentally (Launder and Spalding, 1974), are usually taken as general (Table III).

Table III. Turbulence model constants

	c'_μ	σ_k	σ_ε	$c_{1\varepsilon}$	$c_{2\varepsilon}$
Standard k–ε	0·09	1·0	1·3	1·44	1·92
RNG	0·0845	0·7194	0·7194	1·42	1·68

The standard k–ε model has been shown to perform poorly in flows with significant mean strain, such as occurs in separation zones, and results in an underprediction in separation zone length (Lien and Leschziner, 1994). The turbulence model developed by Yakhot and Orszag (1986), using renormalization group (RNG) theory, results in a model similar to the standard k–ε model, but with different values for the constants, and an extra production term for ε, which was added in a modification proposed by Yakhot *et al.* (1992). These differences result in greater dissipation of turbulence in areas of strong strain, and therefore lead to a reduction in eddy viscosity which improves velocity predictions, for example, by increasing the separation zone length. RNG theory was used to derive the basic turbulence model by systematically removing the smallest scales of turbulence and calculating their effect on the remaining flow scales (Yakhot and Orszag, 1986). A random force (white noise, zero mean) was introduced into the Navier–Stokes equations to drive the turbulent motion. These equations were then averaged over an infinitesimal band of the smallest turbulence scales and the averaging was systematically repeated for successively larger scales until the effects added up to a finite change in viscosity, which was retained. Once the largest scales were removed, the form of the standard k–ε model was obtained with the advantage that the values of the constants were calculated explicitly (Table III). Smith and Reynolds (1992) identified problems with the original derivation of the equation for turbulent dissipation, ε, but the reformulated derivation (Yakhot and Smith, 1992) introduced an additional production term in the equation for ε for which systematic closure was not possible. Yakhot *et al.* (1992) developed a model for this extra term which represented the effect of high strain rates on increasing turbulent dissipation and therefore decreasing viscosity. The constant c_{1e} in the existing term for production of ε was also re-evaluated, and it is this value that is given in Table III. In the existing term, which is the same as that used in the standard k–ε model, the rate of generation of ε is directly related to the rate of production of kinetic energy, P. This assumes spectral equilibrium, in which generation at large turbulent scales is immediately dissipated at smaller turbulent scales. Where strain rates are high, such as at separation and reattachment points, there is a large difference between the scales at which turbulence is generated and dissipated, and there may be an appreciable time lag between generation and dissipation.

In the modified RNG model (Yakhot *et al.*, 1992) the extra production term in the production rate of ε is evaluated using

$$-\rho\alpha\frac{\varepsilon^2}{k} \tag{5}$$

where the parameter α is defined as follows

$$\alpha = c'_\mu\eta^3\frac{(1-\eta/\eta_0)}{(1+\beta\eta^3)} \tag{6}$$

with $\eta_0 = 4{\cdot}8$ and $\beta = 0{\cdot}012$, in which η is the ratio of the time-scale of turbulent strain to that of mean strain, S

$$\eta = S\frac{k}{\varepsilon} \tag{7}$$

where

$$S = \sqrt{2S_{ij}^2} \tag{8a}$$

and

$$S_{ij} = 0{\cdot}5\left(\frac{\partial u_i}{\partial x_j} + \frac{\partial u_j}{\partial x_i}\right) \tag{8b}$$

In Equation (6), η_0 is the fixed point for homogeneously strained turbulent flows and β is evaluated to yield a von Karman constant of 0·4. When the mean strain is weak ($\eta \to 0$), the parameter α, and therefore the extra production term for ε, are small, but when the mean strain rate is strong (large η), the extra production term leads to an increase in turbulent dissipation. This results in a decrease in eddy viscosity and therefore of extraction of momentum from the mean flow. This allows a recirculation zone, for example, to reach a size more comparable with laboratory measurements.

Boundary conditions: wall functions. Since the cell spacing near the solid boundaries is not sufficient to resolve the laminar sublayer, it is necessary to provide a condition for the velocity and turbulence parameters in the cell adjacent to the wall. It is common to use the 'universal' law of the wall based on a shear velocity [Equations (9)–(11)]. However, this is based upon assumptions of a local equilibrium of turbulence that does not hold under conditions of flow separation. For this application, therefore, a non-equilibrium version of the law of the wall is used (Launder and Spalding, 1974). This takes the square root of turbulent kinetic energy, rather than the shear velocity, as the characteristic velocity scale.

The 'universal' law of the wall is

$$\frac{U_{\mathrm{r}}}{U_{\tau}} = \frac{\ln(EY^{+})}{\kappa} \tag{9}$$

with $U_{\tau} = \sqrt{\tau/\rho}$ and $Y^{+} = U_{\tau}Y/\mu$, where U_{r} is the velocity parallel to the wall, U_{τ} is the shear velocity, τ is the bed shear stress, E is the roughness parameter (which in this case is 9·0 for hydraulically smooth walls), Y^{+} is the non-dimensional wall distance, Y is the normal distance to the wall, μ is the laminar viscosity and κ is the von Karman constant (=0·4). Assuming local equilibrium of turbulence, the shear velocity can be related to the local kinetic energy using (Launder and Spalding, 1974)

$$k = \frac{U_{\tau}^{2}}{\sqrt{c'_{\mu}}} \tag{10}$$

The boundary condition for ε is given by

$$\varepsilon = (c'_{\mu})^{3/4}\frac{k^{3/2}}{\kappa Y} \tag{11}$$

However, under conditions of separation, the assumption of local equilibrium [Equation (10)] does not hold, and rather than using U_{τ} as the characteristic velocity scale, the non-equilibrium wall function uses $\sqrt{k}$, and is written as (Launder and Spalding, 1974)

$$\frac{U_{\mathrm{r}} \cdot \sqrt{k}}{U_{\tau}^{2}} = \frac{\ln(0{\cdot}25c'_{\mu}E\sqrt{k}Y)}{0{\cdot}25c'_{\mu}\kappa} \tag{12}$$

where the value of k is taken from the transport Equation (4a) with the diffusion of energy to the wall set to 0. The production term in this equation can be expressed in terms of the shear velocity

$$P = \frac{U_{\tau}^{2} \cdot U_{\mathrm{r}}}{2Y} \tag{13}$$

and the boundary condition for ε is

$$\varepsilon = \frac{(c'_{\mu})^{3/4}k^{3/2}\ \ln(0{\cdot}25c'_{\mu}E\sqrt{k}Y/\mu)}{2kY} \tag{14}$$

If the condition for local equilibrium applies, then Equation (14) reduces to Equation (11) and $P/\varepsilon = 1$.

Boundary conditions: free surface. As surface elevation and depression is a feature of confluence flow dynamics (Best and Reid, 1984; Weerakoon and Tamai, 1989; Bridge, 1993; Rhoads, 1996), the model needs a free surface treatment. Defining the exact contours of the free surface elevation is problematic, and most previous work has used a 'rigid lid' assumption, which prescribes a surface that is usually planar. At the surface, all normal derivatives and velocity resolutes are zero, as for a symmetry plane. The pressure on the surface is not set at zero, but represents the variation in water depth that would occur if the surface were not fixed. Thus, a pressure greater than zero represents superelevation, and a pressure less than zero represents surface depression. This means that the variation in water surface position is accounted for in the pressure gradient term in the momentum equations, but it does not enter the mass continuity equation and this can lead to higher velocity predictions in places where superelevation of the water occurs (Weerakoon and Tamai, 1989), and vice versa for zones of surface depression. At confluences, where streamline curvature may play an important role, significant superelevation of the water surface can occur and representing the effects of spatial variation of water surface elevation is necessary for confluence flow modelling. In the present application, surface depression at the mixing layer is likely, and a method of approximating the free surface displacement is also required.

In this application, this is achieved by representing the deviation from the prescribed planar surface by variable porosities in the surface layer of cells (Spalding, 1985). A porosity is defined for each cell in the top layer of fluid, with the flux across any cell face equal to the porosity multiplied by the area of the face and the velocity component perpendicular to it. The perturbation of the porosity from 1·0 thus represents an effective change in height of the cell. The porosity is calculated so that the pressure deviation is satisfied by this change in height

$$\text{porosity} = 1 + \frac{p}{\rho g h_c} \tag{15}$$

where h_c is the height of the cell. Thus, the superelevation of the water that would occur where the pressure is greater than zero is represented by a volumetric porosity greater than 1·0, and surface depression by a porosity of less than 1·0. This allows an appropriate discharge to pass through the cell such that free surface effects are represented.

Boundary conditions: upstream and downstream. The velocity distribution at the furthest upstream cross-section must be defined. A uniform velocity could have been prescribed, but this would require a large number of cells upstream of the junction to ensure that the flow was fully developed at the junction with respect to the effect of friction at the walls and the bed. To avoid the extra computation that this would require, a separate model was used to calculate the velocity distribution for the upstream cross-sections. The dimensions of the cross-section and the mass flow rate required were defined for each tributary channel, and the fully developed flow for an infinite straight channel of this cross-section were calculated. For this model, the standard k–ε turbulence model [Equations (3) and (4)] and the equilibrium law of the wall [Equations (9)–(11)] were used. The mean axial pressure gradient required to generate the desired mass flow rate is calculated from overall continuity considerations (Pantankar and Spalding, 1972) using

$$\frac{\mathrm{d}p}{\mathrm{d}x} = 0{\cdot}5\,\frac{f\rho(u_{\text{in}})^2}{D} \tag{16}$$

with $f = 1/(1{\cdot}82\log_{10}(Re) + 1{\cdot}64)^2$ for smooth walls, where u_{in} is the cross-sectional average velocity, f is a friction factor and D is the hydraulic radius of the cross-section. The calculated distribution of downstream velocity, u, and the turbulence parameters, k and ε, provided the upstream boundary conditions for each tributary channel.

A boundary condition is also required at the furthest downstream cross-section, for which the pressure is fixed at zero. This means that all other pressure values are defined relative to this, and upstream velocities could occur at this cross-section if negative pressure zones were to extend that far downstream. This is undesirable and was used to indicate that the solution domain should be extended even further downstream. It was also deemed important that secondary flows should be small by the end of the solution domain.

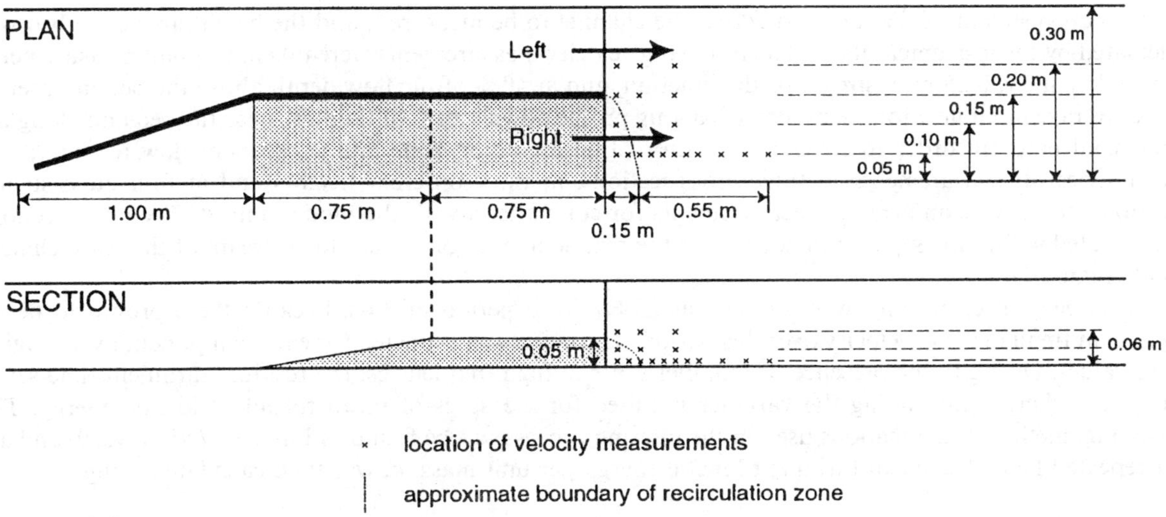

Figure 1. Experimental set-up in the flume with the location of velocity measurements indicated

MODEL VALIDATION

Methods

Model validation involved a number of key stages: (i) mathematical model validation using checks on mass conservation and solution convergence, with the mass and momentum flux residuals reduced to 0·1% of the inlet flux, or at least to near this value with individual values changing by less than 0·1%; (ii) assessment of the grid used for numerical solution to ensure grid independence by undertaking simulations for different grid dimensions and comparing key flow parameters; and (iii) comparison of modelled results with measured values in order to establish confidence in the predictions of the numerical model.

In order to obtain the measured values for this comparison, a simple laboratory experiment was set up in a 0·3-m wide, glass-sided, water-recirculating flume in the Department of Geography, University of Cambridge. An acoustic Doppler velocimeter (ADV) (Kraus *et al.*, 1994) was used to obtain point estimates of three-dimensional velocities and turbulent kinetic energy within the area of interest, and these were compared with output from the corresponding model simulation.

A Perspex dividing plane and step were constructed and fixed into the flume to create two parallel channels, each 0·1475 m wide but with different bed elevations (Figure 1). A gently tapering ramp of 0·75 m long led up to the elevated bed level of 0·05 m in the right-hand channel, and an adjustable plate at the upstream end of the dividing plate allowed the discharge ratio between the two channels to be adjusted. The ADV cannot measure in the top 0·06 m of the flow, and so an overall water depth of 0·15 m was used as this allowed some measurements above the level of the step. The key hydraulic parameters are shown in Table IV.

The ADV was mounted so that the u-velocity was oriented upstream (with subsequent correction of velocities so that further reference to u is positive downstream) and the v-velocity towards the left bank, this orientation is estimated to have been accurate to $\pm 1°$. The vertical velocity was positive upwards, and the level was checked with a spirit bubble on the mounting arm. The mounting structure enabled the sample

Table IV. Hydraulic parameters used for the laboratory assessment of the numerical model

U in left tributary (ms^{-1})	U in right tributary (ms^{-1})	Total discharge ($m^3\ s^{-1}$)	Momentum ratio	Froude no.		Reynolds no.	
				left	right	left	right
0·281	0·297	0·011	1·42	0·23	0·30	28 000	30 000

location downstream of the step and across the channel to be measured, and the height above the bed was calculated by the instrument itself. The following velocity measurements were taken: (i) point measurements in each tributary, 0·25 m upstream of the junction, and at 40% of the flow depth above the bed in order to assess average input flow for the model; (ii) a long profile downstream of the step, near the bed and along the right-hand bank, in order to cover the length of the separation zone and subsequent flow recovery; and (iii) a series of measurements in three cross-sections, at three different heights and at four cross-stream locations, to provide an array of measurements for comparison with the model output. The cross-sections were located within the separation zone, near the reattachment point, and downstream of the reattachment point (Figure 1).

The velocity at each point was sampled at 25 Hz for a period of 30 s, because there proved to be no significant trend in mean velocity over this period. The mean was extracted for each component by averaging these values. Owing to the presence of Doppler noise at high frequencies, the residual turbulent time-series was filtered before calculating the variance required for measures of mean turbulent kinetic energy. The reasoning, method and equations used in the filtering process can be found in Lane *et al.* (in review) and are not repeated here. The mean turbulent kinetic energy per unit mass, k, was then calculated using

$$k = \frac{1}{2}(\overline{u'^2} + \overline{v'^2} + \overline{w'^2}) \tag{17}$$

where u', v' and w' are the filtered turbulent velocity fluctuations around the means for the longitudinal, lateral and vertical components, respectively.

Results

Grid assessment. Figure 2 shows the effect of different grid dimensions on key flow parameters. The grid chosen was the coarsest for which results did not differ appreciably from solutions with finer grids: 70 × 44 × 25 grid cells for a domain of 1 m × 0·3 m × 0·1 m. The grid spacing was finer where velocity gradients were expected to be steep: in the vicinity of the step, near the walls and bed, and in the mixing zone in

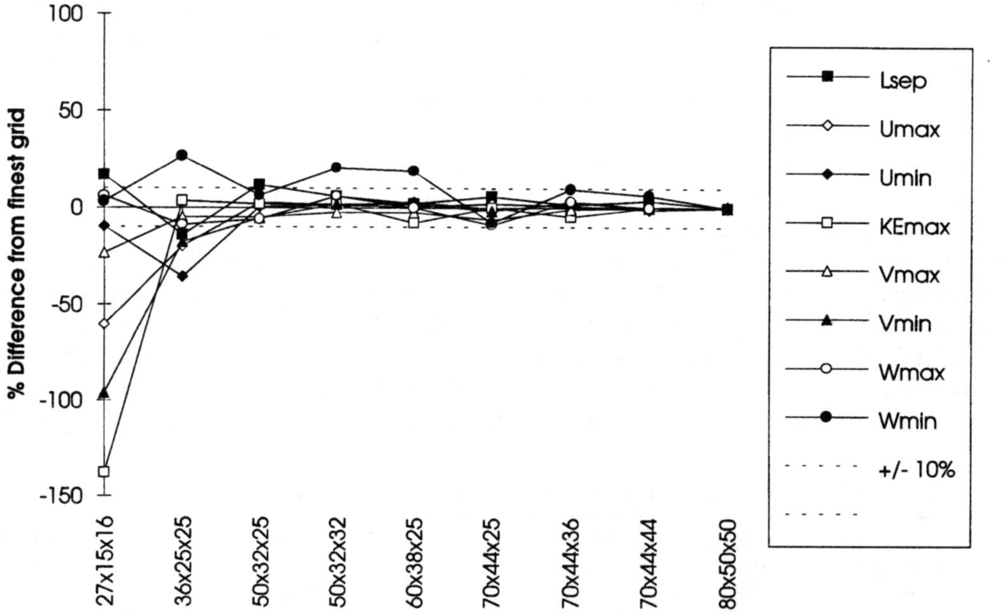

Figure 2. Effect of varying grid dimensions on key maximum and minimum flow parameters. (Lsep: separation zone length, U: downstream velocity, KE: turbulent kinetic energy, V: cross-stream velocity, W: vertical velocity)

Table V. Correlation between model and measured variables

	Correlation coefficient with RNG $k-\varepsilon$ model	Correlation coefficient with standard $k-\varepsilon$ model
Resultant velocity	0·970	0·903
Downstream velocity	0·975	0·918
Cross-stream velocity	0·881	0·444
Vertical velocity	0·845	0·713
Turbulent kinetic energy	0·631	0·688

the centre of the channel. The average grid spacing compares favourably with other similar studies (e.g. Basara and Younis, 1995; Babarutsi *et al.*, 1996; Mayerle *et al.*, 1995). For the laboratory comparison, the number of cells in the vertical was increased to 30 to take into account the extra water depth.

Comparison of the model results with experimental data. In order to compare modelled velocities with those measured using the ADV, the model output was sampled at grid cells equivalent to the measurement locations. Velocities are stored at cell faces and the nearest cell face to the sample point was used.

Table V shows correlations between predicted and observed values of velocity and turbulent kinetic energy for both of the turbulence models used, all of which are statistically significant. Figure 3 shows plots of measured against predicted values for the simulation with the RNG $k-\varepsilon$ turbulence model. The regression line and line of perfect fit are provided for comparison. For the simulation with the RNG $k-\varepsilon$ turbulence model, the equation of the regression lines for downstream velocity and resultant velocity, which it dominates, are not significantly different from the line of perfect agreement (using t-tests on the slope and intercept), and the model explains over 94% of the variation in the measured values. Although the correlations for cross-stream and vertical velocity are lower, they are still reasonable. The magnitude of the predicted values is generally less than that of the measured velocities, but since the absolute values are only a few cms^{-1}, the error is not great. The correlation for turbulent kinetic energy is much lower, and the highest values are underpredicted by about 50%. However, the general range is good, with the regression line not significantly different from the line of perfect agreement. Given the problems with measurement of turbulent kinetic energy (Lane *et al.*, in press), and the success of the model in predicting the mean velocities, the performance of the RNG turbulence model appears to be satisfactory, and is preferable to the standard $k-\varepsilon$ turbulence model.

Turbulence model. Comparison of the results from the RNG model with those from the more standard $k-\varepsilon$ model illustrates the critical importance of careful consideration of appropriate turbulence models in modelling even simple confluence geometries. The improvement in model performance with the RNG version of the $k-\varepsilon$ model was most evident for the cross-stream velocities for which the correlation between predicted and measured values was only 0·444 with the standard $k-\varepsilon$ model, but 0·881 with the RNG version (Table V). The correlation between modelled and measured turbulent kinetic energy is slightly higher with the RNG $k-\varepsilon$ model compared to the standard version and with the standard $k-\varepsilon$ model, the slope of the regression line is 0·692, compared to 1·23 with the RNG version. This implies that the predicted values of turbulent kinetic energy with the standard $k-\varepsilon$ model are generally higher than those observed, and inspection of the actual values shows this is especially true within the recirculation zone. With the RNG model the higher strain rates in the recirculation zone, as expected, resulted in a greater rate of turbulence dissipation and therefore a decrease in turbulent kinetic energy values. The increased dissipation and decrease in turbulent kinetic energy combine to reduce the eddy viscosity within the separation zone, thus promoting the formation of stronger recirculating flow, so explaining the improved flow predictions. It follows that the RNG model is important and necessary for this type of flow problem.

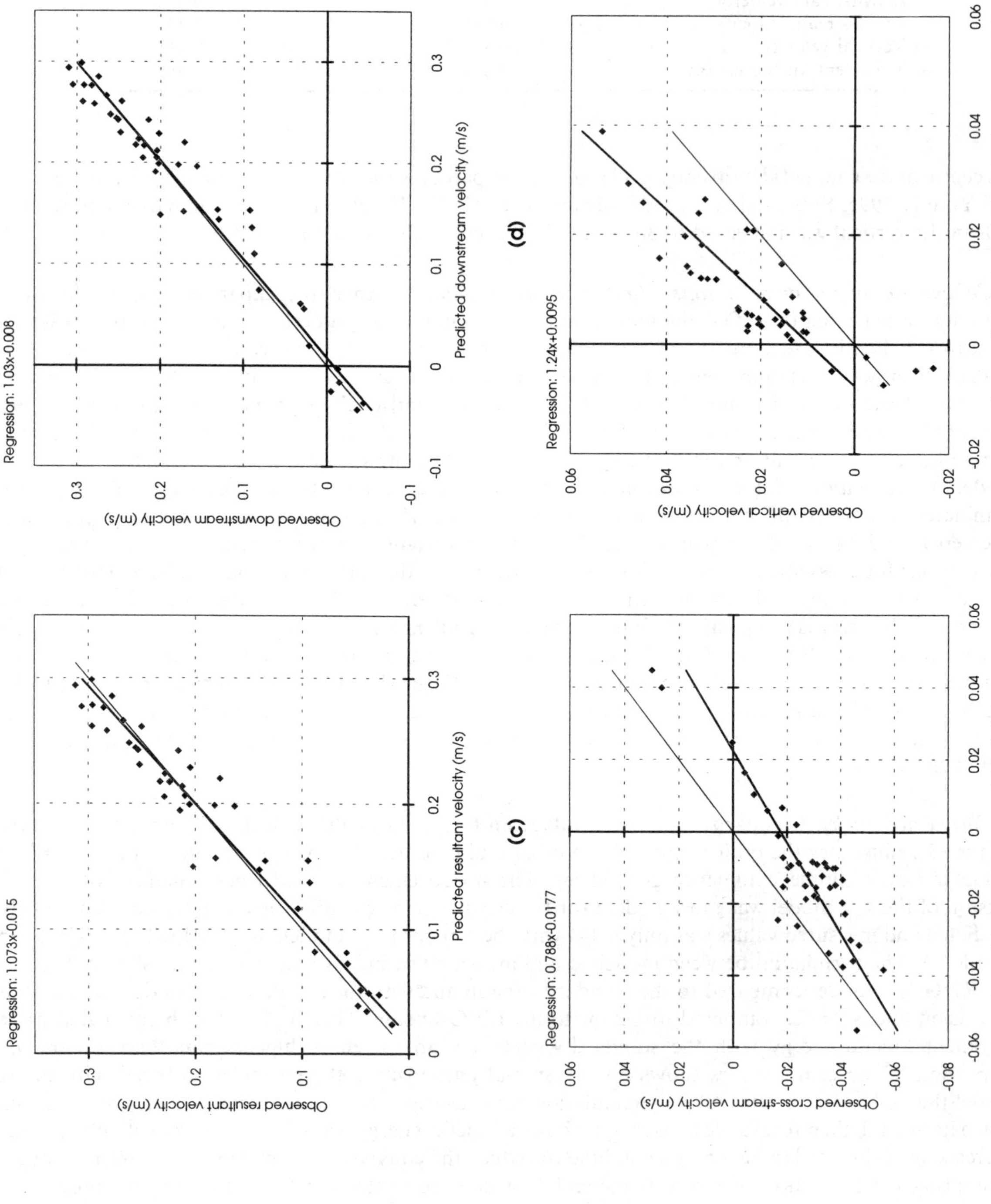
(a)
Regression: 1.073x-0.015
Observed resultant velocity (m/s)
Predicted resultant velocity (m/s)
(b)
Regression: 1.03x-0.008
Observed downstream velocity (m/s)
Predicted downstream velocity (m/s)
(c)
Regression: 0.788x-0.0177
Observed cross-stream velocity (m/s)
Predicted cross-stream velocity (m/s)
(d)
Regression: 1.24x+0.0095
Observed vertical velocity (m/s)
Predicted vertical velocity (m/s)

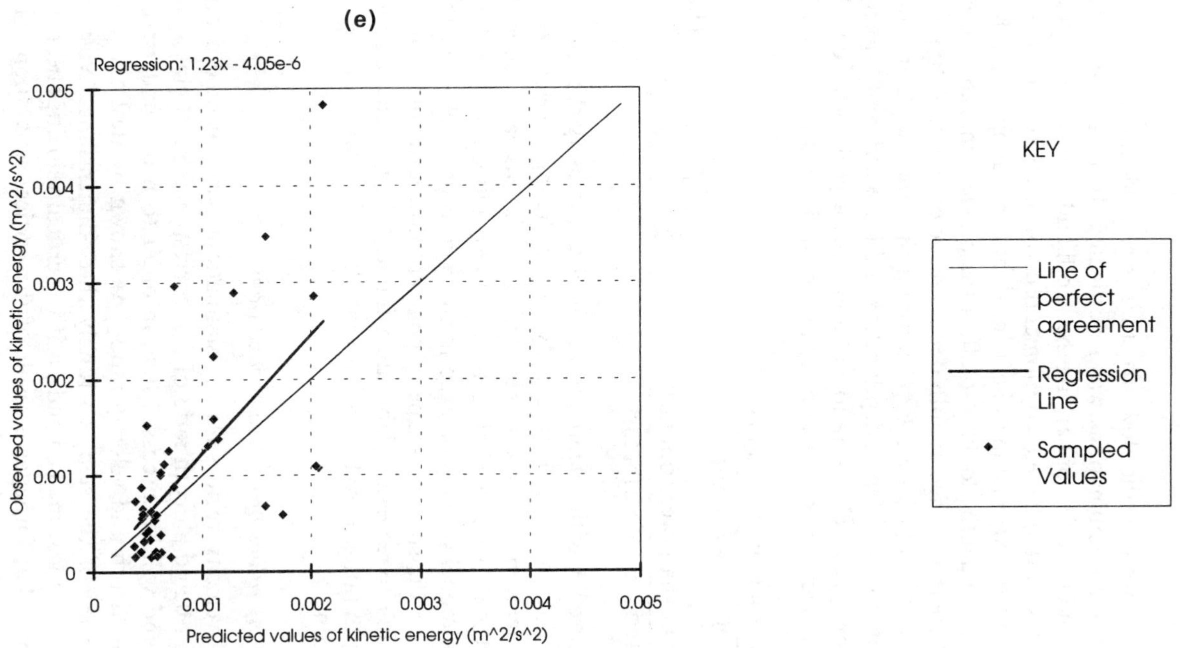

Figure 3. Comparison of model and observed values: (a) resultant velocity; (b) downstream velocity; (c) cross-stream velocity; (d) vertical velocity; (e) turbulent kinetic energy

With the RNG version of the $k-\varepsilon$ turbulence model, the general agreement between model predictions and measured velocity components is very good and this gives sufficient confidence to use the numerical model in simulation mode to investigate the effects of different values of boundary conditions.

NUMERICAL SIMULATION EXPERIMENTS

Experimental design and justification

Having established the ability of the model to reproduce the flow structure measured in a laboratory confluence, the effect of different boundary condition combinations was investigated. In these numerical experiments, junction angle is maintained at zero, and the effect of varying the depth and velocity ratios was considered. Most researchers (e.g. Mosley, 1976; Best, 1988) have used momentum, or discharge ratio, as a controlling variable, with the upstream depth in the tributary channels held constant. This implies that they only examined the velocity ratio, but since a change in depth would also affect the momentum ratio, these results cannot be generalized for the case where there is a change in both depth and velocity ratios. In these simulations, the depth and velocity in the left-hand channel were kept constant and the depth and the velocity in the right-hand channel were varied so that four different depth ratios and three different velocity ratios were considered (Figure 4). Figure 4 also shows the discharge ratio and total discharge associated with each combination of depth and velocity ratio. The depth ratio and velocity ratios are defined by

$$D_R = \frac{\text{depth of shallower tributary}}{\text{depth of deeper tributary}}$$
$$V_R = \frac{\text{velocity in shallower tributary}}{\text{velocity in deeper tributary}} \tag{18}$$

To assess the effect of different total discharge, case B3, with a depth ratio of 0·5 and a velocity ratio of 1·0, was also simulated for a total discharge of twice that shown in Figure 4, by setting the input velocities at 0·6 ms^{-1}. For all these experiments, a maximum depth of 0·1 m was used. This is similar to values used in other laboratory flume experiments (e.g. Best and Roy, 1991; McLelland *et al.*, 1996) and gives a width to depth ratio of 3·0 compared with that of 2·0 in the model verification experiment. This is desirable since the width to depth ratio is likely to have some effect on the size and shape of coherent flow structures, and is generally greater in natural channels than it is possible to recreate in laboratory flumes (e.g. McLelland *et al.*, 1996). To assess the effect of different width to depth ratios, case B3 was also simulated for a width to depth ratio of 6·0 by setting the maximum depth at 0·05 m.

Qualitative appearance of the flow field with different values of boundary conditions

Figure 5 shows typical near-wall and near-bed velocity vectors for the simulations with unequal depth channels. Experiment B3, with a velocity ratio of 1·0 and a depth ratio of 0·5, corresponds most closely to previously reported work (Best and Roy, 1991) and will be discussed first, in order to better understand deviations from this case associated with different depth and velocity ratios. As shown by Best and Roy (1991), the recirculation zone that forms in the lee of the step is strongly three-dimensional. Vertical separation over the step is associated with upstream flow at the bed and positive vertical velocities near the step. This flow is then entrained by the shear layer that forms as a result of the vertical separation. The recirculation associated with horizontal separation created by the splitter plate involves upstream flow along the wall of the shallower channel and towards the centre of the channel at the base of the step. This flow turns downstream and towards the wall of the shallower channel along the edge of the vertical shear layer that forms from the end of the splitter plate and extends downstream, interacting with the shear layer associated with the vertical separation over the step. These flows combine to elongate the separation zone along the wall of the shallower channel and near the bed. The horizontal rotation is around a nodal point (marked 'X' on

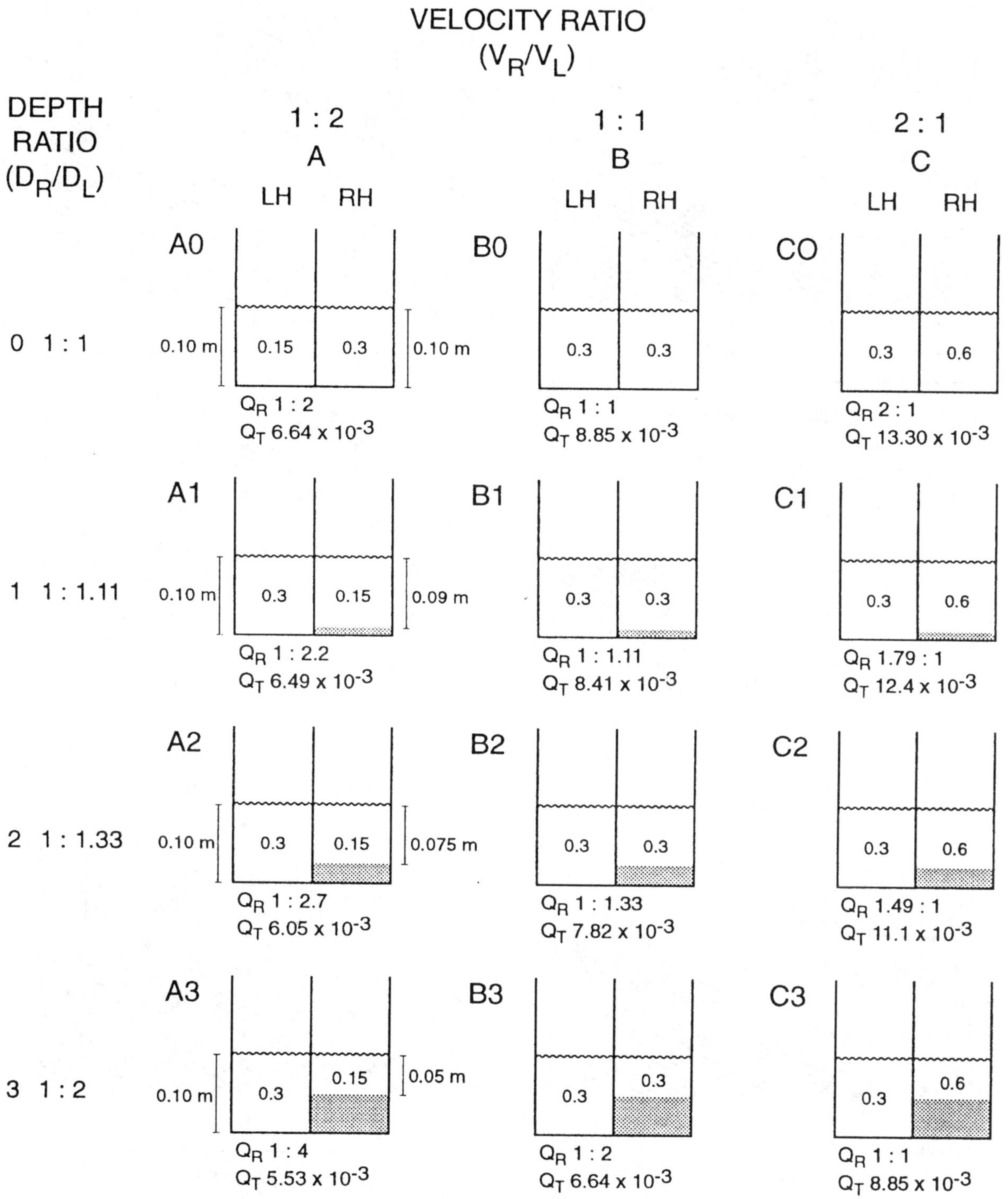

Figure 4. Experimental design for numerical simulations of zero-degree confluence with different depth and velocity ratios. (All velocities in ms^{-1} and discharge in $m^3 s^{-1}$)

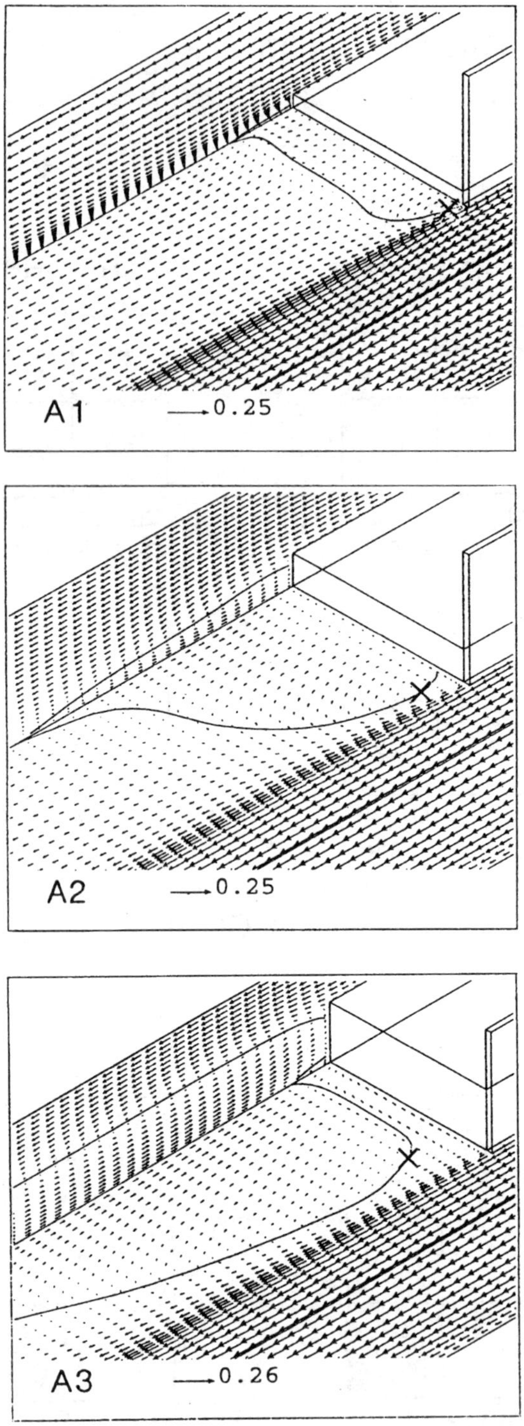

Figure 5a.

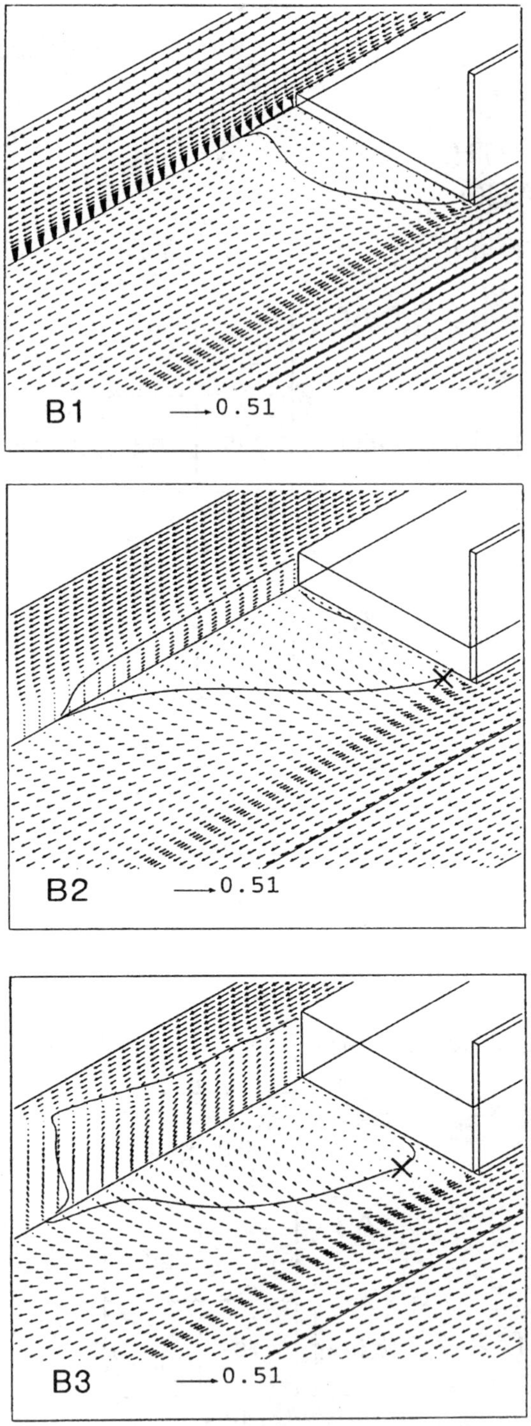

Figure 5b.

Figure 5(a–c). Near-bed and near-wall velocity vectors for unequal depth simulations, with contour of zero downstream velocity. Nodal point of horizontal circulation at bed (see text) marked 'X'. (Note that vector reference length varies between diagrams; value given in ms^{-1})

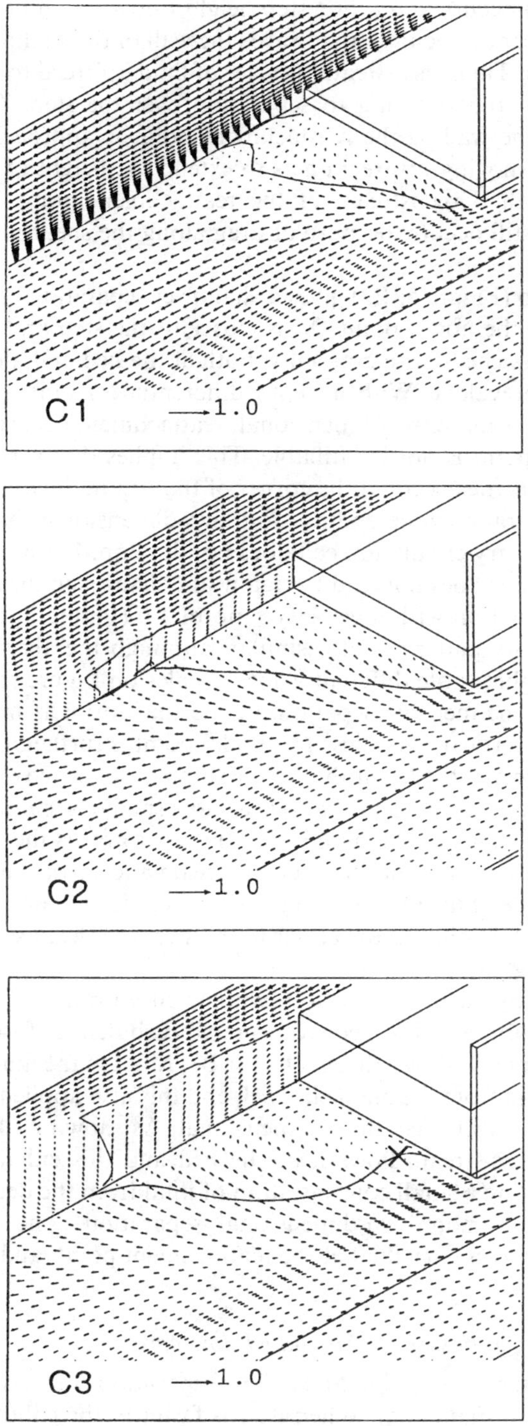

Figure 5c.

Figure 5; B3) which, as reported by Best and Roy (1991) occurs about one step height downstream of the step. Downstream of this nodal point, where the cross-stream velocities are towards the wall of the shallower channel, a streamwise circulation cell develops (Figure 6b). The cross-stream transfer of fluid is forced to rise at the wall and is then entrained in the shear layer that forms owing to separation over the step. This circulation cell becomes smaller and is pushed against the wall as the reattachment point is approached (Figure 6c). The upwelling along the wall of fluid that is moving upstream acts to extend the height of the separation zone at this point, but only close to the wall. Downstream of the reattachment point, the circulation cell persists (Figure 6d), as cross-stream flow still occurs, because of drag along the shear layer, forcing fluid upwards at the wall.

For a fixed velocity ratio of 1·0, a more even depth ratio (Figure 5; B2) gives a shallower separation zone, although the planform shape remains similar. However, the nodal point 'X' is closer to the step and the centre of the channel, and the reattachment point is approached more gradually, both at the bed and along the wall, with cross-stream and vertical velocities less prevalent. With a depth differential of only 10% (Figure 5; B1), the planform shape of the separation zone is more two-dimensional, reattachment occurring along a line more parallel to the step face, and a nodal point is not identifiable. This implies that vertical separation dominates the flow field. High cross-stream velocities occur at the corner of the step in the middle of the channel, but because of the small height of the separation zone and its more two-dimensional shape, fluid transfer across the channel is minimal and the secondary circulation cell does not develop. For a fixed depth ratio of 0·5, an increase in velocity ratio (Figure 5; C3) does not lead to significant changes in the size or shape of the separation zone, compared with the simulation with a velocity ratio of 1·0 (Figure 5; B3), although the nodal point is closer to the step. The velocity pattern is also similar, although the absolute strength of cross-stream and vertical velocities may be higher (note different velocity scales on Figure 5; B3 and C3). However, with very low flow in the shallower channel, the separation zone that forms is much longer (Figure 5; A3), with no significant cross-stream or vertical velocities. The nodal point is further away from the step and a secondary, counter-rotating, eddy forms in the lee of the step. Examination of the other simulations shown on Figure 5 suggests that, in general, a more even depth ratio leads to a smaller separation zone, which becomes more two-dimensional in planform shape as the nodal point moves towards the step and towards the centre of the channel, and is associated with less pronounced cross-stream and vertical flow. Changes in velocity ratio have less effect on the size or shape of the recirculation zone, except for the unusual case of simulation A3. With a higher velocity ratio, the nodal point occurs closer to the step and the absolute strength of cross-stream and vertical velocities may be higher.

The mixing of fluid from each channel was investigated by the use of a 'tracer' with a concentration of 1 in the main channel and 0 in the tributary. The tracer was subject to both convection and turbulent diffusion, and could, for example, represent a soluble pollutant. Figure 7a shows the results for B3. Most of the fluid in the recirculation zone was derived from entrainment of fluid from the main channel, and this also applied for other cases where the circulation in the separation zone was strongly three-dimensional. As shown by Best and Roy (1991), the mixing zone is distorted by the cross-stream flows near the bed and this enhances mixing. The effect of the secondary circulation cell is to bring a small concentration of fluid from the deeper channel to the surface along the tributary wall. For a more two-dimensional separation zone, the recirculating fluid was mostly derived from entrainment of tributary fluid at the reattachment point, and the mixing zone remained fairly vertical (Figure 7b).

Quantitative assessment of the effects of depth and velocity ratios on the simulated flow parameters

The effect of varying the depth and velocity ratios is illustrated in Figure 8. Flow separation is driven by the formation of a low pressure zone in the lee of the step (Figure 8a). When flow is faster in the tributary than in the main channel, the intensity of the low pressure is greater. In general, there is also a slight decrease in pressure with decreasing depth ratio, except for a depth ratio of 0·5 at the highest velocity ratio.

Figure 8b shows the maximum cross-stream velocity towards the shallower channel within the separation zone, and Figure 8c shows the maximum downwards velocity that occurs near the reattachment point. The

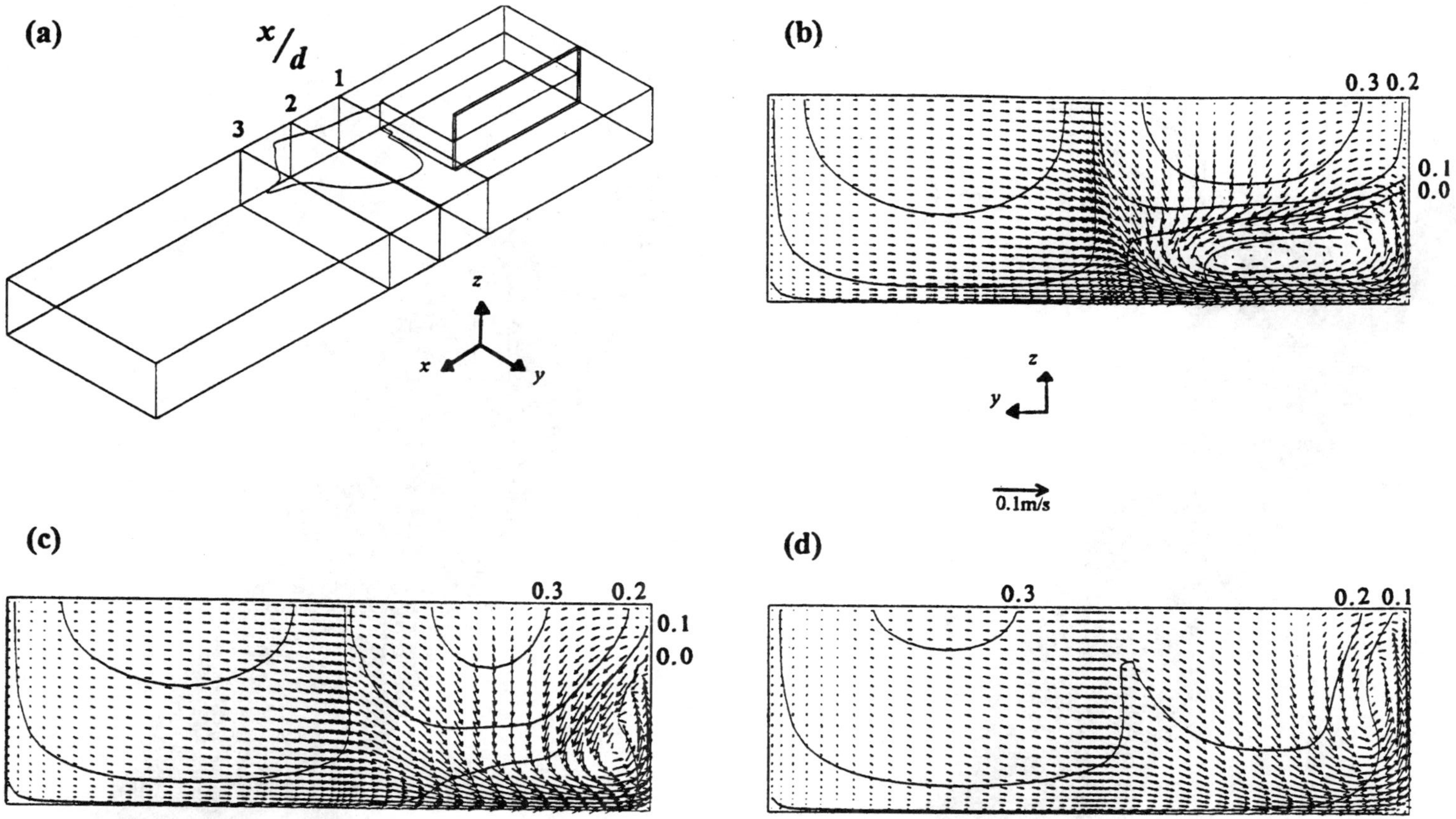

Figure 6. Cross-stream vectors and contours of downstream velocity for case B3: (a) location of cross-sections; (b) $x/d = 1$; (c) $x/d = 2$; (d) $x/d = 3$; where x is the distance downstream from the end of the splitter plate and d is the maximum depth (=0·1 m). All velocities in ms^{-1}

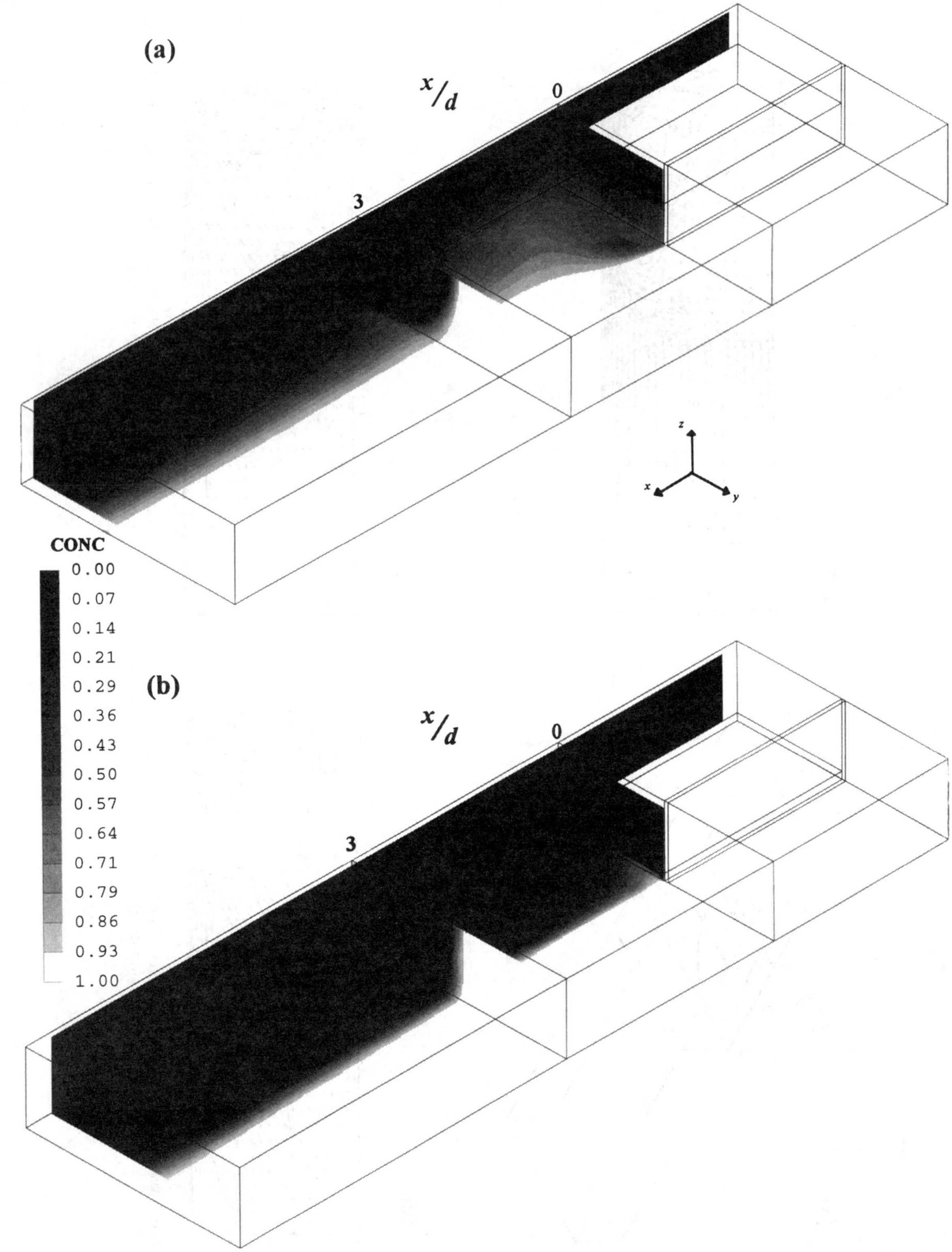

Figure 7. Contours of relative tracer concentration: (a) case B3 (depth ratio = 0·5; velocity ratio = 1·0); (b) case B1 (depth ratio = 0·9; velocity ratio = 1·0)

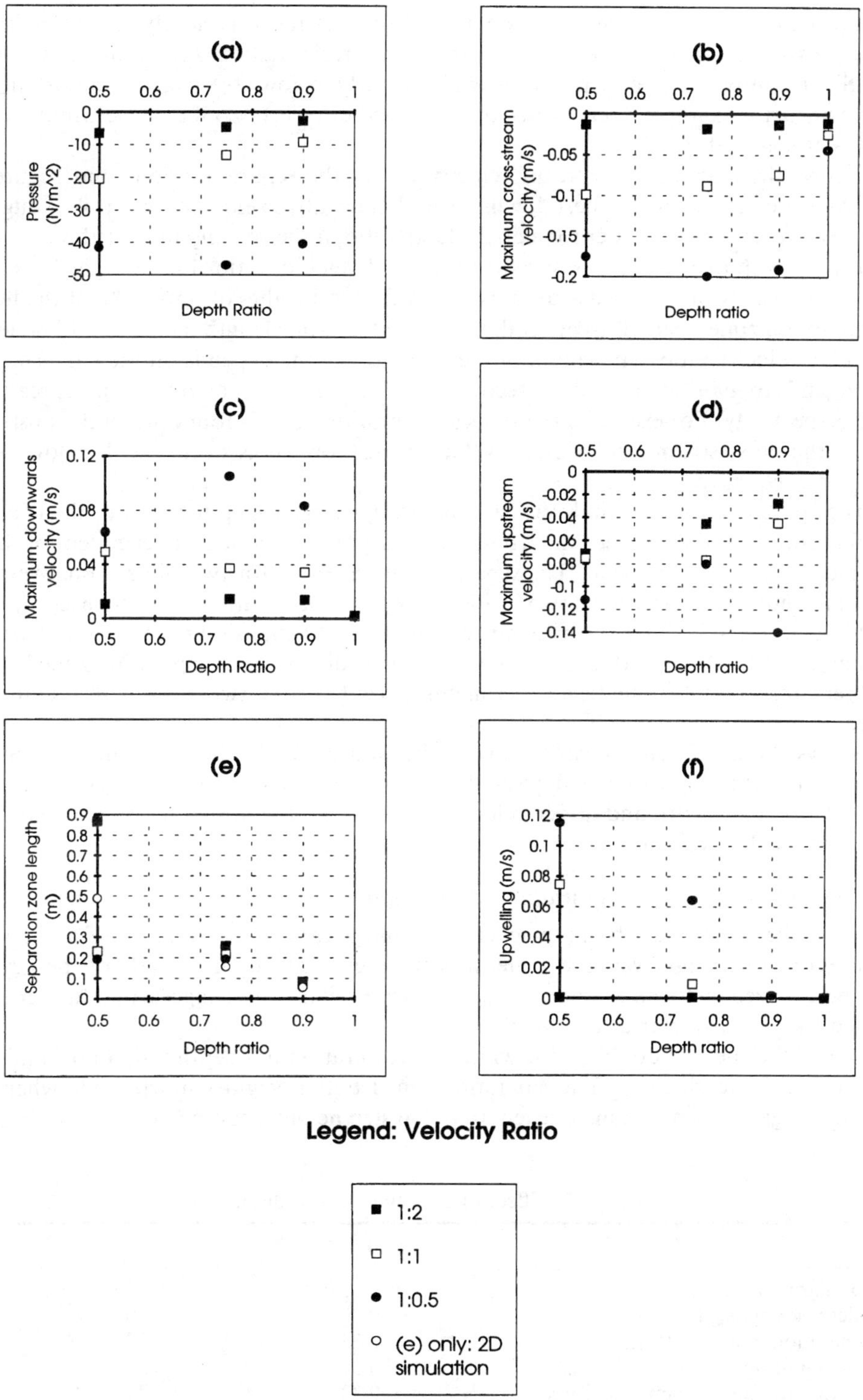

Figure 8. Effect of varying depth and velocity ratios: (a) low pressure minimum below step; (b) maximum cross-stream velocity (+ve towards left-hand bank); (c) maximum vertical velocity (+ve downwards); (d) maximum upstream velocity (+ve downstream); (e) separation zone length; (f) maximum value of upwelling velocity beyond the reattachment point. (Depth and velocity ratios are defined by the value in the shallower tributary divided by the value in the deeper tributary)

general relationship of these quantities to the depth and velocity ratios is closely related to that in Figure 8a, since it is the pressure gradient towards the centre of the recirculation zone that draws fluid across and downwards. Significantly, even a depth ratio only slightly below 1·0 can be important in generating secondary velocities, for example, local cross-stream velocities of up to 33% of the tributary velocity, if the velocity ratio is at least 1·0.

Figure 8d shows the maximum upstream velocity within the separation zone and Figure 8c shows the length of this zone, defined as the furthest downstream that negative velocities occur. Although the upstream velocity (Figure 8d) is generally greater for a given depth ratio, if the velocity in the tributary is faster, there is no consistent trend with decreasing depth ratio for a fixed velocity ratio. Large negative velocities (greater than 20% of the tributary flow velocity) are possible with step heights of only 10% of the flow depth, even though the separation zone is small. In general, the separation zone length increases with a decreasing depth ratio and with the velocity ratio. The length of the separation zone depends on the balance of the upstream velocity, which tends to maintain it, and the secondary velocities, which tend to promote reattachment at the wall and bed, respectively. For example, the longest separation zone is found when the upstream velocity is significant, but the cross-stream and vertical velocities are low. This occurs with a low depth ratio and velocity ratio (both equal to 0·5).

Simulations of pure two-dimensional vertical separation over the step for each of the tributary velocities and step heights considered, show that the length of the separation zone is independent of the velocity over the step, but increases with step height. The separation zone length for two-dimensional vertical separation over the different step heights, is also shown in Figure 8e, and these values, if normalized by step height, are comparable to the results of similar experiments with equivalent expansion ratios (e.g. Babarutsi *et al.*, 1989, 1996; Papadopoulos and Otugen, 1995). In general, the two-dimensional separation zone length is less than that in the three-dimensional simulation experiments, except for the two cases with the highest depth and velocity ratios.

Figure 8f shows the maximum upwards flow at the wall of the shallower channel, downstream of the reattachment point, which is associated with the secondary circulation cell. Upwelling motion is only significant for the smaller depth and larger velocity ratios, when the upwelling velocity can be greater than 20% of the tributary flow velocity.

Effect of varying overall discharge and width to depth ratio

Table VI shows the effect of doubling the overall discharge for a depth ratio of 0·5 and a velocity ratio of 1·0. The separation zone length was the same in both cases, and the magnitude of the key velocity parameters, when normalized by the inlet velocity, u_{in}, are also similar. This suggests that the general conclusions are not affected by the particular discharge chosen.

Table VII shows the effect of doubling the width to depth ratio for a depth ratio of 0·5 and a velocity ratio of 1·0, by reducing the overall depth. The separation zone length is slightly shorter, but, when normalized by the step height, this gives a separation zone length of 7·6 step heights, compared with 4·6 step heights for the

Table VI. Effect of doubling overall discharge

	Standard	$2Q_T$
Discharge ($m^3 s^{-1}$)	$6{\cdot}64 \times 10^{-3}$	0·013
Inlet velocity, u_{in} (ms^{-1})	0·30	0·60
Separation zone length (m)	0·23	0·23
Maximum upstream velocity/u_{in}	0·25	0·30
Maximum cross-stream velocity/u_{in}	0·33	0·39
Maximum downwards velocity/u_{in}	0·16	0·19
Upwelling/u_{in}	0·25	0·24
Average tracer concentration	0·619	0·609

Table VII. Effect of doubling width to depth ratio

	Standard	Width : depth ratio doubled
Width : depth ratio	3 : 1	6 : 1
Maximum depth (m)	0·1	0·05
Separation zone length (m)	0·23	0·19
Separation zone length/step height	4·6	7·6
Maximum upstream velocity (ms^{-1})	0·076	0·074
Maximum cross-stream velocity (ms^{-1})	0·099	0·110
Maximum downwards velocity (ms^{-1})	0·049	0·040
Upwelling (ms^{-1})	0·075	0·055
Average tracer concentration	0·619	0·611

standard case. The other values are not too dissimilar, except for the upwelling velocity, which is reduced. The circulation cell within the recirculation zone is initially circular with a diameter similar to the step height and forms near the centre of the channel, but further downstream it becomes elongated to cover the whole width of the separation zone. As the reattachment point is reached, the recirculation cell is pushed towards the wall of the shallower channel and becomes less elongated.

DISCUSSION

The research presented in this paper has demonstrated that secondary circulation can develop at a channel junction in the absence of planform curvature, under a wide range of combinations of depth and velocity ratios. Considering depth and velocity ratio as the controlling variables has already disaggregated more commonly used variables such as the discharge or momentum ratio. However, depth and velocity ratios can in turn be related to two factors that control the fluid dynamics more directly: (i) the cross-stream pressure gradient; and (ii) the vertical extent of this pressure gradient, which determines the relative depth of flow for cross-stream mass transfer. The former depends primarily upon the velocity ratio and the latter upon the depth ratio. This research has shown that the combination of these two variables determines the rate of cross-stream mass transfer and if both are sufficiently large, then secondary circulation and lateral mixing and mass transfer will be efficient.

For example, a high cross-stream pressure gradient will entrain large cross-stream velocities towards the recirculation zone below a small step, but the cross-stream momentum transfer is limited by the small height of this zone. This configuration does not lead to the development of strong secondary circulation, upwelling or greatly enhanced mixing. However, when the step height and velocity ratio are both large enough that cross-stream momentum transfer leads to significant fluid impingement on the side wall, some fluid will be forced upwards, with the associated flow being re-entrained into the downstream direction in the horizontal shear layer that begins at the step. In such a case, cross-stream velocities will also persist downstream of the reattachment zone, leading to upwelling along the side wall and the associated secondary circulation cell. The stronger the entrainment and subsequent distortion of the mixing layer and secondary circulation, the greater the intensity of mixing of fluid between the two channels.

The rate of cross-stream momentum transfer can also explain the deviation of the separation zone length from that predicted for two-dimensional vertical separation: the entrainment of cross-stream fluid near the bed acts to delay the vertical reattachment, and reattachment occurs slightly further downstream than predicted for two-dimensional separation. However, for the smallest depth ratios with a velocity ratio of at least 1·0, the cross-stream flow is sufficient to force earlier reattachment along the side wall, and the separation zone length is shorter than would occur with two-dimensional vertical separation. In this case, the recirculation zone has a strongly three-dimensional shape.

This research suggests that better understanding of the controls on secondary circulation generation comes from consideration of the pressure gradients that result from particular combinations of boundary condition values. There is the potential to extend this same principle to the investigation of the effect of other controls; for example, it is also the lateral pressure gradient that is important in generating curvature-induced helical cells. Where bed discordance exists in combination with a junction angle (e.g. Best, 1988; Biron *et al.*, 1996a,b), or where the avalanche face is at an angle to the main flow direction (McLelland *et al.*, 1996), then the cross-stream pressure gradient owing to flow separation in the lee of the step is extended further downstream, since the step itself is elongated in that direction. The vertical separation over such a step will include a cross-stream component, and these two factors together are likely to enhance the secondary flow seen in this experiment. For example, observations by Biron *et al.* (1996a,b) where the depth ratio was 0·81, with a velocity ratio of 1·6, showed significant mixing layer distortion and upwelling. The results from the research presented in this paper suggest that some mixing layer distortion and upwelling would occur with such boundary condition values (e.g. Figure 8f), but this may be enhanced by the effect of a 30° junction angle in the experiment investigated by Biron *et al.* (1996a,b). However, this does not consider the interaction of the pressure gradient associated with bed discordance with that owing to streamline curvature. This interaction may intensify or disrupt the secondary circulation cells usually associated with streamline curvature, depending on the exact combinations of boundary condition values. The nature of this interaction is the subject of ongoing research.

Interaction between the effects of junction angle and bed discordance is of great importance to natural channels, which generally exhibit both. However, similarity of the results for different discharge and width to depth ratios is encouraging for the comparison and general application of the results of this, and other similar laboratory-based studies (e.g. McLelland *et al.*, 1996), even though natural rivers have larger width to depth ratios than it is possible to recreate in the laboratory. Other features of natural channels that must be considered when attempting to apply the results of such studies include the fact that natural avalanche faces are rarely vertical, although laboratory experiments by Biron *et al.* (1996a) suggest little difference between the effects of a vertical avalanche face and one at 45°, and the effects of bed roughness, particularly roughness heterogeneities. For example, the scour hole that is part of the explanation of bed discordance may exhibit unusually high bed roughness (Bridge, 1993), as sediment entrainment during scour is partially size selective.

Although, this research has suggested that consideration of pressure gradients is important and useful in understanding the development of coherent flow structures, it is not immediately applicable in the field. Although the nature of the pressure gradient can often be inferred from the expected, or observed, streamline pattern, it is difficult to quantify without numerical modelling, which then also provides full information about the velocity field. Although such a model can be applied to natural confluences, such resources are not always available and it may still be preferable in the field to relate velocity measurements and predictions to measurable quantities such as depth and velocity ratio. Thus, it should be noted that the effect of bed discordance is to promote secondary circulation and enhance mixing, but this is only significant for smaller depth ratios and velocity ratios greater than 1·0. This also gives a context for the application and interpretation of results from laboratory simulations with specific depth and velocity ratios such as those reported in Table I.

However, if the velocity ratio is greater than 1·0, even small differences in the depth of the two channels (for example 10%) can generate significant secondary velocities (for example, local cross-stream velocities and downwards velocities of up to 33% and 13% of the tributary flow velocity, respectively), and reverse flow of more than 20% of the tributary flow velocity. This suggests that, in natural channels, a small step may grow by erosion at the reattachment point because of high velocities and turbulent stresses acting on the bed. Upstream velocities towards the step may maintain the stability of the step as it grows. This feedback would then lead to a greater separation zone length and cross-stream momentum transfer, and could explain the common occurrence of bed discordance at river confluences (Kennedy, 1984).

CONCLUSION

This paper has demonstrated the use of a numerical model to simulate the effects of different combinations of boundary condition values on flow structures in a simple channel confluence. Greater control over the boundary condition values was possible than if the experiment had been conducted in a laboratory flume. In addition, more detailed information about the flow structure could be extracted such as a full three-dimensional network of three orthogonal velocity components covering the whole domain.

Consideration of 12 simulations with different combinations of velocity ratio and depth ratio, for a fixed junction angle of 0°, showed that the combination of these two variables was critical in determining the development of significant secondary circulation. This reiterates the importance of considering bed discordance at confluences, as emphasized by other researchers (e.g. Best and Roy, 1991; Biron *et al.*, 1993b, 1996a,b; Gaudet and Roy, 1995), but specifically when the velocity in the shallower channel is at least as fast as that in the deeper channel. This research has shown that a consideration of the pressure gradients driving the flow can aid understanding of the development of coherent flow structures under different boundary condition combinations. The cross-stream pressure gradient depends primarily on the velocity ratio and initiates cross-stream velocities. However, the depth ratio determines the vertical extent of this pressure gradient, and therefore the extent to which cross-stream velocities lead to significant transfer of fluid in the cross-stream direction, which is required if significant secondary circulation is to form. In this study, strong secondary circulation occurred for a depth differential of 25% or more, as long as the velocity in the shallower tributary was at least as fast as that in the deeper channel.

The use of numerical models in an exploratory fashion can be extended to answer further questions: for example, the consideration of junction angle effects in conjunction with depth and velocity ratios is currently being investigated. An understanding of the individual and combined effects of these controls will aid interpretation of the flow structures seen in more complicated natural channels. The model has also been successfully applied to a natural channel, and results will be published elsewhere. In this case, detailed investigation of the model predictions (pressure gradients, velocities and mixing characteristics) will help to explain existing flow structures. Systematic variation of boundary condition values will enable better prediction of how flow structures change with different bed morphology, roughness or hydrological conditions. Eventual coupling of three-dimensional flow models with sediment transport models could lead to simulation of the feedback between flow and bed topography to model confluence evolution through time.

ACKNOWLEDGEMENTS

This research has been facilitated by NERC grant GR3/09715 (to S.N.L. and K.S.R.), K.F.B. is in receipt of NERC studentship (GT4/94/314/a).

REFERENCES

Andrews, E. D. and Nelson, J. M. (1989). 'Topographic response of a bar in the Green River, Utah to variation in discharge', in Ikeda, S. and Parker, G. (Eds), *River Meandering*, American Geophysical Union Monograph 12. American Geophysical Union, Washington, D.C. pp. 1–50.

Ashmore, P. E. (1993). 'Anabranch confluence kinetics and sedimentation processes in gravel-braided streams', in Best, J. L. and Bristow, C. S. (Eds), *Braided Rivers*, Geological Society Special Publication 75. Geological Society, London. pp. 129–146.

Ashmore, P. E., Ferguson, R. I., Prestegaard, K. L., Ashworth, P. J., and Paola, C. (1992). 'Secondary flow in anabranch confluences of a braided gravel-bed stream', *Earth Surf. Process. Landf.*, **17**, 299–311.

Babarutsi, S., Gonoulis, J., and Chu, V. H. (1989). 'Experimental investigation of shallow recirculating-flows', *J. Hydraul. Engng*, **115**, 906–921.

Babarutsi, S., Nassiri, M., and Chu, V. H. (1996). 'Computation of shallow recirculating flow dominated by friction', *J. Hydraul. Engng*, **122**, 367–372.

Basara, B. and Younis, B. A. (1995). 'Prediction of turbulent flows in dredged channels', *J. Hydraul. Res.*, **33**, 813–824.

Best, J. L. (1987). 'Flow dynamics at river channel confluences: implications for sediment transport and bed morphology', in Etheridge, F. G., Flores, R. M., and Harvey, M. D. (Eds), *Recent Developments in Fluvial Sedimentology*, SEPM Special Publication 39, *Soc. Econ. Paleon. and Mineralogists*, Tulsa, Oklahoma. pp. 27–35.

Best, J. L. (1988). 'Sediment transport and bed morphology at river channel confluences', *Sedimentology*, **35**, 481–498.

Best, J. L. and Reid, I. (1984). 'Separation zone at open channel junctions', *J. Hydraul. Engng*, **110**, 1588–1594.

Best, J. L. and Roy, A. G. (1991). 'Mixing-layer distortion at the confluence of channels of different depth', *Nature*, **350**, 411–413.

Biron, P., Roy, A. G., Best, J. L., and Boyer, C. J. (1993a). 'Bed morphology and sedimentology at the confluence of unequal depth channels', *Geomorphology*, **8**, 115–129.

Biron, P., De Serres, B., Roy, A. G., and Best, J. L. (1993b). 'Shear layer turbulence at an unequal depth channel confidence', in Clifford, N. J., French, J. R., and Hardisty, J. (Eds), *Turbulence: Perspectives on Flow and Sediment Transport*. Wiley, Chichester. pp. 197–214.
Biron, P., Best, J. L., and Roy, A. G. (1996a). 'Effects of bed discordance on flow dynamics at river channel confluences', *J. Hydraul. Engng*, **122**, 676–682.
Biron, P., Roy, A. G., and Best, J. L. (1996b). 'The turbulent flow structure at concordant and discordant open channel confluences', *Exper. Fluids*, **21**, 437–446.
Bridge, J. S. (1993). 'The interaction between channel geometry, water flow, sediment transport and deposition in braided rivers', in Best, J. L. and Bristow, C. S. (Eds), *Braided Rivers*, Geological Society Special Publication 75. Geological Society, London. pp. 13–75.
Bridge, J. S. and Gabel, S. L. (1992). 'Flow and sediment dynamics in a low sinuosity, braided river: Calamus River, Nebraska Sandhills', *Sedimentology*, **39**, 125–142.
Gaudet, J. M. and Roy, A. G. (1995). 'Effect of bed morphology on flow mixing length at river confluences', *Nature*, **373**, 138–139.
Kennedy, B. A. (1984). 'On Playfair's law of accordant junctions', *Earth Surf. Process. Landf.*, **9**, 153–173.
Kraus, N. C., Lohrmann, A., and Cabrera, R. (1994). 'New acoustic meter for measuring 3D laboratory flows', *J. Hydraul. Engng*, **120**, 406–412.
Lane, S. N. 1998. 'Hydraulic modelling in hydrology and geomorphology: a review of high resolution approaches', *Hydrol. Process.*, **12**, 1131–1150 (this issue).
Lane, S. N. and Richards, K. S. 1998. 'High resolution, two-dimensional spatial modelling of flow processes in a multi-thread channel', *Hydrol. Process.*, **12**, 1279–1298 (this issue).
Lane, S. N., Richards, K. S., and Chandler, J. H. (1995). 'Within-reach spatial patterns of process and channel adjustment', in Hicken, E. J. (Ed.), *River Geomorphology*. Wiley, Chichester. pp. 105–130.
Lane, S. N., Biron, P., Bradbrook, K. F., Butler, J. B., Chandler, J. H., Crowell, M. D., McLelland, S. J., and Roy, A. G. (in review). 'Integrated three-dimensional measurement of river channel topography and flow processes using acoustic Doppler velocimetry', *Earth Surf. Process. Landf., Tech. Softw. Bull.*
Launder, B. E. and Spalding, D. B. (1974). 'The numerical computation of turbulent flows', *Comp. Meth. Appl. Mech. Engng*, **3**, 269–289.
Lien, F. S. and Leschziner, M. A. (1994). 'Application of an RNG turbulence model to flow over a backwards-facing step', *Comp. Fluids*, **23**, 983–1004.
Mayerle, R., Toro, F. M., and Wang, S. S. Y. (1995). 'Verification of a three-dimensional numerical model simulation of the flow in the vicinity of spur dikes', *J. Hydraul. Res.*, **33**, 243–256.
McLelland, S. J., Ashworth, P. J., and Best, J. L. (1996). 'The origin and downstream development of coherent flow structures at channel junctions', in Ashworth, P. J., Bennett, S., Best, J. L., and McLelland, S. M. (Eds), *Coherent Flow Structures in Open Channels*. Wiley, Chichester. pp. 459–490.
Mosley, M. P. (1976). 'An experimental study of channel confluences', *J. Geol.*, **84**, 535–561.
Pantankar, S. V. and Spalding, D. B. (1972). 'A calculation procedure for heat, mass and momentum transport in three-dimensional parabolic flows', *Int. J. Heat Mass Transf.*, **15**, 1782.
Papadopoulos, G. and Otugen, M. V. (1995). 'Separating and reattaching flow structure in a suddenly expanding rectangular duct', *ASME J. Fluids Engng*, **117**, 17–23.
Rhoads, B. L. (1996). 'Mean structure of transport-effective flows at an asymmetrical confluence when the main stream is dominant', in Ashworth, P. J., Bennett, S., Best, J. L., and McLelland, S. M. (Eds), *Coherent Flow Structures in Open Channels*. Wiley, Chichester. pp. 459–490.
Rhoads, B. L. and Kenworthy, S. T. (1995). 'Flow structure at an asymmetrical stream confluence', *Geomorphology*, **11**, 273–293.
Richards, K. S. (1980). 'A note on change in geometry at tributary junctions', *Wat. Resour. Res.*, **16**, 241–244.
Shimizu, Y. and Itakura, T. (1989). 'Calculation of bed variation in alluvial channels', *J. Hydraul. Engng*, **115**, 367–384.
Smith, L. M. and Reynolds, W. C. (1992). 'On the Yakhot–Orszag Renormalization group method for deriving turbulence statistics and models', *Phys. Fluids A*, **4**, 364.
Spalding, D. B. (1985). 'The computation of flow around ships with allowance for free-surface and density-gradient effects', *Proceedings of the First Intercontinental Symposium on Maritime Simulation.*
Waterson, N. P. (unpublished).
Weerakoon, S. B. and Tamai, N. (1989). 'Three-dimensional calculation of flow in river confluences using boundary fitted co-ordinates', *J. Hydrosci. Hydraul. Engng*, **7**, 51–62.
Weerakoon, S. B., Kawahara, Y., and Tamai, N. (1991). 'Three-dimensional flow structure in channel confluences of rectangular section', *Proceedings of the 25th IAHR Congress A*. IAHR, Madrid. pp. 373–380.
Yakhot, V. and Orszag, S. A. (1986). 'Renormalization group analysis of turbulence', *J. Sci. Comp.*, **1**, 3.
Yakhot, V. and Smith, L. M. (1992). 'The renormalization group, the eps-expansion and derivation of turbulence models', *J. Sci. Comp.*, **7**, 35.
Yakhot, V., Orszag, S. A., Thangam, S., Gatshi, T. B., and Speziale, C. G. (1992). 'Development of a turbulence model for shear flow by a double expansion technique', *Phys. Fluids A*, **4**, 1510–1520.

8

FLOOD ROUTING BASED ON DIFFUSION WAVE EQUATION USING MIXING CELL METHOD

V. P. SINGH,* GUANG-TE WANG AND D. D. ADRIAN

Department of Civil and Environmental Engineering, Louisiana State University, Baton Rouge, LA 70803, USA

ABSTRACT

A one-dimensional non-linear diffusion wave equation is derived from the Saint Venant equations with neglect of the inertia terms. This non-linear equation has no general analytical solution. Numerical schemes are therefore employed to discretize the space and time axes and convert the differential equation to difference form. In this study, the mixing cell method is used to convert the diffusion wave equation to difference form, in which the difference term can be eliminated by selecting an optimal space step size Δx when time step size Δt is given. When the time step size $\Delta t \rightarrow 0$, the space step size $\Delta x = Q/(2S_0BC_k)$ where Q is discharge, S_0 is bed slope, B is channel width and C_k is kinematic wave celerity, which is the same as the characteristic length proposed by Kalinin and Milyukov. The results of application to two cases show that the mixing cell and linear channel flow routing methods produce hydrographs that are in agreement with the observed flood hydrographs.

Hydrol. Process., Vol. 11, 1881–1894 (1997)

(No. of Figures: 2 No. of Tables: 1 No. of Refs: 10)

KEY WORDS characteristic length; flood routing; diffusion wave equation; mixing cell method; numerical scheme

INTRODUCTION

Flood wave movement is a complex unsteady flow and can be described by the one-dimensional form of the Saint Venant equations of continuity and momentum, respectively,

$$\frac{\partial A}{\partial t} + \frac{\partial Q}{\partial x} = 0 \tag{1}$$

$$S_f = S - \frac{\partial y}{\partial x} - \frac{1}{g}\frac{\partial v}{\partial t} - \frac{v}{g}\frac{\partial v}{\partial x} \tag{2}$$

where A is the cross-sectional area of flow, g is the gravitational acceleration, v is the average velocity of the flow cross-section, Q is the discharge, $Q = Av$, y is the water flow depth, S is channel bed slope, S_f is the frictional slope, t is the time and x is the distance along the flow direction. For rectangular channels, $A = By$, B is the top width. Equations (1) and (2) are highly non-linear and their closed form analytical solution does not exist, except under highly simplified conditions. Therefore, numerical methods are the only resort for solving them. Numerical methods convert these equations into difference form that may then be solved for water depth and velocity at finite incremental values of x and t. The numerical methods can be grouped into six categories: (1) finite difference schemes that solve the characteristic grid; (2) explicit finite difference schemes for the characteristic equations using a rectangular x–t grid; (3) explicit finite difference schemes for the original equations using a rectangular grid; (4) implicit finite difference schemes for the original equations using a rectangular grid; (5) finite element methods; and (6) boundary fitted coordinate methods. Explicit schemes use *explicit* equations for y and v at a given x–t point, while implicit schemes use sets of simultaneous equations to find y and v at several x values simultaneously for a given t value. A good discussion of these methods is given in Singh (1996).

* Correspondence to: V. P. Singh.

In each category numerical methods have been developed to a level satisfying the accuracy requirements of most stream flow routing applications. The methods using a characteristic grid, although suited to a wide range of unsteady flow problems, require a large amount of interpolation and place a heavy demand on computer resources. Programming simplicity tends to favour the explicit schemes, but in these schemes the time step size, Δt, is limited by the Courant stability condition to

$$\Delta t \leqslant \frac{\Delta x}{v + (gy)^{1/2}}$$

Implicit methods are not subject to this stability condition. These methods are the only schemes suited to deal with large variations in flow characteristics between adjacent channel sections. For routing computations of slowly varying floods in natural rivers with typical durations of several days, implicit methods thus appear to be the most efficient.

In this study, the mixing cell method is used to solve the diffusion wave equations which are a simple form of the Saint Venant equations. The theory of the mixing cell method is based on the concentration of mass for water and solute and the assumption of complete mixing within a cell. This method has been used in modelling migration of reactive and non-reactive solute in soil water, groundwater and surface water (Van Ommen, 1985; Appelo and Willemsen, 1987; Bajracharya and Barry, 1992). Wang and Chen (1996) employed this method for describing advection–dispersion phenomena with an optimal time interval, $\Delta t \to 0$. The resulting model is an n-order ordinary differential equation which is similar to the continuous channel reach routing model proposed by Kalinin and Milyukov (Dooge, 1973).

THEORETICAL BASIS FOR DIFFUSION WAVE FLOOD ROUTING

Unsteady flow in an open channel can be represented by the St Venant Equations (1) and (2). For many flow situations, the inertia terms (local and convective acceleration) can be considered much smaller than friction slope and pressure gradient, i.e.

$$\left|\left(S_{\mathrm{f}} + \frac{\partial y}{\partial x}\right)\right| \gg \left|\left(\frac{1}{g}\frac{\partial v}{\partial t} + \frac{v}{g}\frac{\partial v}{\partial x}\right)\right|$$

and can, therefore, be neglected. Furthermore, experimental and field observations show that in most flood routing cases the convective acceleration and the local acceleration are of the same order of magnitude but opposite in sign, thus counter-balancing each other. Equation (2) then reduces to

$$S_{\mathrm{f}} = S - \frac{\partial y}{\partial x} \tag{3}$$

If the assumption is made that the friction slope can be determined as in steady, uniform flow (e.g. Chezy's equation), then

$$S_{\mathrm{f}} = \frac{v^2}{C^2 R} \tag{4}$$

where R is the hydraulic radius, and C is Chezy's roughness coefficient. Substituting Equation (4) into Equation (3), one gets

$$v = CR^{1/2}\left(S - \frac{\partial y}{\partial x}\right)^{1/2} \tag{5a}$$

or

$$Q = ACR^{1/2}\left(S - \frac{\partial y}{\partial x}\right)^{1/2} \tag{5b}$$

For shallow, wide rectangular channels, $R \approx y$ and Equation (5b) becomes

$$\frac{\partial y}{\partial x} = S - \frac{Q^2}{C^2B^2y^3} \tag{6}$$

If the Chezy coefficient and bed slope, S, are assumed constant, Equation (6) can be easily combined with Equation (1). To that end, differentiating Equation (6) with respect to time, yields

$$\frac{\partial^2 y}{\partial t \partial x} = \frac{3Q^2}{B^2C^2y^4}\frac{\partial y}{\partial t} - \frac{2Q}{B^2C^2y^3}\frac{\partial Q}{\partial t} \tag{7}$$

Equation (1) can be written as

$$\frac{\partial y}{\partial t} = -\frac{1}{B}\frac{\partial Q}{\partial x} \tag{8}$$

Differentiating Equation (8) with respect to x, one gets

$$\frac{\partial^2 y}{\partial x \partial t} = -\frac{1}{B}\frac{\partial^2 Q}{\partial x^2} \tag{9}$$

On substituting Equation (9) into Equation (7), the result is

$$-\frac{1}{B}\frac{\partial^2 Q}{\partial x^2} = \frac{3Q^2}{B^2C^2y^4}\frac{\partial y}{\partial t} - \frac{2Q}{B^2C^2y^3}\frac{\partial Q}{\partial t} \tag{10}$$

Substituting Equation (8) into Equation (10), one obtains

$$\frac{\partial Q}{\partial t} = \frac{BC^2y^3}{2Q}\frac{\partial^2 Q}{\partial x^2} - \frac{3Q}{By}\frac{\partial Q}{\partial x} \tag{11}$$

Equation (11) is a highly non-linear one-dimensional second-order diffusion wave equation which includes two unknown variables y and Q, and requires one initial and two boundary conditions for its solution.

MIXING CELL METHOD

Standard mixing cell method

The objective of solving Equation (11) is to determine the water depths and discharges at all points in the channel reach for all time. There is no known general analytical solution for this equation. Numerical schemes are therefore employed to discretize the space and time axes and convert the differential equation to difference form. The structure of Equation (11) suggests the use of the mixing cell method in which the diffusion term can be eliminated by selecting an optimal space interval Δx when time interval Δt is given. By applying the Taylor series expansion to the first-order terms of Equation (11) at (i, k) expanding forward in time and backward in space and neglecting the terms of third and higher order, we get

$$\frac{\partial Q}{\partial t} = \frac{Q_i^{k+1} - Q_i^k}{\Delta t} - \frac{\Delta t}{2}\frac{\partial^2 Q}{\partial t^2} \tag{12}$$

$$\frac{\partial Q}{\partial x} = \frac{Q_i^k - Q_{i-1}^k}{\Delta x} + \frac{\Delta x}{2}\frac{\partial^2 Q}{\partial x^2} \tag{13}$$

In order to express the second term on the right-hand side of Equation (12), we introduce the kinematic wave model, i.e.

$$\frac{\partial Q}{\partial t} + C_{\mathrm{k}}\frac{\partial Q}{\partial x} = 0 \tag{14}$$

where $C_{\mathrm{k}} = 3Q/(2By)$, and is the celerity of the kinematic wave. The last term in Equation (12) can be expressed by employing the kinematic wave model as:

$$\frac{\partial^2 Q}{\partial t^2} = \frac{\partial}{\partial t}\left(\frac{\partial Q}{\partial t}\right) = \frac{\partial}{\partial t}\left(-C_{\mathrm{k}}\frac{\partial Q}{\partial x}\right) = -\frac{\partial C_{\mathrm{k}}}{\partial t}\frac{\partial Q}{\partial x} + C_{\mathrm{k}}\frac{\partial C_{\mathrm{k}}}{\partial x}\frac{\partial Q}{\partial x} + C_{\mathrm{k}}^2\frac{\partial^2 Q}{\partial x^2} \tag{15}$$

C_{k} can be expressed as

$$C_{\mathrm{k}} = aQ^b \tag{15a}$$

where a and b are parameters. Differentiating Equation (15a) with respect to time t and space x, one gets

$$\frac{\partial C_{\mathrm{k}}}{\partial t} = abQ^{b-1}\frac{\partial Q}{\partial t} \tag{15b}$$

$$\frac{\partial C_{\mathrm{k}}}{\partial x} = abQ^{b-1}\frac{\partial Q}{\partial x} \tag{15c}$$

From Equations (15b) and (15c), we find that $-\partial C_{\mathrm{k}}/\partial t = C_{\mathrm{k}}\partial C_{\mathrm{k}}/\partial x$. Therefore, Equation (15) becomes

$$\frac{\partial^2 Q}{\partial x^2} = 2C_{\mathrm{k}}\frac{\partial C_{\mathrm{k}}}{\partial x}\frac{\partial Q}{\partial x} + C_{\mathrm{k}}^2\frac{\partial^2 Q}{\partial x^2} \tag{15d}$$

For an actual flood event, the variation of C_{k} is usually small along the profile section. The example in the next section shows that the average velocity during the flood event at the upstream section is 2·914 m/s and at the downstream section is 3·057 m/s, and the channel reach is 247 km long. For this event, $\mathrm{d}C_{\mathrm{k}}/\mathrm{d}x = 5{\cdot}7 \times 10^{-7}$. Therefore,

$$2C_{\mathrm{k}}\left|\frac{\partial C_{\mathrm{k}}}{\partial x}\frac{\partial Q}{\partial x}\right| \ll C_{\mathrm{k}}^2\left|\frac{\partial^2 Q}{\partial x^2}\right|$$

and this term can be neglected and Equation (15) becomes

$$\frac{\partial^2 Q}{\partial t^2} = C_{\mathrm{k}}^2\frac{\partial^2 Q}{\partial x^2} \tag{15a}$$

Therefore, Equation (12) becomes

$$\frac{\partial Q}{\partial t} = \frac{Q_i^{k+1} - Q_i^k}{\Delta t} - \frac{\Delta t}{2}C_{\mathrm{k}}^2\frac{\partial^2 Q}{\partial x^2} \tag{16}$$

Substituting Equations (16) and (13) into Equation (11), the result is:

$$\frac{Q_i^{k+1} - Q_i^k}{\Delta t} = \frac{C_\mathrm{k}}{\Delta x}(Q_{i-1}^k - Q_i^k) + \left(\frac{\Delta t}{2}C_\mathrm{k}^2 - \frac{\Delta x}{2}C_\mathrm{k} + \frac{3C^2y^2}{4C_\mathrm{k}}\right)\frac{\partial^2 Q}{\partial x^2} \tag{17}$$

To eliminate the second-order term of Equation (17), let

$$\frac{\Delta t}{2}C_\mathrm{k}^2 - \frac{\Delta x}{2}C_\mathrm{k} + \frac{3C^2y^2}{4C_\mathrm{k}} = 0 \tag{18a}$$

This leads to

$$\Delta x = \Delta t C_\mathrm{k} + \frac{3C^2y^2}{2C_\mathrm{k}^2} \tag{18b}$$

Using Equation (18*b*), Equation (17) reduces to

$$Q_i^{k+1} = Q_i^k + \frac{\Delta t C_\mathrm{k}}{\Delta x}(Q_{i-1}^k - Q_i^k) \tag{19}$$

Equation (19) is a mixing cell form of Equation (11). When Δt is given, Δx can be calculated from Equation (18*b*).

Crank–Nicolson method

An explicit form of the Crank–Nicolson method is proposed which is more accurate than the standard form. In this scheme the advective term is approximated as the average of the backward difference approximations at the (i, k) and $(i, k+1)$ grid points as in the conventional Crank–Nicolson scheme (Noye, 1982). The scheme is given by

$$Q_i^{k+1} = Q_i^k + \frac{\Delta t C_\mathrm{k}}{\Delta x}\left(\frac{Q_{i-1}^{k+1} + Q_{i-1}^k}{2} - \frac{Q_i^{k+1} + Q_i^k}{2}\right) \tag{20}$$

Re-arranging Equation (20), we get

$$Q_i^{k+1} = \left(\frac{2 - C_\mathrm{r}}{2 + C_\mathrm{r}}\right)Q_i^k + \left(\frac{C_\mathrm{r}}{2 + C_\mathrm{r}}\right)(Q_{i-1}^{k+1} + Q_{i-1}^k) \tag{21}$$

where $C_\mathrm{r} = \Delta t C_\mathrm{k}/\Delta x$ and expresses the ratio of the kinematic wave celerity to the computational grid celerity. Equation (21) can be easily solved numerically. The solution procedure is then identical to that outlined above for the standard method.

Linear channel routing method

Without lateral inflow. Let $\mu = BC^2y^3/(2Q)$ and $C_\mathrm{k} = 3Q/(2By)$, then Equation (11) becomes

$$\frac{\partial Q}{\partial t} = \mu\frac{\partial^2 Q}{\partial x^2} - C_\mathrm{k}\frac{\partial Q}{\partial x} \tag{22}$$

If μ and C_k are constant, Equation (22) reduces to a linear diffusion wave equation. The initial and boundary conditions for Equation (22) can be expressed as:

$$Q(x, 0) = 0 \quad (x \geqslant 0) \tag{23}$$

$$Q(0, t) = Q_0(t) \quad (t \geqslant 0) \tag{24}$$

$$\left.\frac{\partial Q(x, t)}{\partial x}\right|_{x=\infty} = 0 \tag{25}$$

Equation (22) has been solved using Laplace transform, separation of variables, characteristic variables, transformed variables, etc. The solution is usually in terms of exponential functions and hyperbolic functions, error functions, modified Bessel functions or harmonic functions (Dooge, 1973; Singh, 1996). The solution here is derived using the mixing cell method. Equation (22), with the initial and boundary conditions given by Equations (23)–(25), can be transformed to an n-order linear differential equation by using the mixing cell concept (Wang and Chen, 1996). Thus

$$k^n \frac{d^n Q(t)}{dt^n} + k^{n-1} C_n^{n-1} \frac{d^{n-1} Q(t)}{dt^{n-1}} + \cdots + C_n^0 Q(t) = Q_0(t) \tag{26a}$$

which can be written in terms of linear operators as

$$Q(t) = \frac{Q_0(t)}{(1 + kD)^n} \tag{26b}$$

where $k = 2\mu/C_k^2$ and $C_n^j = n(n-1)(n-2)\ldots(n-j+1)/j!$. Equation (26) is a zero state condition where the output discharge response depends only on the input. If $Q_0(t) = Q_0\,\delta(t)$, the output discharge of Equation (26) is

$$Q(t) = \frac{Q_0}{k\Gamma(n)} \mathrm{e}^{-t/k} \left(\frac{t}{k}\right)^{n-1} \tag{27}$$

The boundary condition given by Equation (24) can be written in terms of a unit step function (Singh, 1988; Wang and Wu, 1983) as

$$Q_0(t) = I_0 u(t) + (I_1 - I_0)u(t - \Delta t) + \cdots + (I_m - I_{m-1})u(t - m\Delta t) - I_m[t - (m+1)\Delta t] \tag{28}$$

With the expression of inflow of the upper cross-section, the solution of Equation (26) can be obtained (Singh, 1988; Wang and Wu, 1983; Wang and Chen, 1995) as

$$Q(t) = \sum_{j=0}^{m} A_j u(t - j\Delta t)\left[1 - \sum_{i=0}^{n-1} (i!)^{-1} \left(\frac{(t - j\Delta t)}{k}\right)^i \exp\left(-\frac{(t - j\Delta t)}{k}\right)\right] \tag{29}$$

where $A_0 = I_1 - I_0$, $A_1 = I_2 - I_1, \ldots, A_{m-1} = I_m - I_{m-1}$, $A_m = -I_m$.
Usually, the initial condition in a channel reach is not zero; therefore, Equation (23) becomes

$$Q(x, 0) = Q_v \quad (x \geqslant 0) \tag{30}$$

For the non-zero state response, the output response should include the non-zero state and input response. Therefore, the output hydrograph can be estimated as (Wang and Chen, 1995)

$$\begin{aligned} Q(t) = {} & \sum_{j=0}^{m} A_j u(t - j\Delta t)\left[1 - \sum_{i=0}^{n-1} (i!)^{-1} \left(\frac{t - j\Delta t}{k}\right)^i \exp\left(-\frac{t - j\Delta t}{k}\right)\right] \\ & + C_v \sum_{i=1}^{n} \frac{1}{\Gamma(i)} \exp\left(-\frac{t - j\Delta t}{k}\right)\left(\frac{t - j\Delta t}{k}\right)^{i-1} \end{aligned} \tag{31}$$

With lateral input discharge. For lateral inflow discharge, the ordinary differential equation for the first channel reach is

$$k\frac{\partial Q(t)}{\partial t} + Q(t) = Q_0(t) \tag{32a}$$

For convenience, the first term on right-hand side in Equation (31) may be expressed as $kDQ(t)$. Therefore, Equation (31) becomes

$$Q(t) = \frac{Q_0(t)}{1 + kD} \tag{32b}$$

For the second channel reach, Q_1 is the inflow, Q_{1a} is the lateral inflow discharge and Q_2 is the output discharge, given as

$$Q_2 = \frac{Q_1 + Q_{1a}}{1 + kD} = \frac{Q_0(t)}{(1 + kD)^2} + \frac{Q_{1a}}{1 + kD} \tag{33}$$

For the nth channel reach, Q_{n-1} is the inflow, Q_{2a} is the lateral inflow discharge and Q_n is the output discharge $Q(t)$ given by

$$Q(t) = \frac{Q_0(t)}{(1 + kD)^n} + \frac{Q_{1a}(t)}{(1 + kD)^{n-1}} + \cdots + \frac{Q_{(n-1)a}}{1 + kD} \tag{34a}$$

or

$$Q(t) = \frac{Q_0(t)}{(1 + kD)^n} + \sum_{i=1}^{n-1} \frac{Q_{(n-i)a}}{(1 + kD)^i} \tag{34b}$$

If $Q_0, Q_{1a}, \ldots, Q_{(n-1)a}$ are unit step functions, i.e. $Q_{1a}, Q_{2a}, \ldots, Q_{(n-1)a}$ can be expressed by Equation (28), under zero condition, the solution of Equation (34*b*) is given as

$$\begin{aligned} Q(t) = {} & \sum_{j=0}^{m} Q_0 u(t - j\Delta t)\left[1 - \sum_{i=0}^{n-1} (i!)^{-1}\left(\frac{t - j\Delta t}{k}\right)^i \exp\left(-\frac{t - j\Delta t}{k}\right)\right] \\ & + \sum_{L=1}^{n-1}\sum_{j=0}^{m} Q_{(n-L)a} u(t - j\Delta t)\left[1 - \sum_{i=0}^{L} (i!)^{-1}\left(\frac{t - j\Delta t}{k}\right)^i \exp\left(-\frac{t - j\Delta t}{k}\right)\right] \end{aligned} \tag{35}$$

Equation (35) is the general equation for linear channel routing with lateral inflows. In practice there is only one lateral inflow in the whole channel reach routing. For instance, if the lateral inflow is in the third characteristic reach, Equation (34) reduces to

$$Q_n = \frac{Q_0}{(1 + kD)^n} + \frac{Q_{2a}}{(1 + kD)^{n-2}} \tag{36}$$

and Equation (35) reduces to

$$\begin{aligned} Q_n = {} & \sum_{j=0}^{m} Q_0 u(t - j\Delta t)\left[1 - \sum_{i=0}^{n-1} (i!)^{-1}\left(\frac{t - j\Delta t}{k}\right)^i \exp\left(-\frac{t - j\Delta t}{k}\right)\right] \\ & + \sum_{j=0}^{m} Q_{2a} u(t - j\Delta t)\left[1 - \sum_{i=0}^{n-3} (i!)^{-1}\left(\frac{t - j\Delta t}{k}\right)^i \exp\left(-\frac{t - j\Delta t}{k}\right)\right] \end{aligned} \tag{37}$$

ANALYSIS OF THE MIXING CELL METHOD

Consistency analysis. Applying the Taylor series expansion to the first-order terms in Equation (11) one gets

$$\frac{\partial Q}{\partial t} = \frac{Q_i^{k+1} - Q_i^k}{\Delta t} - \frac{\Delta t}{2}\frac{\partial^2 Q}{\partial t^2} + 0\left(\frac{\Delta t^2}{3!}\right) \tag{38}$$

$$\frac{\partial Q}{\partial x} = \frac{Q_i^k - Q_{i-1}^k}{\Delta x} + \frac{\Delta x}{2}\frac{\partial^2 Q}{\partial x^2} + 0\left(\frac{\Delta x^2}{3!}\right) \tag{39}$$

With the kinematic wave equation introduced to express $\partial^2 Q/\partial t^2$ by $C_k^2\ \partial^2 Q/\partial x^2$, Equation (38) becomes

$$\frac{\partial Q}{\partial t} = \frac{Q_i^{k+1} - Q_i^k}{\Delta t} - \frac{\Delta t}{2} C_k^2 \frac{\partial^2 Q}{\partial x^2} + 0\left(\frac{\Delta t^2}{3!}\right) \tag{40}$$

Substituting Equations (39) and (40) into Equation (11) and arranging, one gets

$$\frac{Q_i^{k+1} - Q_i^k}{\Delta t} = \frac{C_k}{\Delta x}(Q_{i-1}^k - Q_i^k) + \left(\frac{\Delta t}{2} C_k^2 - \frac{\Delta x}{2} C_k + \frac{3C^2 y^2}{4C_k}\right)\frac{\partial^2 Q}{\partial x^2} + 0\left(\frac{\Delta t^2}{3!}, \frac{\Delta x^2}{3!}\right) \tag{41}$$

The diffusion term $\partial^2 Q/\partial x^2$ can be eliminated by selecting an optimal Δx. If $\Delta x = \Delta t C_k + 3C^2 y^2/(4C_k^2)$ is imposed, Equation (41) reduces to Equation (19). Equation (41) expresses that Equation (19) has a truncation error of $0(\Delta t^2/3!, \Delta x^2/3!)$. Thus, the mixing cell method is more accurate than the finite difference method.

Truncation error. Equations (19) and (21) have a truncation error of $(\Delta x^2, \Delta t^2)$. To minimize the error in Equation (19), we can limit the time interval to as small a value as desired. Therefore, when $\Delta t \to 0$, Equation (18*a*) reduces to

$$-\frac{\Delta x}{2} C_k^2 + \frac{3C^2 y^2}{4} = 0 \tag{42}$$

which yields

$$\Delta x = \frac{Q}{S_0 B C_k} = \frac{2y}{3S_0} \tag{43}$$

Relation to the Kalinin–Milyukov method. Kalinin and Milyukov (1957) proposed the concept of characteristic length in which the relationship between storage of the channel reach and the discharge of the lower cross-section shows a straight line. Based on this concept, the formula estimating the characteristic length was obtained as

$$\zeta = \frac{Q_0}{S_0}\left(\frac{\partial H}{\partial Q}\right) = \frac{Q}{S_0 B C_k} \tag{44}$$

where ζ is the characteristic length. Equations (43) and (44) are the same but are derived using different ways and concepts.

Based on the mixing cell concept a linear flood diffusion wave equation is reduced to an n-order ordinary differential equation when $\Delta t \to 0$. The parameter of the linear channel flow routing model is $k = C_k^2/(2\mu)$. However, the routing method of Kalinin and Milyukov (1957) was derived based on water balance and storage release equations in each characteristic length. In this study, the inflow process is expressed by using a unit step function and the Laplace transform method is used to obtain Equations (31) and (35) which can be used directly to calculate the flood hydrograph. For the linear routing method of Kalinin and Milyukov, the S curve was obtained by integrating the routing model and a unit hydrograph of finite duration was derived from the S curve by displacing it by the required amount of lag, subtracting the ordinates and then normalizing the volume. Then, the discharge for a given inflow was obtained by numerical convolution.

Linear flow routing as a special case. Based on Equations (43) or (44), the characteristic length Δx depends on water depth and S_f; and Δx is variable. Therefore, the linear routing model is a special case. Actually, if we take the limit of Δt, then

$$\lim_{\Delta t \to 0} \frac{Q_i^{k+1} - Q_i^k}{\Delta t} = \frac{dQ_i}{dt} \tag{45}$$

Substituting Equations (43) and (45) into Equation (17), gives

$$\frac{dQ_i}{dt} = \frac{C_k}{\Delta x}(Q_{i-1} - Q_i) = \frac{3S_f C_k}{2y}(Q_{i-1} - Q_i) \tag{46}$$

When $i = 1$, Q_{i-1} becomes inflow expressed by I. Thus, Equation (46) reduces to

$$\frac{dQ_i}{dt} = \frac{3S_f C_k}{2y}(I - Q_1) \tag{47-1}$$

where Q_1 is the output discharge from the first channel characteristic reach, the length of which is Δx. For the second channel reach, Q_1 is the input discharge and Q_2 is the output discharge, expressed as

$$\frac{dQ_2}{dt} = \frac{3S_f C_k}{2y}(Q_1 - Q_2) \tag{47-2}$$

In a similar manner, for the nth channel reach, Q_{n-1} is the inflow discharge, and Q_n is the output discharge from the nth reach expressed as

$$\frac{dQ_n}{dt} = \frac{3S_f C_k}{2y}(Q_{n-1} - Q_n) \tag{47n}$$

Equations (47-1)-(47–n) are a number of n first-order non-linear ordinary differential equations.

As mentioned above, Δx changes with discharge and the size of the space step is not equal. Therefore, the mixing cell method requires a large amount of interpolation.

Equation (11) is a one-dimensional non-linear diffusion wave equation. The mixing cell method cannot be used to solve this equation numerically because of the non-linearity of the first-order and second-order spatial derivative terms. Fortunately, the kinematic wave model is introduced to replace the term $\partial^2 Q/\partial t^2$ by $C_k^2 \partial^2 Q/\partial x^2$. This makes the mixing cell method solve the numerical diffusion wave with more accuracy and simplicity than the finite difference method.

EXAMPLE APPLICATIONS

The following examples are taken to illustrate applications of the mixing cell method and continuous linear channel flow routing method. The first example is to use the mixing cell methods and the linear channel routing method to the Yuanling–Wangjiahe River reach, and the second is to apply the mixing cell methods to the Longjie–Qiaojia River reach.

Table I. Inflow and outflow data (m^3/s)

Time (hours)	Yuanling station	Wangjiahe station	Longjie station	Qiaojia station
0	1950		2490	
3	1960		4655	
6	1970		10 100	
9	1980		13 550	
12	1990		13 050	2410
15	2080		11 070	2410
18	2330	2050	8850	5950
21	2910	2050	6670	10 300
24	4840	2150	5380	11 400
27	6750	2350	5000	11 200
30	6870	3450	4640	9750
33	6140	5710	4300	8300
36	5210	6650	4065	6760
39	4560	6400	3935	5670
42	4140	5900	3830	4910
45	3840	5250	3745	4410
48	3570	4710		
51	3330	4350		
54	3100	4000		
57	2870	3700		
60	2650	3400		
63	2480	3200		
66	2350	2980		
69	2240	2780		
72	2140	2570		

First example

The Yuanling–Wangjiahe River reach is 112 km long and has a mean bed slope $S_0 = 0{\cdot}0004$. In August 1968 a flood event was measured at two stations and is shown in Table I. Based on hydraulic elements of the two cross-sections, the relationships between kinematic wave celerity C_k, characteristic reach length ζ and inflow Q were determined by using the least-squares method:

$$C_k = 0{\cdot}047 Q^{0{\cdot}5}$$

and

$$\zeta = 7{\cdot}2886 Q / C_k$$

Standard mixing cell method. In this example, $\Delta t = 3600$ s, the first interval inflow is 1950 m^3/s, the second interval flow is 1960 m^3/s, third interval flow is 1970 m^3/s, and so on. The initial value is: $t = 0$, $Q = 20{\cdot}0$ m^3/s.

For the first time interval:

(1) For the first reach,

$$\begin{aligned}\Delta x &= \Delta t \times C_k + \zeta = 3600 \times 0{\cdot}047 \times 1950^{0{\cdot}5} + 7{\cdot}2886 \times 1950/C_k \\ &= 7471{\cdot}7 + 6849{\cdot}5 = 14\,321{\cdot}2 \text{ m}\end{aligned}$$

Substituting these values into Equation (19), one gets

$$Q_1^2 = Q_1^1 + \frac{\Delta t \times C_k}{\Delta x}(Q_0^1 - Q_1^1) = 20 + 0{\cdot}5217 \times 1930 = 1026{\cdot}9 \text{ m}^3/\text{s}$$

(2) For the second reach,

$$Q_2^2 = Q_2^1 + \frac{\Delta t \times C_k}{\Delta x}(Q_1^1 - Q_2^1) = 20{\cdot}0 + 0 = 20{\cdot}0 \text{ m}^3/\text{s}$$

Similarly, $Q_3^2 = Q_4^2 = \cdots = 20{\cdot}0 \text{ m}^3/\text{s}$.

For the second time interval:

(1) For the first channel reach

$$\Delta x = \Delta t \times C_k + \zeta = 3600 \times 0{\cdot}047 \times 1960^{0{\cdot}5} + 7{\cdot}2886 \times 1960/2{\cdot}08 = 14\,358{\cdot}9 \text{ m}$$

Substituting these values into Equation (19), one gets

$$Q_1^3 = Q_1^2 + \frac{\Delta t \times C_k}{\Delta x}(Q_0^2 - Q_1^2) = 1026{\cdot}9 + \frac{7490{\cdot}8}{14\,358{\cdot}9}(1960 - 1026{\cdot}9)$$
$$= 1026{\cdot}9 + 486{\cdot}8 = 1513{\cdot}7 \text{ m}^3/\text{s}$$

(2) For the second channel reach

$$\Delta x = \Delta t \times C_k + \zeta = 3600 \times 0{\cdot}047 \times 1026{\cdot}9^{0{\cdot}5} + 7{\cdot}2886 \times 1026{\cdot}9/C_k$$
$$= 5422{\cdot}1 + 4669{\cdot}9 = 10\,392 \text{ m}$$

(3) For the third channel reach

$$Q_2^3 = Q_2^2 + \frac{\Delta t \times C_k}{\Delta x}(Q_1^2 - Q_2^2) = 20 + 0{\cdot}52173 \times 1007{\cdot}1 = 545{\cdot}3 \text{ m}^3/\text{s}$$
$$Q_3^3 = Q_3^2 + \frac{\Delta t \times C_k}{\Delta x}(Q_2^2 - Q_3^2) = 20{\cdot}0 \text{ m}^3/\text{s}$$

In the same way, we calculate flows for the third, fourth, etc., time interval. The result is shown in the Figure 1.

Crank–Nicolson method. For the Crank–Nicolson method, the procedure is the same as the standard mixing cell method. The result is also presented in the Figure 1 for comparison.

Linear channel routing method. The characteristic channel reach length Δx and the celerity of the kinematic wave can be obtained based on the data of the observed water level, discharge and cross-sectional average velocity: $\Delta x = 10{\cdot}18$ km and $C_k = 3{\cdot}327$ m/s.

The number of characteristic channel reaches is estimated as

$$N = \frac{L}{\zeta} = \frac{112}{10{\cdot}18} = 11$$

and the time parameter k_ζ expressed as

$$k_\zeta = \frac{2\mu}{C_k^2} = \frac{\zeta}{C_k} = \frac{10{\cdot}18}{3{\cdot}327} = 0{\cdot}85h$$

Substituting these values into Equation (28), one gets

$$Q(t) = \sum_{j=0}^{m} A_j U(t - j\Delta t)\left[1 - \sum_{i=0}^{10}(i!)^{-1}\left(\frac{t - j\Delta t}{0{\cdot}85}\exp - \frac{t - j\Delta t}{0{\cdot}85}\right)\right]$$

where $m = 27$, $\Delta t = 3$ h. The discharge values computed are shown in Figure 1.

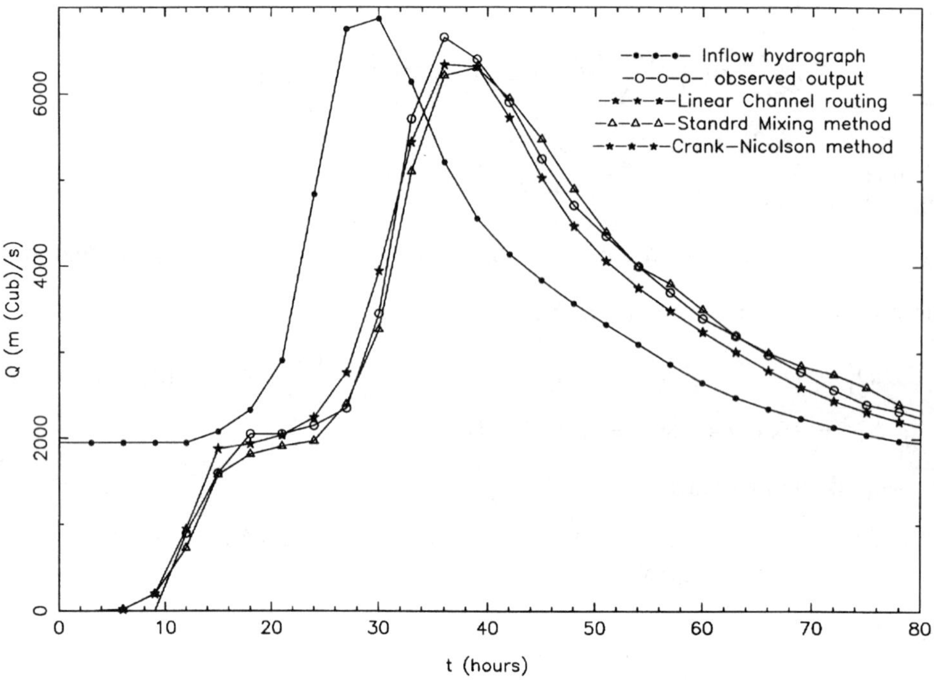

Figure 1. Comparison of calculated and observed hydrographs for Yuanling and Wangjiahe stations

Second example

The Longjie–Qiaojia River reach is located upstream of the Yangtze River in China. The length of the reach is 247 km and the mean bed slope is 0·00112. On 18–20 June 1967 a flood event was measured at the two stations and is shown in Table I. Based on hydraulic elements of the two cross-sections, the relationships between kinematic wave celerity C_k, characteristic reach length ζ and inflow Q were determined by using the least-squares method

$$C_k = 0{\cdot}18Q^{0{\cdot}36}$$

and

$$\zeta = 4150{\cdot}0 + 0{\cdot}85Q$$

Standard mixing cell method. In this example, $\Delta t = 3600$ s, the first interval inflow is 2490 m^3/s, the second is 2510 m^3/s, the third is 2530 m^3/s, and so on. The initial value is $t = 0$, $Q = 20{\cdot}0$ m^3/s.

For the first interval time

(1) For the first reach

$$\begin{aligned}\Delta x &= \Delta t \times C_k + \zeta = 3600 \times 0{\cdot}18 \times 2490^{0{\cdot}36} + 4150 + 0{\cdot}85 \times 2490\\ &= 12\,021{\cdot}7 + 6266{\cdot}5 = 18\,288{\cdot}2 \text{ m}\end{aligned}$$

Substituting these values into Equation (19) gives

$$Q_1^2 = Q_1^1 + \frac{\Delta t \times C_k}{\Delta x}(Q_0^1 - Q_1^1) = 20{\cdot}0$$

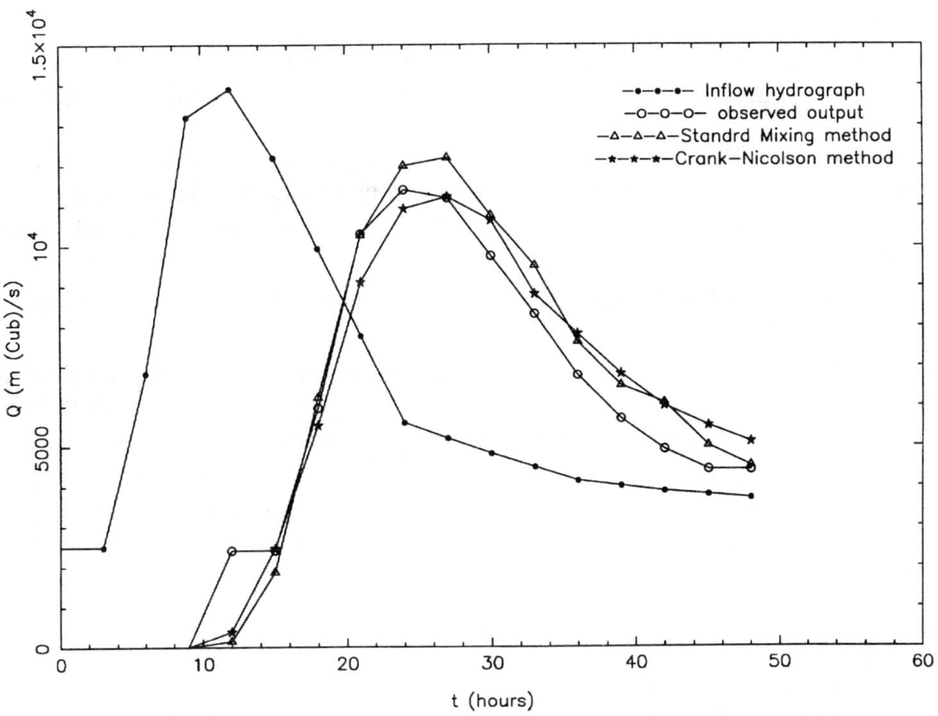

Figure 2. Comparison of calculated and observed hydrographs for Longjie and Giaojia stations

(2) For the second reach,

$$Q_2^2 = Q_2^1 + \frac{\Delta t \times C_k}{\Delta x}(Q_1^1 - Q_2^1) = 20{\cdot}0 + 0 = 20{\cdot}0\ \mathrm{m^3/s}$$

Similarly, $Q_3^2 = Q_4^2 = \cdots = 20{\cdot}0\ \mathrm{m^3/s}$.

In the same way, we calculate flows for the second, third, fourth, etc. time intervals. The results of the computation are shown in Figure 2. For the Crank–Nicolson method, the procedure is the same as the standard mixing cell method. The results of the computation are also shown in Figure 2 for comparison. For the linear channel reach routing method, the calculation was not carried out because the channel reach is too long and the number of characteristic reaches is more 200.

Discussion of results

The mixing cell method and the linear channel flow routing model compare well with the observed flood hydrographs for two flood events. Figure 1 shows that the hydrographs calculated by both standard and Crank–Nicolson mixing cell methods are close to the observed flood hydrographs. The standard mixing cell method is more accurate than the Crank–Nicolson mixing cell method. Figure 1 shows that the mixing cell method can be used to solve the non-linear flood diffusion wave equation and to predict the flood hydrograph. Example 1 illustrates that the procedure is easy and simple because the difficulties of the second-order derivative are avoided. Figure 1 demonstrates that the results of the linear channel routing model almost exactly match the observed flood hydrograph. It is clear that the predictions of the mixing cell and linear channel routing model are in agreement with the observed flood hydrographs. Figure 2 also shows that the hydrographs calculated by both standard and Crank–Nicolson mixing cell methods are close to the observed flood hydrographs. However, the Crank–Nicolson method is more accurate than the standard mixing cell method in this case.

CONCLUSIONS

Based on the mixing cell concept, the mixing cell and linear channel routing methods were developed for solving the diffusion wave equation. Based on application of these methods the following conclusions can be drawn.

1. The structure of the non-linear diffusion wave equation suggests the use of the mixing cell method in which the diffusion term can be eliminated by calculating an optimal space size Δx. Therefore, the mixing cell method only contains the first-order finite difference form. This not only saves computation time but also avoids numerical dispersion.

2. The algebraic formulae of the mixing cell method for channel flood routing are explicit, and the formulae are simple to use and programme.

3. The linear channel flood routing model can be used to solve flood routing with multiple tributaries in a natural river. When the tributary flows enter any characteristic reach of the main channel, the linear channel routing model with lateral input discharge can be used to obtain the output discharge.

4. The mixing cell method and the linear channel routing method proposed in this study do not need optimization of parameters from the observed input and output data. The cross-sectional hydraulic elements such as water depth, width, average velocity and so on can be used to determine their parameters.

REFERENCES

Appelo, C. A. J. and Willemsen, A. 1987. 'Geochemical calculations and observations on salt water intrusions. 1. A combined geochemical/mixing cell method', *J. Hydrol.*, **94**, 313–330.

Bajracharya, K. and Barry, D. A. 1992. Mixing cell models for nonlinear nonequilibrium single species adsorption and transport', *Wat. Resour. Res.*, **29**(5), 1405–1413.

Dooge, J. C. I. 1973. 'Linear theory of hydrologic systems', US Dept Agric., Res. Serv., Tech. Bull. No. 1468. U.S. Department of Agriculture, Agricultural Research Service, Washington, D.C.

Kalinin, J. P. and Milyukov, B. I. 1957. 'On the computation of unsteady flow in open channels', *Meteor. Gidrol. Zh.*, **12**, 12–18.

Noye, J. 1982. 'Finite difference methods for partial differential equations, in Noye, J. (Ed.), *Numerical Solutions of Partial Differential Equations*. North-Holland, New York. 647 pp.

Singh, V. P. 1988. *Hydrologic System: Rainfall–Runoff Modeling*. Prentice Hall, Englewood Cliffs.

Singh, V. P. 1996. *Kinematic Wave Modeling in Water Resources: Surface–Water Hydrology*. John Wiley and Sons, New York.

Van Ommen, H. C. 1985. 'The mixing cell concept applied to transport of non-reactive and reactive components in soils and groundwater', *J. Hydrol.*, **78**, 210–213.

Wang, G.-T. and Chen, S. 1996. 'A new model describing convective-dispersive phenomena derived by using the mixing-cell concept', *Appl. Math. Model.*, in press.

Wang, G.-T. and Wu, K. 1983. 'The unit-step function response for several hydrological conceptual models', *J. Hydrol.*, **62**, 119–128.

9

INVESTIGATING TWO-DIMENSIONAL, FINITE ELEMENT PREDICTIONS OF FLOODPLAIN INUNDATION USING FRACTAL GENERATED TOPOGRAPHY

P. D. BATES[1],* M. HORRIT[2] AND J.-M. HERVOUET[3]

[1]*Department of Geography, University of Bristol, Bristol, BS8 1SS, UK*
[2]*Environmental Science Systems Centre, University of Reading, Whiteknights, Reading, UK*
[3]*Laboratoire National d'Hydraulique, EDF DER, 6, Quai Watier, Chatou 78401, France*

ABSTRACT

Two-dimensional, finite element hydraulic models have been developed to simulate river flood flows at high spatial and temporal resolutions over river reach lengths of 1–60 km. Such models have been shown to be capable of simulating bulk flow; however, model redesign to predict spatially distributed hydraulic variables has been constrained by lack of suitable topographic and hydraulic data. Here we begin this development process using a hypothetical river channel/floodplain domain where the topographic surface is parameterized using scaling information derived from a fractal analysis of a real floodplain DTM. This is used to test the relative effect of the boundary friction calibration, numerical model grid resolution, topography sampling error and floodplain relative height on model predictions of outflow discharge, inundation extent and local hydraulic variables. Simulations indicate that model calibration is the dominant factor affecting the above three quantities. Moreover, model sensitivity to spatially uniform change is shown to be simple for bulk flow and inundation extent but spatially complex for local hydraulics. The study has a number of implications for model calibration and set-up procedures, as well as indicating the need to develop a new suite of analysis techniques for this class of model.

KEY WORDS floodplains; hydraulic modelling; finite elements; fractals

CURRENT RESEARCH PROBLEMS IN PREDICTING FLOODPLAIN INUNDATION

In this paper we address the numerical simulation of free surface flows in compound channels. For such flows, construction of accurate predictive models remains a fundamental research problem. Such studies are not only of importance to engineers and planners engaged in the water resources field, but are also an important prerequisite for modelling investigations of floodplain sediment transport (James, 1985; Pizzuto, 1987; Marriot, 1992; Falconer and Chen, 1996; Walling *et al.*, 1996), contaminant transport (Marron, 1989; Burt *et al.*, 1993), floodplain ecology (Jolly, 1996; Richards *et al.*, 1996) and catchment hydrology (Beven, 1985; Abbott *et al.*, 1986; Beven *et al.*, 1987), which either require hydraulic data as an input or incorporate a submodel for hydraulic processes. For all these tasks we require a model capable of predicting spatially distributed estimates of hydraulic variables such as water depth and flow velocity.

---

* Correspondence to: P. D. Bates, Department of Geography, University of Bristol, Bristol, BS8 1SS, UK

Contract grant sponsor: UK NERC.
Contract grant number: GR3/09925.

In recent years increasingly sophisticated questions have been asked about the spatial patterns and dynamics of flow and associated sediment transport through complex, channel–floodplain systems. This has led to the development and widespread use of two-dimensional models for reach-scale hydraulic problems (Gee *et al.*, 1990; Baird *et al.*, 1992; Bates *et al.*, 1992; Feldhaus *et al.*, 1992) in order to improve the simulation of floodplain inundation over previous one-dimensional approaches. However, Bates *et al.* (1994, 1995) have shown that, whilst it has been possible to prove that such models can be successful in simulating bulk hydraulic variables such as discharge and stage hydrographs after appropriate calibration, a convincing validation of the distributed water depth, flow velocity and inundation extent predicted by such models has not, thus far, been obtained. In reality, data sets capable of fully validating high resolution, two-dimensional flow models do not exist. Moreover, detailed testing of models has also been hampered by a general lack of understanding regarding the dominant controls on flood prediction. Multiple adjustable parameters, complex data requirements and boundary conditions all lead to considerable difficulties in interpreting those modelling results that we do obtain. Carefully constructed sensitivity analyses can, however, help to overcome this problem and provide a baseline understanding of system response to various modelling choices.

While previous sensitivity studies have individually identified the importance of model calibration (Bates *et al.*, 1996), topographic specification (Lane *et al.*, 1994a,b; Bates and Anderson, 1996; Walling *et al.*, 1996) and the resolution of the numerical model discretization (Bates *et al.*, 1996), no comprehensive attempt has yet been made to quantify their relative importance, or to provide a strategy for field data collection for two-dimensional, finite element models. Moreover, at present, it is unknown whether effects such as these are more or less important areas for future research than, for example, further development of numerical solution techniques or the methodology used for sensitivity analysis. Indeed, there is a general lack of information available concerning how such models should be applied. The focus of this paper is therefore to set forth an analytical framework for exploring the sensitivity of two-dimensional hydraulic models to a range of scale and parameter choices. Specifically, we present an investigation of calibration, topography and mesh resolution effects on two-dimensional finite element model predictions in order to achieve a more clearly prioritized strategy for model development and redesign.

RESEARCH DESIGN

Sensitivity analyses for two-dimensional hydraulic models may be conducted using either field data sets or hypothetical, yet realistic, test reaches. Using actual field data includes real-world complexity and avoids problems of scaling, but measurement errors may be complex and the data can contain significant gaps or spatial heterogeneities. A better solution may therefore be to employ hypothetical domains in order to generate a controlled, data-rich framework for model investigation. However, a problem arises owing to the degree of simplification usually employed in the construction of such domains. Previous investigations (Bates *et al.*, 1996) have used simple planar floodplain topographies and straight or sinusoidal channels with uniform slope. Planar surfaces bear little relation to known river long profiles and floodplain micro-topographic variations (Walling *et al.*, 1996) which may affect the frictional energy losses though the reach. In effect this form of drag will be subsumed within the friction term and one may overestimate the significance of this parameter. Consequently, we cannot be confident that such results can be easily generalized to other, more realistic, geometries or models. Thus, a clear need exists to define a method of developing synthetic model topography with realistic scaling properties in order to provide a more robust framework for sensitivity analyses.

At present very little research has been conducted on the development of synthetic topographies, although a number of techniques do present themselves. For the purposes of this paper three potential methods exist, which rely on various amounts of seed data for success. At the most basic level a methodology could be developed that has no reliance on field data but merely uses an appreciation of the typical length scales of topographic variation for particular environments. Each typical length scale may then be transformed into a

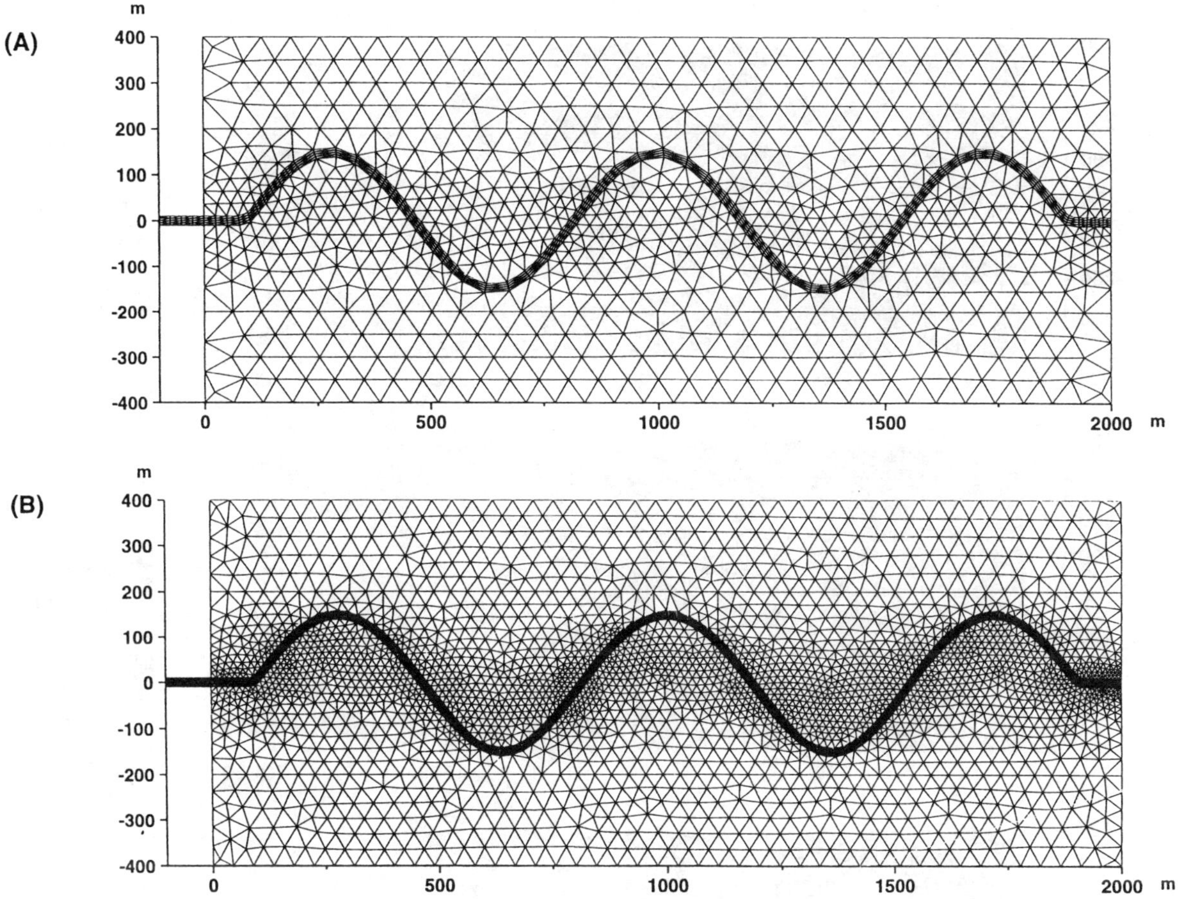

Figure 1. Finite element discretizations developed for the river channel/floodplain domain. Mesh A consists of 2284 elements and 1199 nodes. Mesh B consists of 7310 elements and 3746 nodes

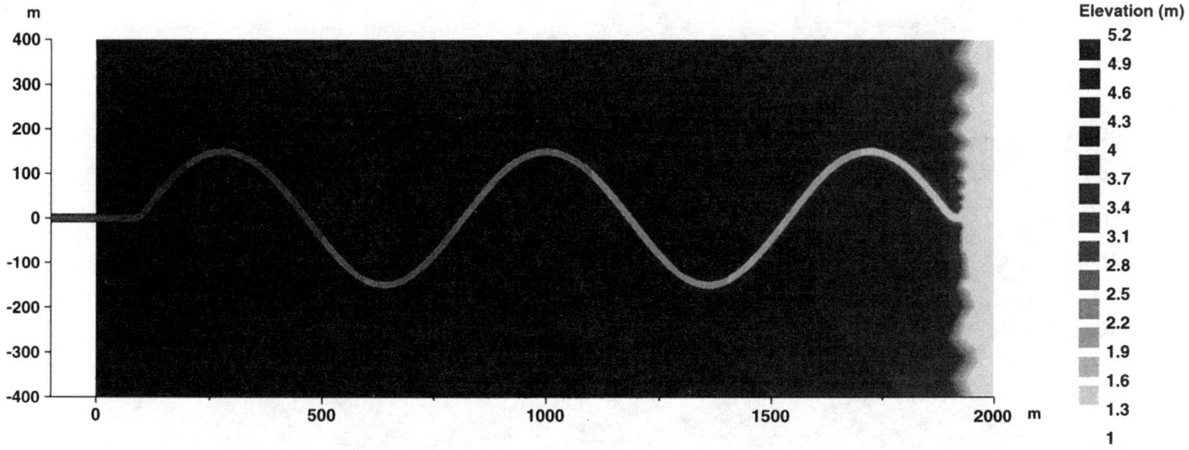

Figure 2. Initial planar floodplain topography prior to perturbation using a fractal scaling analysis of real topography

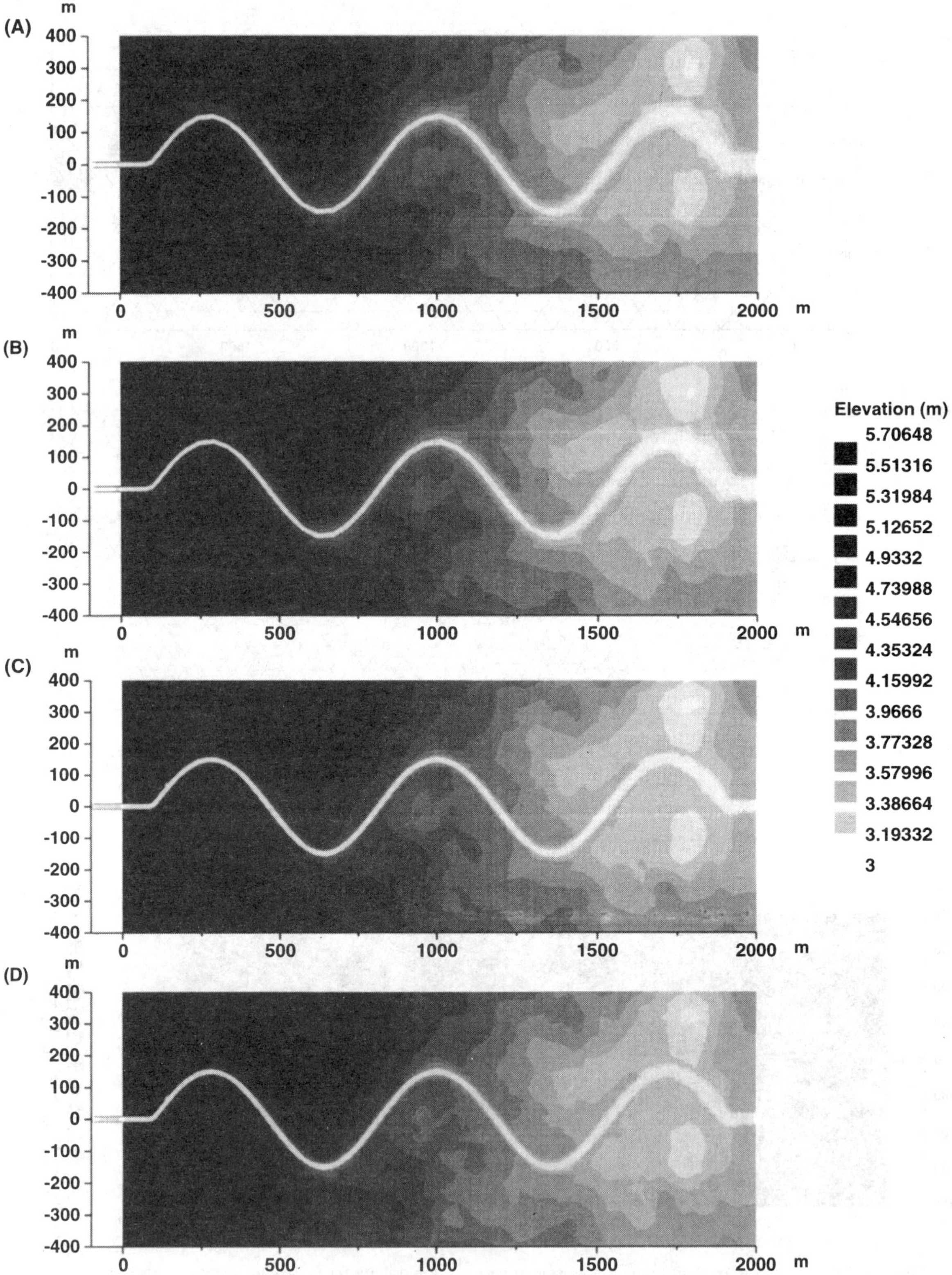
(A)
(B)
(C)
(D)
m
400
300
200
100
0
-100
-200
-300
-400
0
500
1000
1500
2000
Elevation (m)
5.70648
5.51316
5.31984
5.12652
4.9332
4.73988
4.54656
4.35324
4.15992
3.9666
3.77328
3.57996
3.38664
3.19332
3

single sine function with given amplitude and wavelength. The individual sine functions may then be combined to create pseudo-realistic variations in a topographic surface. Secondly, kriging techniques (Gambolati and Volpi, 1979; Hughes and Lettenmaier, 1981; Clarke, 1994) could be used to interpolate realistic topographic surfaces stochastically from sparse field surveys. While these approaches are relatively simple, the potential for bias owing to subjectivity or lack of data is relatively high. A better, but more data-dependent, method involves statistical analysis of scaling in topographic data to develop laws of self-similarity in terrain information that can categorize terrain heterogeneity over a broad range of scales (see, for example, Culling and Datko, 1987). This is the approach taken here, where we have used an analysis of the fractal geometry of real floodplain terrain to generate a synthetic DTM (digital terrain model) for hydraulic model parameterization using the method given by Turcotte (1992).

A digital terrain model of a 12 km reach of the floodplain of the River Stour, Dorset, UK, was used as the seed data for the procedure. This is a small river, approximately 10 m in width, meandering within a floodplain belt up to 700 m wide. The DTM was generated from height and contour information contained in the UK Ordnance Survey 1:10 000 series maps of the area, converted into a 10 × 10 m grid, and approximately 25 surveyed cross-sections of the channel and floodplain collected by the UK Environment Agency. As the surveyed cross-sections were expected to be accurate to within a few cm these were taken as 'ground truth'. Channel bank and bed slopes were linearly interpolated between these surveyed cross-sections to generate the channel-long profile, and, finally, the Ordnance Survey map data were used to 'fill in' the remaining floodplain regions. A series of post-processing steps were then developed to identify and correct points where the map data were not consistent with the ground survey. Errors were expected to show up as steep gradients at the boundary between each data set; however, no such circumstances were noted. The resulting DTM was estimated to have a worst case vertical accuracy of ± 30 cm, although this will improve towards each surveyed cross-section. In terms of typically available field data, this DTM is of relatively high resolution and accuracy and has been used to develop model applications with reasonable success (Bates *et al.*, 1996). It was also felt that it compared favourably with coarse scale floodplain topographic features and would therefore be suitable for an analysis of the fractal geometry at the grid resolution of typical reach-scale finite element models (*c.* 30–50 m).

To undertake the fractal analysis, features with a length scale of less than 200 m were analysed by determining the variogram for lags from 0 to 200. These features were found to have a fractal dimension of 2·4 and a variance of 0·4. This is higher than the values found by Culling and Datko (1987). However, the variogram method used did not address such issues as anisotropy and spatial variations in the correlation surface, which would be required for a more complete analysis. A regular grid of dimensions 512 × 128 was constructed and a height at each node set to a random value using a normal distribution and unit variance. A fast Fourier transform was performed on the grid of random numbers and each component in Fourier space multiplied by a factor representing a function of the radial wavenumber, k. For large k, corresponding to wavelengths less than 200 m the function is

$$k^{-(4-D)} \tag{1}$$

where D is the fractal dimension required. For small k, the function tends to 1. These data were transformed back into real space using a further fast Fourier transform and the resulting data set was normalized to give the required variance and a ground resolution of approximately 10 × 10 m. This generated a random perturbation at each grid point with an rms height of 0·223 m which could be used to generate real topographic variation at a range of length scales and which was comparable with the DEM quality identified in

Figure 3. Floodplain topographic surfaces used in the study. These were generated using a fractal perturbation imposed on to the planar floodplain topography shown in Figure 2 to produce two digital terrain models at 10 m × 10 m and 20 m × 20 m resolution, which were then sampled onto the two finite element mesh discretizations in Figure 1. Shown are (A) low resolution mesh and 10 m × 10 m DTM; (B) low resolution mesh and 20 m × 20 m DTM; (C) high resolution mesh and 10 mx 10 m DTM; and (D) high resolution mesh and 20 m × 20 m DTM this study. The 512 × 128 DTM was then degraded to give a second grid of random perturbations with a ground resolution of approximately 20 × 20 m.

These two perturbation surfaces were then used to generate a series of topographic surfaces for use with a two-dimensional, finite element model of compound channel flow in a hypothetical channel–floodplain reach. The domain used consisted of a 20 m wide, 2 m deep, sinuous channel flowing down the centre of a floodplain domain 2000 m long and 800 m wide. Thus, the dimensions of the model used were similar to those of the River Stour on which we conducted the fractal analysis of floodplain geometry. Bank-full discharge was approximately 12 $m^3 s^{-1}$. Two finite element discretizations of the domain were constructed at different mesh resolutions. These consisted of 2284 and 7310 elements (1199 and 3746 nodes), respectively, thereby giving average element sizes of approximately 58·7 and 30·5 m (see Figure 1). The higher resolution mesh therefore represents a 310% increase in mesh density and a 50% reduction in element size over the smaller mesh. Finally, the gradient in element size from the channel to the boundary of the domain was steeper in the higher resolution mesh. Although this may potentially be an important mesh quality criterion in its own right, its impact was not analysed in the present study for the sake of simplicity.

Floodplain topography was assumed to be planar in the initial configuration for each finite element discretization (see Figure 2). The field of random perturbations generated using the fractal analysis was then used to generate microtopographic variations on to this initial floodplain at each nodal point. The combination of two mesh resolutions and two topography resolutions gave a total of four models in all, thus enabling controlled testing of the relative impact of mesh and topographic resolution on model results. Figure 3 also demonstrates that a fractal analysis method is capable of producing apparently realistic synthetic floodplain topographies with a high level of structural complexity, which compare well with the microtopography noted in high resolution surveys of real floodplains (Walling *et al.*, 1996; Nicholas and Walling, 1998 this issue). Despite appearing similar, variations in height of the topographic surface at individual points are of the order of ±0·11 m, approximately equivalent to the rms error. This can be seen in Figure 4 where we plot the difference surface between the 10 × 10 and 20 × 20 DTMs when sampled on to the low and high resolution finite element meshes.

The two-dimensional hydraulic model, TELEMAC-2D (Hervouet, 1989, 1993), was then used to simulate a dynamic medium-sized flood event passing through each of the four domains shown in Figure 3. The TELEMAC-2D code was selected since it has undergone considerable validation (Hervouet and Janin, 1994; Hervouet and Van Haren, 1996) against analytical solutions and a range of fluvial flow problems including floodplain inundation (Bates *et al.*, 1994, 1995, 1996). The code solves the second-order partial differential equations for depth-averaged fluid flow derived from the full three-dimensional Navier–Stokes equations (the shallow water equations) and is fully described by Stewart *et al.* (1998 this issue) to which the reader is referred for further information. Boundary conditions for the model consisted of a prescribed dynamic discharge at the upstream inflow and a fixed water surface elevation at the downstream outflow. This combination was chosen in preference to a freely varying water surface elevation at the lower boundary in order to obtain a well-posed problem under the theory of characteristics. For real applications a dynamic water surface elevation would be prescribed at this boundary from gauging station data, although this was obviously not available in this case. A dynamic lower boundary condition could have been assumed; however, it was felt that this would have led to additional complications in the interpretation of results. Test simulations showed that any enhancement of water surface slopes only occurred within an extremely localized region less than 50 m from the lower boundary and that the upstream water levels were insensitive to this simplifying assumption. Moreover, as the overall range in water surface elevation (approximately 60 cm) during the flood event was relatively small, the effect of this boundary condition approximation could be expected to be minimal.

The event hydrograph used in the simulations was scaled from real discharge data to account for the dimensions of the hypothetical domain and to ensure significant, but not total, floodplain inundation. The event lasted approximately 19·5 hours in real time, with a peak discharge of 40 $m^3 s^{-1}$ and a total volume of approximately 1 728 000 m^3. This was then discretized into 35 000 time-steps of 2 s duration. Separate

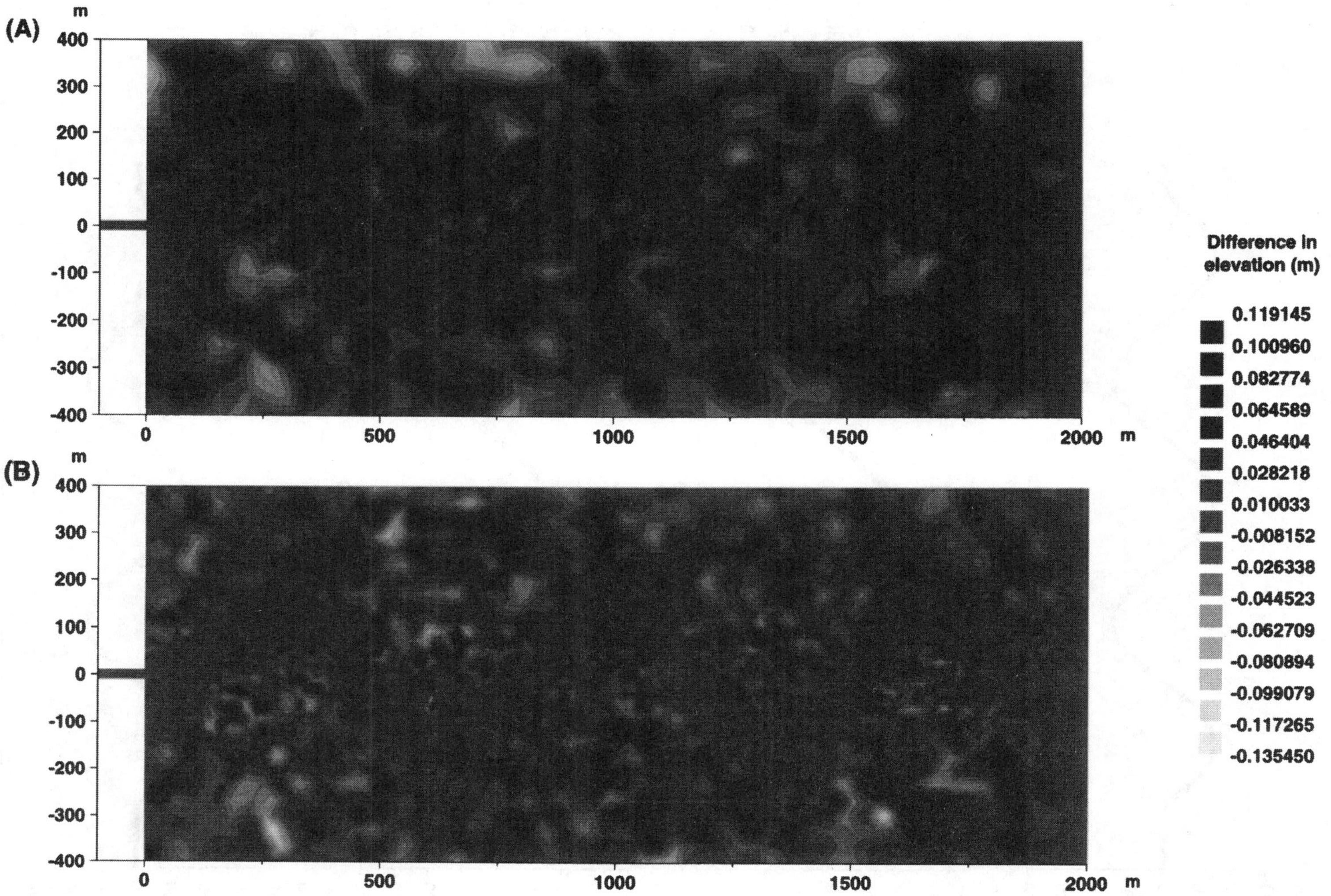

Figure 4. Difference in elevation generated by changing mesh DTM resolution from 10 m × 10 m to 20 m × 20 m for the (A) low resolution and (B) high resolution finite element mesh discretizations shown in Figure 1

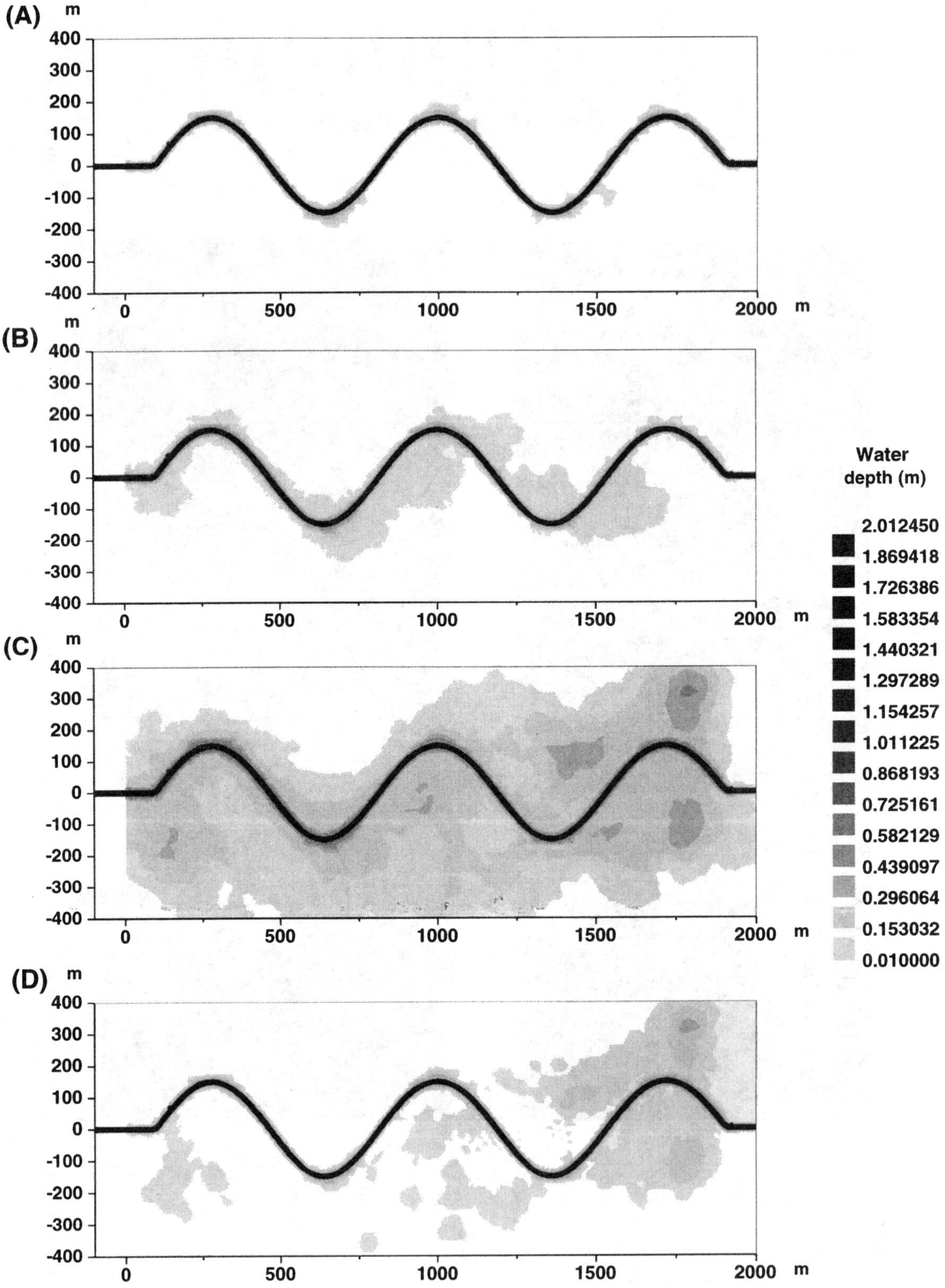

(A)
(B)
(C)
(D)
m
400
300
200
100
0
-100
-200
-300
-400
0
500
1000
1500
2000
Water depth (m)
2.012450
1.869418
1.726386
1.583354
1.440321
1.297289
1.154257
1.011225
0.868193
0.725161
0.582129
0.439097
0.296064
0.153032
0.010000

Table I. Simulations realized in the sensitivity analysis

Simulation number	Mesh resolution (m)	DTM resolution	Friction coefficients	Raised floodplain
1	Low	10	0·021 0·056	No
2 (control simulation)	Low	10	0·025 0·067	No
3	Low	10	0·031 0·083	No
4	Low	20	0·021 0·056	No
5	Low	20	0·025 0·067	No
6	Low	20	0·031 0·083	No
7	High	10	0·021 0·056	No
8	High	10	0·025 0·067	No
9	High	10	0·031 0·083	No
10	High	20	0·021 0·056	No
11	High	20	0·025 0·067	No
12	High	20	0·031 0·083	No
13	Low	10	0·025 0·067	Yes
14	Low	20	0·025 0·067	Yes
15	High	10	0·025 0·067	Yes
16	High	20	0·025 0·067	Yes

Manning's n friction factors were applied in the main channel and floodplain with respective values of 0·025 and 0·067 chosen for the control simulation on each mesh resolution/topography resolution combination. Each simulation was then repeated using two different friction calibration sets: one below the control value ($n_{channel} = 0{\cdot}021$ and $n_{floodplain} = 0{\cdot}056$) and one above ($n_{channel} = 0{\cdot}031$ and $n_{floodplain} = 0{\cdot}083$). This represented a $\pm 20\%$ range in roughness and was chosen to be approximately equivalent to the field estimation error for this parameter. A further four simulations were conducted to assess the effect of errors in the relative relief between the main channel and floodplain on model results, and hence the effect of systematic rather than random topographic perturbations. To achieve this, each mesh/topography combination was reconfigured by increasing the height of each floodplain node by 0·1 m. Simulations were then performed using the control friction parameterization ($n_{channel} = 0{\cdot}025$ and $n_{floodplain} = 0{\cdot}067$). Each simulation commenced from steady-state, bank-full discharge (12 $m^3\ s^{-1}$) and continued after the passage of the floodwave until near-bank-full conditions again existed along the entire reach. This gave a total of 16 simulations in all, the details of which are summarized in Table I.

RESULTS

Typical results from the simulations are given in Figures 5 and 6. These were examined in three ways: in terms of the effect on the reach outflow hydrograph; the predicted inundation extent; and the local hydraulics. As expected, increasing the roughness delayed and attenuated the reach outflow hydrograph (Figure 7a). In the case of the low resolution, finite element mesh parameterized with a 10×10 m DTM, the 40% increase in roughness caused an attenuation in peak flow of 1 $m^3\ s^{-1}$ (or 2·5%) and a delay in its arrival of 3000 s. Whilst small, such effects are significant when one considers the scale of the domain. Altering the mesh resolution, and hence the accuracy of the finite element approximation, has a similar magnitude effect on flow volume and peak flow magnitude, but has no effect on the timing of the flood peak (Figure 7b). Topographic sampling error (Figure 7b) has no discernible effect on the outflow hydrograph in the simulations conducted; however, this is not unexpected as random sampling errors will average out at the domain scale.

Figure 5. Time-series of predicted water depths during a small event for the high resolution mesh and 10 m × 10 m DTM (Figure 3C). Shown are (A) bank-full, steady-state flow at the beginning of the simulations; (B) initial flooding; (C) peak inundation extent; and (D) water ponded on floodplain at the end of the simulation after passage of the floodwave

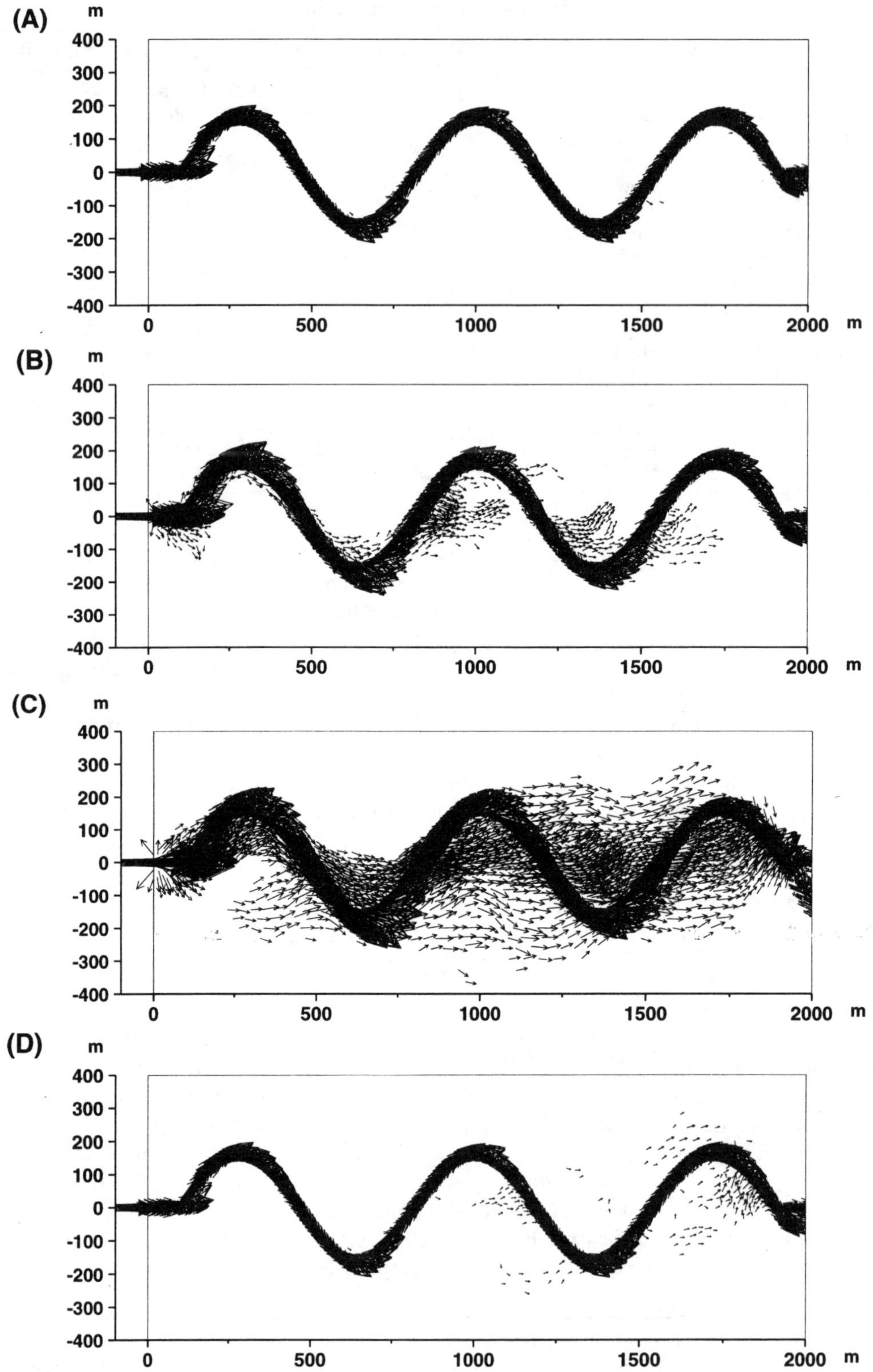
(A)
m
400
300
200
100
0
-100
-200
-300
-400
0
500
1000
1500
2000
m
(B)
(C)
(D)

Floodplain height (Figure 7c) appears only to affect the bulk outflow in the early stages of the flood. The onset of floodplain inundation will be earlier in the control model topography, with a longer period of channel-only flow occurring when the floodplain is raised. Studies have shown that when out-of bank flow occurs there is a significant reduction in the floodwave speed (Knight and Shiono, 1996), probably caused by the interaction that develops between main channel and floodplain flows and the relative roughness effects (topography and friction) associated with the floodplain surface. Wave speeds are at a theoretical maximum at two-thirds of bank-full flow, before decreasing to a minimum at a particular overbank flow with a shallow floodplain depth. After this point the compound channel begins to behave increasingly as a single unit (Ervine and Baird 1982) and the wave speed gradually increases. Hence, for the raised floodplain geometry, wave speeds will be higher at the beginning of the simulation compared to the control, resulting in a more rapid transfer of water to the downstream outflow. After this initial phase, wave speeds in each configuration equilibrate and the flood routing properties of each geometry are similar. This mechanism would seem to account for the more rapid increase in outflow discharge noted for these simulations in Figure 7c.

Inundation extent boundaries are shown for peak flow conditions in Figure 8a to 8d. Figure 8a shows that increasing the friction by 40% generated a simple and significant change in the predicted inundation extent field. As expected, this change increased inundation extent over the entire domain. In places this caused a lateral extension of the inundation boundary of 200 m, equivalent to half the floodplain width. Changing the mesh resolution (Figure 8b) or DTM resolution (Figure 8c) had only a marginal effect (mostly less than 50 m) on the predicted inundation extent field, with both positive and negative effects in the case of mesh resolution. Further tests showed even less variability on both the rising and falling limb of the hydrograph. In both cases the insensitivity of inundation extent to these variables is likely to be a result of their interaction with the local topography. While the rms error of the perturbations generated by the analysis was 0·223 m, the distribution is random, with many places showing smaller perturbations. Clearly, the inundation extent boundary will only show a significant change where it coincides with one of the larger topographic perturbations, which will, in part, be a product of their frequency of occurrence resulting from the fractal analysis of real floodplain terrain. The minor effect of these two parameters on inundation extent was therefore not unexpected, owing to the relatively low probability that the inundation boundary will intersect with a significant topographic perturbation. In addition, raising the floodplain surface by 0·1 m had only a minor effect on the inundation extent predicted by the model at peak flow (Figure 8d). This would appear to be contradictory. However, raising the topography also causes an upward shift in the level of the water surface of slightly less than 0·1 m, as one would expect from conservation principles. Thus, the depth of water on the floodplain is similar in the control and raised floodplain simulations and the zero depth contour remains in approximately the same location. This is not true of the channel, as the base of channel elevations remains constant leading to an increase in depth. The effect of raising the floodplain surface was therefore neutral, although a greater inundation extent can be noted at the start of the control simulation. This is again owing to the lower channel capacity of this geometry and the resulting earlier onset of floodplain inundation.

The bulk flow and inundation extent results demonstrate that the effect of changing topography is small relative to calibration effects for the range of values tested here. We are therefore able to begin to establish the relative dominance of these factors for predicting a number of different quantities. Moreover, it implies that inundation extent predictions are not greatly affected by microtopographic variations of this scale and that in constructing hydraulic models to simulate this quantity it is overall topographic form that is of greater importance. It should be noted, however, that friction and topography are not independent parameters and, furthermore, that this relationship is scale dependent, since the form drag associated with microtopography at subgrid scales forms part of the friction parameterization in a given model application. Thus, one way to represent the effect on flow routing of subgrid-scale microtopography would be to increase the magnitude of the friction parameter. In this study increasing the amount of form drag explicitly included in the model by

Figure 6. Time-series of predicted velocity vectors during a small event for the high resolution mesh and 10 m × 10 m DTM (Figure 3C) for the times shown in Figure 5. Note the complex flow paths around topographic obstacles in (C)

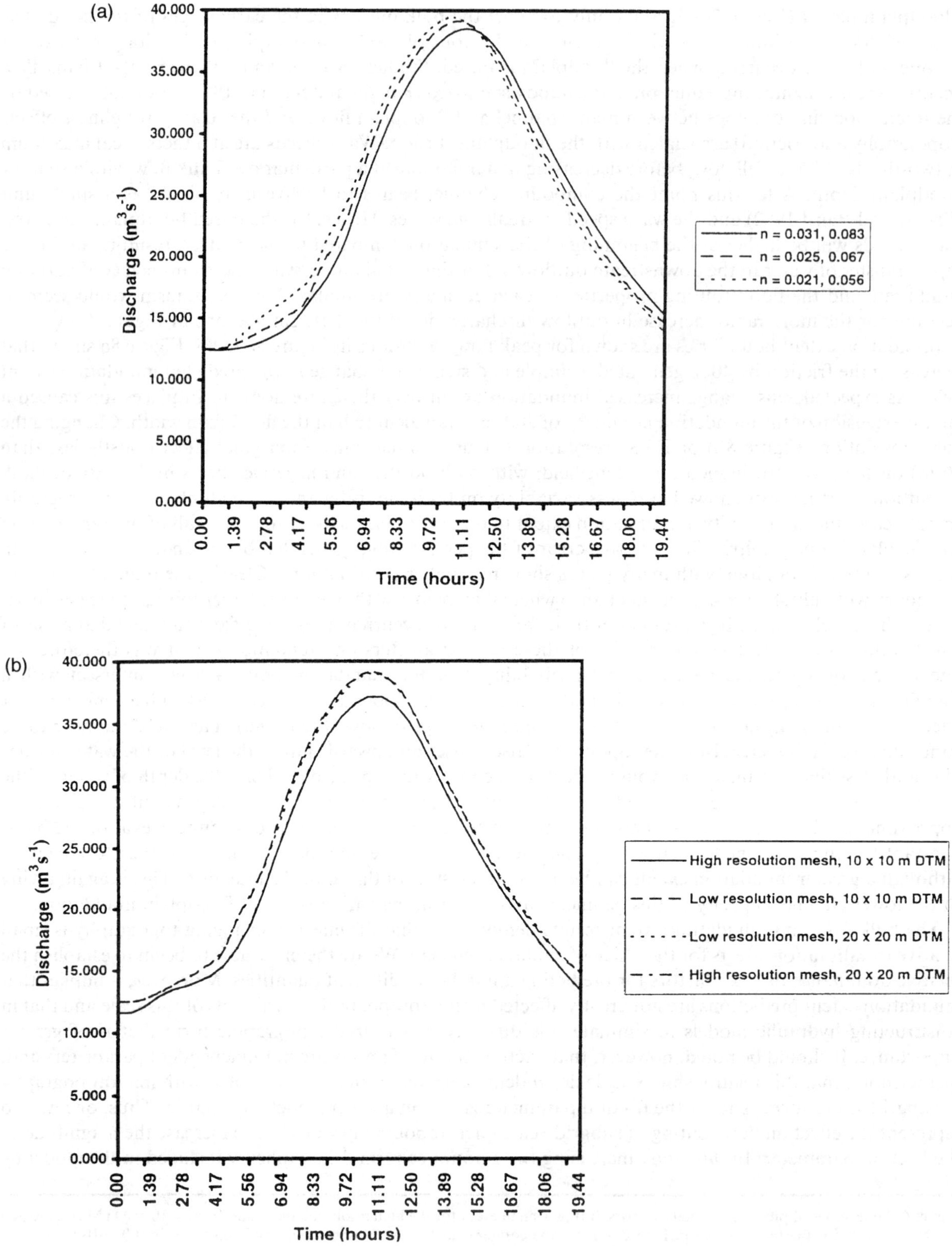
(a)
Discharge (m^{3}s^{-1})
40.000
35.000
30.000
25.000
20.000
15.000
10.000
5.000
0.000
0.00 1.39 2.78 4.17 5.56 6.94 8.33 9.72 11.11 12.50 13.89 15.28 16.67 18.06 19.44
Time (hours)
n = 0.031, 0.083
n = 0.025, 0.067
n = 0.021, 0.056
(b)
Discharge (m^{3}s^{-1})
40.000
35.000
30.000
25.000
20.000
15.000
10.000
5.000
0.000
0.00 1.39 2.78 4.17 5.56 6.94 8.33 9.72 11.11 12.50 13.89 15.28 16.67 18.06 19.44
Time (hours)
High resolution mesh, 10 x 10 m DTM
Low resolution mesh, 10 x 10 m DTM
Low resolution mesh, 20 x 20 m DTM
High resolution mesh, 20 x 20 m DTM

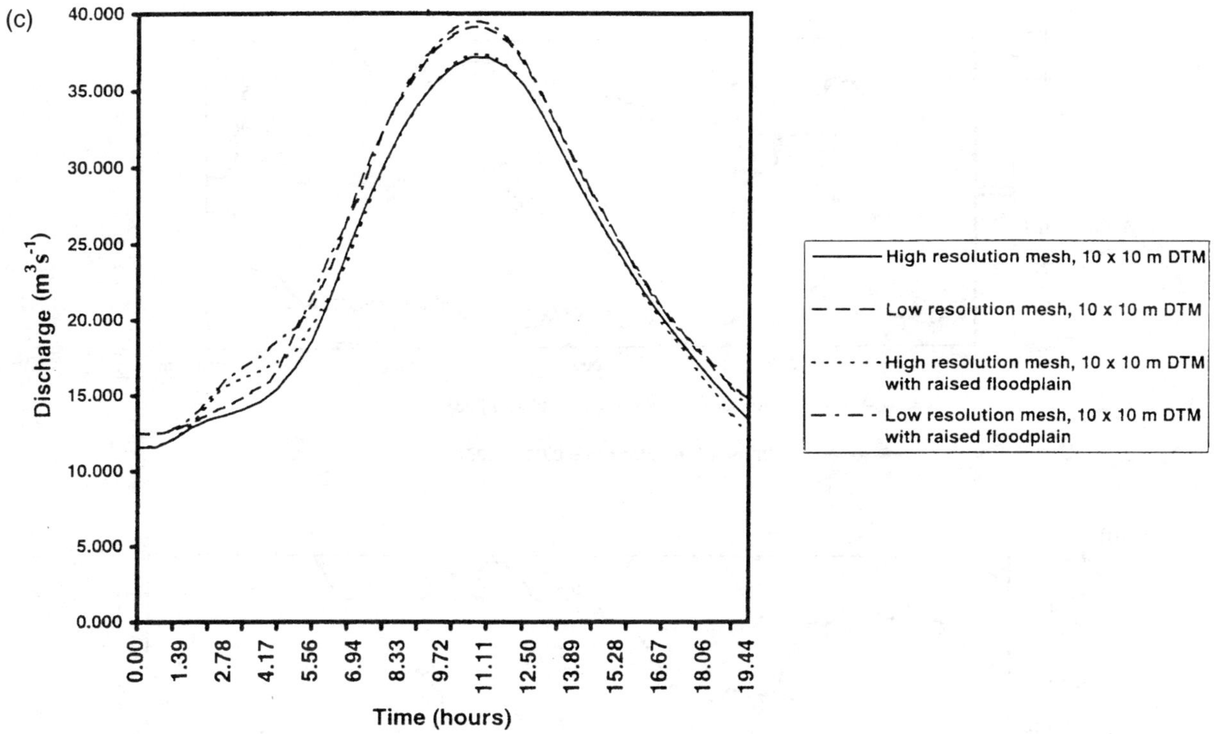

Figure 7(a–c). Model predictions of discharge at the downstream reach outflow showing the effect of varying, (a) model calibration for the low resolution mesh and 10 m × 10 m DTM; (b) mesh and DTM resolution for a friction parameterization of $n = 0{\cdot}025$ and 0·067; and (c) floodplain relative height for a friction parameterization of $n = 0{\cdot}025$ and 0·067

increasing the DTM resolution had a relatively minor effect on the flow. These experiments therefore begin to shed some light on this scaling process and suggest that little of the friction term can be attributed to microtopographic effects not represented by the mesh. However, this observation is dependent on both the application scale and the degree to which the topographic reconstruction used here reproduces actual topography and therefore requires further investigation.

A very different pattern emerges when one considers the results for spatial distributions of hydraulic variables such as water depth and flow velocity (Figure 9a to 9f). Figure 9a and 9b shows the effect of a 40% increase in friction on water depth and flow velocity, respectively, for the low resolution mesh and 10 m × 10 m DTM at peak flow. Unlike inundation extent, where the model response was unidirectional and spatially uniform, here a uniform change in friction caused a spatially variable water depth difference surface. Some dynamic effects are also observed, with increased friction in certain areas leading to a reduction in water depth. Thus, with dynamic and highly non-linear models a simple, spatially uniform change in a particular parameter may have a complex, spatially distributed effect on predicted variables. This is consistent with findings from other recent hydraulic modelling studies (Lane *et al.*, 1994b; Miller, 1995; Bates and Anderson, 1996) and implies that simple calibration techniques, which fail to take account of such behaviour, may be inappropriate for two-dimensional, finite element models applied to fully distributed problems. Changes generated by the friction perturbation are in the range −0·03 to 0·22 m for depth and −0·14 to 0·07 ms^{-1} for velocity. These changes are significant in terms of typical floodplain hydraulic conditions, where water depths are typically less then 1 m and velocities less then 1 ms^{-1}. The velocity difference surface is particularly interesting as it displays lateral zoning within the floodplain belt and distinct velocity decreases at the apex of channel bends. Despite the fact that both floodplain and channel friction

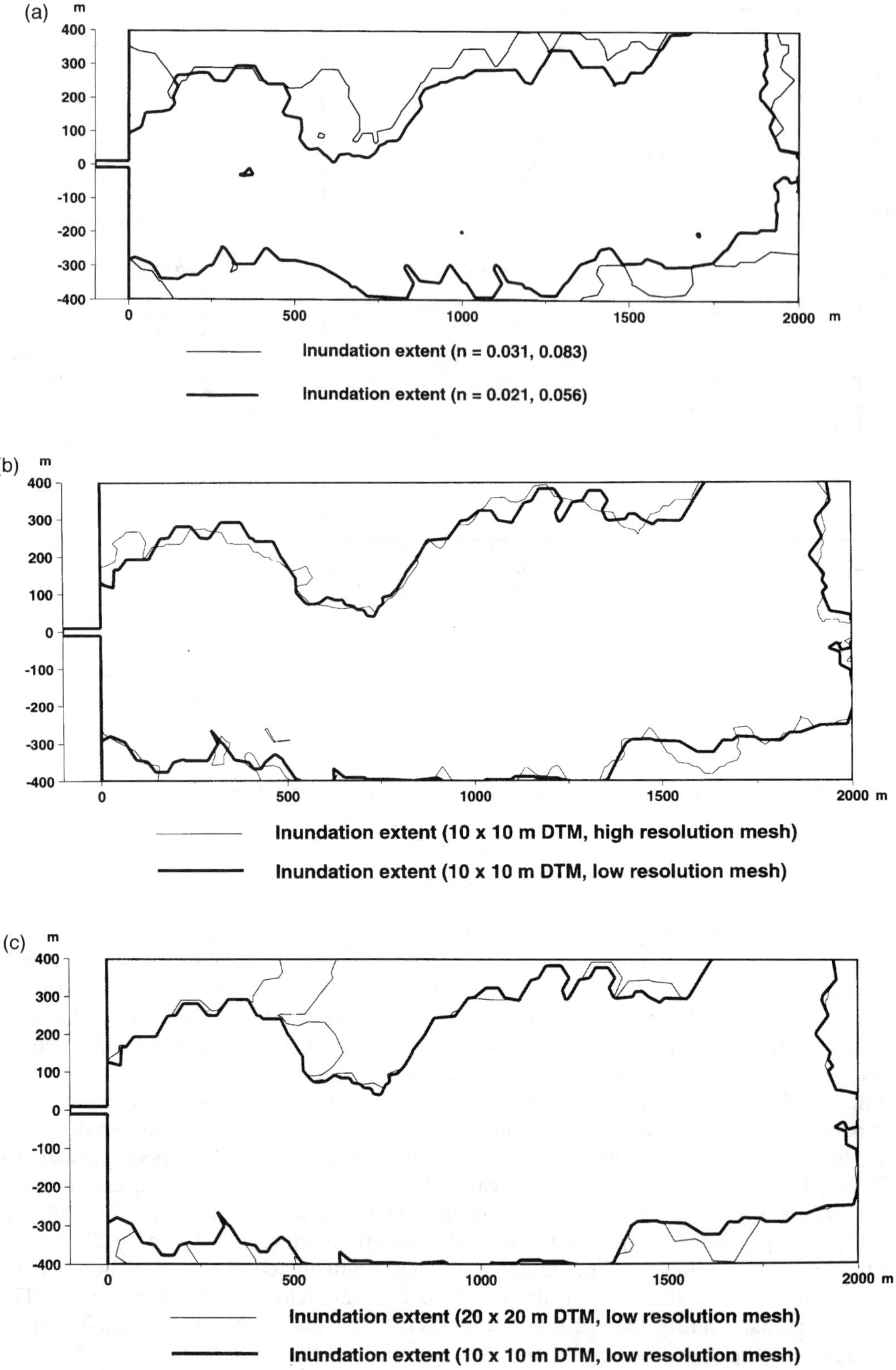
(a)
m
400
300
200
100
0
-100
-200
-300
-400
0
500
1000
1500
2000 m
Inundation extent (n = 0.031, 0.083)
Inundation extent (n = 0.021, 0.056)
(b)
m
400
300
200
100
0
-100
-200
-300
-400
0
500
1000
1500
2000 m
Inundation extent (10 x 10 m DTM, high resolution mesh)
Inundation extent (10 x 10 m DTM, low resolution mesh)
(c)
m
400
300
200
100
0
-100
-200
-300
-400
0
500
1000
1500
2000 m
Inundation extent (20 x 20 m DTM, low resolution mesh)
Inundation extent (10 x 10 m DTM, low resolution mesh)

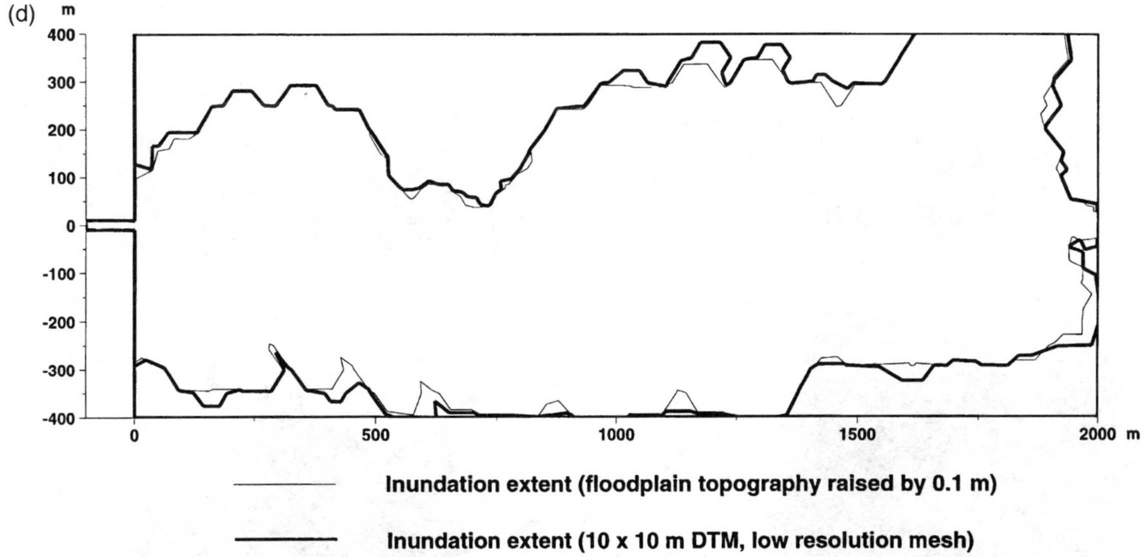

Figure 8. Model predictions of inundation extent showing the effect of varying, (a) model calibration from $n = 0{\cdot}021$ and $0{\cdot}056$ to $n = 0{\cdot}031$ and $0{\cdot}083$ for the low resolution mesh and 10 m × 10 m DTM; (b) mesh resolution for a friction parameterization of $n = 0{\cdot}025$ and $0{\cdot}067$; (c) DTM resolution for a friction parameterization of $n = 0{\cdot}025$ and $0{\cdot}067$; and (d) floodplain relative height for a friction parameterization of $n = 0{\cdot}025$ and $0{\cdot}067$

have been increased, a noticeable band of increased velocity occurs at the outside edge of the floodplain. This can be explained by the higher friction in the channel reducing channel conveyance, which is countered by a greater transfer of water on to the floodplain.

In terms of changing mesh resolution, it was not possible to generate exact difference plots because of the varying number of points in each mesh. Interpolating data onto a regular grid was rejected as a solution to this problem since this would have resulted in the loss of the very information that we wished to compare. However, analysis of plots of water depth at peak flow showed that the high resolution mesh enabled much greater spatial resolution of local hydraulics, although the broad pattern remained reasonably similar. Changing DTM resolution produced a similar complex response in water depth and flow velocity predictions (Figure 9c and 9d). Unlike calibration changes, however, channel depths and velocities were largely unaltered: the magnitude of the changes caused was confined to a smaller range and the spatial pattern was far more sparse and discrete. This probably reflects the fact that only floodplain areas were topographically perturbed, although it is clear that the effect on local hydraulics of topographic changes of this magnitude is much less than that of changing boundary friction. As one would expect from the previous analyses of bulk flow and inundation extent, floodplain height (Figure 9e and 9f) only significantly affected channel velocities and depths as it is only here that the increase in water surface elevation leads to an increase in depth. This has some knock-on effects on floodplain hydraulics owing to a change in the intensity of the shear layer between main channel and floodplain flows, although these differences are mostly less than 0·02 m for depth and 0·03 ms^{-1} for velocity. Floodplain depths and velocities do, however, increase at the far downstream end of the reach.

At the far downstream end it appears that, as the flow is subcritical, the fixed water surface elevation used as a downstream boundary condition exerts an influence on the local water depths and velocities. Hence, in the control simulation the water surface elevation is higher than the local friction slope. In this region the water surface in the control and raised floodplain topographies is at approximately the same elevation; however, the topography difference between the two leads to a difference in the water depth. As a consequence, local velocities are also affected.

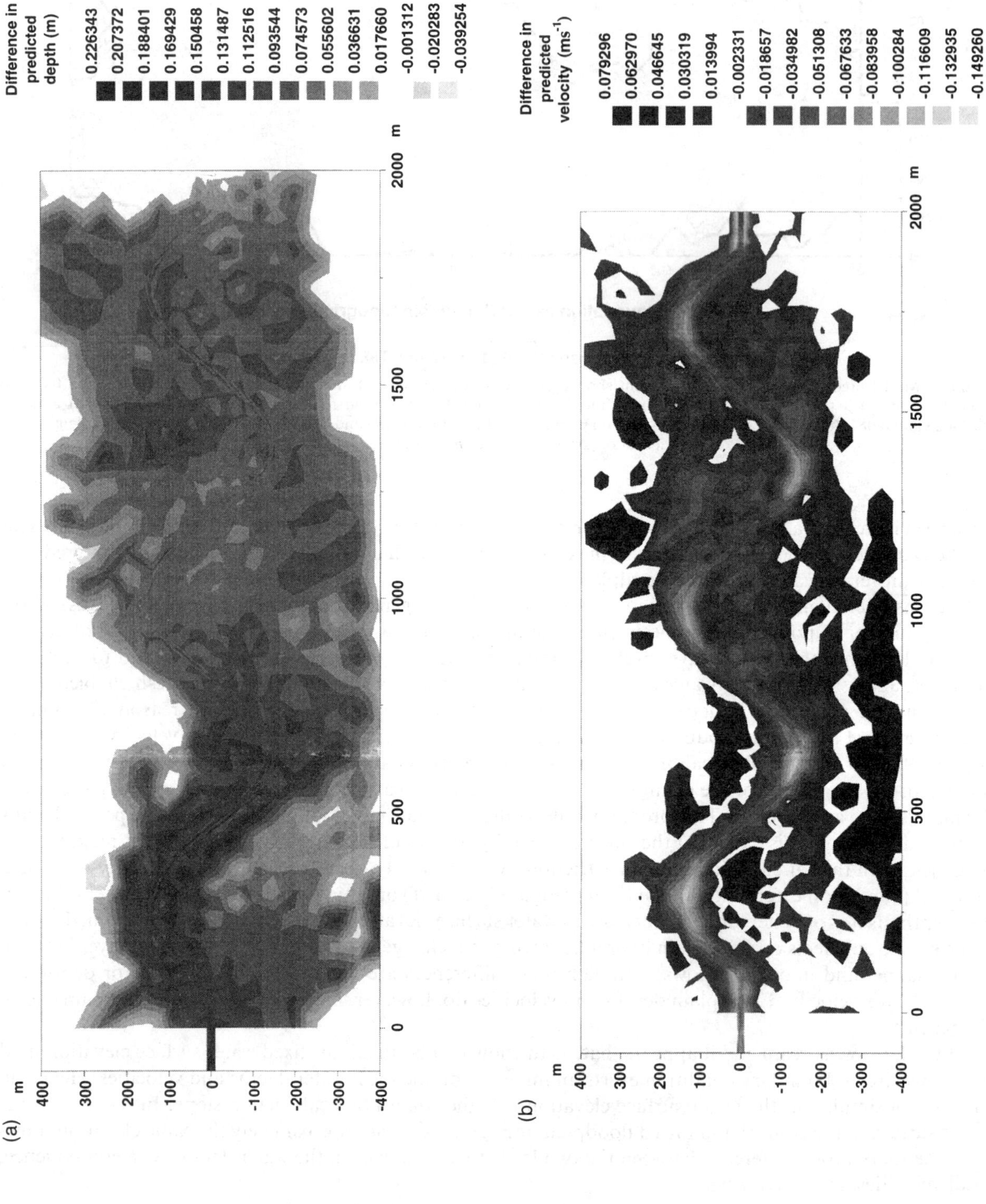
(a)
Difference in predicted depth (m)
0.226343
0.207372
0.188401
0.169429
0.150458
0.131487
0.112516
0.093544
0.074573
0.055602
0.036631
0.017660
-0.001312
-0.020283
-0.039254
m
400
300
200
100
0
-100
-200
-300
-400
0
500
1000
1500
2000 m
(b)
Difference in predicted velocity (ms-1)
0.079296
0.062970
0.046645
0.030319
0.013994
-0.002331
-0.018657
-0.034982
-0.051308
-0.067633
-0.083958
-0.100284
-0.116609
-0.132935
-0.149260
m
400
300
200
100
0
-100
-200
-300
-400
0
500
1000
1500
2000 m

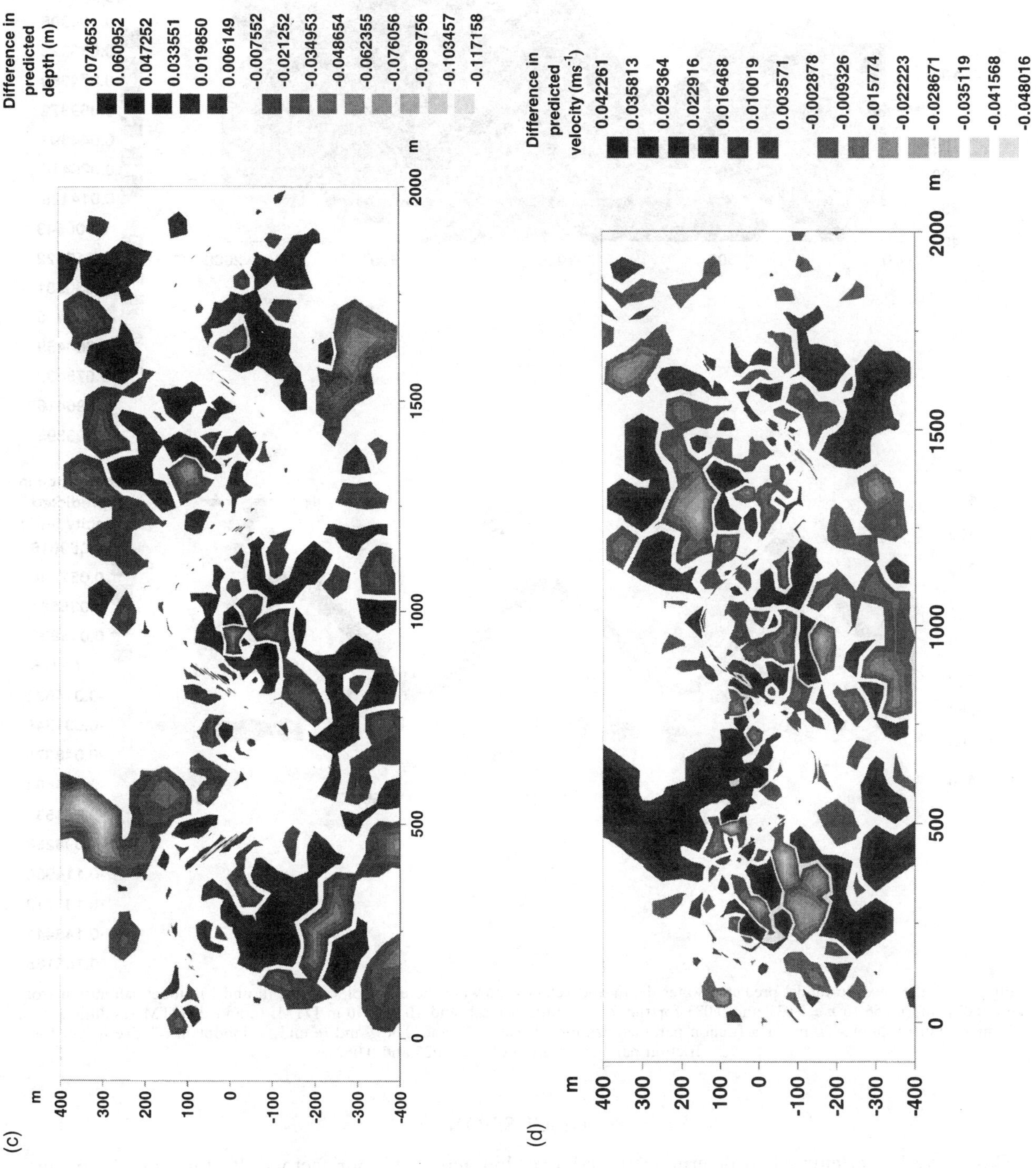
(c)
Difference in predicted depth (m)
0.074653
0.060952
0.047252
0.033551
0.019850
0.006149
-0.007552
-0.021252
-0.034953
-0.048654
-0.062355
-0.076056
-0.089756
-0.103457
-0.117158
m
400
300
200
100
0
-100
-200
-300
-400
0
500
1000
1500
2000 m
(d)
Difference in predicted velocity (ms-1)
0.042261
0.035813
0.029364
0.022916
0.016468
0.010019
0.003571
-0.002878
-0.009326
-0.015774
-0.022223
-0.028671
-0.035119
-0.041568
-0.048016
m
400
300
200
100
0
-100
-200
-300
-400
0
500
1000
1500
2000 m

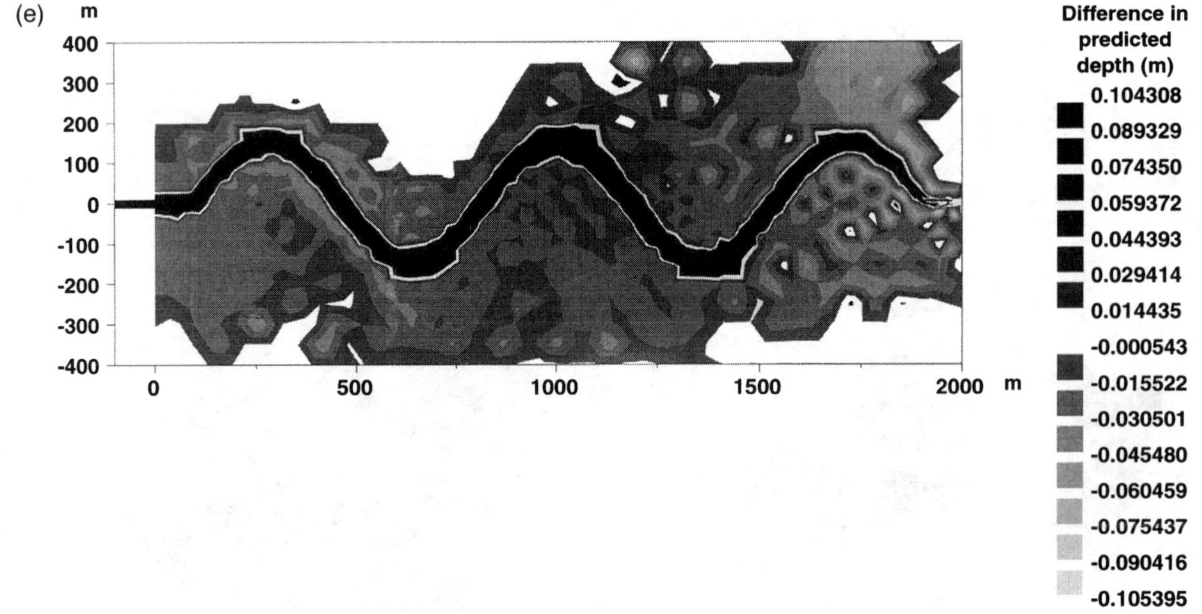

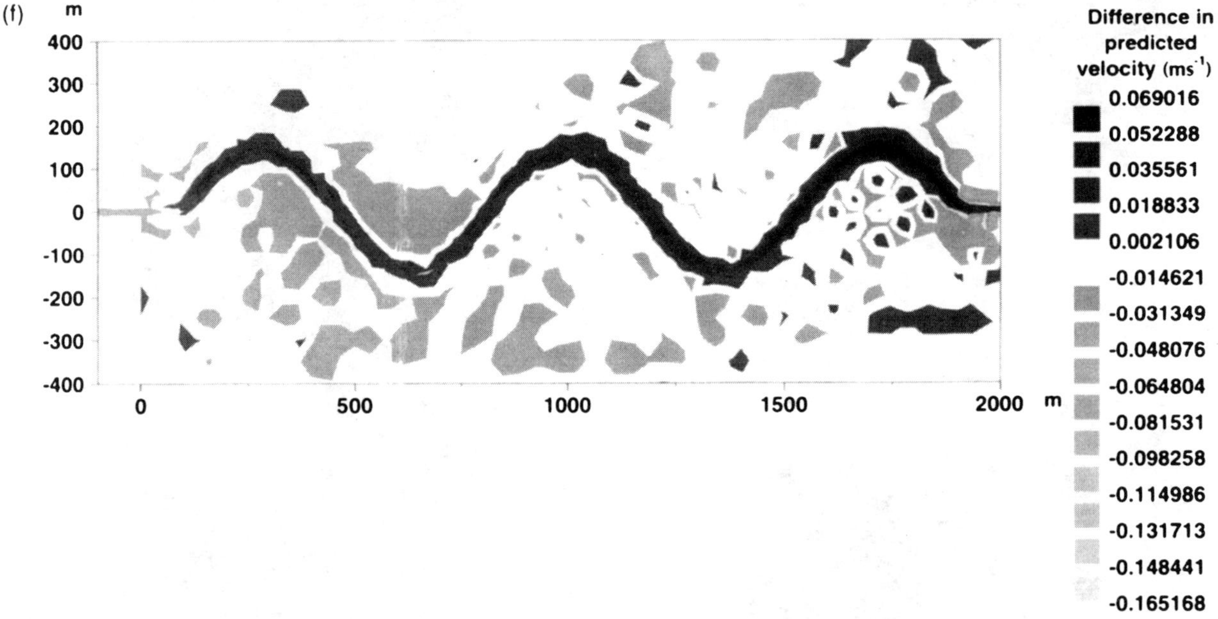

Figure 9. Differences in model predicted water depth and velocity showing the effect of varying, (a and b) model calibration from $n = 0{\cdot}021$ and $0{\cdot}056$ to $n = 0{\cdot}031$ and $0{\cdot}083$ for the low resolution mesh and 10 m × 10 m DTM; (c and d) DTM resolution from 10 m × 10 m to 20 m × 20 m for a friction parameterization of $n = 0{\cdot}025$ and $0{\cdot}067$; and (e and f) floodplain relative height for a friction parameterization of $n = 0{\cdot}025$ and $0{\cdot}067$

DISCUSSION

This study has attempted to determine the relative dominance of four factors affecting inundation predictions from a two-dimensional, finite element hydraulic model: calibration, mesh resolution, topography sampling error and floodplain relative height, in order to further the development of this class of model. This has been achieved using an hydraulic model specifically adapted to floodplain flows and parameterized using a novel fractal analysis technique for generating synthetic topography. Examination of the hydraulic calculations performed by the finite element model demonstrated it to be capable of providing a realistic simulation of dynamic flood water depths (Figure 5) and velocity vectors (Figure 6). In particular, the model is capable of simulating a complete range of inundation states, commencing with dry floodplains and ultimately returning to this state after substantial inundation, albeit with some water ponded in floodplain hollows generated by the fractal topography at the end of the simulation. Velocity vectors demonstrate the model to be capable of simulating flow spilling out from the channel and across meander loops, prior to re-entering, or crossing over, the channel flow on the downstream side of the meander. This has been noted as an important momentum exchange mechanism in compound channel flume studies (Sellin *et al.*, 1993; Ervine *et al.*, 1994) which would be impossible to represent using a one-dimensional model. Significant floodplain flow structures driven by the imposed topographic perturbations may also be discerned. Figure 6b, for example, shows flow paths diverting around topography during the initial flooding phase. The results, summarized in Table II, have been analysed in terms of the predicted outflow hydrograph, inundation extent and local hydraulics. The results show that for the range of parameters studied, calibration is the dominant factor affecting model predictions.

It has been demonstrated that model response to uniform parameter change can be spatially complex. While it would appear straightforward to calibrate such models to predict bulk flow and inundation extent using the relationships implied by Figures 7a–c and 8a–d, calibrating a model for local hydraulics is a considerably more difficult proposition. We have demonstrated that topographic specification, mesh resolution and calibration procedure should all vary depending on the particular model application attempted. If the aim is simply to predict bulk flows and inundation extent then relatively simple mesh resolution and topography may be appropriate, with research effort focused on calibration issues. If local hydraulics are of relevance then mesh/topography resolution should be commensurate with the expected process variability. In this case resolution issues attain greater relative importance although, as noted above, complex calibration problems remain. Such statements are, however, limited to the specific case of two-dimensional models and do not necessarily imply that the processes that the model represents are *equally* insensitive to topographic variation. A better understanding is required of such calibration and resolution effects, particularly as these may be linked through the operation of grid square effective parameters (cf. Beven, 1989). By changing the mesh resolution we may also change effective parameter values, thereby affecting the calibration despite an apparently static parameterization prescribed by the modeller.

Lastly, the issue of topographic specification for floodplain hydraulic modelling is also emergent from this research. While we have identified the effect of topography on local hydraulics, the scale and accuracy of topographic information needed to achieve particular simulation results is not clear. In particular, there is a need to conduct a detailed examination of threshold effects in the DTM and mesh resolution and their resulting effect on the model. This study has also indicated that small-scale topography does not have a

Table II. Summary of model responses to changes in friction calibration, mesh resolution, DTM resolution and relative floodplain height

Factor	Effect on bulk flows		Effect on inundation extent	Effect on local hydraulics
	Magnitude	Timing		
Friction calibration	Yes	Yes	Yes	Yes
Mesh resolution	Yes	No	Minor	Yes
DTM resolution	No	No	Minor	Minor
Floodplain height	Minor impact at start of simulation	Minor impact at start of simulation	Negligible impact apart from at start of simulation	Significant impact in channel, minor impact elsewhere

large-scale impact on the model results for simple river/floodplain configurations, yet this requires further testing under more realistic conditions if it is to be confirmed.

CONCLUSIONS

This study has highlighted the potential for using fractal-generated topography to parameterize hydraulic models in hypothetical domains as a means of advancing our understanding how to develop models to simulate complex flows. This is particularly true of environments such as floodplains where data availability is often poor. There remains a need, however, to extend these concepts from hypothetical to real applications. This will be a major research effort, requiring both high resolution topography and hydraulic data. Studies such as the one presented here would appear to be an important element in the design of such a research programme. In addition we have further confirmed the viability of two-dimensional, finite element methods applied to floodplain flow problems. Such models have been shown to be capable of representing significant flow complexity, although their analysis here has highlighted a number of new research issues. In particular, complex model response to uniform parameter change has been noted. There is increasing awareness that standard model calibration and analysis techniques may be inappropriate for certain classes of model and particular types of model application. Spatially distributed, dynamic simulations using highly non-linear equation sets may thus require a new research methodology and the development of more sophisticated analysis techniques if their potential to make accurate spatially distributed predictions is to be fully exploited.

ACKNOWLEDGEMENTS

This research has been made possible with the support of Laboratoire National d'Hydraulique, France, the UK Natural Environment Research Council (Grant No. GR3/09925) and the UK Ordnance Survey for the provision of DTM data. Matthew Horritt was supported by a studentship funded by NRSC Ltd. The authors are grateful to Professor Malcolm Anderson (University of Bristol) and Dr Andrew Miller (University of Maryland) for helpful review comments on a draft version of this paper.

REFERENCES

Abbott, M. B., Bathurst, J. C., Cunge, J. A., O'Connell, P. E., and Rasmussen, J. 1986. 'An introduction to the European Hydrological System — Systeme Hydrologique Europeen, "SHE", 2: structure of a physically based distributed modelling system', *J. Hydrol.*, **87**, 45–59.

Baird, L., Gee, D. M., and Anderson, M. G. 1992. 'Ungauged catchment modelling. II. Utilization of hydraulic models for validation', *Catena*, **19**, 33–42.

Bates, P. D. and Anderson, M. G. 1996. 'A preliminary investigation into the impact of initial conditions on flood inundation predictions using a time/space distributed sensitivity analysis', *Catena*, **26**, 115–134.

Bates, P. D., Anderson, M. G., Baird, L., Walling, D. E., and Simm, D. 1992. 'Modelling floodplain flows using a two-dimensional finite element model', *Earth Surf. Process. Landf.*, **17**, 575–588.

Bates, P. D., Anderson, M. G., and Hervouet, J.-M. 1994. 'Computation of a flood event using a two-dimensional finite element model and its comparison to field data', in P. Molinaro and L. Natale (Eds), *Modelling Flood Propagation over Initially Dry Areas*. American Society of Civil Engineers, New York. pp. 243–256.

Bates, P. D., Anderson, M. G., and Hervouet, J.-M. 1995. 'An initial comparison of two 2-dimensional finite element codes for river flood simulation', *Proc. Inst. Civil Eng., Water Maritime Energy*, **112**, 238–248.

Bates, P. D., Anderson, M. G., Price, D., Hardy, R., and Smith, C. 1996. 'Analysis and development of hydraulic models for floodplain flow', in M. G. Anderson, D. E. Walling, and P. D. Bates (Eds), *Floodplain Processes*. John Wiley and Sons, Chichester. pp. 215–254.

Beven, K. J. 1985. 'Distributed models', in M. G. Anderson and T. P. Burt (Eds), *Hydrological Forecasting*. Wiley, New York. pp. 405–435.

Beven, K. J. 1989. 'Changing ideas in hydrology: the case of physically based distributed models', *J. Hydrol.*, **105**, 157–172.

Beven, K. J., Calver, A., and Morris, E. M. 1987. 'The Institute of Hydrology distributed model', *Institute of Hydrology Report no. 98*. Institute of Hydrology, Wallingford.

Burt, T. P., Heathwaite, A. L., and Trudgill, S. T. 1993. *Nitrates: Processes, Patterns and Management*. John Wiley and Sons, Chichester.

Clarke, R. T. 1994. *Statistical Modelling in Hydrology*. John Wiley and Sons, Chichester. 412 pp.
Culling, W. E. H. and Datko, M. 1987. 'The fractal geometry of soil covered landscape', *Earth Surf. Process. Landf.*, **12**, 369–385.
Ervine, D. A. and Baird, J. I. 1992. 'Rating curves for rivers with overbank flow', *Proc. Inst. Civil Eng. Part 2*, **73**, 465–472.
Ervine, D. A., Sellin, R. J., and Willets, B. B. 1994. 'Large flow structures in compound meandering channels', in W. R. White and J. Watts (Eds), *2nd International Conference on Hydraulics*. John Wiley and Sons, Chichester. pp. 459–469.
Falconer, R. A. and Chen, Y. 1996. 'Modelling sediment transport and water quality processes on tidal floodplains', in M. G. Anderson, D. E. Walling, and P. D. Bates (Eds), *Floodplain Processes*. John Wiley and Sons, Chichester. pp. 361–398.
Feldhaus, R., Hòttges, J., Brochhaus, T., and Rouvé, G. 1992. 'Finite element simulation of flow and pollution transport applied to a part of the River Rhine', in R. A. Falconer, K. Shiono, and R. G. S. Matthews (Eds), *Hydraulic and Environmental Modelling; Estuarine and River Waters*. Ashgate Publishing, Aldershot. pp. 323–334.
Gambolati, G. and Volpi, G. 1979. 'A conceptual deterministic analysis of the kriging technique in hydrology', *Wat. Resour. Res.*, **15**, 625–629.
Gee, D. M., Anderson, M. G., and Baird, L. 1990. 'Large scale floodplain modelling', *Earth Surf. Process. Landf.*, **15**, 512–523.
Hervouet, J.-M. 1989. 'Comparison of experimental data and laser measurements with computational results of the TELEMAC-2D code (shallow water equations)', in C. Maksimovic and M. Radojkovic (Eds), *Computational Modelling and Experimental Methods in Hydraulics (HYDROCOMP '89)*. Elsevier, Amsterdam. pp. 237–342.
Hervouet, J.-M. 1993. 'Validating the numerical simulation of dam-breaks and floods', in *Advances in Hydro-Science and Engineering*, Volume 1, Part A. Washington, D.C. pp. 754–761.
Hervouet, J.-M. and Janin, J.-M. 1994. 'Finite element algorithms for modelling flood propagation', in P. Molinaro and L. Natale (Eds), *Modelling Flood Propagation over Initially Dry Areas*. American Society of Civil Engineers, New York. pp. 102–113.
Hervouet, J.-M. and Van Haren, L. 1996. 'Recent advances innumerical methods for fluid flows', in M. G. Anderson, D. E. Walling, and P. D. Bates (Eds), *Floodplain processes*. John Wiley and Sons, Chichester. pp. 183–214.
Hughes, J. P. and Lettenmaier, D. P. 1981. 'Data requirements for kriging estimation and network design', *Wat. Resour. Res.*, **17**, 1641–1650.
James, C. S. 1985. 'Sediment transfer to overbank sections', *J. Hydraul. Res.*, **23**, 435–452.
Jolly, I. D. 1996. 'The effects of river management on the hydrology and hydroecology of arid and semi-arid floodplains', in M. G. Anderson, D. E. Walling, and P. D. Bates (Eds), *Floodplain Processes*. John Wiley and Sons, Chichester. pp. 577–609.
Knight, D. W. and Shiono, K. 1996. 'River channel and floodplain hydraulics', in M. G. Anderson, D. E. Walling, and P. D. Bates (Eds), *Floodplain Processes*. John Wiley and Sons, Chichester. pp. 139–181.
Lane, S. N., Chandler, J. H., and Richards, K. S. 1994a. 'Developments in monitoring and modelling small-scale river bed topography'. *Earth Surf. Process. Landf.*, **19**, 349–368.
Lane, S. N., Richards, K. S., and Chandler, J. H. 1994b. 'Application of distributed sensitivity analysis to a model of turbulent open channel flow in a natural river channel', *Proc. R. Soc. London, Series A*, **447**, 49–63.
Marriot, S. 1992. 'Textural analysis and modelling of a flood deposit: River Severn, UK', *Earth Surf. Process. Landf.*, **17**, 687–697.
Marron, D. C. 1989. 'The transport of mine tailings as suspended sediment in the Belle', *IAHS Publ.*, **184**, 19–26.
Miller, A. J. 1995. 'Valley morphology and boundary conditions influencing spatial patterns of flood flow', in J. E. Costa, A. J. Miller, K. W. Potter, and P. R. Wilcock (Eds), *Natural and Anthropogenic Influences in Fluvial Geomorphology: the Wolman Volume*, American Geophysical Union Geophysical Monograph 89. American Geophysical Union, Washington, D.C. pp. 57–81.
Nicholas, A. and Walling, D. E. 1998. 'Numerical modelling of floodplain hydraulics and suspended sediment transport and deposition', *Hydrol. Process.* **12**, 1339–1355 (this issue).
Pizzuto, J. E. 1987. 'Sediment diffusion during overbank flows', *Sedimentology*, **34**, 301–317.
Richards, K. S., Hughes. F. M. R., El-Hames, A. S., Harris, T., Pautau, G., Peiry, J.-L., and Girel, J. 1996. 'Integrated field, laboratory and numerical investigations of hydrological influences on the establishment of riparian tree species', in M. G. Anderson, D. E. Walling, and P. D. Bates (Eds), *Floodplain Processes*. John Wiley and Sons, Chichester. pp. 611–635.
Sellin, R. H. J., Ervine, D. A., and Willets, B. B. 1993. 'Behaviour of meandering two stage channels', *Proc. Inst. Civil Eng., Water Maritime Energy*, **101**, 99–111.
Stewart, M. D., Bates, P. D., Price, D. A., and Burt, T. P. 1998. 'Modelling the spatial variability in floodplain soil contamination during flood events to improve chemical mass balance estimates', *Hydrol. Process.*, **12**, 1233–1255 (this issue).
Turcotte, D. L. 1992. *Fractal Models in Geology and Geophysics*. Cambridge University Press, Cambridge. 221 pp.
Walling, D. E., He, Q., and Nicholas, A. P. 1996. 'Floodplains as sediment sinks', in M. G. Anderson, D. E. Walling, and P. D. Bates (Eds), *Floodplain Processes*. John Wiley and Sons, Chichester. pp. 399–440.

10

INTEGRATING REMOTE SENSING OBSERVATIONS OF FLOOD HYDROLOGY AND HYDRAULIC MODELLING

P. D. BATES,[1]* M. S. HORRITT,[2] C. N. SMITH[1] AND D. MASON[2]

[1]*Department of Geography, University of Bristol, Bristol BS8 1SS, UK*
[2]*Environmental Science Systems Centre, University of Reading, Whiteknights, Reading RG6 6AB, UK*

ABSTRACT

The further development of two-dimensional finite element models of river flood flow is currently constrained by a lack of data for rigorous parameterization and validation. Remote sensing techniques have the potential to overcome a number of these constraints thereby allowing a research design for model development. This is illustrated with reference to a case study of a two-dimensional finite element model applied to the Missouri River, Nebraska and compared with a synchronous Landsat TM image of flood inundation extent. The case study allows research needs for the integration of hydraulic modelling and remote sensing to be defined.

Hydrol. Process., Vol. 11, 1777–1795 (1997)

(No. of Figures: 7 No. of Tables: 0 No. of Refs: 85)

KEY WORDS floodplain flow; hydraulic modelling; finite elements; remote sensing; edge detection

INTRODUCTION

In this paper we address the numerical simulation of free surface flows in compound channels. For such flows construction of accurate prediction models remains a fundamental research problem in hydrology and hydraulics. Such studies are not only of importance to engineers and planners engaged in the water resources field, but are also an important prerequisite for modelling investigations of floodplain sediment transport (James, 1985; Pizzuto, 1987; Marriot, 1992; Falconer and Chen, 1996; Walling *et al.*, 1996), contaminant transport (Marron, 1989; Burt *et al.*, 1993), floodplain ecology (Jolly, 1996; Richards *et al.*, 1996) and catchment hydrology (Beven, 1985; Abbott *et al.*, 1986; Beven *et al.*, 1987) which either require hydraulic data as an input or incorporate a submodel for hydraulic processes. For all these tasks we require a model capable of accurately predicting spatially distributed estimates of hydraulic variables such as water depth and flow velocity. A fundamental constraint to the development of such a model is the lack of suitable calibration and validation data sets. Whilst it is now possible to develop high space/time resolution prediction schemes, current data capture techniques do not allow the utility of such developments to be fully assessed. The purpose of this paper is to review recent progress in hydraulic modelling and examine the extent to which recent advances in the remote sensing of flood hydrology have the potential to overcome current data constraints, thereby leading to an integration of remote sensing and modelling methodologies.

REVIEW OF RECENT PROGRESS IN FLOOD HYDRAULIC MODELLING

Progress in hydraulic modelling over the last decade has led to considerable improvements in our ability to simulate river flooding problems. The impetus for this progress has come from a number of fields and incorporates improvements in process understanding, mathematical and numerical developments and

---

* Correspondence to: P. D. Bates.

Contract grant sponsor: NERC.
Contract grant number: GR3/09925.

available computational power. Large-scale flume studies (see, for example, Knight and Shiono, 1996; Sellin and Willetts, 1996) have demonstrated the importance of such mechanisms as momentum transfer between main channel and floodplain flows (Sellin, 1964; Knight and Demetriou, 1983) and flow across meander loops (Toebes and Sookey, 1967; Kiely, 1990) in compound channels. These studies have indicated that while in-channel flows may be satisfactorily described by a one-dimensional representation, out-of-bank flows incorporate significant three-dimensional effects which invalidate this class of model (Knight and Shiono, 1996). A need has therefore arisen to develop higher resolution hydraulic models (two- and three-dimensional) to account for the impact of these processes during river flood events. Concurrently, developments in computational fluid dynamics (CFD) have led to a number of techniques that have enabled this need to be met. Stable numerical techniques for free surface flow problems have been developed. For example, the Streamline Upwind/Petrov–Galerkin technique developed by Brookes and Hughes (1982) eliminates spurious oscillations from numerical solutions of the Navier–Stokes equations without adding artificial diffusivity. Element-by-element matrix storage techniques (Carey and Jiang, 1986; Hervouet, 1992; Binley and Beven, 1993) have reduced computer storage requirements and improved efficiency, allowing spatially complex higher dimensional problems to be simulated. Lastly, and of critical importance to flood inundation problems, algorithms have been developed to simulate dynamically moving inundation extent boundaries (Lynch and Gray, 1980; King and Roig, 1988). Thus, the last decade has seen a significant stimulus to the development of two-dimensional, or higher, numerical models for flood inundation problems, while at the same time a series of technical advances has rendered this goal attainable.

Two- and three-dimensional modelling approaches have now been developed for application to a number of free surface flow problems. Such models solve some form of the Navier–Stokes equations, such as the St Venant or shallow water equations, in conjunction with a number of approximations to represent such processes as turbulence and boundary friction (see, for example, Hervouet and Van Haren, 1996; Lane, Chapter 1). An appropriate discretization of the domain geometry and topography is developed to define a network of computational nodes where the above equations can be solved by some numerical procedure. Codes have been developed using a variety of numerical methods including finite difference (Zeilke and Urban, 1981), finite element (Gee *et al.*, 1990; Bates *et al.*, 1992) and finite volume (Lane *et al.*, 1994). This solution in space is then projected forward in time using a further numerical procedure, typically a finite difference scheme, to give a dynamic simulation. Given appropriate boundary conditions, estimates of water depth and flow velocity may thus be obtained at each computational node at each time step. In particular, a series of studies has focused on the development of hydraulic models for long reach flood inundation problems. Until recently, application of hydraulic models to free surface flow problems was characterized by three specific attributes; the scale of application was small (0·5–2 km) and focused on the analysis of detailed flow patterns, the main channel was rarely resolved separately from the floodplain and the entire domain was inundated during the simulation. However, most floodplain inundation problems of practical interest occur at long reach scales (10–60 km) and involve a dynamic flood inundation boundary. As a consequence there is a need to represent dry areas within the computational domain. A number of models have therefore been developed that are capable of application at this scale and that incorporate algorithms to account for wetting and drying processes (see, for example, Gee *et al.*, 1990; Baird *et al.*, 1992; Feldhaus *et al.*, 1992; Leclerc *et al.*, 1990; Bates *et al.*, 1995). Given the scale of the problem, the above approaches all adopt a two-dimensional finite element approximation to the flow field, which enables complex topography to be represented with a minimum number of computational points.

Such studies have been largely successful in providing realistic simulations of flood inundation phenomena, which, when compared to the available field observations, show a reasonable correspondence. For example, Bates *et al.* (1996) describe the application of the TELEMAC-2D model to a 12 km reach of the River Stour, Dorset, UK. A finite element mesh consisting of 12 000 linear triangles, with a minimum size of 10 m, and 7000 computational nodes (see Figure 1) was developed for this reach. Topography was ascribed from 25 channel cross-sections surveyed by the UK Environment Agency and a 10 m × 10 m digital terrain model (DTM) of the floodplain developed by the UK Ordnance Survey from height information on

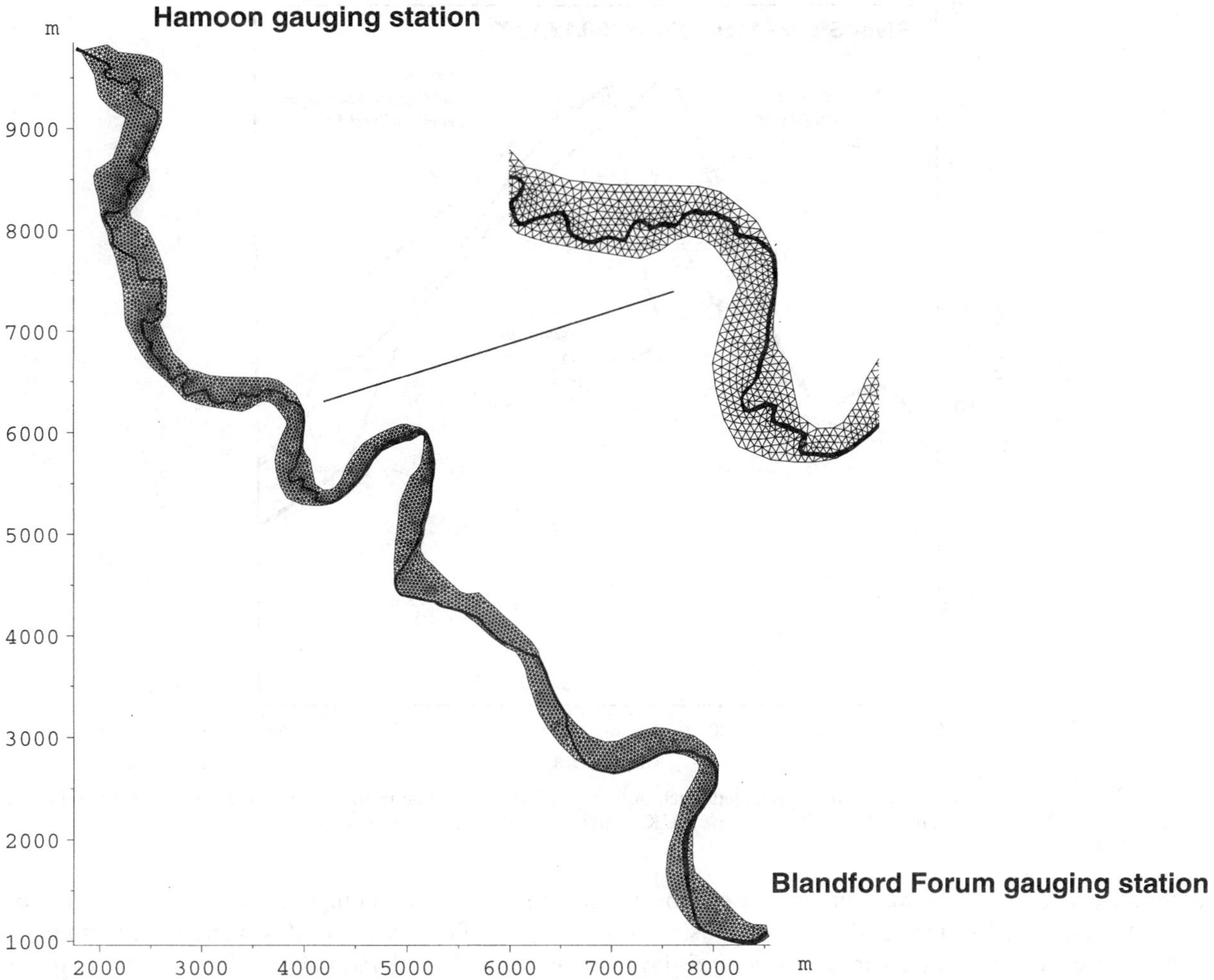

Figure 1. Finite element mesh consiting of 12 000 linear triangles and 7000 computational nodes developed for a 12 km reach of the River Stour, Dorset, UK

its 1:10 000 map series. A one in four-year recurrence interval flood event, which took place over four days around 20 December 1993, was then simulated for this computational domain using boundary condition data provided by flow gauging stations at the up- and downstream extremities of the reach. The simulation began at below bankfull discharge and continued until flow data indicated that within-bank flow had once again been attained along the whole reach. The event was discretized into 55 800 time steps of 4 seconds duration, for which the model predicted water depth and depth-averaged velocity at each computational node. A reasonable correspondence was obtained between observed and model-predicted downstream discharge (see Figure 2). Moreover, analysis of the predicted inundation extent patterns (Figure 3) shows the model to be capable of predicting the advance and recession of the flood wave across complex topography. While the model is able to commence the simulation with a fully dry floodplain and return to that state after the passage of the flood peak, no field observations of flood inundation extent were available to validate this aspect of the model predictions. Model behaviour is therefore consistent with known process behaviour and, while the lack of validation data is apparent, initial results are extremely encouraging.

Such simulations are, however, by no means straightforward and in developing an appropriate flood simulation model a number of issues have to be considered. First, the model must represent a complex set of

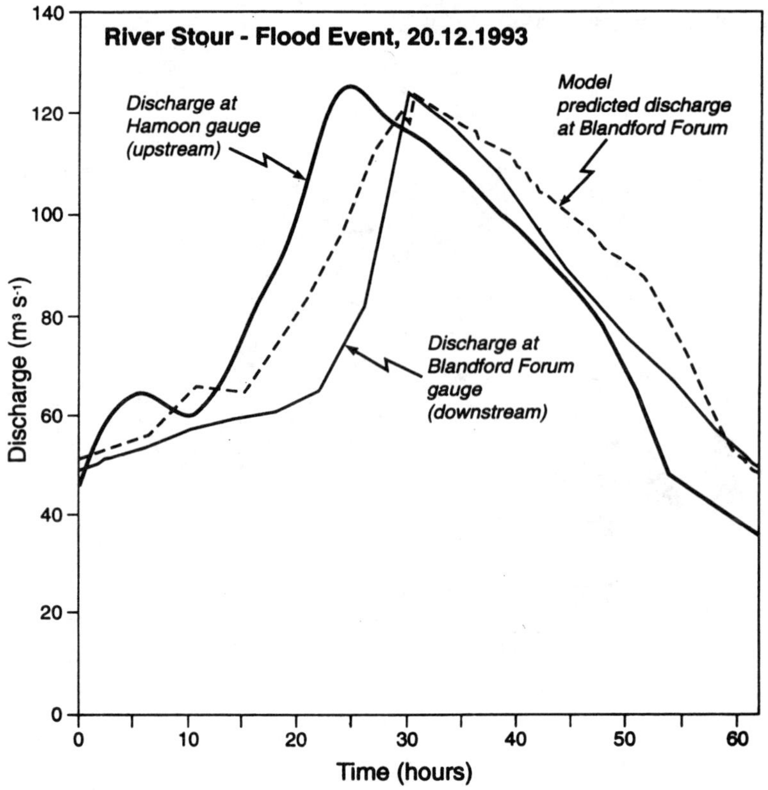

Figure 2. Comparison of observed and model-predicted reach outflow discharge for a one in four-year recurrence interval flood event on the River Stour, Dorset, UK, which occurred on 20 December 1993

hydraulic processes operating over a range of time and length scales. These range from turbulence at micro-scales, for example 1–10 cm and 1–10 seconds, to the scale of the flood wave itself, which may be hundreds of kilometres in length and persist for several days (see Bates and Anderson, 1993, for a more complete discussion). The level of detail at which processes can be represented is dependent on the scale of the simulation. A compromise therefore exists between processes representation and scale, with the modeller being required to make a number of subjective decisions regarding which processes are relevant for particular applications. Secondly, field observations have shown that floodplains possess a complex micro-topography that may have a significant effect on both local hydraulic and sediment transport processes (Walling *et al.*, 1986, 1992) as well as the inundation extent (Bates and Anderson, 1996) for a particular recurrence interval flood. Given this complexity of process and initial condition specification, hydraulic models must simplify actual physical processes. This, in conjunction with the impossibility of obtaining complete *a priori* knowledge of parameters such as boundary friction in distributed models, means that recourse must inevitably be made to calibration procedures in any practical application to achieve a correspondence between model predictions and field data.

Although calibration is a necessity, it may lead to a number of potentially significant constraints on our ability to develop further this class of model. In effect the numerical procedure attempts to solve a system of simultaneous equations where there are many more unknowns than equations. The objective behind estimating these unknowns is an attempt to obtain a match with some test data set. However, given that data availability is limited, particularly when compared with the number of data points that are unknown, the degrees of freedom present within the modelling procedure are such that the solution obtained may not be unique. Thus, more than one set of calibration parameters may fit the available data. This is particularly true

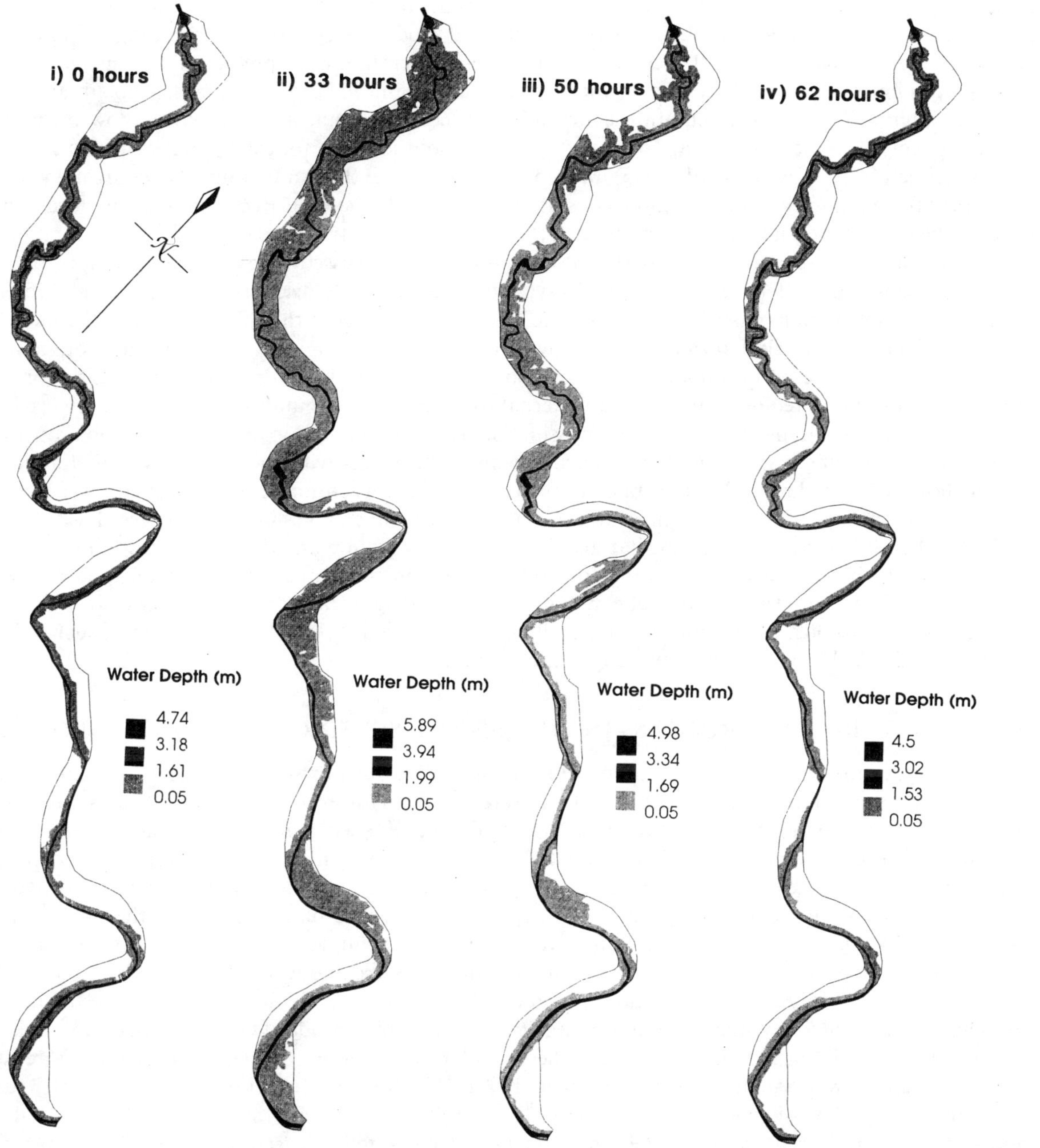

Figure 3. Inundation extent patterns predicted by a two-dimensional finite element hydraulic model for the River Stour flood event shown in Figure 2

of inundation extent predictions. Available calibration and validation data for river reaches predominantly consist of observations of bulk flow properties such as discharge. These are typically available only at the reach outflow, as in the River Stour example above. It may be entirely possible to develop a number of calibrations for this problem that match the available downstream discharge record but give different predictions of flood inundation extent. For example, Bates *et al.* (1996) studied a hypothetical channel/

floodplain domain using a Monte Carlo procedure to generate boundary friction and turbulent viscosity parameters for input into a two-dimensional finite element model. They showed that a range of calibrations could be obtained that gave equal acceptability in terms of their ability to predict the downstream outflow hydrograph, but gave inundation extents varying between 80 and 93% of the total floodplain area. Thus, incorrect internal process representation can produce a 'correct' result at the reach outflow owing to an inability to constrain the calibration procedure to a unique solution (see, for example, Beven, 1989; Konikow and Bredehoeft, 1992; Fawcett *et al.*, 1995). Moreover, the results of this study show that we are unlikely to be able to calibrate complex hydraulic models using hydrograph data alone. Recent modelling developments have enabled the potential for prediction to far outstrip our ability to collect sufficient calibration and validation data. For example, the River Stour simulation discussed above generated approximately $1{\cdot}17 \times 10^9$ items of data. This compares with an available data set of 248 discharge measurements. Moreover, these data are not spatially distributed and we are thus unable to utilize the full potential of the model.

A lack of spatially distributed calibration and validation data therefore currently constrains our ability to develop further high-resolution flood hydraulic models. In order to determine a research design for future model development we require sufficient data internal to the model computational domain to achieve both a calibration that constrains the number of possible solutions as far as possible and a validation of this prediction model using a split sample approach. Owing to problems with the collection of hydraulic data, such as flow velocity, during flood events, an obvious focus for attention should be direct observation of flood inundation extent. Not only is this variable sensitive to small changes in water surface elevation owing to shallow floodplain bed gradients, it has also been shown to require an additional set of approximations over and above those needed to solve the controlling equations. Accurate prediction of flood inundation extent is therefore a good test of model capabilities and is of significant practical interest. In this context remote sensing methods, both airborne and satellite, have significant potential for acquiring such information at a spatial scale commensurate with model resolution.

REVIEW OF RECENT PROGRESS IN THE REMOTE SENSING OF FLOOD HYDROLOGY

Ground observations of flood events can provide valuable data such as stage and discharge readings from gauging stations and inundation extent from high water marks. However, such measurements are time and resource intensive and may be difficult to obtain during flood events when installations may be destroyed and sampling may become dangerous. It is also relatively difficult to obtain large area synoptic observations of hydraulic properties, creating a significant disparity with the scale at which recently developed hydraulic models can operate. Remote sensing, from both space and airborne platforms, offers the opportunity to collect spatially distributed data rapidly and over large areas (Schultz, 1988) without the need for costly ground surveys. This data may then be used as an input to hydraulic and hydrological models in a variety of ways. For example, multi-spectral data can be used to classify land use, which may be used to estimate better the spatially distributed boundary friction parameters for hydraulic models. Ground-based radar has been used to monitor rainfall in order to provide the boundary conditions for hydrological flood forecasting models (Schultz, 1988). Satellite radar altimetry on board GEOSAT has been used to measure water surface elevation levels of the Amazon River (Koblinsky *et al.*, 1993). Ultimately, this could replace gauged stage data, although, at present, accuracy (± 1 m) and frequency (satellite overpasses every 17 days) are not sufficient for most applications while the altimeter's large radar footprint for significant returns (0·2–2 km radius) effectively limits the technique to only a few rivers. Remote sensing can even be used to determine sedimentological properties. For example, Yamagata *et al.* (1988) used Landsat TM data to estimate water turbidity, while (Currey, 1977) used Landsat-visible imagery and aerial colour photography to determine suspended sediment concentrations. Despite this variety of uses the most common hydraulic property studied by remote sensing is the inundation extent field.

Three types of differences between land and water can be detected by remote sensing: differences in emissivity (passive), differences in the reflection of natural radiation (also passive) and differences in the reflection of satellite-generated radiation (active). In terms of emissivity, field experiments (Schmugge, 1988) using passive microwave radiometry have shown a correlation with soil water content. This has enabled the difference in brightness temperature, as measured by AVHRR (Advanced Very High Resolution Radiometer) thermal infrared channels, to produce flood inundation extent maps (Barton and Bathols, 1988). The technique can only be used at night as it is based on the assumption that the land surface cools more quickly than water. AVHRR also suffers from a poor spatial resolution (1 km) and cannot penetrate cloud cover.

Passive techniques based on reflectance have also been employed. For example, Rango and Salomonson (1974) found that near-infrared (0·8–1·1 μm) images from the Landsat MSS could be used to distinguish between dry soil and vegetation (good infrared reflectors) and wet soil, stressed, damaged or wet vegetation and standing water (poor infrared reflectors) with a resolution of *c.* 80 m. Since the technique also detects wet soil, it can be used to map the inundated area after the flood wave has passed, even though the area of standing water may have receded. Thus, although the precise timing of the satellite overpass is less critical with this technique, it may only be able to detect maximum inundation extent rather than a time-series of inundation states as predicted by hydraulic models. A further problem is that, again, systems operating in the visible, near-infrared or thermal infrared wavebands are incapable of penetrating cloud cover.

Active reflectance techniques such as synthetic aperture radar (SAR) have considerable potential for overcoming the limitations of the above methods. SAR is an imaging microwave system that gives high ground resolution (*c.* 30 m) and is capable of both cloud cover penetration and 24-hour operation. In the simplest model of SAR imaging of flooded areas, smooth water acts as a specular reflector (i.e. it has a low radar back-scattering coefficient) and returns little energy back to the satellite or aircraft. Land areas are rougher and have higher back-scattering coefficients. Thus, water appears dark in the image, with land brighter. The situation is, however, complicated as the water surface may be roughened by wind or broken by protruding vegetation, the image may be degraded by speckle and the sideways-looking radar system can produce complex returns from topography that may introduce distortion into the images. To date, much of the literature on delineating flood boundaries using SAR adopts a qualitative approach. Tholey (1995) used ERS-1 images combined with a geographical information system (GIS) to carry out land use impact studies. Bonansea (1995) compared ERS-1 imagery to SPOT data, also using a GIS, while others (Badji and Dautrebande, 1995; Kannen, 1995; Noyelle *et al.*, 1995) have used multi-temporal SAR scenes, one of which observes a flood, to contribute the red, green and blue channels of a colour image in order to aid visual interpretation.

The use of SAR imagery compares favourably with other remote sensing systems. Imhoff *et al.* (1987) compared Landsat MSS and SIR-B images with a control set of colour and infrared aerial photographs of monsoon flooding in Bangladesh. Multi-spectral data have the advantage that data from all bands can be used in the land cover classification process, whereas only simple schemes (spatial filtering, density slicing) can be applied to black-and-white SAR images. In spite of this, SAR data processed with a simple density slicing and threshold technique provided a better identification of the flooded area (85% correct) compared with Landsat (64% correct) when compared with an air photograph control (Biggin and Blyth, 1996).

Of the complications noted above, the effects of flooded vegetation are well explored. Multiple reflections from the water surface and upright vegetation enhance back-scattering, giving flooded vegetation a bright return on the image. Inundated forests have been identified (Richards *et al.*, 1987) with high returns in the L-band (*c.* 20 cm), but this effect is reduced at shorter wavelengths (*c.* 6 cm) owing to increased volume. scattering in the canopy. Ormsby *et al.* (1985) found that the X-band (*c.* 3 cm) gave bright returns for flooded marshland but no back-scatter enhancement in forests. Henderson (1995) found that flooded forest and swamp gave high returns, and Ramsey (1995) found an inverse relationship between SAR returns and the depth of marsh flooding. Solomon (1993) found that the appearance of rivers in tropical forest areas varied depending on the orientation of the river with respect to the radar look direction, with the water acting as a specular reflector when looking along the river but giving higher returns when looking across, as a result of multiple water/vegetation reflection. Wang *et al.* (1995) have modelled these interactions mathematically, their results predicting that for C-band SAR at an incidence angle of approximately 20° (the characteristics of ERS-1), flooding should increase back-scatter by approximately 2·6 dB. As a further complication Giacomelli *et al.* (1995) has shown that SAR back-scatter can also increase with soil moisture content for unflooded areas. The speckle inherent in SAR imagery means that often traditional image processing techniques (as surveyed in Sonka *et al.*, 1993) are inappropriate and new approaches have to be developed. Despite this, a number of methods are currently available that could be applied to inundation extent detection. For example, White (1994) developed a simulated annealing algorithm that adopted a probabilistic approach for speckle reduction. Given an image, a model of how the SAR turns back-scatter cross-sections on the ground into image grey values and a number of assumptions about how the cross-section varies, the method searches for the most likely (maximum *a posteriori* probability) cross-section. Results show that images are smoothed (despeckled) without loss of spatial resolution. Oliver *et al.* (1994) used local image texture to map clearings in rain forest. *K*-distributed image statistics are assumed and a measure based on the order parameter (related to the variance) is used to calculate local texture. This measure is then smoothed using a simulated annealing algorithm. Similar techniques were used by Beattie and Elder (1995) on side-scan sonar images, the acoustic analogue of SAR. Finally, distortion owing to topography (layover and shadowing) is not significant on low gradient floodplains as problems are most evident where terrain slopes approach the radar look angle of 20° (Tilley and Bonwit, 1989), although automatic classification and segmentation algorithms may misclassify regions of radar shadow as floods (Henderson, 1995).

INTEGRATING HYDRAULIC MODELLING AND REMOTE SENSING

Maps of flood extent derived from remotely sensed data have so far only been used to calibrate and assess simple models and empirical relationships in hydraulic modelling. Imhoff *et al.* (1987) and Ramamoorthi (1990) derived approximate linear relationships between inundated area (as measured from SAR imagery) and gauged water levels for monsoon floods in Bangladesh and India. These empirical schemes need to be calibrated and extension beyond the observed range to extreme floods is not viable. Smith *et al.* (1995) used SAR imagery to calibrate a simple power law model relating width of glacial braided streams to discharge, and Moll and Overmars (1990) outlined a scheme for the calibration of a one-dimensional flow model using Landsat TM data. Here, river levels predicted by the model were converted to inundation extent via empirical relationships and compared with remotely sensed observations.

The simple models described above do not consider complex flow processes such as momentum transfer between channel and floodplain, known to be critical to the accurate simulation of flood inundation phenomena (Knight and Shiono, 1996). Two-dimensional models can incorporate such flow processes, but to date very few studies have compared direct flood inundation extent predictions from two, or higher, dimensional hydraulic models with remote sensing observations. Bates *et al.* (1996) compare inundation extent prediction from a two-dimensional finite element model of an 11 km reach of the River Culm, Devon, UK for a one in one-year recurrence interval flood with an estimate of the maximum inundated area obtained from air and ground photography and post-event surveying of floodplain deposits. Various mesh resolution and numerical solver techniques were tested with similar results: the model overpredicted inundation extent by approximately 20%. Reasons for this might include the rather simple representation of the floodplain topography as broadly planar, but no definitive conclusions were drawn as the data did not provide uniform coverage and were subject to a degree of error. When this was taken into account, model predictions were on the upper limit of the observed data error band.

A need therefore exists for integration of full spatial coverage SAR inundation extent data with high resolution hydraulic modelling parameterized using appropriate topographic data sources. Thus far this has only been attempted for Landsat TM and SPOT data. Figure 4 shows a finite element domain developed for a 60 km reach of the Missouri River along the South Dakota–Nebraska border between the gauging stations at Gavins Point Dam and Maskell. The model consists of 10 000 elements and 6000 computational nodes, giving an average element size of approximately 90 m. This is a channel-only reach up to 2 km wide but containing three permanent islands and a number of sandbanks that may become inundated at high flows. A

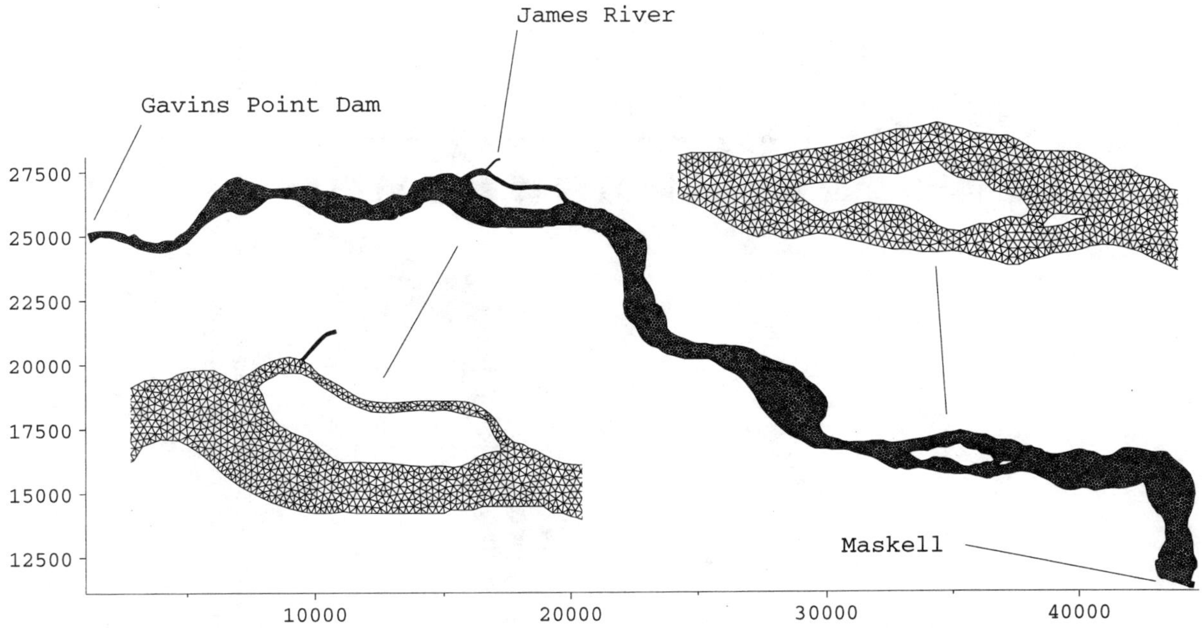

Figure 4. The finite element mesh, consisting of 10 000 elements and 6000 nodes, for the modelled reach of the Missouri River from Gavins Point Dam to Maskell, Nebraska

single tributary, the James River, enters from the north and is gauged 33 km upstream from the confluence at Scotland, South Dakota. A high quality topographic data set (Figure 5) was obtained from boat surveys of the channel bed cross-sections taken approximately every 150 m and island topography was estimated from USGS maps. Boundary conditions for the numerical model were taken from gauging station information, with a simple one-dimensional finite difference hydraulic model used to route the James River inflow to the confluence point. For this reach a single Landsat TM image taken on 6 June 1994 was available with synchronous boundary condition information. The resolution of this image is approximately 30 m, which compares favourably with the resolution of the finite element model.

A simulation over the domain shown in Figure 4 with appropriate boundary condition data was made using the two-dimensional finite element model TELEMAC-2D. Close inspection of the flow data within several days of the image revealed that there was very little ($<2\%$) variation in the data set. This enabled constant inflows, of 905 m^3 s^{-1} at Gavins Point and 45 m^3 s^{-1} from the James River, and a constant downstream stage value at Maskell to be used to simulate the flow regime at the time of the image. The model was run using a 4 second time step until a steady state was achieved where the inflow equals the outflow, taking into account a mass conservation error of less than 1%. Figure 6 shows a comparison of the available satellite image to the model predictions. The two do compare quite well in large-scale features, the model representing the large sandbanks and channel boundaries especially well (Figure 7). The model does not, however, pick up many of the smaller scale features that can be seen on the Landsat image. Such features are either smaller than the mesh resolution of the model or were overlooked during the survey of the topography, These problems were not unexpected as, first, the topography above the water line in the model is only crudely represented from USGS maps, hence small-scale features could not be included, and, secondly, the resolution of the satellite image is greater than that of the model, so smaller scale features are bound to be depicted.

While this data set and comparison with two-dimensional model results is unique, the success was largely owing to the scale of the reach and the fact that the flow regime is highly regulated. This allowed an assumption of steady-state flow to be made which both simplifies the modelling problem and gives a larger 'window' in which to obtain a cloud-free image. Moreover, the resolution of Landsat TM data (30 m) would

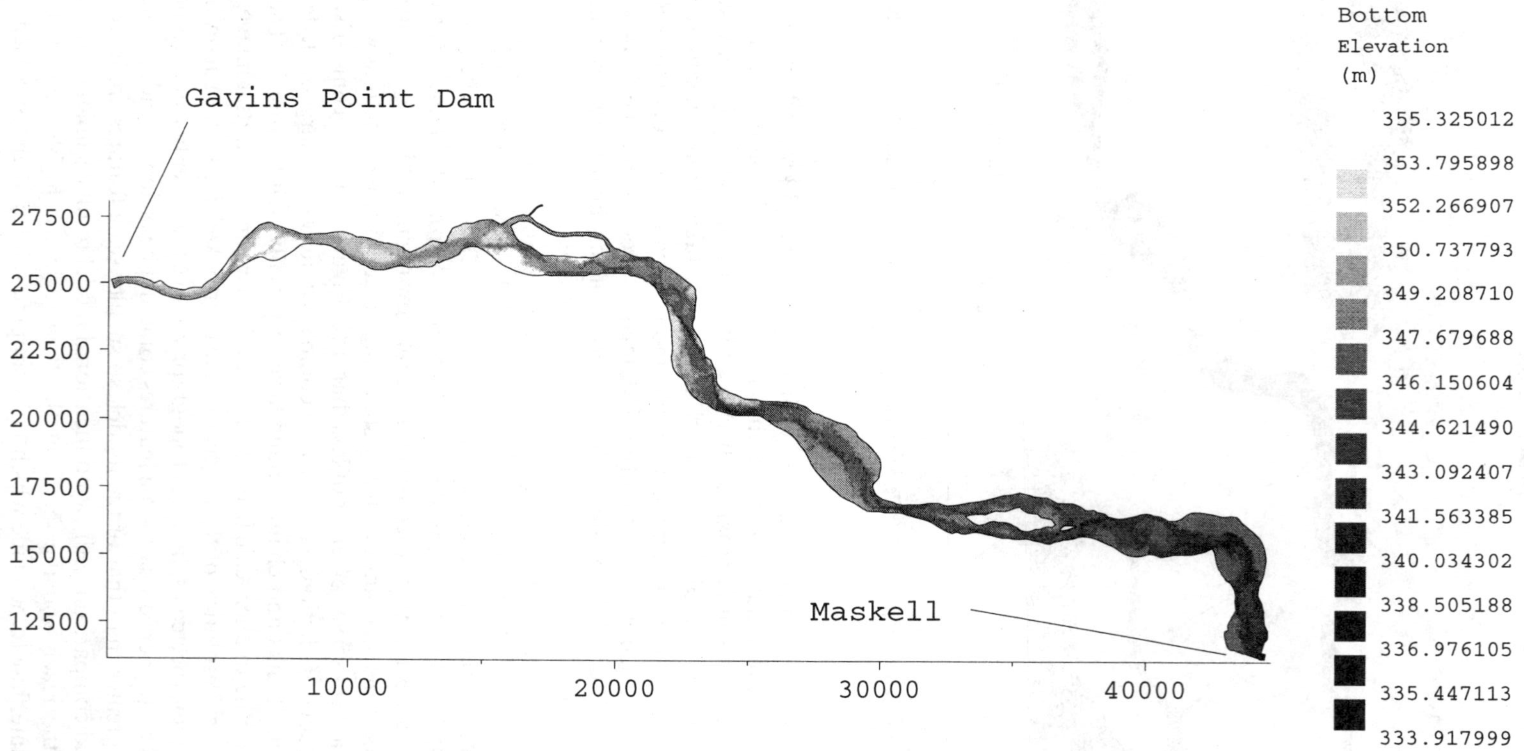

Figure 5. The high quality topographic data set, obtained from boat surveys of bed cross-sections, as applied on to the Missouri River model. The raw data points were interpolated on to the computational nodes using a downstream directional method between cross-sections to give the complete topography shown

(a)

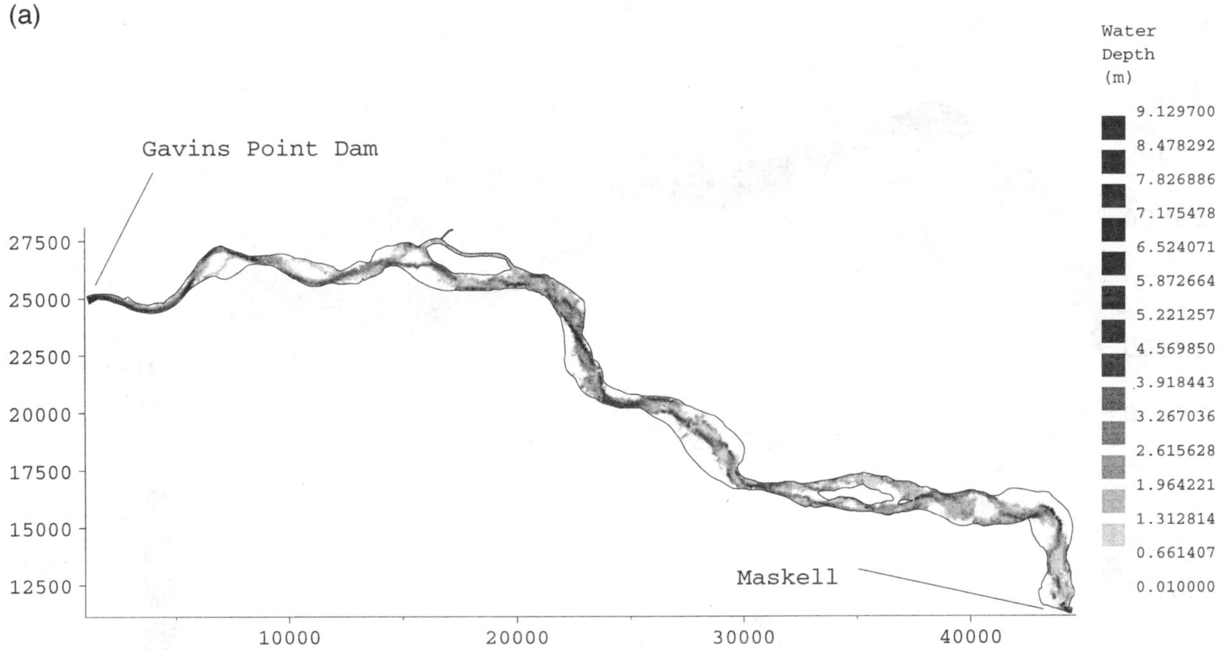

(b)

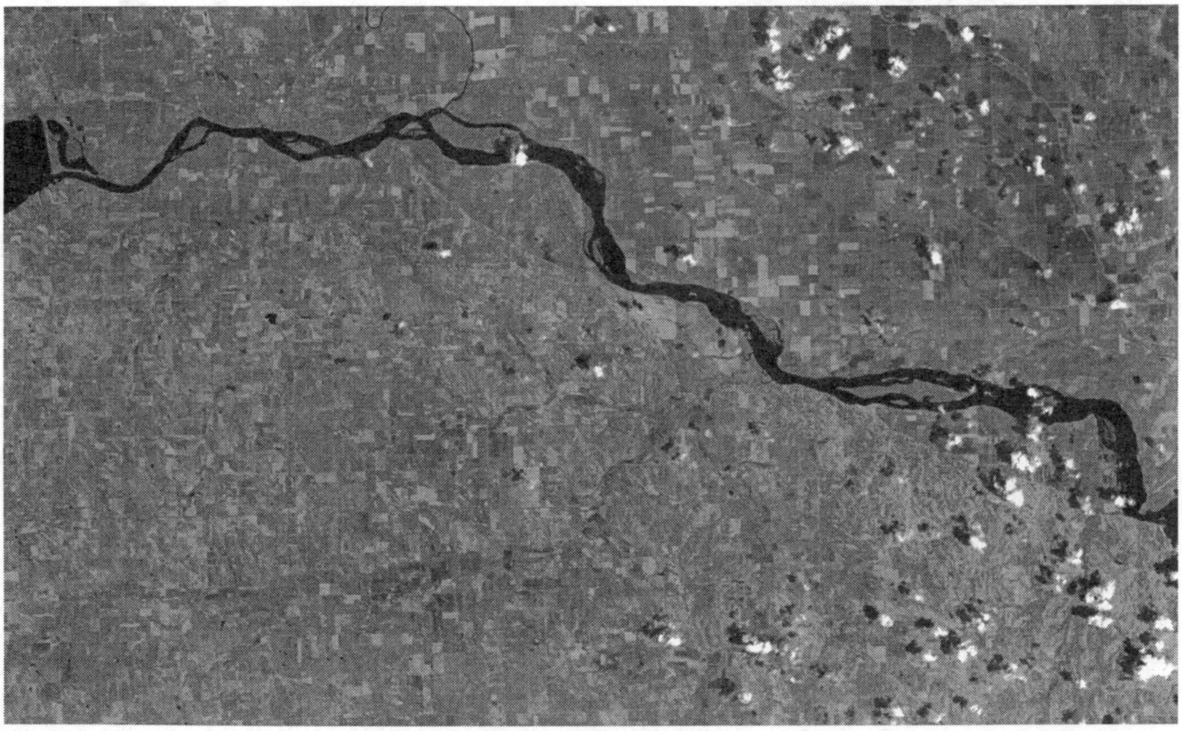

Figure 6. (a) Predicted water depths from the full Missouri River model for 6 June 1994, the data of the Landsat TM image. (b) Landsat TM image of the modelled Missouri River reach, 6 June 1994

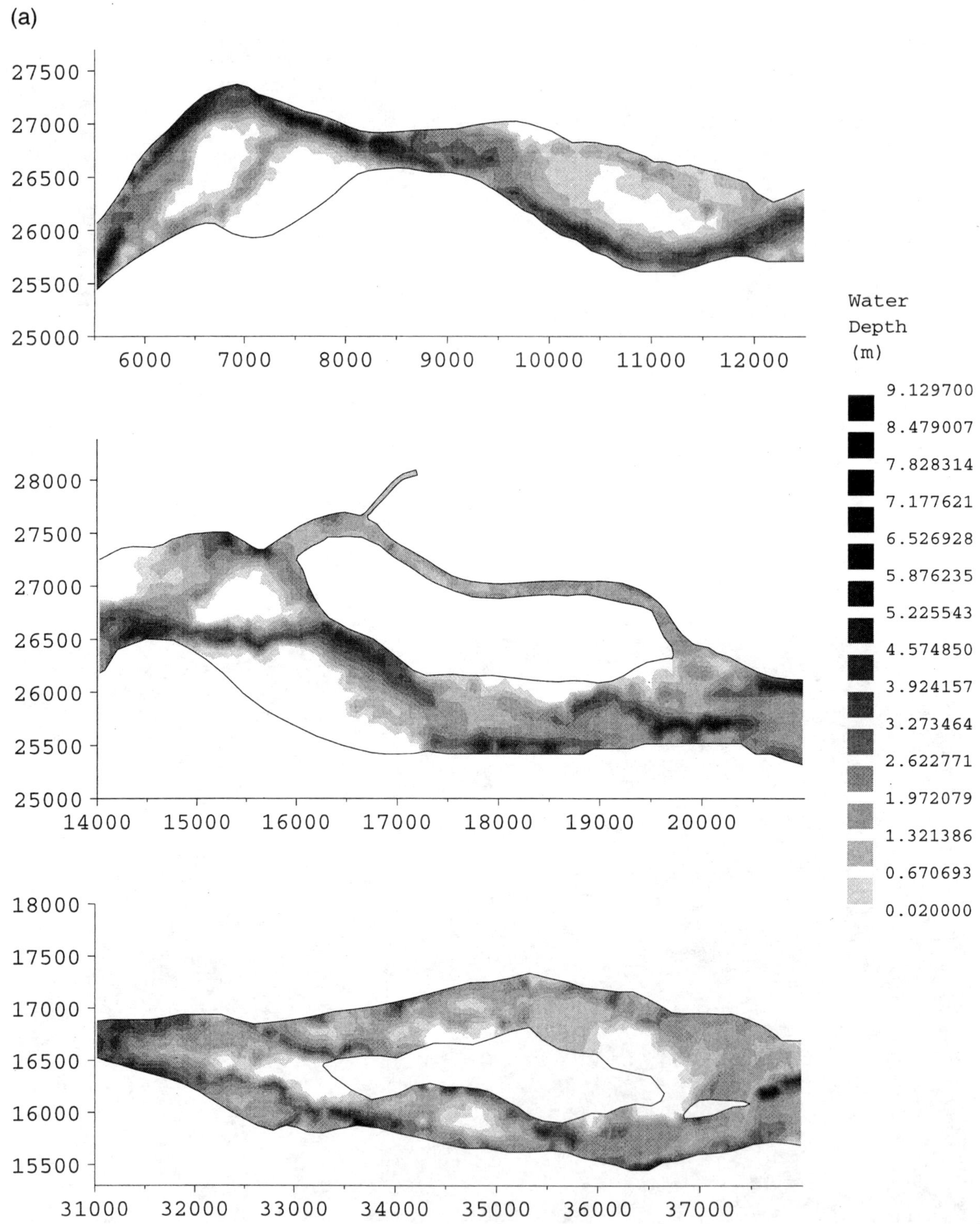

Figure 7. (a) Model-predicted water depths for three subsections of the plot shown in Figure 6a. (b) Corresponding subsections of the Landsat TM image shown in Figure 6b

(b)

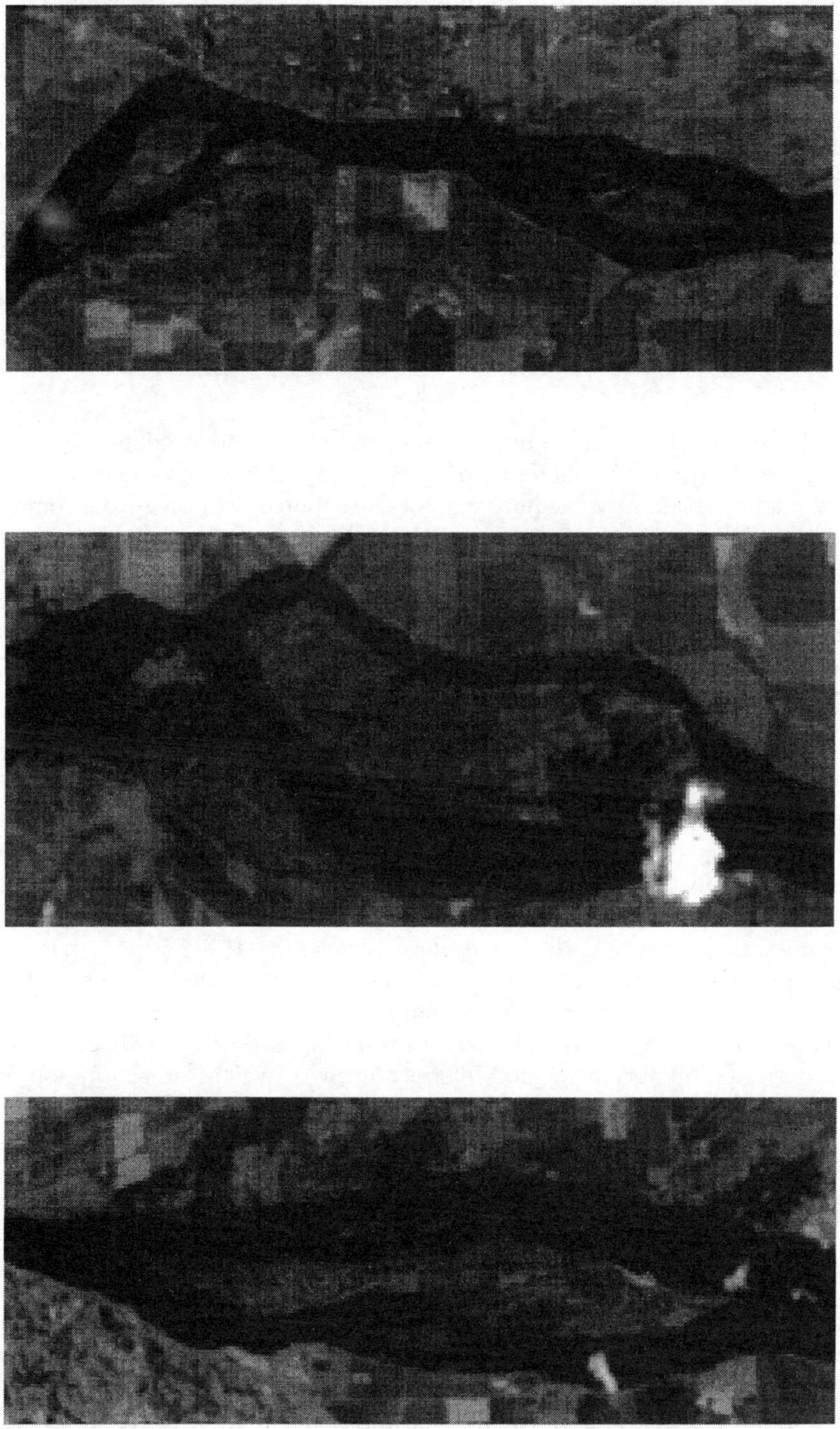

seem to be relevant to large-scale river problems. In the case of the River Stour model discussed above, average element resolution is of the order of 10 m for which Landsat TM data would be somewhat coarse.

RESEARCH NEEDS

The Missouri River case study demonstrates the potential for integrating flood hydraulic modelling and remote sensing data capture for flood inundation extent, while highlighting the need to move to the use of full spatial coverage SAR data identified in the review section of this paper, in order to obtain better spatial resolution and an all-weather capability. Although it is clear that such research is a priority, a number of prerequisite research needs are apparent from the evidence discussed above. These require investigation prior to the development of an integrative research design for enhancing both model calibration and validation procedures and the remote sensing of flood hydraulic properties.

Hydraulic modelling

This paper has demonstrated a need to improve the calibration and validation of hydraulic models to enable robust, spatially distributed predictions to be obtained. It has been suggested that remote sensing techniques now offer an opportunity to acquire spatially distributed data on a scale commensurate with, or in excess of, model resolution, thus facilitating model development and redesign. At present, however, no methodology exists for the calibration and validation of distributed models using distributed data. Standard calibration procedures for hydraulic models are typically based on an attempt to match model predictions with an observed discharge hydrograph, with simple sensitivity analysis techniques (McCuen, 1973, 1974) used to provide data on the effect of parameter perturbation on model behaviour. This is usually achieved using a high degree of spatial lumping in the parameter surface to simplify the calibration problem. For example, in the model of the River Stour discussed above, Bates *et al.* (1996) used only two values for boundary friction: one applied to the channel and one to the floodplain. It has recently been realized (Lane *et al.*, 1994) that such calibration and sensitivity analysis techniques were developed in hydrology and hydraulics for the previous generation of lumped models and may fail to capture the complexity of high resolution spatially distributed schemes, which are based on highly non-linear equations applied using dynamic boundary conditions. For example, Bates and Anderson (1996) have demonstrated the complex response in terms of inundation extent prediction of a two-dimensional finite element hydraulic model to small, uniform changes in simulation initial conditions. Lane *et al.* (1994) demonstrated similarly complex spatial responses to parameter variation for a two-dimensional finite volume model of flow in a proglacial stream. At present, knowledge concerning even the simplest case of a model of spatially distributed response to uniform parameter variation is largely unknown. Introducing a more spatially complex parameter space will invariably increase the number of degrees of freedom with which the calibration may operate. The development of effective calibration and validation methodologies that optimize a spatially distributed parameter surface against spatially distributed data should therefore be a focus for future research. Recent progress in this field has centred on inverse modelling techniques (see, for example, Tarantola, 1987; Feddes, 1995) which use probability theory to examine the range of parameter sets that produce an acceptable solution of a problem when compared with given objective criteria. Such approaches to parameter estimation have recently been applied to hydraulic models. Romanowicz *et al.* (1996) used a Bayesian approach to the inverse problem in conjunction with a Monte Carlo procedure for parameter selection to perform a full spatial optimization of 12 boundary friction parameters in a simple hydraulic of the River Culm, Devon, UK. 'Observed' data for the Bayesian procedure consisted of 12 stage hydrographs taken from a two-dimensional finite element model of this reach (Bates *et al.*, 1992) which was assumed to represent ground truth. Barros (1996) presented an iterative minimization algorithm for the optimization of friction parameters in the two-dimensional finite element model, TELEMAC-2D. Again, stage hydrograph measurements were used to test the procedure which, theoretically, could provide a full, spatially distributed optimization. Such methods would appear to have considerable potential for development to allow spatially distributed data to be used in the optimization, although computational costs are considerable.

With the development of such a methodology, three further research issues relating to the prediction of flood inundation extent will require investigation. First, to date, no studies have adequately explored the relationship between the calibration, model topographic specification and the finite element mesh resolution. All these factors can be expected to have an effect on the model's ability to predict inundation extent; however, their relative dominance is poorly understood. Such information would enable research priorities to be assessed and would give benchmark information for assessing model developments. For example, much research in hydraulic modelling is focused on improving numerical solution algorithms (see, for example, Molinaro and Natale, 1994). Initial research (Bates *et al.*, 1996) has, however, shown that the impact of numerical improvements may be of the same order as mesh resolution effects. Secondly, there is scope for further development and testing of algorithms to simulate dynamic wetting and drying processes in two-dimensional schemes. Such algorithms are central to flood inundation prediction, yet all involve some degree of compromise in the flow predictions and still require comparative testing against a high quality data set. In particular, there is a need to extend wetting and drying algorithms to three-dimensional schemes. Hervouet and Van Haren (1996) illustrate problems with current approaches in three dimensions with reference to the TELEMAC-3D code. Here a two-dimensional mesh is extended to three dimensions by specifying a finite number of layers between the bed and the free surface. This gives a system of prismatic finite elements for which a solution of the three-dimensional shallow water equations can be obtained. A problem occurs where the water depth tends to zero, since, here, elements are 'smashed', thus generating instability. Partially dry elements can be excluded from the solution but this results in a loss of mass conservation. Flood inundation problems are therefore currently restricted to two-dimensional solutions. Lastly, approaches to turbulence closure modelling in hydraulic schemes continue to attract much interest and debate (see, for example, Rodi, 1980; Younis, 1996). The development of robust solutions applicable at a scale commensurate with reach length numerical models is a priority here. The current restriction to two-dimensional approaches is also relevant, as a number of higher order turbulence closure schemes strictly only apply to three-dimensional problems.

Remote sensing

Having identified SAR as a potentially useful remote sensing system for the mapping of flood extent, it remains to identify the requirements for a data set to validate its use. The high spatial resolution of SAR (12·5 m pixel size for ERS-1 PRI data) means that imagery from other sensors (e.g. Landsat TM, 30 m resolution) is of insufficient resolution for validation, and the satellites' orbits preclude the possibility of obtaining coincident scenes. Other systems' failure to operate at night or when the floodplain is obscured by cloud cover would also make such a data set difficult to obtain. Given these problems with space-borne sensors other than SAR, aerial photography would seem to be the way forward to validating the identification of the flood boundary from ERS-1 imagery. Such a data set has been obtained by the Institute of Hydrology, Wallingford, UK (Biggin and Blyth, 1996) and comprises aerial photographs coincidental (to within *c.* 3 hours) with an ERS-1 SAR scene of a flood of the River Thames, UK, on 4 December 1992, along with pre- and post-flood SAR scenes. Because of the presence of low cloud and the pressures of covering a reach of *c.* 100 km in a matter of hours, the photographs were shot obliquely to gain more coverage. Even so, this type of data set is in itself rare, and possibly unique within the UK. The 12·5 m pixel size of ERS-1 imagery compares well with the spatial resolution inherent in the most detailed flood modelling studies to date, e.g. the River Stour investigation discussed above. This sets the resolution required for the validation at *c.* 10 m, a more accurate study than in Blyth *et al.* (1993). Delineation of the flood boundary to within 10 m from the aerial photo set is relatively straightforward. Ground control points such as field boundaries and pylons are easily identifiable to <10 m on the photographs and on scanned 1:25 000 UK Ordnance Survey maps. An orthographic transform can then be used to convert image coordinates into map coordinates. Thus a map-referenced flood boundary with an accuracy of *c.* 10 m can be produced. Areas where the SAR response may be confused, such as flooded vegetation and buildings can also be identified from the photographs. Thus, the data set also provides the opportunity to study the water–vegetation–microwave interactions discussed above for land cover typical to many temperate floodplain environments.

SAR segmentation to within *c.* 1 pixel presents more of a problem: how to reduce the effects of speckle but maintain spatial resolution? The simulated annealing techniques discussed previously (White, 1994; Oliver *et al.*, 1994) achieve this at the expense of computational complexity and speed, and also require post-processing (edge detection linking and vectorization) to achieve results compatible with the aerial data. Active contour models (snakes) (Cohen, 1991; Williams and Shah, 1992) have recently gained favour in image segmentation applications because of their ability to link detected edges broken by noise, to produce smooth region boundaries. The snake method uses a dynamic curvilinear contour to search the image space until it settles upon image region boundaries. The technique has been applied to SAR images of the North Sea shoreline on the UK's east coast (Mason *et al.*, 1995). Ivins and Porrill (1994) present a statistical snake that does not require an initial edge detection stage, instead operating directly on the image statistics that can be calculated from pixels along the length of the snake. Thus, by looking at a number of pixels, the effects of speckle can be reduced without a loss of resolution perpendicular to the shoreline. Such active contour models are usually cast as an energy minimization problem:

$$E = \frac{1}{2}\int \lambda \left| \frac{\partial \boldsymbol{u}}{\partial s} \right|^2 \mathrm{d}s + \frac{1}{2}\int \gamma \left| \frac{\partial^2 \boldsymbol{u}}{\partial s^2} \right|^2 \mathrm{d}s + \int P(I(\boldsymbol{u}))\, \mathrm{d}s \tag{1}$$

where $\boldsymbol{u}(s)$ is the contour position as a function of the parameter $s(s \in [0, 1])$ and $I(\boldsymbol{u})$ is the image. The first two terms represent the snake's tension and stiffness, the parameters λ and γ being used to adjust the importance of length and curvature in the model, and the third is an image energy function. The continuous curve $\boldsymbol{u}(s)$ is represented by a series of linear segments.

Such a formulation presents problems: estimation of tension and curvature measures for a finite difference contour, parameter (λ, γ, size of linear segments) estimation, energy minimization of algorithm design and determining the initial contour (seeding). Curvature estimation is well explored in Williams and Shah (1992), but parameter estimation in previous work has been conducted on an *ad hoc* basis, by trial and error. The solution may lie in the consideration of the relationship between image geometry and statistics. Curvature of typical flood shorelines can vary widely, dependent on the local floodplain environment: purely topographic considerations tend to produce smooth shorelines, whereas in reality man-made features (roads, fields) will produce sharp corners. Similarly, the contrast in image statistics will depend on the floodplain land use (grass, arable land, trees) and the meteorological conditions (wind-roughening of the water surface). This makes parameter estimation difficult, and it may not be possible to select global parameter values that ensure that the shoreline is always a strong minimum of the contour energy functional. The choice of a minimization algorithm is important for two reasons: computational efficiency and solution stability. In dynamic programming, the energy gradient is used as a force that moves the snake (e.g. Cohen, 1991; Ivins and Porrill, 1994), which iteratively descends the energy slope into local minima. Williams and Shah (1992) use a greedy algorithm that considers all possible moves of a point on the snake and chooses the one with the most negative change in E. This could easily be extended into a simulated annealing scheme, but that would introduce the extra question of what annealing schedule to use. The stability problem is addressed in Cohen (1991), for snakes operating on edge-detected images, and in Ivins and Porrill (1994), but this approach assumes statistics vary smoothly from one region to another over a distance of several pixels. This is not the case for SAR, because the surface back-scattering coefficient can change significantly over a distance of 10 m, causing image statistics to vary sharply. The stability of solutions around the minimum of the energy function is all the more important in this situation, because of the need to isolate edges to within 1 pixel. The choice of initial contour position is vital to snakes operating on edge-detected images: if the snake starts in the wrong place, it will settle on edges not associated with the flood shoreline. This is especially problematic for typical UK floodplains where the land image is composed of fields and is therefore far from homogeneous.

For the statistical snake the initial contour position is less critical (it will expand away from areas with statistics different to the flood area) but requires that the flooded area image statistics are known, and so a training area is needed. An initial contour near the flood boundary would be advantageous in reducing the number of iterations needed for the snake to reach equilibrium on the shoreline. Previous active contour model studies have concentrated on detected edges and image intensity in the energy formulation, but the use of a statistical snake means more information can be used. Image texture (local variance) is a well-known, useful property of SAR imagery (Oliver *et al*., 1994), and many previous qualitative studies have used multi-temporal SAR scenes for flood detection. An active contour model operating on local image intensity and texture over two or three image planes simultaneously is potentially a very powerful SAR segmenting tool.

CONCLUSIONS

The integration of high resolution hydraulic modelling with remote sensing data capture techniques for flood hydrology variables has the potential to improve substantially our understanding of floodplain flow processes. While it will be necessary to determine how errors inherent in each procedure interact and combine in order to comment on the robustness of the approach, remote sensing techniques confer, for the first time, an ability to collect spatially distributed data on a scale and extent commensurate with recently developed flood inundation models. Moreover, this development will allow fundamental research issues in terms of both model calibration and remote sensing image processing to be addressed.

ACKNOWLEDGEMENTS

This research has been made possible with the support of the Laboratoire National d'Hydraulique, France, the UK Natural Environment Research Council, Grant No. GR3/09925, and the US Army Corps of Engineers. The authors are grateful for provision of data by the UK Ordnance Survey, the UK Environment Agency and the UK Institute of Hydrology. Matthew Horritt was supported by a studentship funded by NRSC Ltd and Christopher Smith by a UK Natural Environment Research Council Studentship.

REFERENCES

Abbott, M. B., Bathurst, J. C., Cunge, J. A., O'Connell, P. E., and Rasmussen, J. 1986. 'An introduction to the European Hydrological System — Système Hydrologique European, "SHE", 2: structure of a physically based distributed modelling system', *J. Hydrol.*, **87**, 45–59.

Badji, M. and Dautrebande, S. 1995. Characterisation of flood inundated areas and delineation of poor internal drainage soil using ERS-1 SAR imagery', *Proceedings, 1st ERS Thematic Working Group Meeting on Flood Monitoring, ESA–ESRIN, Frascati, Italy.*

Baird, L., Gee, D. M., and Anderson, M. G. 1992. 'Ungauged catchment modelling. II. Utilization of hydraulic models for validation', *Catena*, **19**, 17–31.

Barros, E. 1996. 'Estimation de parametres dans les equations de Saint-Venant', *Laboratoire National d'Hydraulique report no. HE-43/96/034/A*. Electricité de France, Chatou, France. 214 pp.

Barton, I. J. and Bathols, J. M. 1989. 'Monitoring floods with AVHRR', *Rem. Sens. Environ.*, **30**, 89–94.

Bates, P. D. and Anderson, M. G. 1993. 'A two dimensional finite element model for river flood inundation', *Proc. R. Soc. Lond., Ser. A*, **440**, 481–491.

Bates, P. D. and Anderson, M. G. 1996. 'A preliminary investigation into the impact of initial conditions on flood inundation predictions using a time/space distributed sensitivity analysis', *Catena*, **26**, 115–134.

Bates, P. D., Anderson, M. G., Baird, L., Walling, D. E., and Simm, D. 1992. 'Modelling floodplain flows using a two-dimensional finite element model', *Earth Surf. Process. Landf.*, **17**, 575–588.

Bates, P. D., Anderson, M. G., and Hervouet, J.-M. 1995. 'An initial comparison of two 2-dimensional finite element codes for river flood simulation. *Proc. Inst. Civil Eng. Wat. Marit. Energy*, **112**, 238–248.

Bates, P. D., Anderson, M. G., Price, D., Hardy, R., and Smith, C. 1996. Analysis and development of hydraulic models for floodplain flows', in Anderson, M. G., Walling, D. E., and Bates, P. D. (Eds), *Floodplain Processes*. John Wiley and Sons, Chichester. pp. 215–254.

Beattie, R. S. and Elder, S. C. 1995. 'Side scan sonar image restoration using simulated annealing and iterative conditional modes. *Proc. Inst. Acoust.*, **17**, 161–167.

Beven, K. J. 1985. 'Distributed models', In Anderson, M. G. and Burt, T. P. (Eds), *Hydrological Forecasting*. Wiley, New York. pp. 405–435.

Beven, K. J. 1989. 'Changing ideas in hydrology: the case of physically based distributed models', *J. Hydrol.*, **105**, 157–172.

Beven, K. J. Calver, A., and Morris, E. M. 1987. 'The Institute of Hydrology distributed model', *Institute of Hydrology report no. 98*. Institute of Hydrology, Wallingford.

Biggin, D. S. and Blyth, K. 1996. 'A comparison of ERS-1 satellite radar and aerial photography for river flood mapping. *J. Chart. Inst. Wat. Eng. Manage.*, **10**, 59–64.

Binley, A. and Beven, K. 1993. 'Three dimensional modelling of hillslope hydrology', in Beven, K. and Moore, I. (Eds), *Terrain Analysis and Distributed Modelling in Hydrology*. John Wiley and Sons, Chichester. pp. 107–120.

Blyth, K., Biggin, D. S., and Ragab, R. 1993. ERS-1 SAR for monitoring soil moisture and river flooding', *Proceedings, 2nd ERS-1 Symposium, Hamburg, Germany*.

Bonansea, E. 1995. 'Mapping 1994 floods in Piedmonte Region, Italy: an example of remote sensing and GIS application', *Proc. 1st ERS Thematic Working Group Meeting on Flood Monitoring, ESA–ESRIN, Frascati, Italy*.

Brookes, A. N. and Hughes, T. J. R. 1982. 'Streamline Upwind/Petrov Galerkin formulations for convection dominated flows with particular emphasis on the incompressible Navier–Stokes equations', *Comp. Meth. Appl. Mech. Eng.*, **32**, 199–259.

Burt, T. P., Heathwaite, A. L., and Trudgill, S. T. 1993. *Nitrates: Processes, Patterns and Management*. John Wiley and Sons, Chichester.

Carey, G. F. and Jiang, B. N. 1986. 'Element-by-element linear and non-linear solution schemes', *Communications in Applied Numerical Methods*, **2**, 145–153.

Cohen, L. D. 1991. 'On active contour models and balloons', *CVGIP: Image Understand.*, **53**, 211–218.

Currey, B. 1977. 'Identifying water flood movement', *Rem. Sens. Environ.*, **6**, 51–61.

Falconer, R. A. and Chen, Y. 1996. 'Modelling sediment transport and water quality processes on tidal floodplains', in Anderson, M. G., Walling, D. E., and Bates, P. D. (Eds), *Floodplain Processes*. John Wiley and Sons, Chichester. pp. 361–398.

Fawcett, K. R., Anderson, M. G., Bates, P. D., and Jordan, J.-P. 1995. 'The importance of internal validation in the assessment of physically based distributed models', *Trans. Inst. Br. Geogr.*, **NS20**, 248–265.

Feddes, R. A. 1995. 'Remote sensing-inverse modelling approach to determine large scale effective soil hydraulic properties in soil-vegetation-atmosphere systems', in Feddes, R. A. (Ed.), *Space and Time Scale Variability and Interdependencies in Hydrological Processes*. Cambridge University Press, Cambridge. pp. 33–42.

Feldhaus, R., Höttges, J., Brockhaus, T., and Rouvé, G. 1992. 'Finite element simulation of flow and pollution transport applied to a part of the River Rhine', in Falconer, R. A., Shiono, K., and Matthews, R. G. S. (Eds), *Hydraulic and Environmental Modelling, Esturine and River Waters*. Ashgate Publishing, Aldershot. pp. 323–334.

Gee, D. M., Anderson, M. G., and Baird, L. 1990. 'Large scale floodplain modelling', *Earth Surf. Process. Landf.* **15**, 512–523.

Giacomelli, A., Bacchiega, U., Troch, P. A., and Mancini, M. 1995. Evaluation of surface soil moisture distribution by means of SAR remote sensing techniques and conceptual hydrological modelling', *J. Hydrol.*, **166**, 445–459.

Henderson, F. M. 1995. 'Environmental factors and the detection of open surface water using X-band radar imagery', *Int. J. Rem. Sens.*, **16**, 2423–2437.

Hervouet, J.-M. 1992. 'Element-by-element methods for solving shallow water equations with FEM', *Proceedings, 9th International Conference on Computational Methods in Water Resources, Denver, USA*.

Hervouet, J.-M. and Van Haren, L. 1996. 'Recent advances in numerical methods for fluid flows', in Anderson, M. G., Walling, D. E., and Bates, P. D. (Eds), *Floodplain Processes*. John Wiley and Sons, Chichester. pp. 183–214.

Imhoff, M. L., Vermillion, C., Story, M. H., Choudhury, A. M., Gafoor, A., and Polcyn, F. 1987. 'Monsoon flood boundary delineation and damage assessment using space borne imaging radar and Landsat data', *Photogram. Eng. Rem. Sens.*, **53**, 405–413.

Ivins, J. and Porrill, J. 1994. 'Statistical snakes: active region model', *Proc. 5th British Machine Vision Conference*, pp. 377–386.

James, C. S. 1985. 'Sediment transfer to overbank section', *J. Hydraul. Res.*, **23**, 435–452.

Jolly, I. D. 1996. 'The effects of river management on the hydrology and hydroecology of arid and semi-arid floodplains', in Anderson, M. G., Walling, D. E., and Bates, P. D. (Eds), *Floodplain Processes*. John Wiley and Sons, Chichester. pp. 577–609.

Kannen, A. 1995. 'Use of SAR data for flood mapping and monitoring in Thuringen, Germany', *Proceedings, 1st ERS Thermatic Working Group Meeting on Flood Monitoring, ESA–ESRIN, Frascati, Italy*.

Kiely, G. 1990. 'Overbank flow in meandering compound channels: the important mechanisms', in White, W. R. (Ed.), *International Conference on River Flood Hydraulics*. John Wiley and Sons, Chichester. pp. 207–217.

King, I. P. and Roig, L. C. 1988. 'Two-dimensional finite element models for floodplains and tidal flats', in Niki, K. and Kawahara, M. (Eds), *Proceedings, International Conference on Computational Methods in Flow Analysis, Okayama, Japan*, pp. 711–718.

Knight, D. W. and Demetriou, J. D. 1983. Floodplain and main channel interaction', *Proc. Am. Soc. Civil Eng. J. Hydraul. Eng.* **109**, 1073–1092.

Knight, D. W. and Shiono, K. 1996. 'River channel and floodplain hydraulics', in Anderson, M. G., Walling, D. E., and Bates, P. D. (Eds), *Floodplain Processes*. John Wiley and Sons, Chichester. pp. 139–181.

Koblinsky, C. J., Clarke, R. T., Brenner, A. C., and Frey, H. 1993. 'Measurement of river level variations with satellite altimetry', *Wat. Resour. Res.* **29**, 1839–1848.

Konikow, L. F. and Bredehoeft, J. D. 1992. 'Ground-water models cannot be validated', *Adv. Wat. Resour.* **15**, 75–83.

Lane, S. N. 1997. 'Hydraulic modelling in geomorphology and hydrology: a review of high resolution approaches', this volume.

Lane, S. N., Richards, K. S., and Chandler, J. H. 1994. 'Application of distributed sensitivity analysis to a model of turbulent open channel flow in a natural river channel', *Proc. R. Soc. Lond. Ser. A*, **447**, 49–63.

Leclerc, M., Belemare, J.-F., Dumas, G., and Dhatt, G. 1990. 'A finite element model of estuarian and river flows with moving boundaries', *Adv. Wat. Resour.*, **13**, 158–168.

Lynch, D. R. and Gray W. G. 1980. 'Finite element simulations of flow deforming regions', *J. Comput. Phys.*, **36**, 135–153.

Marriot, S. 1992. 'Textural analysis and modelling of a flood deposit: River Severn, UK', *Earth Surf. Process. Landf.*, **17**, 687–697.

Marron, D. C. 1989. 'The transport of mine tailings as suspended sediment in the Belle', *IAHS Publ.* **184**, 19–26.

Mason, D. C., Davenport, I. J., Robinson, G. J., Flather, R. A., and McCartney, B. S. 1995. 'Construction of an inter-tidal digital elevation model by the "water-line" method', *Geophys. Res. Lett.*, **22**, 3187–3190.

McCuen, R. H. 1973. 'The role of sensitivity analysis in hydrologic modelling', *J. Hydrol.*, **18**, 37–53.

McCuen, R. H. 1974. 'A sensitivity and error analysis of procedures used for estimating evaporation', *Wat. Resour. Bull.* **10**, 484–498.

Molinaro, P. and Natale, L. 1994. *Modelling of Flood Propagation over Intially Dry Areas*. American Society of Civil Engineers, New York. 373 pp.

Moll, J. R. and Overmars, J. F. M. 1990. 'River hydrometry by pattern recognition in remote sensing. *Proc., International Symposium on Remote Sensing in Water Resources*. International Association of Hydrogeologists Publication, The Netherlands. pp. 443–452.

Noyelle, J., Delimiere, S., and Marinelli, L. 1995. 'Identification of flooded areas in the Rhone', *Proceedings, 1st ERS Thematic Working Group Meeting on Flood Monitoring, ESA–SRIN, Frascati, Italy.*

Oliver, C. J., Blake, A., and White, R. G. 1994. 'Optimum texture analysis of synthetic aperture radar images', *Alg. Synth. Aper. Radar Imag.*, **2230**, 389–398.

Ormsby, J. P, Blanchard, B. J., and Blanchard, A. J. 1985. 'Detection of lowland flooding using active microwave systems', *Int. J. Rem. Sens.*, **5**, 317–328.

Pizzuto, J. E. 1987. 'Sediment diffusion during overbank flows', *Sedimentology*, **34**, 301–317.

Ramamoorthi, A. S. 1990. 'Flood damage forecasting by satellite remote sensing', *Proceedings, International Symposium on Remote Sensing in Water Resources.* International Association of Hydrogeologists Publication, The Netherlands. pp. 433–442.

Ramsey, E. W. 1995. 'Monitoring flooding in coastal wetlands by using radar imagery and ground based measurements', *Int. J. Rem. Sens.* **16**, 2495–2502.

Rango, A. and Salmonson, V. V. 1974. 'Regional flood mapping from space', *Wat. Resour. Res.*, **10**, 473–484.

Richards, J. A., Woodgate, P. W., and Skidmore, A. K. 1987. 'An explanation of enhanced radar backscattering from flooded forests', *Int. J. Rem. Sens.*, **8**, 1093–1100.

Richards, K. S., Hughes. F. M. R., El-Hames, A. S., Harris, T., Pautau, G., Peiry, J.-L., and Girel, J. 1996. 'Integrated field, laboratory and numerical investigations of hydrological influences on the establishment of riparian tree species', in Anderson, M. G., Walling, D. E., and Bates, P. D. (Eds), *Floodplain Processes.* John Wiley and Sons, Chichester. pp. 611–635.

Rodi, W. 1980. *Turbulence Models and their Application in Hydraulics.* International Association for Hydraulics Research, Delft, The Netherlands. 104 pp.

Romanowicz, R., Beven, K., and Tawn, J. 1996. 'Bayesian calibration of flood inundation models', in Anderson, M. G., Walling, D. E., and Bates, P. D. (Eds), *Floodplain Processes.* John Wiley and Sons, Chichester. pp. 333–360.

Schmugge, J. 1988. 'Microwave approaches in hydrology', *Photogram. Eng. Rem. Sens.*, **46**, 495–507.

Schultz, G. A. 1988. 'Remote sensing in hydrology', *J. Hydrol.*, **100**, 239–265.

Sellin, R. J. 1964. 'A laboratory investigation into the interaction between flow in the channel of a river and that over its floodplain', *La Houille Blanche*, **7**, 793–802.

Sellin, R. H. J. and Willetts, B. B. 1996. 'Three dimensional structures, memory and energy dissipation in meandering compound channel flow. in Anderson, M. G., Walling, D. E., and Bates, P. D. (Eds), *Floodplain Processes.* John Wiley and Sons, Chichester. pp. 255–297.

Smith, L. C., Isacks, B. L., Forster, R. R., Bloom, A. L., and Preuss, I. 1995. 'Estimation of discharge from braided glacial rivers using ERS-1 SAR: first results', *Wat. Resour. Res.*, **31**, 1325–1329.

Solomon, S. I. 1993. 'Methodological considerations for the use of ERS-1 SAR imagery for the delineation of river networks in tropical forest areas', *Proceedings, 1st ERS-1 Symposium, ESA–ESRIN, Frascati, Italy*, SP-359, pp. 595–600.

Sonka, M., Vaclav, H., and Boyle, R. 1993. *Image Processing, Analysis and Machine Vision.* Chapman and Hall, London. 555 pp.

Tarantola, A. 1987. *Inverse Problems Theory, Methods for Data Fitting and Model Parameter Estimation.* Elesevier, The Netherlands.

Tholey, N. 1995. 'Monitoring flood events with remote sensing data: an example of ERS-1's contribution to flood events in northern and southern France regions', *Proceedings, 1st ERS Thematic Working Group Meeting on Flood Monitoring, ESA–ESRIN, Frascati, Italy.*

Tilley, D. G. and Bonwit, K. S. 1989. 'Reduction of layover distortion in SAR imagery', *Rem. Sens. Environ.* **27**, 211–220.

Toebes, G. H. and Sookey, A. A. 1967. 'Hydraulics of meandering rivers with floodplains', *Proc. Am. Soc. Civil Eng. J. Waterways Harbours Div.*, **93**, 213–226.

Walling, D. E., Bradley, S. B., and Lambert, C. P. 1986. 'Conveyance loss of suspended sediment within a floodplain system', *IAHS Publ.*, **159**, 119–132.

Walling, D. E., He, Q., and Nicholas, A. P. 1996. 'Floodplains as suspended sediment sinks', in Anderson, M. G., Walling, D. E., and Bates, P. D. (Eds), *Floodplain Processes*, John Wiley and Sons, Chichester, 399–440.

Walling, D. E., Quine, T. A., and He, Q. 1992. 'Investigating contemporary rates of floodplain sedimentation', in Petts, G. E. and Carling, P. A. (Eds), *Lowland floodplain rivers: a geomorphological perspective*, John Wiley and Sons, Chichester, 165–184.

Wang, Y., Hess, L. L., Filoso, S., and Melack, J. M. 1995. 'Understanding the radar backscattering from flooded and non–flooded Amazonian forests: results from canopy backscatter modelling', *Rem. Sens. Environ.*, **54**, 324–332.

White, R. G. 1994. 'A simulated annealing algorithm for radar cross-section estimation and segmentation', *Applic. Artif. Neural Net. V*, **2243**, 231–240.

Williams, D. J. and Shah, M. 1992. 'A fast algorithm for active contours and curvature estimation', *CVGIP: Image Understand.*, **55**, 14–26.

Yamagata, Y., Wiegand, C., Akiyama, T., and Shibayama, M. 1988. 'Water turbidity and perpendicular vegetation indices for paddy rice flood damage analyses', *Rem. Sen. Environ.*, **26**, 241–251.

Younis, B. 1996. 'Progress in turbulence modelling for open channel flows', in Anderson, M. G., Walling, D. E. and Bate, P. D. (Eds), *Floodplain Processes.* Wiley and Sons, Chichester. pp. 299–332.

Zeilke W. and Urban W. 1981. *Two Dimensional Modelling of Rivers with Flood Plains*, Numerical Modelling of River Channel and Overland Flow for Water Resources and Environmental Applications. IAHR, Delft, The Netherlands.

SECTION III

HYDROLOGICAL MODELLING

11

AN INTEGRATED, PHYSICALLY BASED MODEL FOR ARID REGION FLASH FLOOD PREDICTION CAPABLE OF SIMULATING DYNAMIC TRANSMISSION LOSS

A. S. EL-HAMES* AND K. S. RICHARDS

Department of Geography, University of Cambridge, Cambridge CB2 3EN, UK

ABSTRACT

This paper outlines a physically based model developed for prediction of both arid region flash floods and associated transmission loss. It is based on coupling numerical solutions of the St Venant equations and the kinematic wave equation for channel and overland flow routing, respectively, with the Richards' equation for the determination of infiltration. These hydrological components are integrated in one comprehensive model for the above-mentioned purposes. The robustness of the model and its components have been tested and it has then been applied to a medium-size catchment in an arid region. The results indicate that reasonable simulations can be achieved with a minimum of calibration, but point to problems of representing input rainfall conditions with the necessary time and space resolutions.

KEY WORDS arid region; flash flood; catchment; flood routing; transmission loss; wadi bed

INTRODUCTION

Rainfall–runoff modelling in warm, low-latitude, arid region catchments is rendered challenging because of environmental heterogeneity in space and time, which results in the production of irregular and extreme flood events. These floods cause a significant annual loss in life and property because of the increasing population of these regions and the frequent lack of long-term planning. Such floods are rare and of very variable magnitude in time and space, and thus flood prediction by methods that rely on extensive rainfall–runoff data rarely prove satisfactory. Also, the difficulties of data collection, owing to the low population density and inaccessibility of these areas, limits the reliability of the data record for such environments (El-Hames and Richards, 1994). Extreme climatic variation from year to year, and variation in annual precipitation, increase the problems of constructing probabilistic models. As a result, flood prediction for wadi channels is plagued with both risks and uncertainties, and requires complex techniques for successful achievement. Arid region rainfall may occur in one of two seasons, one during winter and the other during summer. Winter rainfall events often cover large areas and have a long duration per storm, while summer rainfall events are more localized and are of shorter duration and higher intensity. These localized, high intensity storms produce 'flash floods' in arid areas, where interception capacity is almost negligible. Arid region floods are characterized by steep rising limbs, which indicate a short time to peak often in the form of 'bores' advancing on a dry bed. This creates another difficulty for flood prediction since most of the available physically based numerical techniques cannot handle sudden changes in flood depth, which cause kinematic shocks owing to numerical instability.

* Correspondence to: Dr A. S. El-Hames, Department of Geography, University of Cambridge, Downing Place, Cambridge CB2 3EN, UK.

A general solution to the problem of hydrological modelling in arid zones thus requires exploitation of developments in physically based numerical modelling techniques. Representation of the hydrological components of rainfall, infiltration, runoff and (particularly important in these regions) transmission losses, all need to be combined in such models. Flood prediction models that deal with large catchment areas usually include an overland flow component, and invariably require a channel routing component. In arid region catchments rainfall usually takes place over small portions of catchments. Accordingly, the runoff produced over the upstream areas where rainfall occurs must be translated to downstream sites through the main channel and its tributaries, which all may have a thick alluvial fill into which transmission loss occurs. In this case, comprehensive methods for flow routing are needed, such as those that employ a detailed solution of the St Venant equations, but these must be coupled with an infiltration model to deal with transmission losses. Infiltration in such models may be treated using simple relationships (e.g. Wooding 1965), or may employ more rigorous physically based methods such as Richards' equation (e.g. Smith and Woolhiser, 1971), the Green–Ampt formula (e.g. Akan, 1985) or Philip's equation (e.g. Field and Williams, 1987).

In this paper the development of an integrated, physically based rainfall–runoff model for arid region flash-flood prediction is discussed. The complete model is composed of three major submodels. The first calculates excess rainfall by employing the Richards' equation to simulate one-dimensional infiltration. The second then solves for overland flow routing by using the kinematic wave theory; this is coupled to the first model to simulate the effects of ponded head of runoff. The third major submodel then consists of the full solution of the St Venant equation to solve for channel flow routing. This is also coupled with Richards' equation to account for infiltration losses into channel beds. The model is thus able to provide simulation of dynamic transmission loss driven by the ponded head in the wadi channel as the hydrograph passes downstream.

MODEL COMPONENTS

The complete catchment-scale model consists of hillslope and channel flow submodels that employ the three major components identified above, coupled together in one comprehensive model to account for catchment rainfall–runoff production and flood routing applicable to flood transmission in arid and semi-arid regions. The processes simulated therefore include determination of infiltration over hillslopes and into channel beds, overland flow routing and channel flow routing. The development of each component is described in more detail below, and the overall structure is shown in Figure 1.

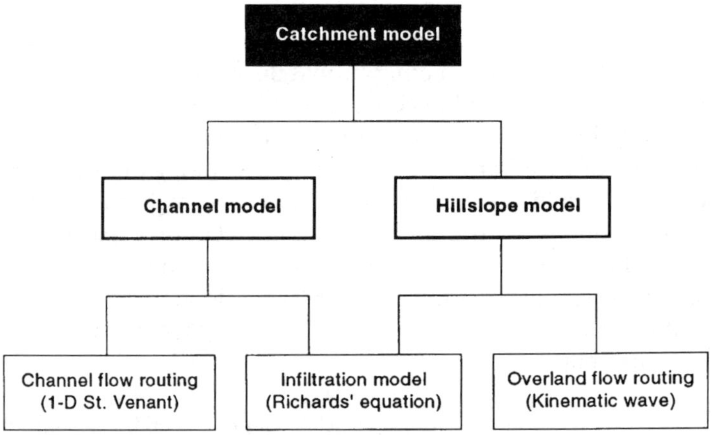

Figure 1. Integrated model components

Infiltration component

This component was developed initially to account for excess rainfall on slopes, a critical component of any rainfall–runoff model. However, it also represents a good tool for calculation of transmission loss into channel beds when coupled with a channel routing method. A variety of techniques is available for infiltration calculation, such as the Green and Ampt model or the Philip equation. Given the complexity of the soil profiles in the study area (and the sediment stratigraphy in wadi beds in the south-western region of Saudi Arabia), and the need to simulate infiltration through these vertically heterogeneous layers, a more generalized method had to be used. Thus, the equation of flow for unsteady saturated–unsaturated flow in the one-dimensional vertical direction, in the form of Richards' equation, was employed as follows

$$C(\psi)\frac{\partial\psi}{\partial t} = \frac{\partial}{\partial z}\left(K(\psi)\frac{\partial\psi}{\partial z} + K(\psi)\right) \tag{1}$$

Here, C is the specific moisture capacity, ψ is the pressure head, t is time, K is saturated hydraulic conductivity and z is the depth of the soil profile. This procedure has been tested by a number of investigators on dry and partially wet soils (e.g. Giesel *et al.*, 1973; Haverkamp *et al.*, 1977; Redinger *et al.*, 1984; Ross, 1990). The success of the method in handling both dry and deep soil columns, and the results produced, are both very encouraging, and justify considering this technique as particularly suited to the problem of infiltration and runoff prediction, since employed soil characteristics in these studies are similar to those measured in the field in the study area. The Crank–Nicholson forward and backward finite difference method was used to solve Equation (1) in the model developed here. This method is second-order accurate in space and time and unconditionally stable (Press *et al.*, 1989), and results in a set of non-linear equations that can be presented in tridiagonal matrix form. The advantage of tridiagonal matrices is that computer storage is not required for the full $N \times N$ matrix, but only for the non-zero components stored as vectors. These matrices were solved by the Gaussian elimination technique. The infiltration rate (f) was calculated from Darcy's law as the flux entering the upper boundary, or as the volumetric soil moisture content difference between two successive time increments, as follows

$$f(t) = \sum_{j=1}^{m}(\theta_j^{i+1} - \theta_j^{i})\frac{\Delta z}{\Delta t} \tag{2}$$

where j and i are the spatial and temporal nodes, respectively, m is the number of vertical nodes and θ is the volumetric soil moisture content. The upper boundary condition was defined as a pressure head, representing the ponded head of water propagating over the soil surface, or as a consequence of the water wave translated downstream over the wadi bed in the case of the transmission loss calculations.

Overland flow routing component

This component is required to route the excess rainfall produced over hillslopes downslope towards river channels. Overland flow routing is slightly different from channel routing analysis because of the steep slopes, which result in shallow water flow propagating rapidly over the surface, mainly influenced by gravitational forces. Applying the full momentum and continuity equations to describe such cases may result in numerical instability, since these equations cannot handle numerically either very shallow flow or initially dry cases, owing to singularity problems. However, overland flow routing problems may be treated using the kinematic wave method, firstly employed comprehensively by Henderson and Wooding (1964) and Wooding (1965). This technique has been found to be an accurate tool when applied to thin sheet flow in overland flow problems, when the energy slope may be approximated as the surface slope and the rest of the momentum equation terms are neglected. Furthermore, its numerical stability is relatively, relaxed, which makes it a convenient technique to describe the rapid change in flow depth, occurring in arid region flash-flood cases. Either the Manning or Darcy–Weisbach equations can be used in this model to link the velocity with the continuity equation. The resultant kinematic flow equation, in the conservation form and utilizing the Darcy–Weisbach friction factor (F) may be written as

$$\frac{\partial h}{\partial t} + (8g)^{1/2} \frac{\partial}{\partial x}\left(\frac{s_0^{1/2} h^{3/2}}{F^{1/2}}\right) - q - r + f = 0 \tag{3}$$

where h is the flow depth, g is the gravitational acceleration, x is the horizontal distance, s_0 is the surface slope, q is the lateral inflow per unit width and r is the rainfall intensity. Equation (3) was solved numerically by applying the single-step Lax–Wendroff scheme for the second-order approximation, as explained by Kibler and Woolhiser (1970), Rovey *et al.* (1977) and Yao-Huang *et al.* (1978). This method was adopted because of its relatively relaxed stability criterion and its ability to handle difficult cases. The model solves for the overland flow problem in the form of one-dimensional unsteady flow and allows for variable geomorphological and hydraulic parameters. If there is no flow contributing to the slope at its upper limit, a zero boundary condition is simulated, but flow may also be generated as a result of rainfall, lateral inflow or both. Lateral inflow may enter any segment of the plane representing part of a hillslope, throughout the total length of the slope.

Channel flow routing component

Different methodologies are available to solve the channel routing problem, ranging from the Muskingum method to black-box regression models. These techniques vary in terms of their assumptions, the degree of difficulty in their solution and application and in the conceptual basis of their solutions. Other more rigorous methods require a physically based approach, such as the full solution of the conservation of mass and momentum equations for channel flow (the St Venant equations), or may adopt a simplified form or partial solution of these equations, such as the kinematic wave or diffusion methods. Selecting the channel routing technique to be applied depends on the sort of problem under study, the type of output required and the degree of accuracy needed in the output. However, the most comprehensive and reliable method is based on a full solution of the St Venant equations. This method is used here to cope with the rapid temporal change of depth at a point relative to the spatial change in water surface elevation. The usual forms of the unsteady flow equations in one-dimensional flow, incorporating information on rainfall intensity and the infiltration rate, and for non-prismatic channels, are

$$\frac{\partial Q}{\partial x} + \frac{\partial A}{\partial t} - q - w(r - f) = 0 \tag{4}$$

$$\frac{\partial v}{\partial t} + v\frac{\partial v}{\partial x} + g\frac{\partial h}{\partial x} + \frac{vq}{wh} + g(s_{\mathrm{f}} - s_0) - \frac{v}{h}(r - f) = 0 \tag{5}$$

where Q is the discharge, A is the channel cross-sectional area, v is the flow velocity, s_{f} is the friction line slope and w is the cross-section width. The mathematical approximation of these equations utilizes Taylor's expansion, and the numerical treatment employs the Lax–Wendroff scheme; this being adopted since it is flexible in application to relatively complicated hydraulic channel routing problems, and permits variable channel geometries. A significant advantage of this method was found to be that it does not require excessively large computer memory, or a large number of iterations.

During floodwave translation downstream over an initially dry bed, infiltration occurs into the bed to produce the phenomenon of transmission loss. The forms of the unsteady flow equations used above [equations (4) and (5)] include an infiltration term, and the rate of infiltration is simulated using the same one-dimensional solution of the Richards' equation as outlined in the first part of this section. This employs a pressure head boundary condition at the surface, which is defined by the flow routing predictions. Thus, transmission losses are estimated dynamically as a function of the head of water imposed by the passing floodwave, in a simulation that couples routing and infiltration procedures explicitly and physically for the first time. Existing methods for estimating transmission loss have usually been based on empirical, regression-based approaches (e.g. Lane, 1982; Walters, 1990).

AN INTEGRATED, PHYSICALLY BASED CATCHMENT RAINFALL–RUNOFF MODEL

The developed and integrated catchment model thus consists of two major submodels (for hillslope and channel flow), each of which is composed of two of the components outlined above. The first major submodel (Figure 1) solves for excess rainfall produced on slopes and accounts for the overland runoff routing problem. This has been achieved by coupling Equations (1) and (3) using a technique similar to that of Akan and Yen (1981). Infiltration rate values obtained at the previous time step from equations (1) and (2) are substituted into Equation (3). Equation (3) then calculates the piezometric head at each computational node throughout the hillslope profile. These values are employed as the upper boundary condition for the infiltration model in the next time step (as a ponded head). These steps are repeated until the total period of simulation is complete. The outputs of this first submodel are then treated as inputs to the second major submodel in the form of lateral inflow coming from the hillslopes surrounding the main channels. The infiltration and rainfall parameters are readily introduced into the kinematic wave theory by adding an infiltration and rainfall factor (*fr*), which is

$$fr = \frac{\Delta t}{2}([r_j^{i+1} - r_j^i) - (f_j^{i+1} - f_j^{i+1})] \tag{6}$$

where Δt is the time increment, and superscript i and subscript j refer to the time and space nodes, respectively. This has to be added to the equation that results from expansion of equation (3) numerically by the Lax–Wendroff scheme.

The second major submodel (Figure 1) couples Equations (1), (4) and (5), in a similar way to that described above, to account for channel routing while considering losses owing to infiltration into the channel bed during hydrograph propagation (i.e. transmission losses). The case of channel flow routing is not as straightforward as overland flow, since a different infiltration and rainfall factor has to be added to the resultant *depth* equation, after the numerical expansion of equation (4). This is

$$fr = -\Delta t(f - r) + \frac{\Delta t^2}{4\Delta x}\left[(f - r)(v_{j+1}^{i+1} - v_{j-1}^{i+1}) + \frac{v}{w}(f - r)(w_{j+1}^{i+1} - w_{j-1}^{i+1}) - \frac{2\Delta x}{\Delta t}(f_j^{i+1} - f_j^i) + \frac{2\Delta x}{\Delta t}(r_j^{i+1} - r_j^i)\right] \tag{7}$$

Similarly, another infiltration and rainfall factor, which may be written as

$$fr = -\Delta t\frac{v}{h}(r - f) + \frac{\Delta t^2}{4\Delta x}(r - f)\left\{\left[\left(\frac{g}{h} - \frac{2v^2}{h^2}\right)(h_{j+1}^* - h_{j-1}^*) - \left[2\Delta x\frac{g}{h}(s_f^* - s_o)\right] + \frac{v^2}{wh}(w_{j+1}^* - w_{j-1}^*)\right.\right.$$
$$\left.\left. - \frac{v}{h}(v_{j+1}^* - v_{j-1}^*) - 4\Delta x\frac{vq}{wh^2}\right]\right\} + \frac{\Delta t^2}{4\Delta x}\left[2\Delta x\frac{v}{h}\left(\frac{r_j^* - r_j - f_j^* - f_j}{\Delta t}\right)\right] \tag{8}$$

must be added to the resultant *velocity* equation after expanding Equation (5) numerically. Here the superscript (*) refers to the new time increment ($i + 1$). A mass balance check routine was built in, and automatically operated to calculate input, storage and output volumes in order to ensure accurate mass balance tracking and to reduce losses resulting from numerical dissipation.

When the rainfall event being simulated takes place over the headwater slopes, the first submodel must be employed. Generation of excess rainfall and propagation of floodwater over the slopes is accounted for, the losses are calculated and deducted from the overland flow in real time and the calculated outflow hydrographs at slope bases are stored as data files. These files are then transferred to the second submodel as input hydrographs to the channel in the form of lateral inflow, with consideration being taken of the location along the channel of each lateral inflow, as well as the timing of inflow to the main channel. Once these data have been read by the second submodel, this accounts for channel routing and transmission loss and calculates the output hydrograph at the catchment outlet.

TESTING, CALIBRATING AND APPLYING THE MODEL TO CATCHMENT-SCALE PROBLEMS

Testing the hillslope submodel

The robustness of the hillslope submodel was tested by using evidence from a number of published sources. One of the tests was against the empirical data and the model KININF generated by Lima (1992). This model also consists of an overland flow routing model employing the kinematic wave theory and incorporating an infiltration method based on Richards' equation. This test exploited empirical data from a small experimental soil plot of 1 m × 20 m, with an average surface slope of 0·02, as employed by Lima (1992) with soil data obtained from Mein and Larson (1973) for a Columbia sandy loam soil. A variable rainfall intensity was utilized by Lima and the same rainfall values were simulated in this test. Good agreement between the results obtained from Lima's work (both observed and simulated runoff) and the simulations obtained using the developed model can be seen in Figure 2. This submodel could not be calibrated for any real case in the field area described below because of an absence of overland flow information. Measured values of soil and sediment characteristics in the study area were, however, used throughout the simulations described below. The only parameter employed in this submodel that could not be measured directly was the Manning roughness coefficient, and this had to be partly calibrated. It is recommended from previous work in the field area that this parameter ranges in value between 0·03 and 0·05, and the upper limit of this range was found to give reasonable results when the full model was implemented, as shown in the following sections.

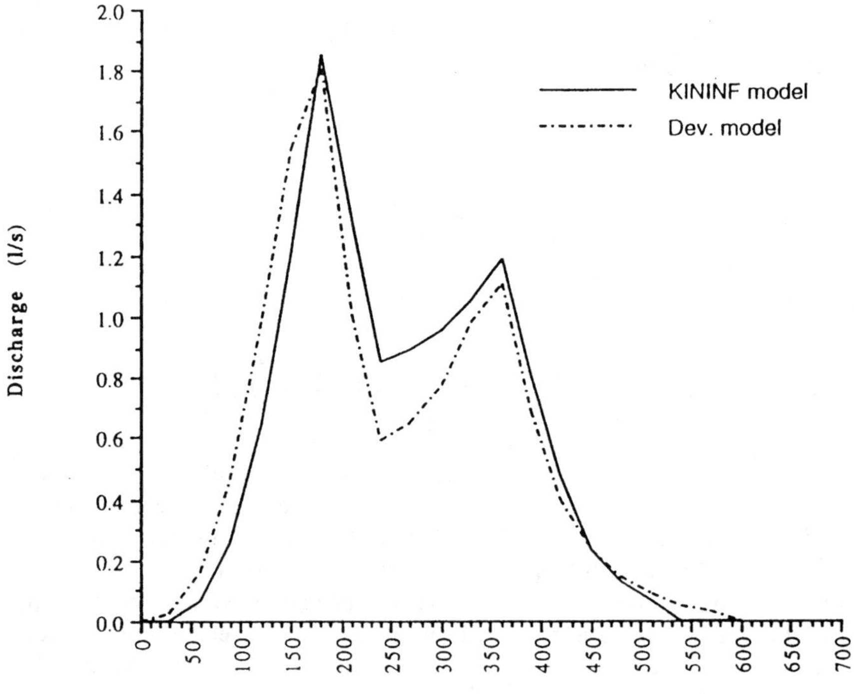

Figure 2. Submodel 1, results from test problem 3

Testing and calibration of the channel routing submodel

The accuracy of the channel routing submodel was investigated by studying observed flood cases in an arid region where transmission loss is known to have taken place, but where no lateral inflow or precipitation input occurred in the reach experiencing transmission loss during floodwave propagation. Data and information about the area chosen for these tests were mainly collected from Abdulrazzak *et al.* (1989) and from unpublished reports prepared by the Dames & Moore consulting company (1988) describing the catchment under study, which is that of Wadi Tabalah, an arid catchment in south-western Saudi Arabia (see Figure 4 later). Three runoff stations are installed along the main channel of this wadi, namely, B413, B412 and B405 numbered from upstream to downstream. The transmission loss simulations were carried out between stations B412 and B405. The length of the reach between these stations is 22·7 km, and this was divided into two reaches that are different from each other in their physical and geomorphological characteristics (e.g. wadi channel width, slope and bed sedimentology). These reaches are defined by the channel cross-section at station B412 at the upstream end, a channel cross-section located at 14·5 km downstream of B412 and the cross-section of station B405, downstream. These cross-sections are irregular in shape, but a rectangular shape was employed in these tests in order to accord with the simplifying assumptions adopted in the mathematical model. The surveyed width of the upstream cross-section at B412 is 110 m, the middle cross-section is 800 m wide and the downstream cross-section is 178 m wide. Channel bed slope is 0·0053 and 0·0034 in the upper reach and lower reach, respectively. The whole study basin (upstream catchment) is described in more detail in the next section.

The wadi bed sediment stratigraphy was found to be similar throughout in the upper 5 m, and consists of sand and silt. The main soil layer was found to have a porosity of 0·3 and a hydraulic conductivity of about 0·5 cm min^{-1} from laboratory experiments and pumping tests. Sediment logs indicate that this layer underlies a surface clogging layer of silt in the upper 20 cm, which has an hydraulic conductivity an order of magnitude lower. The hydraulic conductivity of this layer has a very marked effect on the transmission loss into the channel bed.

An initial soil moisture distribution in the unsaturated zone was obtained from the Dames & Moore (1988) soil moisture studies. For events where this information is not available, and where there was no major rainfall in the preceding 10 days or more, the distribution was constructed based on experience gained from neutron probe measurements in the field, as follows. In the top 30 cm of the soil, moisture content was found to be at the residual value of 0·03 and increases to reach the field capacity of 0·08 at a depth of 150 cm. Below this depth, moisture content increased rapidly to reach 0·22 at 200 cm. From this depth downwards, moisture content increased gradually to reach the saturated moisture content value (0·30) at the water table. Accordingly, field-measured climatological and hydrological data were employed for all required model data except for the Manning roughness coefficient, as in the hillslope submodel whose testing is described above.

Two events (10 April 1985 and 7 April 1986) were used for calibration and testing of the channel routing submodel, and two parameters (the Manning coefficient and the clogging layer hydraulic conductivity) were adjusted to obtain the best fit. In all the simulations employed here these parameters were used within the recommended range of 0·03–0·05 and 0·03–0.10 cm min^{-1} for the Manning roughness and the top layer hydraulic conductivity, respectively, and these gave good results. The simulated events were generated by rainfall above station B412, and the downstream area did not receive any rainfall during these events. Flood hydrographs, initiated in the upstream areas, passed station B412 and reached station B405. This makes the events convenient for analysis of combined routing and transmission loss. The results of these tests and comparisons with the observed hydrographs are shown in Figure 3. They show good agreement with observed hydrographs in terms of the resulting outflow volumes, flood peak magnitudes and times to peak, which indicates the applicability of this model to prediction of arid land flood routing over initially dry beds, and transmission losses. It can be seen from these test cases that the transmission loss average rate was about 16 000 m^3 per kilometre, which is mainly a function of sediment characteristics (especially the clogging layer hydraulic conductivity), the initial moisture content and the depth of the unsaturated zone.

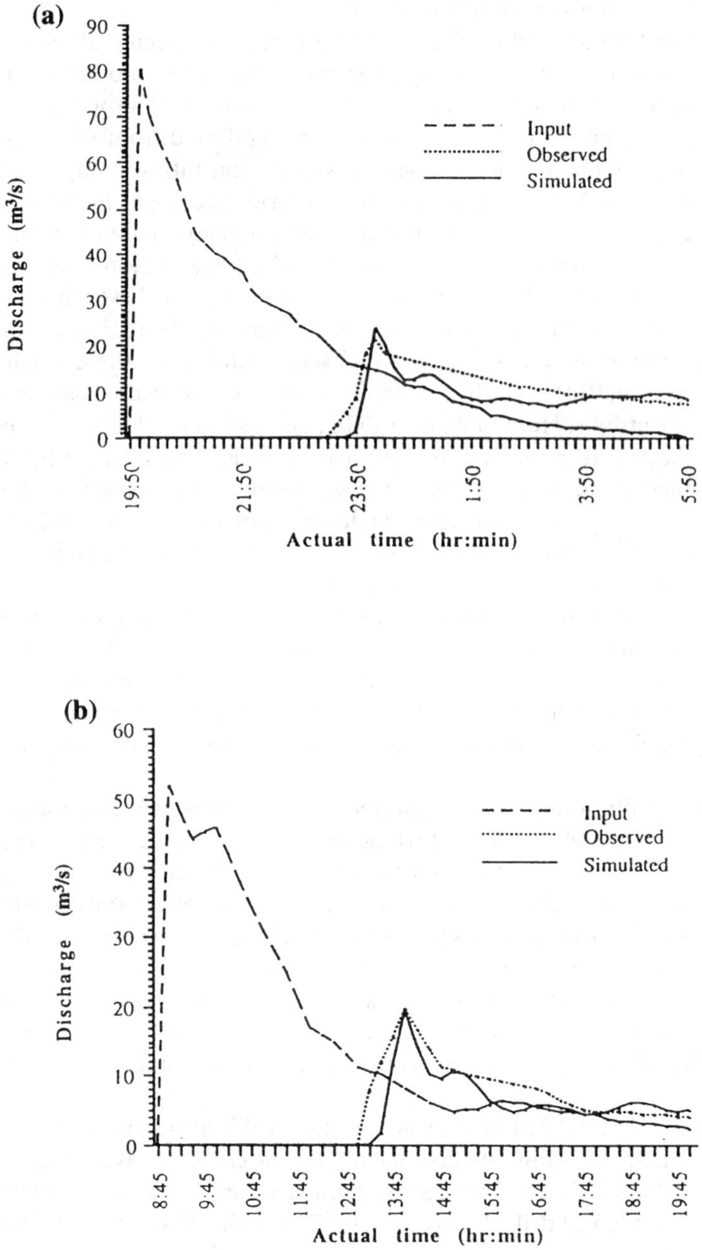

Figure 3. Transmission loss simulation. Events of (a) 10 April 1985 and (b) 7 April 1986

Application of the model to catchment-scale problems

The complete model (which consists of the two major submodels) was tested for three full rainfall–runoff cases in Wadi Tabalah. The need to study the hazard resulting from flash-floods in this area was the initial reason for the development of the model. The unique climatic and topographic characteristics of the area, as well as the availability of relatively detailed hydrological and climatological networks, made it suitable for applying such a model. The study basin is the upper Tabalah catchment (above station B413) which has an

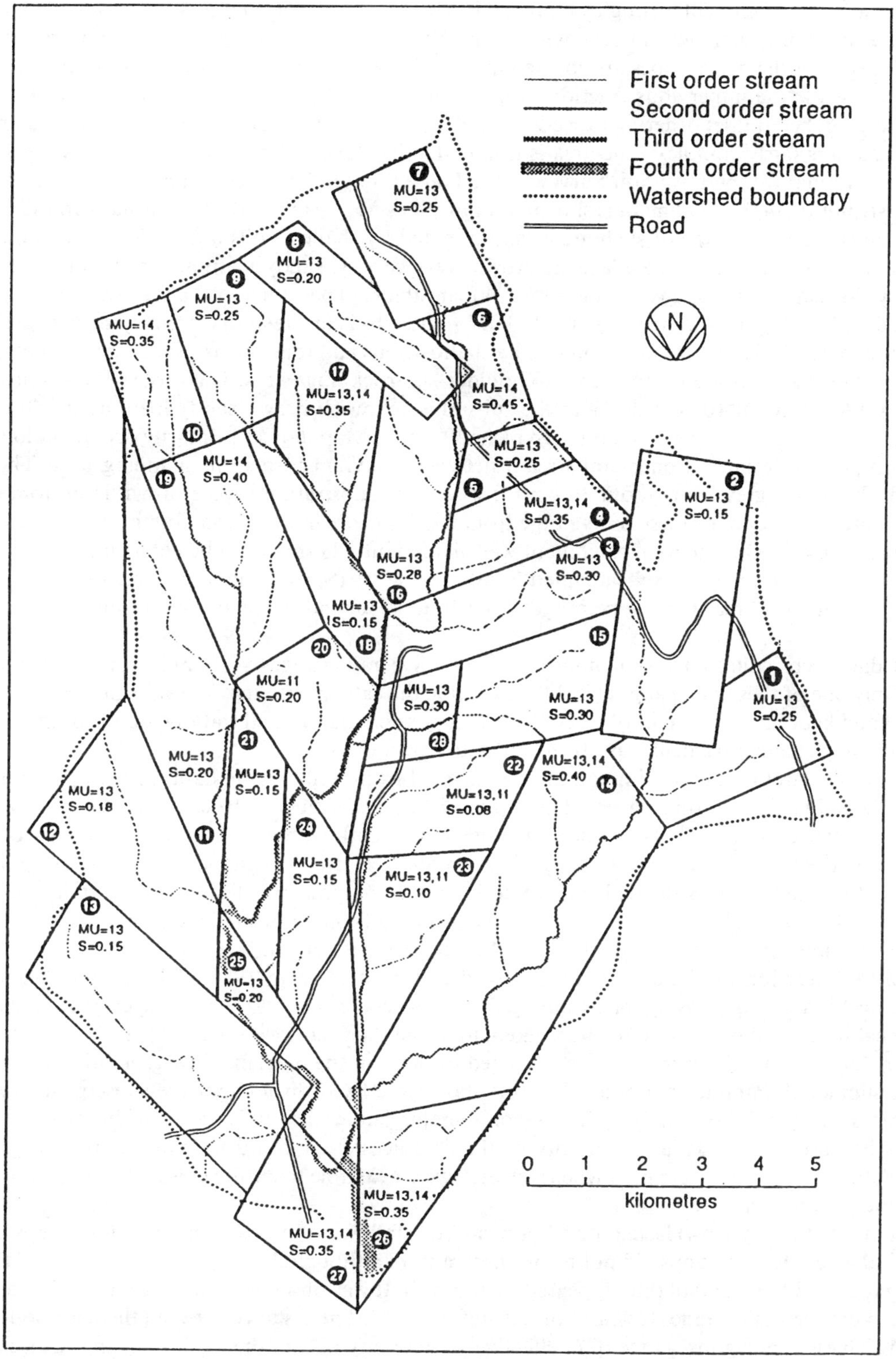

Figure 4. Dividing the upstream subcatchment of the Tobalah Basin

area of 170 km^2. The catchment is roughly elliptical in shape, as shown in Figure 4. The wadi channels in this basin are steeply sloping, narrow and deep with steep valley walls; the surrounding mountain ridges are also steep. About 80% of the basin slopes are covered by gravel and cobbles, and alluvium only represents 7% of the total area, while the impervious fraction ranges from 20 to 30%. The main channel is a fourth-order stream (following the Strahler ordering procedure) and has a total length of 27 km. Thus, the catchment is of medium size and is fairly complex. The area is dominated by three soil mapping units following standards used by the National Cooperative Soil Survey in the United States. These are mapping units 13, 14 and 11 and their distribution in the catchment is shown in Figure 4 as MU. Map unit 13 is normally found on rolling to hilly uplands. The landscape of such areas is dominated by shallow soils and rock outcrops on igneous and metamorphic rocks. The surface layer is brown, very cobbly, sandy loam about 10 cm thick, and the subsoil is yellowish brown sandy loam about 40 cm thick. The slopes of areas with this map unit (abbreviated by S in Figure 4) range from 0–30%. Map unit 14 is on steep to extremely steep side slopes on mountain uplands. The landscape is dominated by shallow soils and rock outcrops. A high percentage of the soil surface (85–95%) is covered with gravel- to boulder-size rock fragments. The surface layer is brown, very cobbly, sandy loam about 10 cm thick and the subsoil is yellowish brown sandy loam about 40 cm thick. Slopes of areas having this mapping unit range from 30–75%. Map unit 11 primarily occurs on low alluvial terraces along drainage lines. Ephemeral stream network alluvium forms this mapping unit. The surface layer is light brownish grey loam about 30 cm thick and the substratum is pale brown loam to a depth of 230 cm or more. The slopes of these units range from 0–2%. More information about the study area can be found in El-Hames (1993). The total catchment was divided into 28 smaller subcatchments according to the stream network structure. Each subcatchment's average slope (S) and mapping unit (MU) is shown in Figure 4. Overland flow areas have slope ranges from 10 to 45%, and the main stream slopes range from 1·2 to 1·7%.

Rainfall data were collected on an hourly basis by five rain gauges scattered over an area of about 560 km^2. Only one of these rain gauges (B237) was inside the catchment area. The rain gauges were installed in late 1984 and kept in operation until early 1988. It is possible that not all stations were working in a given event, which may create a particular problem for data extrapolation.

The soil profile employed in all simulations was derived from well logging data. In the upland areas, the soil column was divided into three layers. The top layer has a thickness of 10 cm and a saturated hydraulic conductivity of 0·025 cm min^{-1}. The middle layer is 15 cm depth with a hydraulic conductivity of 0·21 cm min^{-1}. Finally, the bottom layer has a depth of 25 cm and a hydraulic conductivity of 0·5 cm min^{-1}. In a similar way, the alluvium was divided into three layers, with similar hydraulic conductivity values. The depths of these layers were 20, 30 and 170 cm for the top, middle and bottom layers, respectively. Field data showed that the average water table level for 1984 and 1985 was 230 cm below the surface.

The events selected for simulation tests were those occurring on 14 May 1984, 25 January 1985 and 5 April 1985. Partly calibrated cases (using the Manning roughness coefficient and the top layer hydraulic conductivity as the calibrated parameters) were undertaken, in which the results were acceptable after only the first or second run. Adjustment of parameters was restricted to those in the overland flow generation routine in the first major submodel. Parameters obtained at this stage were fixed throughout the subsequent simulation. Beyond this stage (i.e. when applying the second major submodel) it is very difficult to adjust input parameters because of the spatial complexity of the channel network and its input–output system. Field measured data for these tests were used whenever they were available to ensure close similarity to the natural characteristics of the area.

The first event (14 May 1984) lasted for 4 hours and rainfall was only recorded by rain gauge B237 inside the simulated area. Other stations did not record any rainfall although they were in operation. Further away from rain gauge B237 the rainfall thus appeared to diminish. It was observed that the centre of this storm was concentrated between rain gauge B237 and runoff station B413. The distance between them is about 2200 m. By applying Thiessen polygons, gauge B237 effectively represents 54% of the total basin area. Since the other stations did not record any rainfall, rainfall intensity was assumed to be zero at the boundary of the effective area of rain gauge B237. Rainfall intensities were linearly distributed with respect to distance from the centre of the storm according to the following equation

$$ARI = \left(1 - \frac{ESD}{TOD}\right) ARI_{\max} \tag{9}$$

where ARI is the estimated rainfall intensity, ESD is the measured distance from the centre of storm to the centre of the subcatchment under simulation, TOD is the total distance from the centre of the storm to the boundary, where the rainfall is zero, and $ARI_{\max}$ is the maximum rainfall intensity. This gives a non-uniform rainfall over each subcatchment.

The initial soil moisture distribution for this event was approximated to match field measurements as follows. In the hillslope areas, moisture content was found to be 0.17 $cm^3\ cm^{-3}$ in the top layer, increasing gradually to 0·28 $cm^3\ cm^{-3}$ at a depth of 50 cm, with impervious rock just below. In the channel alluvium, moisture content was measured ranging from a value of 0.18 $cm^3\ cm^{-3}$ in the top 30 cm, reducing to 0.10 $cm^3\ cm^{-3}$ at 65 cm. This latter value was constant until a depth of 100 cm, then increased to 0·28 $cm^3\ cm^{-3}$ at the lowest depth.

In the second event (25 January 1995), rainfall continued for 7 hours recorded by station B237, and the weighted average rainfall was assumed to have covered the whole area homogeneously owing the absence of other rain gauge records. It is not known whether the other stations were in operation and did not record any rainfall (which would nullify this assumption) or whether different rainfall intensities occurred at these sites. Initial soil moisture distribution for this event was assumed to be the same as for the previous event. This is a typical residual moisture distribution after a lengthy dry period in these soils and sediments, except that, in the hillslope areas, initial moisture content in the surface layer was set at 0.12 $cm^3\ cm^{-3}$ to match the observed data.

The assumed rainfall distribution, however, resulted in a poor simulation in this second event, in which the simulated runoff volume was greater than that of the observed hydrograph (Figure 5b). However, it was observed that the simulated runoff in this event for subcatchment 14 (Figure 4), which has a channel 9000 m long, gave both a peak discharge and total runoff volume very similar to that recorded by runoff station B413. This implies that the event only involved rainfall over part of the catchment, and that the assumption of complete rainfall coverage in this event was inappropriate. Clearly, therefore, the need for more reliable rainfall data is crucial. A more dense rain gauge network and shorter rainfall recording time is required for better results.

In the third event (5 April 1985), rainfall continued for 7 hours with a rainfall pattern involving two rainfall intensity peaks. The first peak of 7·4 mm hr^{-1} took place in the second hour, and the second peak of 6·7 mm hr^{-1} occurred in the seventh hour. Rainfall was found to cover the whole area. The initial soil moisture distribution for this event was again approximated to measured data prior to the storm. In the hillslope areas the moisture content was 0.10 $cm^3\ cm^{-3}$ at the surface layer, and reached 0.12 $cm^3\ cm^{-3}$ at 10 cm, then 0.16 $cm^3\ cm^{-3}$ at a depth of 20 cm. It then remained constant for the next 22 cm, then increased to 0·20 $cm^3\ cm^{-3}$ at the base of the soil. For the channel alluvium, slightly lower values than those used in tests 1 and 2 above were applied.

The results of simulating events 1 and 3 are displayed in Figure 5a and c. These show very good agreement between simulated peak flows, initial time of rise, time to peak and runoff volumes, except that in the third event the model produced two peaks while there are three peaks in the observed data. This might be attributed to the hourly recording of rainfall, which is insufficiently detailed to represent all the significant rainfall bursts that occurred. However, the general behaviour and magnitude of the results are very encouraging taking into consideration that the simulations were done without extensive calibration and by initial selection of physically realistic parameter values.

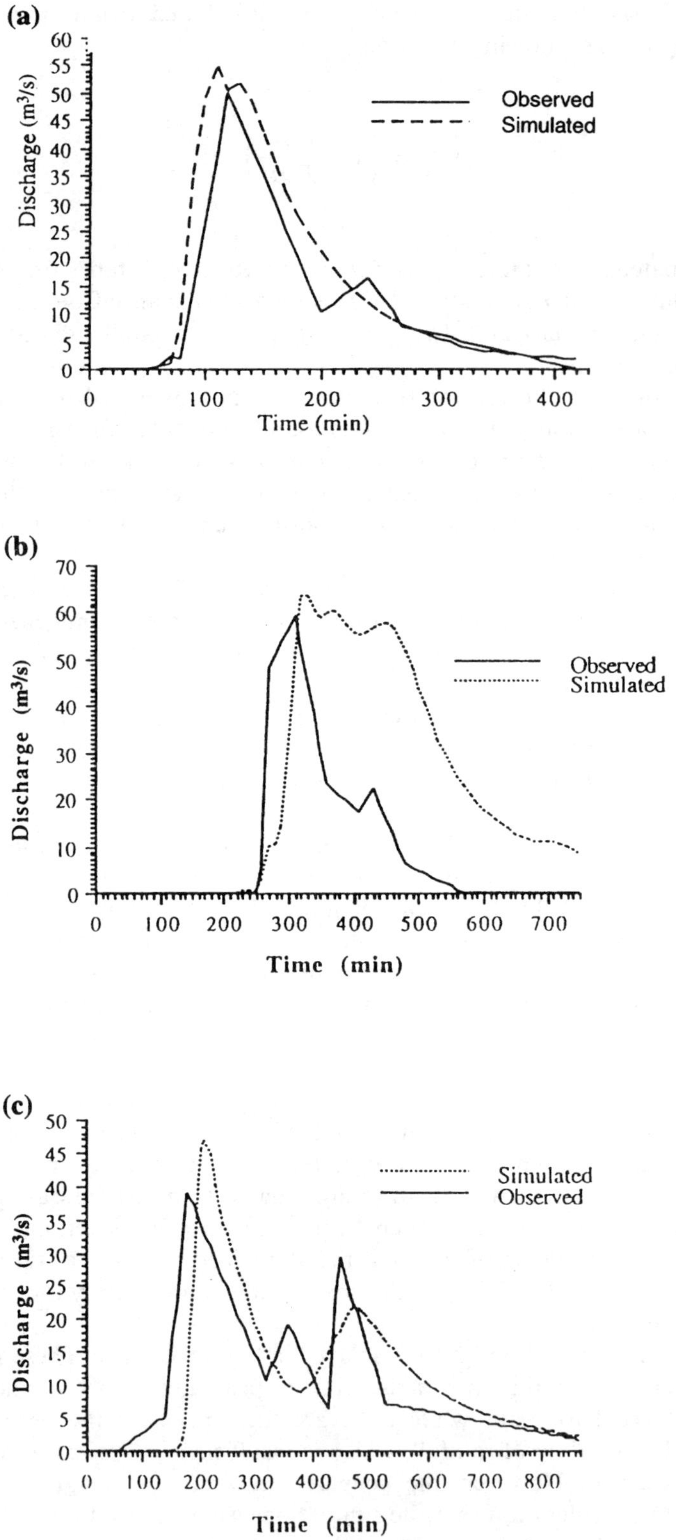

Figure 5. Output hydrographs at station B143. (a) 14 May 1984, (b) 25 January 1885, (c) 5 April 1985

CONCLUSION AND RECOMMENDATIONS

The development of a physically based, semi-distributed model, suitable for arid region flood prediction, has been discussed in this paper. Three physically based techniques to account for the main hydrological components were coupled together to create an integrated model. Both the kinematic wave theory and the full solution of the St Venant equations are utilized here, in order to account for flood routing over slopes and down wadi channels, respectively. An infiltration method based on the Crank–Nicholson numerical scheme to solve Richards' equation was coupled with the above flood routing techniques. The complete model has been tested in an arid region catchment, and the results suggest that it should be a good tool for runoff prediction in such medium-sized (*c.* 200 km^2) catchments. It has been observed from sensitivity analysis conducted in association with these test cases that three of the input parameters are the most effective factors in terms of affecting the output hydrograph's shape and volume. These are the initial soil moisture content, the soil saturated hydraulic conductivity and the Manning's roughness coefficient. Palacios-Velez and Cuevas (1992), in testing their model SHIFT, found that the same parameters were the most crucial ones. Initial soil moisture content defines the hydrograph starting time as well as time to peak, hydraulic conductivity influences the runoff volume, and the roughness coefficient affects the hydrograph shape. It has also been observed that the developed model is sensitive to rainfall bursts and hence hourly rainfall may often be an insufficient resolution of record in arid regions. Furthermore, the simulation for the event of 25 January 1985 illustrates the difficulty in calibrating models such as this in the absence of reliable spatially distributed rainfall data. These simulations show that up to 88% of rainfall volume may be lost as a result of infiltration into channel beds in catchments of this size. Thus, the model was found to be an appropriate tool for transmission loss simulation in arid wadis. Indeed, it is believed to be the first physical model capable of simulating dynamic transmission loss during hydrograph translation. In spite of the scarcity of hydrological information and the difficulty of data collection in such an environment, reasonable data were acquired. The variation of catchment and channel physical properties was overcome by dividing the basin into a reasonable number of roughly homogeneous units.

It is noted that this type of model requires a very long CPU time to complete a simulation. The main cause of this long time requirement is the infiltration component. This makes an average of 10 iterations per time increment because of the existence of a ponded head (flooding water) as the upper boundary condition. The channel routing component makes about 4 iterations per time increment in extreme cases. Supposing that the channel is divided into 10 segments, this means that about of 1 440 000 iterations are required to complete a simulation for only one branch of the channel network. Thus, it is not practical to use this model on personal computers.

In spite of the difficulty of performing a full calibration of the model, and the restriction imposed on the quality of data used, the model has proved its capability to produce results with a fairly high degree of accuracy, and with minimal calibration. The evidence from this application suggests that, if experience is gained with such a model in gauged arid catchments, and the effective parameters can be identified for different soil conditions, it should be possible to set-up the model for use in ungauged basins as a prediction tool for medium-size catchments.

REFERENCES

Abdulrazzak, M. J., Sorman, A. U. and Alhames, A. S. 1989. 'Water balance approach under extreme arid conditions — a case study of Tabalah basin, Saudi Arabia', *Hydrol. Process.*, **3**, 107–122.
Akan, O. A. 1985. 'Similarity solution of overland flow on pervious surface', *J. Hydraul. Engng*, **111**, 1057–1067.
Akan, O. A. and Yen, B. C. 1981. 'Mathematical model of shallow water flow over porous media', *ASCEJ. Hydraul. Div.*, **107**, 479–494.
Dames and Moore, 1988. 'Wadi Tabalah, Land classification survey', *Final Report*, vol. C9, unpublished report.
El-Hames, A. S. 1993. 'A physically-based model for the prediction of flood hydrographs in arid zone catchments', *PhD Thesis*, Cambridge University.
El-Hames, A. S. and Richards, K. S. 1994. 'Progress in arid lands rainfall–runoff modelling', *Progr. Phys. Geog.*, **18**, 343–365.
Field, W. G. and Williams, B. J. 'A generalized kinematic catchment model', *Wat. Resour. Res.*, **23**, 1693–1696.

Giesel, W., Renger, M. and Strebel, O. 1973. 'Numerical treatment of the unsaturated water flow equation: comparison of experimental and computed results', *Wat. Resour. Res.*, **9**, 174–177.
Haverkamp, R., Vauclin, M., Touma, J., Wierenga, P. J. and Vachaud, G. 1977. 'A comparison of numerical simulation models for one-dimensional infiltration', *Soil Sci. Soc. Am. J.*, **41**, 285–293.
Henderson, F. M. and Wooding, R. A. 1964. 'Overland flow and groundwater flow from steady rain of finite duration', *J. Geophys. Res.*, **69**, 1531–1540.
Kibler, D. F. and Woolhiser, D. A. 1970. 'The kinematic cascade as a hydrologic model', *Hydrology Papers*, no. 39. Colorado State University, Fort Collins, Colorado.
Lane, L. J. 1982. 'Distributed model for small semiarid watersheds', *ASCE J. Hydraul. Div.*, **108**, 1114–1131.
Lima, J. T. 1992. 'Model KININF for overland flow on previous surfaces', in Parsons, A. J. and Abrahams, A. D. (Eds), *Overland Flow*. University College London, London. pp. 69–88.
Mein, R. G. and Larson, C. L. 1973. 'Modeling infiltration during a steady rain', *Wat. Resour. Res.*, **9**, 384–394.
Palacios-Velez, O. L. and Cuevas-Renaud, B. 1992. 'SHIFT: a distributed runoff model using irregular triangular facets', *J. Hydrol.*, **134**, 35–55.
Press, W. H., Flannery, B. P., Teukolsky, S. A. and Vetterling, W. T. 1989. '*Numerical Recipes. The Art of Scientific Computing*. Cambridge University Press, Cambridge. pp. 637–660.
Redinger, G. J., Campbell, G. S., Saxton, K. E. and Papendick, R. I. 1984. 'Infiltration rate of salt mulches: measurement and numerical simulation', *Soil Sci. Am. J.*, **48**, 982–986.
Ross, P. J. 1990. 'Efficient numerical methods for infiltration using Richards' equation', *Wat. Resour. Res.*, **26**, 279–290.
Rovey, E. W., Woolhiser, D. A. and Smith, R. E. 1977. 'A distributed kinematic model of upland watersheds', *Hydrology Papers*, no. 93. Colorado State University, Fort Collins, Colorado. 36 pp.
Smith, R. E. and Woolhiser, D. A. 1971. 'Overland flow on an infiltrating surface', *Wat. Resour. Res.*, **7**, 899–913.
Walters, M. O. 1990. 'Transmission losses in arid regions', *J. Hydraul. Engng*, **116**, 129–138.
Wells, S. G. 1980. 'Geomorphic controls of alluvial deposition in the Sonoran desert, southwestern Arizona', in Doehring, D. O. (Ed.), *Geomorphology in Arid Regions*. George Allen & Unwin, London. pp. 27–50. A proceedings volume of the 8th Annual Geomorphology Symposium held at the State University of New York, at Binghamton.
Woodring, R. A. 1965. 'A hydraulic model for the catchment-stream problem, I. Kinematic wave theory', *J. Hydrol.*, **3**, 254–267.
Yao-Huang, Wu, Yevjevich, V. and Woolhiser, D. A. 1978. 'Effects of surface roughness and its spatial distribution on runoff hydrographs', Hydrology Papers, no. 96. Colorado State University, Fort Collins, Colorado.

12

MODELLING THE SPATIAL VARIABILITY IN FLOODPLAIN SOIL CONTAMINATION DURING FLOOD EVENTS TO IMPROVE CHEMICAL MASS BALANCE ESTIMATES

M. D. STEWART,[1]* P. D. BATES, D. A. PRICE[1] AND T. P. BURT[2]

[1]*Department of Geography, University of Bristol, University Road, Bristol, BS8 1SS, UK*
[2]*Department of Geography, University of Durham, Durham, DH1 3LE, UK*

ABSTRACT

The spatial distribution of pollutants entering floodplain soils during a flood event will be controlled by the interactions between flood hydraulics, floodplain topography, pollutant concentration, soil conditions, suspended sediment size and chemical reactions between the pollutants, water and soil. These interactions produce spatial variations in pollution build-up that are difficult to sample using current field-based techniques. Coupled hydraulic and hydrological models provide an ideal framework in which to explore these interactions and to improve current, reach-scale, chemical budget calculations. This paper presents a distributed model of floodplain pollution transport which for the first time allows a prediction of the spatial distribution of deposited and infiltrated pollutants within river reaches over a range of scales (1–60 km). The model is shown to be able to predict many of the features of floodplain pollutant build-up that are observed in the field. Using data from hypothetical examples it is suggested that the model can be used in the development of site-specific field sampling strategies.

KEY WORDS distributed modelling; floodplain pollution; chemical mass balance; hydraulic modelling

INTRODUCTION

In this paper we address the derivation of realistic chemical mass balance estimates for riparian environments as an essential first step in the investigation of hydrologically mediated floodplain chemical processes. To date, field sampling strategies have failed to capture the essential hydrological and chemical variability of floodplain river reaches (Berndtsson, 1990; Macklin *et al.*, 1994; Taylor, 1996). For example, mass balances are typically calculated as the loss occurring between gauging stations that may be up to 40 km apart. Such measurements cannot be used to disaggregate the complex chemical processes occurring within floodplain environments as they merely represent their integrated effect. Also, these measurements cannot be easily reconciled with process investigations that have typically focused on small-scale studies (e.g. Lewin *et al.*, 1983; Taylor, 1996). A need therefore exists to investigate hydrological and chemical processes at the river reach (1–60 km) scale, yet retain a high degree of spatial resolution in order to capture the known reach-scale variability. High resolution distributed hydraulic models provide one methodology for the solution of this research problem, having recently been developed to simulate floodplain flows over scales of up to 10 km, with grid resolutions of less then 10 m (Bates *et al.*, 1995, 1996). However, while such a core modelling platform is available there is currently no existing model that integrates hydraulics, sediment and pollutant

* Correspondence to: Dr M. D. Stewart, Department of Geography, University of Bristol, University Road, Bristol BS8 1SS, UK.

Contract grant sponsors: NERC and Electricité de France, Laboratoire National d'Hydraulic.
Contract grant number: GR3/09925.

transport and floodplain hydrology at this scale. The objective of this paper is to describe such an integrated model and demonstrate its potential for investigating floodplain hydrology and chemistry at the reach scale.

The floodplain soils of lowland river systems can act as major stores for a range of pollutants (e.g. heavy metals, nitrate, phosphate). Pollutant storage within floodplains is an important sink component in chemical mass balance studies (Christopherson and Wright, 1981; Elder, 1985; Berndtsson, 1990; Jain, 1996), and has important effects on river water quality and the floodplain environment (Lewin *et al.*, 1983; Haycock and Burt, 1993; Ross, 1994). It has also been shown that contaminants in floodplain soils can be used as stratigraphical markers and indicators of historical pollution levels (Macklin and Lewin, 1989; Macklin *et al.*, 1994).

Pollutants can reach floodplain soils through a number of pathways (rainfall, surface and subsurface runoff from adjacent hillslopes; Burt, 1992), although the river itself is often the most important source of contaminants (Bradley and Lewin, 1982; Lewin and Macklin, 1987), with river-borne pollutants becoming trapped on the floodplain during flood events as a result of the deposition of contaminated sediment and the infiltration of polluted waters. Hence, the spatial distribution of pollutants entering the floodplain during a flood event will be predominantly controlled by flood hydraulics, floodplain topography, concentration of the pollutants, soil conditions, suspended sediment size and chemical reactions between the pollutants, water and soil (Bradley and Lewin, 1982; Lewin and Macklin, 1987; Bird, 1987; Macklin, 1996). The complex interaction of these factors will produce wide spatial variability in the location of pollution sinks on the floodplain surface, both parallel and perpendicular to the river (Macklin *et al.*, 1994). It has become increasingly obvious that there are problems in producing field sampling strategies that are able to represent this spatial variability adequately (Taylor, 1996). Berndtsson (1990) illustrated that integrating the mean pollutant content from at-a-point floodplain soil samples across the entire floodplain did not produce a value that was large enough to explain depositional losses predicted from mass balance calculations. This was because the floodplain pollution was concentrated in depositional niches that were not sampled in the study. If the major niches of pollutant build up are not knowingly sampled then integrating an average value across the floodplain will not express the spatial variation or the total value of pollutant accumulation. The sizes of the dominant niches that can be sampled will depend on the scale of the river reach being considered. On small-scale reaches (1 km), niches produced by local variations in topography or friction (1–10 m) will be important, whereas on longer reaches (40 km) the effect of larger scale (100 m) niches will have to be considered. Hence, there is a clear need to develop a model that will allow the prediction of the location and areal extent of these niches on the scale of long (1–60 km) river reach scales in order to revise chemical mass balance estimates and allow the significance of these niches relative to the entire floodplain to be assessed.

The use of flood hydraulic modelling has been proposed as a tool for prediction of the location of deposited contaminated suspended sediment (Grimshaw *et al.*, 1976; Bird, 1987; Leenaers, 1989). However, there have been few studies that relate field measurements to flood hydraulics (Graf, 1990) or that attempt to consider the pollution deposited during individual flood events (Bradley and Lewin, 1982; Bird, 1987; Graf, 1990). Also there have been few studies that attempt to model the spatial variability of floodplain soil pollution. A number of these studies have used simple, one-dimensional models, e.g. downstream exponential decay models (Wolfenden and Lewin, 1978; Marcus, 1987; Lewin and Macklin, 1987), or overbank diffusion models (Pizzuto, 1987). Detailed field studies, such as Graf (1990), have illustrated that these models cannot represent the actual spatial variability seen in the field. However, recent advances in the numerical modelling of the hydraulics of river flood events means that there now exists the potential to produce simulations of the interaction of hydraulics, topography and soil pollutant distribution in floodplain environments using a two-dimensional approach (Bates and Anderson, 1993; Bates *et al.*, 1994, 1995).

This paper addresses the development of such a model through the investigation of three research issues. First, we extend current two-dimensional hydraulic and sediment transport models to consider infiltration into floodplain soils, thereby giving an innovative representation of floodplain hydrology. To date, such linkages have not been made and the effect of infiltration during flood events on floodplain hydrology and chemical budgets has not been assessed. Investigation of the effect of inclusion of this process forms the second objective of the paper. Lastly, we discuss the potential of the model to be used as a tool in the planning of field sampling strategies.

DETAILS OF THE MODELLING PROCEDURE

The model developed in this study consists of three components: a two-dimensional finite element hydraulic model, a two-dimensional finite element contaminant transport model and a one-dimensional finite difference soil hydrology model. This suite of submodels allows the integration of detailed hydraulic predictions with transport equations of dissolved and suspended load sediment, providing a calculation of the spatial distribution of infiltrated and deposited contaminants across a floodplain. This allows, for the first time, an hydraulics-based approach that is able to simulate the contamination of floodplain soils by dissolved and suspended pollutants during flooding events, and on a scale broadly commensurate with the size of niches of pollutant build up. Example simulations illustrate the potential of the modelling technique to enable the calculation of reach-scale sediment budgets and the development of field sampling strategies. We now describe each model component and the developed linkage mechanisms.

Two-dimensional finite element hydraulic model

Depth-averaged calculations of river hydraulics are based on the TELEMAC-2D two-dimensional finite element model (Hervouet, 1989, 1993) adapted specifically for long-reach floodplain studies (Bates *et al.*, 1994, 1995). Specific modifications to enable application to floodplain environments include a method to simulate flow over partially wet elements, new numerical algorithms particularly suited to floodplain finite element discretizations and improvements to the mass conservation properties of the model for this class of application (Bates *et al.*, 1994). The model solves the second-order partial differential equations for depth-averaged fluid flow derived from the full three-dimensional Navier–Stokes equations. This gives an equation set consisting of an equation for mass continuity and two force–momentum equations. These equations, called the shallow water or Saint-Venant equations, are given in non-conservative form as

$$\frac{\partial h}{\partial t} + \vec{u}.\overrightarrow{\text{grad}}\,(h) + h\text{div}(\vec{u}) = 0 \tag{1}$$

$$\frac{\partial u}{\partial t} + \vec{u}.\overrightarrow{\text{grad}}\,(u) + g\frac{\partial h}{\partial x} - \frac{1}{h}\text{div}\left[v_{\text{t}}\vec{h}, \overrightarrow{\text{grad}}\,(u)\right] = S_x - g\frac{\partial z_{\text{b}}}{\partial x} \tag{2}$$

$$\frac{\partial v}{\partial t} + \vec{u}.\overrightarrow{\text{grad}}\,(v) + g\frac{\partial h}{\partial y} - \frac{1}{h}\text{div}\left[v_{\text{t}}h, \overrightarrow{\text{grad}}\,(v)\right] = S_y - g\frac{\partial z_{\text{b}}}{\partial y} \tag{3}$$

where u and v are the velocity components in the x and y Cartesian directions, h is the depth of flow, z_{b} is the bed elevation, v_{t} is the turbulent viscosity, S_x and S_y are the source terms, g is the gravitational acceleration and t is the time.

The model solves for the three unknowns u, v and h. Turbulence is treated by Reynolds averaging of the governing equations. The resultant Reynolds stresses are added to the governing equations to represent the increased internal shear stress on mean flow produced by velocity fluctuations. Evaluation of these poses a number of difficulties as they represent additional unknowns, but with no additional equations to solve them. Hence, a turbulence model is introduced, based upon the Boussinesq approximation. The Reynolds stresses are assumed to be the product of the depth–mean velocity gradient and an exchange coefficient, ε, dimensionally similar to the coefficient of molecular viscosity, μ, and termed the eddy viscosity. This eddy viscosity term may then be given from established values or determined from a first-order k–ε turbulence closure model (see Lane, 1998, this issue).

To solve the governing equations, a fractional step method is employed (Marchuk, 1975) whereby advection terms are solved initially, separately from propagation, diffusion and source terms, which are solved together in a second step. This is achieved for a space discretization consisting of linear triangular or bilinear quadrilateral elements with three or four nodes per element, respectively. For the advection step several schemes may be used, with the method of characteristics chosen here for the momentum equation. To ensure mass conservation, the Streamline Upwind Petrov–Galerkin (SUPG) method (Brookes and Hughes, 1982) has been applied for the advection of the water depth (h) in the continuity equation. This is particularly suited to floodplain finite element problems because of the need for complex mesh discretizations. In such situations standard Galerkin methods may produce spurious oscillations in the solution which can only be eliminated by severe mesh refinement, often in areas where such resolution is not required. Further refinement of an already complex mesh may be an inefficient use of computing resources, hence the need for an alternative formulation. According to the SUPG technique, standard Galerkin weighting functions are modified by adding a streamline upwind perturbation which can be used to develop oscillation-free solutions regardless of mesh refinement. The second step (propagation) makes use of an implicit time discretization and solves the resulting linear system with a conjugate gradient-type method. In addition, the code makes significant savings in both computational time and storage requirements through the use of an element by element solution technique. Here, the matrices of the linear system are stored in their elementary form without full assemblage. Closure of this initial boundary value problem is then achieved by specification of inflows and outflows to the computational domain, usually at the upstream and downstream extremities of the reach, and the provision of an initial estimate of water depth and velocity at each computational node.

The model incorporates an additional modification to simulate areas of low lateral bed slope, such as floodplains or tidal flats. If this option is not implemented for floodplain areas, partially wet elements are retained within the solution domain and the model interpolates a spurious non-zero, lateral, free water surface across the element. For such an element the driving term in the momentum equation is

$$\frac{\partial u}{\partial t} = -g\frac{\partial z_{\mathrm{ws}}}{\partial x} \tag{4}$$

where z_{ws} is the free surface elevation.

Other terms, for example friction, diffusion, etc., still act on such an element but have been discarded for clarity. In dry elements, a fictitious free surface elevation is computed and will cause the numerical scheme to overpredict flow velocities. The object of the modification for floodplain areas is therefore to provide a better approximation of the water surface slope. First, a check is made for each element to test for the existence of a partially dry element. If the bed elevation for a particular node, $z_{\mathrm{b}2}$ with depth h_2, is greater than the water surface elevation ($z_{\mathrm{b}1} + h_1$) at one of the other nodes in that element, then a new water surface elevation (z_{ws}) for the dry node is defined in the momentum equation as

$$z_{\mathrm{ws}2} = z_{\mathrm{b}1} + h_2 \tag{5}$$

This can then be used to develop more realistic flow velocities in floodplain areas. Moreover, as only the momentum equation is modified, mass conservation properties are not affected.

Pollutant and suspended sediment transport model

The transport of dissolved pollutants is modelled using the two-dimensional, depth-averaged tracer model (SUBIEF) described by Moulin and Ben Slama (1998, this issue). Similar approaches for the transport of pollutants have been used by Falconer (1984, 1986) and Falconer and Liu (1994). This model solves an advection–diffusion equation driven by the hydrodynamics produced by the TELEMAC-2D model for one or several tracers, which can either be dissolved or in suspension. In the latter case the particle is prescribed a size and the model calculates the settling of sediment from suspension, erosion from the bed or resuspension of deposited sediment according to the hydrodynamic conditions. The advection–diffusion equation is given in non-conservative form as

$$\frac{\partial c}{\partial t} + \vec{U}.\overrightarrow{\mathrm{grad}}\, c = \frac{1}{h}\mathrm{div}\left(\vec{K}.h, \overrightarrow{\mathrm{grad}}\, c\right) + \frac{Q_{\mathrm{e}} - Q_{\mathrm{d}}}{h} + S \tag{6}$$

where c is the concentration of tracer, K is a dispersion coefficient, U is the flow velocity [the vector (u, v) from Equations (2) and (3)], S is a source term and Q_d and Q_e are the depositional flux and erosional flux calculated by the Krone and Partheniades formulae, respectively

$$Q_d = W_c.c.\left[1 - \left(\frac{u^*}{u_d^*}\right)^2\right] \tag{7}$$

$$Q_e = M\left[\left(\frac{u^*}{u_e^*}\right)^2 - 1\right] \tag{8}$$

where W_c is the settling velocity, M is the Partheniades constant, u^* is the friction velocity at the bottom and u_d^* and u_e^* are the critical settling and erosion velocities, respectively. The model therefore takes water depth and flow velocity from the hydrodynamic code described above and uses this to calculate the advection $(\vec{U}.\overrightarrow{\text{grad}})$ and the diffusion $[1|h\text{div}(\vec{K}.h\,\overrightarrow{\text{grad}}\,c)]$ of the tracer.

The equations assume that the pollutants do not change or modify the flow. Moreover, the two-dimensional approximation requires that the pollutants are well mixed through the water column such that there is no stratification of tracer concentration. The equations are solved using a finite element scheme based on the method of characteristics with linear interpolation, and using the same space discretization as that employed for the hydrodynamic model. Boundary conditions are a specification of the tracer concentration at each inflow boundary and the initial tracer concentration in the domain at the beginning of the simulation. At the outflow boundary one may prescribe either a concentration for the tracer or let its value vary freely. This latter option was used for all the model simulations reported in this paper. Interactions between tracers can also be modelled and Moulin and Ben Slama (1998, this issue) give the example of a dissolved metal interacting with fluvially transported suspended sediment. However, in this paper the simple case of conservative pollutants is considered.

Infiltration model

A one-dimensional infiltration model facilitates the relaxation of the zero flux boundary that is present at the interface between the flood flow and the subsurface soil moisture store in all current river floodplain hydraulic schemes. Linkage between the hydraulic and hydrology components is unidirectional, with water allowed to infiltrate into the floodplain according to soil moisture conditions. However, return flow is not considered in this paper. The modelling technique is discussed fully in Anderson and Howes (1985) with applications and calibration in Anderson (1983) and Brooks and Anderson (1995). Similar techniques have also been utilized and tested by Geisel *et al.* (1973) and Haverkamp *et al.* (1977). Water movement in the saturated zone is restricted to one-dimensional vertical flow described using Darcy's law

$$q = K\Delta H/L \tag{9}$$

where q is the flux density, K is the hydraulic conductivity and $\Delta H/L$ is the head drop per unit flow distance.

Darcy's law, conceived for saturated flow, was extended by Richards to unsaturated flow by restating the hydraulic conductivity as a function of the matrix suction head (ψ)

$$q = -K(\psi)\nabla H \tag{10}$$

where ∇H is the hydraulic head, gradient, which may include both suction and gravitational components. This formulation does not take account of hysteresis within the soil water behaviour.

We may extend the combined flow equation (8) by rewriting it as

$$\frac{\partial\theta}{\partial t} = \frac{\partial K(\theta)}{\partial z} - \frac{\partial}{\partial z}\left[K(\theta)\frac{\partial\psi}{\partial z}\right] \tag{11}$$

where θ is the soil moisture content, K is the unsaturated hydraulic conductivity and z is the vertical elevation.

Whilst there is no universally valid method for the *a priori* prediction of K based upon the soil suction, certain methods are available based on Poiseuille's law. In particular, the Millington–Quirk method (1959) codified by Jackson (1972) has substantial acceptance (Hillel, 1980) and is thus used here

$$K_i = K_s(\theta_i/\theta_s)\left[\frac{\sum_{j=i}^{m}((2j-1-2i)\psi_j^{-2}}{\sum_{j=1}^{m}((2j-1)\psi_j^{-2})}\right] \tag{12}$$

where K_i is K at moisture θ_i, θ_s is the saturated soil moisture content, m is the number of increments of θ between $\theta = 0$ and $\theta = \theta_s$ and ψ_i is the soil suction at the midpoint of each θ increment.

This scheme is solved using a forward difference, explicit, block-centred, finite difference scheme for a soil column beneath each computational node on the hydraulic model finite element mesh. Because of the ability of the hydraulic scheme to predict dry areas within the computational domain, the top boundary condition for the soil water model can be provided by either hydraulic or atmospheric inputs. If the node is dry, evaporation or rainfall drive the model and are used to update the soil moisture status. If the node is inundated, the infiltration rate is no longer supply limited and is set to its maximum. Calculation of the maximum infiltration rate for particular soil conditions gives the nodal vertical water flux from the hydraulic to the soil water model. This forms a sink term in the hydraulic model continuity equation and can also be used to update surface flow calculations at each time step. Hitherto, zero flux boundary conditions have been relaxed by the hydraulic model source and floodplain/channel sink models described above.

As the infiltration model solves the Richards' equation over a soil column, there is the potential to consider the passage of pollutants through the entire soil column. However, in the examples that follow the pollutants are assumed to be trapped in the upper layers of the soil, with no leaching or transport of the pollutants to groundwater aquifers. Over the short time-scale of the flood event simulations this is thought to be a reasonable approximation, assuming that the metal fractions are tightly bonded to the suspended sediments, and even on longer time-scales, work by Bradley and Cox (1987) has shown that many heavy metals become trapped in the upper few centimetres of the soil.

SIMULATION EXAMPLES

In this section the integrated model is illustrated through two simple examples that simulate hypothetical pollution events down reaches of the River Culm and River Severn, England. These simulations highlight the capabilities of the model in simulating the interactions between flood hydraulics, floodplain topography, soil conditions and water-borne pollutants, for reaches on different scales. The River Culm example considers a detailed topographical representation of a small (1 km) reach, while the River Severn example considers a longer (40 km) reach. These are chosen because they reflect two spatial scales of importance to floodplain pollution studies. The small scale is similar to that of detailed field sampling studies where the spatial pattern of pollutants is described by a finite number of samples (e.g. Macklin *et al.*, 1994; Taylor, 1996). The longer reach is of a scale frequently used in chemical mass balance studies, where a mass balance is calculated from data obtained at gauging stations (e.g. Bird, 1987; Jain, 1996). In each of the simulations there is a necessary trade-off between the number of computational points in the solution domain and the available computational limits. This will result in differences in the scale of the pollutant niches that each simulation can predict, with the Culm example allowing a consideration of small-scale variation in pollutant build up, and the Severn example only allowing the prediction of large-scale depositional niches. These differences are similar to those seen in field studies of reaches of different scales; as discussed previously. For both reaches

Table I. Physical constants used in simulations

Constant	Culm	Severn
Channel friction (Manning's n)	0·025	0·0167
Floodplain friction (Manning's n)	0·083	0·1
Pollutant dispersion coefficient ($m^2\ s^{-1}$)	0·4	0·4
Partheniades constant	0·0	0·0
Velocity diffusivity ($m^2\ s^{-1}$)	10·0	2·0
Critical velocity for settling ($m\ s^{-1}$)	0·1	0·1

the effect of the infiltration component of the model is isolated through a comparison between runs that do and do not include infiltration. The examples are then used to discuss the application of the model to chemical budget calculation and the development of field sampling strategies. The physical constants used in the simulations are listed in Table I. The friction values are taken from Chow (1959) and previous calibration studies. Owing to the lack of other published data, the values for pollutant dispersion and velocity diffusivity are taken from previous studies using TELEMAC-2D and SUBIEF (Moulin and Gailhard, 1996), and the values for the critical velocity of settling are calibrated from mass balance studies and are within the range of previously published values (Yalin, 1977).

River Culm example

Model set-up. This simulation considers the floodplain contamination produced by two large, hypothetical, pollution events entering a small reach (1 km) of the River Culm, England, during a single-peaked, once in one year flood event. The reach is a detailed section of the River Culm finite element mesh that was used with TELEMAC-2D by Bates *et al.* (1994, 1995), as shown in Figure 1. This reach is highly susceptible to flooding, with 90% of the floodplain inundated during the peak of the one-in-one year event. The finite element mesh contained 3570 nodes which allowed detailed floodplain topography to be considered (see Figure 2). The floodplain topography was defined from a field survey presented in Walling *et al.* (1992), which provided topographical constraint for the mesh that was superior to that used in Bates *et al.* (1994, 1995), and to the detail available from 1 : 10 000 topographic maps of the region. The hydraulics used for this simulation were based on the calibration studies of Bates *et al.* (1994, 1995) which used the same flood event.

Each of the hypothetical pollution events entered the reach through a point source, which discharged directly into the river channel. There is no attempt to model any specific pollutant in this experiment; however, it is assumed that the pollutant can exist in a dissolved state, and also adsorbed on to suspended sediment, with no reaction between the two phases. The pollution events discharged 0·25 $m^3\ s^{-1}$ of water contaminated with 200 mg l^{-1} of dissolved pollutant 200 mg l^{-1} of contaminated suspended sediment. Each pollution event lasted 14 400 s, with one event during the rising limb of the flood hydrograph and the second event during the peak of the flood. The background discharge of the point source was 0·12 $m^3\ s^{-1}$ with 5 mg l^{-1} each of dissolved pollutant and suspended sediment.

For the infiltration component of the model the simplest possible case of a constant soil type across the entire floodplain was used. The effects of spatially varied soil types on infiltration is discussed elsewhere (Price (1997)). The soil type chosen was a silty loam, a common soil type on floodplains in the south-west of England. The soil had a saturated hydraulic conductivity of $1{\cdot}25 \times 10^{-6}$ ms^{-1} and the initial water content of the soil at the onset of each simulation was at 60% saturation, thought to be typical of dry conditions. Other variables that could affect the activity of pollutants entering the soil, such as the pH (Frenet, 1981; Webb and Walling, 1985), were assumed to be constant.

Results. The hydrograph and chemical budgets (chemographs) for the simulation are shown in Figure 3. The shape of both chemographs illustrates the passage of the two pollution events through the reach. Both

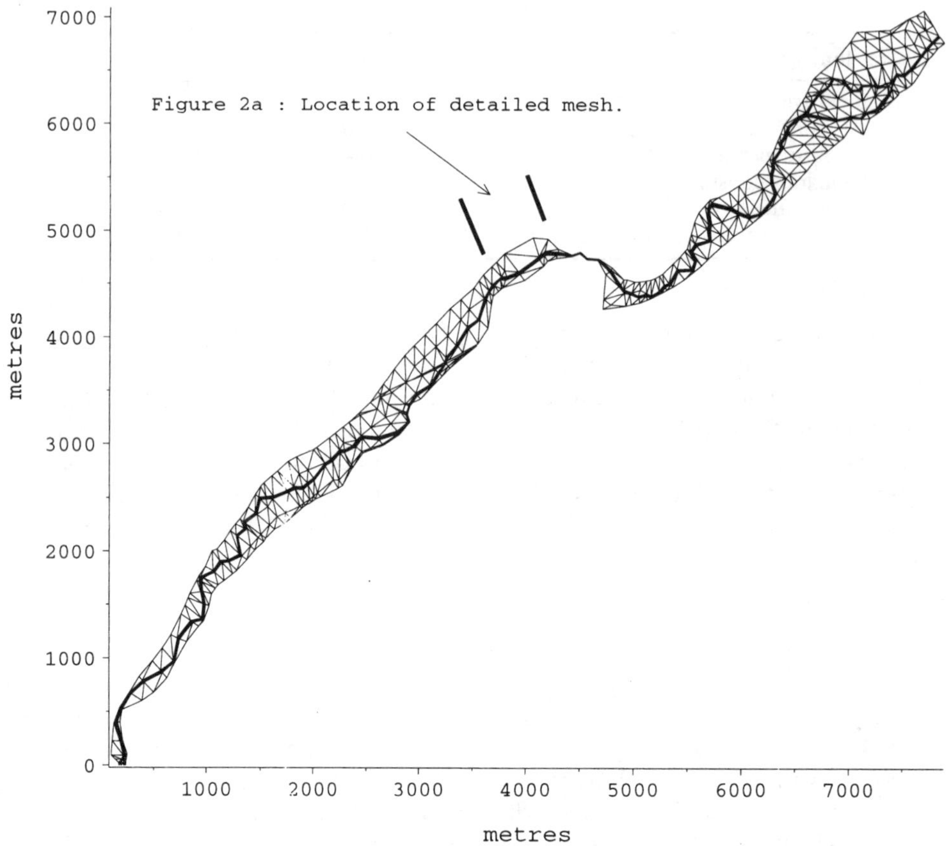

Figure 1. Location of the detailed River Culm finite element mesh, within the mesh used by Bates *et al.* (1994, 1995)

chemical budgets show good mass conservation. The quality of the chemical mass balances produced in SUBIEF are dependent on the water mass balances produced in TELEMAC-2D. In this example, the water mass balance only showed a volumetric loss of 3% over the entire simulation.

It is notable from Figure 3 that less suspended sediment and dissolved pollutants are discharged from the reach during the first pollution event than during the second event. This illustrates the response of the model to the changing hydraulic and hydrological conditions between the two events. During the first pollution event more pollutant is infiltrated into, and deposited on, the floodplain than is the case during the second pollution event. More pollutant is deposited during the first event because of the lower velocities that occur during the rising limb of the hydrograph, compared with the high velocities during the peak of the flood when the second event occurs. Also, during the second pollution event some of the pollutant deposited during the first pollution event is eroded and transported out of the reach. Infiltration is also lower during the second pollution event as a result of the soil becoming progressively saturated throughout the flood. These effects on the outflow of pollutant are compounded by the fact that some of the pollutant is still resident within the reach at the end of the first event and contributes to the second pollution peak.

The development of the pollutant plume formed during the experiment is illustrated in Figure 4. The figure shows the suspended sediment plume during and after the first pollution event, and illustrates the passage of the plume through the entire reach. The width of the plume is controlled by the inundation extent and the diffusivity of the plume. As similar values for diffusivity are used for the two pollutant plumes, the plume for the dissolved pollutant is very similar to that for the suspended sediment. It is notable that the core of the plume does not remain in-channel for the whole of the event, but that during the period of peak flow it leaves

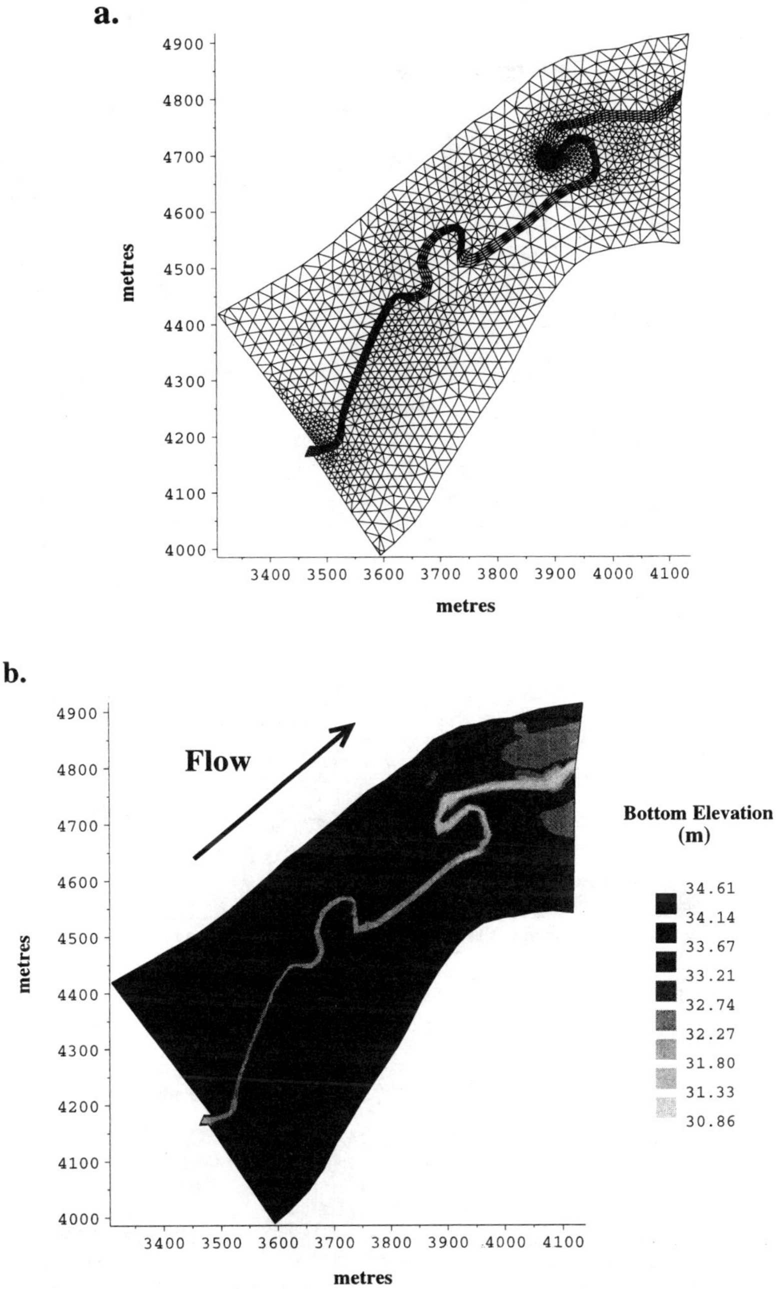

Figure 2. (a) Finite element mesh of the detailed River Culm reach. The mesh is composed of triangular elements and 3750 nodal points. More elements are present at regions where there is complex topography. (b) Contour map of the floodplain topography used in the simulation. Topographic data are from a survey by Walling *et al.* (1992)

the channel and lies on the inundated floodplain to the north of the channel. An appreciation of such a complex hydraulic response is only possible through a two-dimensional distributed approach.

The pattern of water infiltrated into the floodplain during the event is shown in Figure 5a. As a constant soil type was chosen for all of the floodplain, and because large areas of the floodplain were inundated for

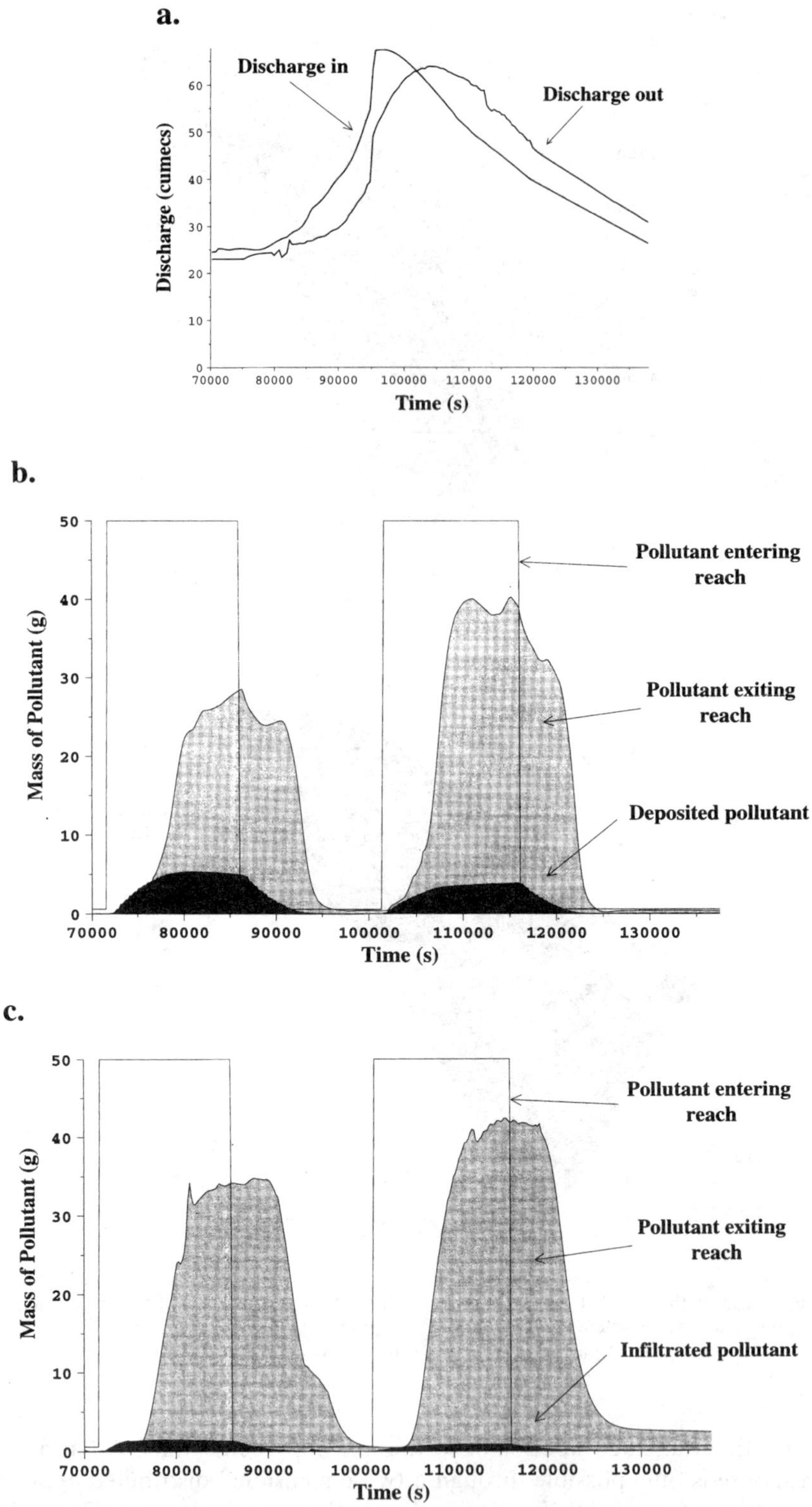

Figure 3. Results for the River Culm simulation. (a) Inflow and outflow flood hydrographs for the once in one year event. (b) Suspended sediment chemograph. (c) Dissolved pollutant chemograph

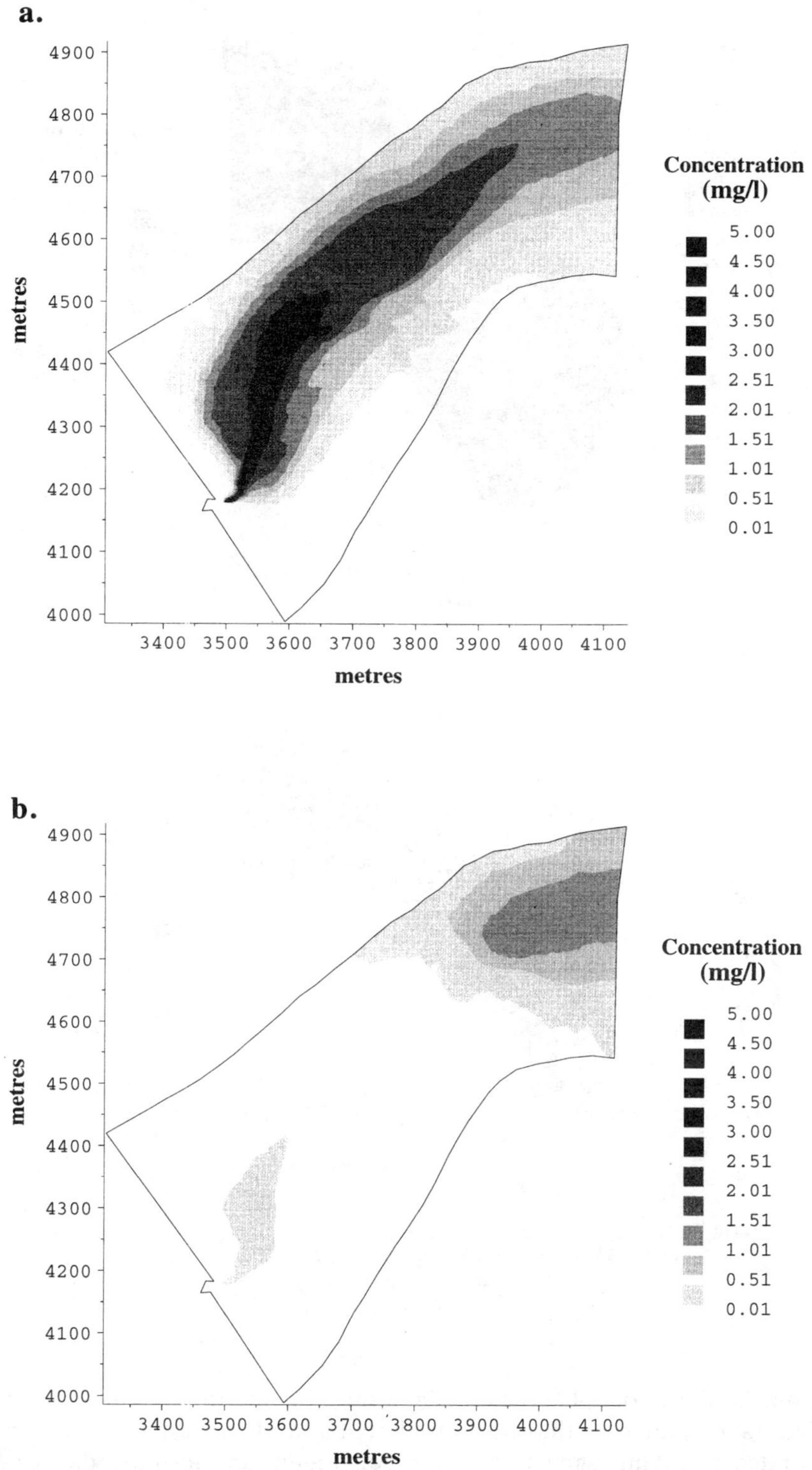

Figure 4. Development of the suspended sediment plume during the first pollution event. (a) Well-developed plume forming during high pollutant outflow. (b) Situation soon after the end of the pollution event when most of the pollutant has passed through the reach; some pollutant is trapped on the floodplain, with low levels of background pollution discharging from the source

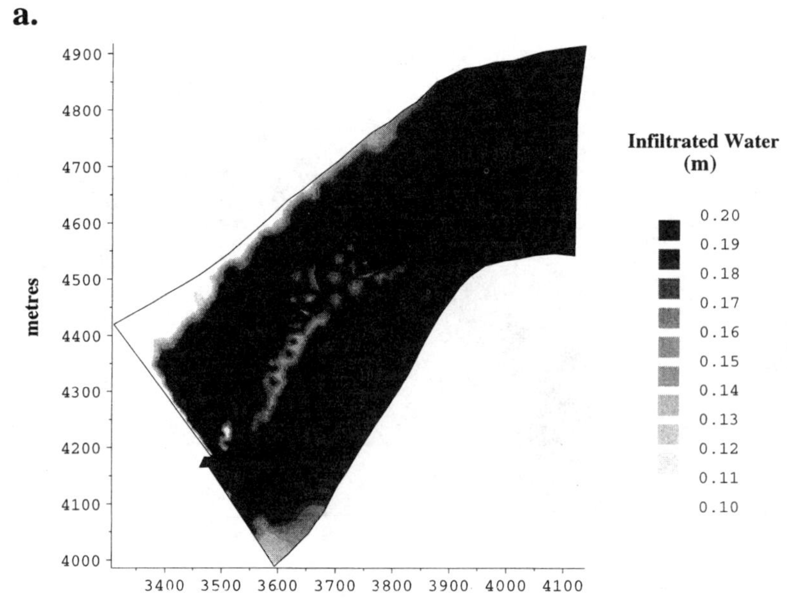

Figure 5a

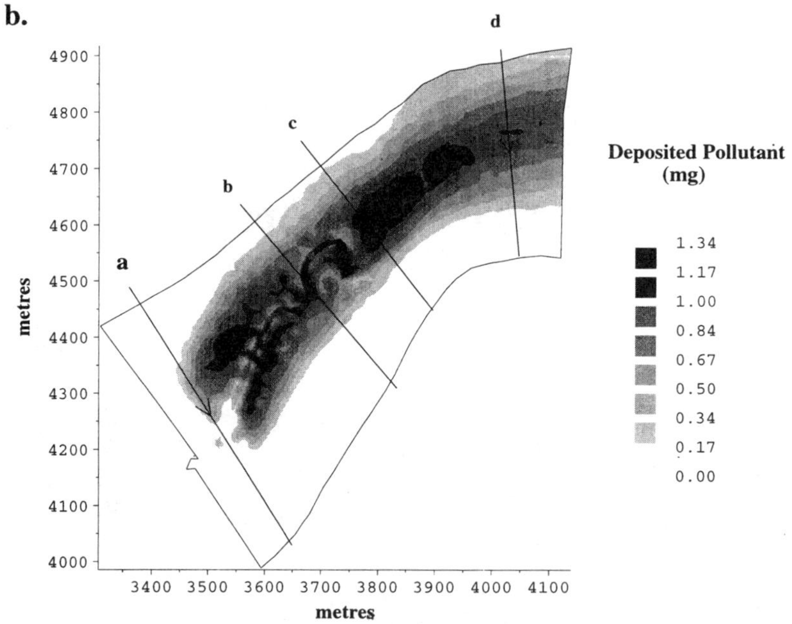

Figure 5b

most of the event, much of the floodplain has similar values for the total infiltrated water. However, the interaction of the infiltration pattern with the dissolved pollutant plume results in more spatial variation in the pattern of infiltrated pollutant (shown in Figure 5c). Figure 5b illustrates the pattern of deposited suspended sediment. The spatial distributions of the pollutants remaining within the reach illustrate that the model predicts distinct niches for the build up of both the infiltrated and deposited pollutant. Hence the

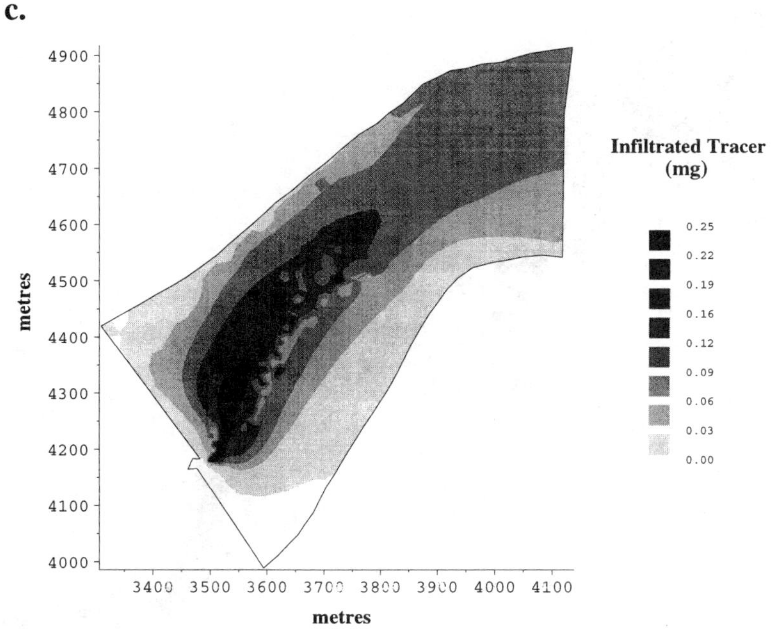

Figure 5c

Figure 5a–c. Distributed outputs from the model. (a) Spatial distribution of total infiltrated water following the flood event. (b) Spatial distribution of deposited suspended sediment-borne pollutant The positions of the cross-sections (a–d) used in Figure 10 are also shown. (c) Spatial distribution of infiltrated pollutant

spatial distributions of both pollutants are dominated by strong localizations of concentration, with 50% of the infiltrated pollutant occurring in 20% of the surface area, and 50% of the deposited suspended sediment occurring in only 10% of the surface area. Although the pattern of infiltrated pollutant may be strongly affected by the assumption of equivalent soil conditions across the floodplain, the example does clearly illustrate that the spatial distribution of infiltrated pollution and deposited pollution can be very different. The distribution of the deposited suspended sediment is predominantly controlled by the hydraulics, the topography and the nature of the sediment particles, while the spatial distribution of the infiltrated pollutant is predominantly controlled by the soil conditions and the plume extent. In the example presented here, the pattern of infiltrated pollutant is seen to be strongly influenced by the location of the pollutant source, with the concentrations attenuating rapidly downstream. By contrast the pattern of deposited suspended sediment is more clearly influenced by local hydraulic effects, with the highest concentrations reflecting the position of hydraulic niches.

These results, from a single simulation, are presented to illustrate the model linkages and the potential of the modelling approach. The effect of changing the model parameterization will, of course, affect the model predictions. Bates *et al.* (1996) have illustrated the effect of topographic representation, friction and turbulent viscosity on the hydraulic predictions, while the effect of using different soil types on floodplain infiltration is discussed by Price (1997). Furthermore, a simulation run without an infiltration component did not produce a noticeable difference in the development of the pollutant plume, the flooding extent or in the mass of deposited suspended sediment. This was because of the low percentage of water removed (2·5%) during the flood event. Further studies of model parameterization and calibration must form major future research efforts, but are beyond the scope of this paper.

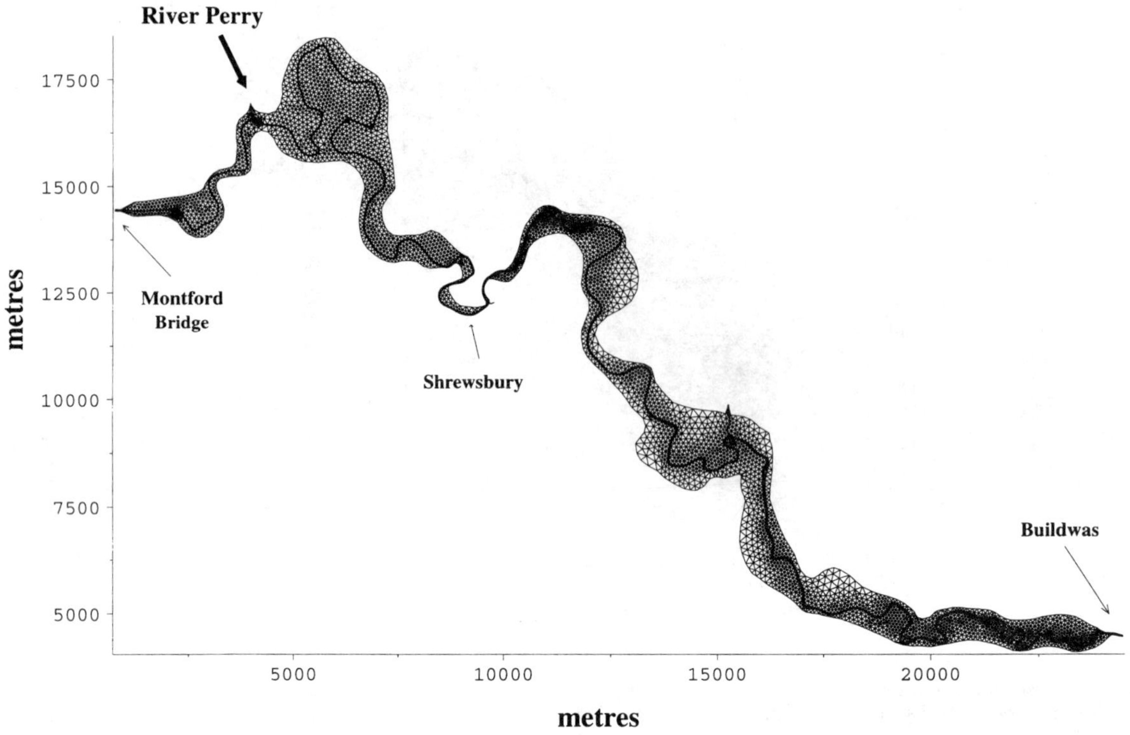

Figure 6. River Severn finite element mesh. The mesh contains 12 000 elements and 6500 nodal points

River Severn example

Model set-up. The second example simulates the effect of the discharge of a polluted tributary into a 40 km reach of the River Severn in Shropshire, England, during an approximately one-in-one year flood event which inundates 40% of the floodplain. The reach lies between the Montford Bridge and Buildwas gauging stations, and contains a number of wide floodplain sections. The finite element mesh has 6500 nodal points and over 12 000 elements (Figure 6), allowing a reasonably detailed representation of the floodplain topography. The topography is taken from a 10 m × 10 m DTM (digital terrain model), produced from the Ordnance Survey 1:10 000 series maps of the region. This simulation illustrates that the linked model is applicable to large-scale reaches, with complex finite element meshes, which are at the current limit of available computer power. The hydraulics of the model are calibrated by comparing the model outflow to a known outflow at Buildwas, by the adjustment of the modelled floodplain and channel friction coefficients (see Bates *et al.*, 1995). The fit of the predicted to the actual hydrograph is shown in Figure 7. The simulation provided reasonable water and pollutant mass balances, considering the length and complexity of the reach.

The simulation considers the entry of a polluted tributary into the main channel of the River Severn, during a one-in-one year flood event. The event lasted 250 000 s, which was divided into 4-s time steps. The tributary, the River Perry, contains 50 mg l^{-1} of dissolved pollutant, and has a discharge that ranges from 8·0 to 9·8 m^3 s^{-1}. In contrast, the River Severn at Montford Bridge has a zero concentration of pollutant and a discharge during the event of between 175·6 and 318·2 m^3 s^{-1}. Although the chemistry of no specific pollutant is modelled in this example, a value of 50 mg l^{-1} is equivalent to the EC recommended maximum level of nitrate in rivers. As with the River Culm example, a constant soil type of silty loam is used for each of the nodal points.

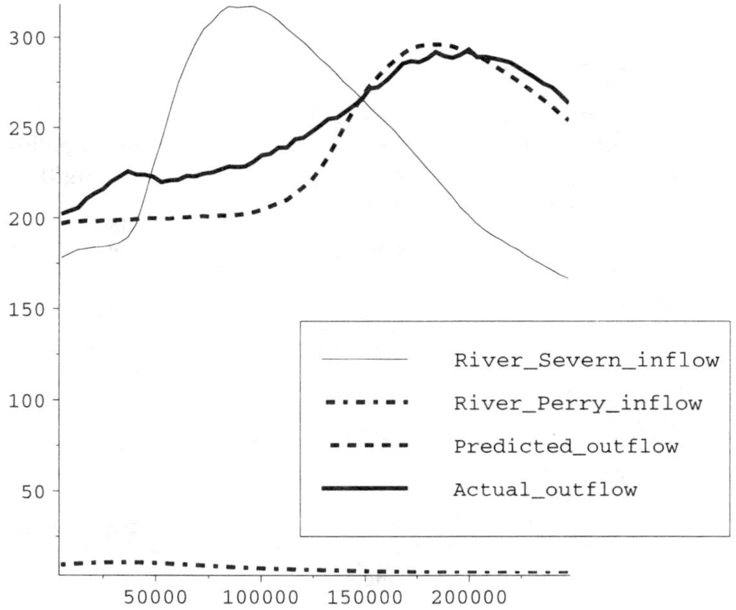

Figure 7. Inflow and outflow hydrographs for the one-in-one year flood event for the River Severn simulation. There is a poor fit between the predicted and observed outflow discharges near the start of the simulation as true steady state conditions did not exist in the reach at the beginning of the real flood event

Results. Figure 8 illustrates the development of the pollutant plume throughout the flood event. There is an obvious rapid dilution of the plume as it enters the main River Severn. At the end of the flood event some of the pollutant plume has begun to pass out of the end of the reach at Buildwas, 25 km downstream of the tributary inflow. However, most of the pollutant still lies within the reach.

The spatial distribution of infiltrated water into the floodplain is shown in Figure 9. This pattern also reflects the maximum inundation extent of the flood. The pattern of infiltrated pollutant is similar to the extent of the pollutant plume, shown in Figure 8. Unlike the River Culm example, there is greater spatial variation in the amount of infiltrated water on the floodplain. The variable pattern of infiltrated water illustrates the importance of the changing extent of flood inundation on the spatial pattern of infiltration and infiltrated pollutant on the floodplain. It is notable that smaller floods would be expected to concentrate high concentrations of pollutant close to the channel, while large flood events may result in more dilute concentrations being distributed across a wider stretch of the floodplain. In total, 4% of the flood water was infiltrated into the floodplain, and, as with the Culm example, running a simulation without infiltration did not result in noticeable changes in the flood hydraulics.

IMPLICATIONS FOR FLOODPLAIN CHEMICAL BUDGETS AND FIELD SAMPLING STRATEGIES

Owing to its inherent spatial variability, floodplain pollution presents a modelling problem that is perfectly suited to a distributed approach. No other methodology allows the integration of small-scale sediment and pollutant deposition processes over the reach scale. However, there are problems associated with this class of model (Beven, 1989), especially in terms of parameterization and calibration. In this study, calibration of the full integrated model has not been attempted, because of a lack of suitable field data. However, each component of the integrated model has been independently validated (Bates *et al.*, 1995; Brooks and Anderson, 1995, Moulin and Gailhard, 1996) and, importantly, the hydraulics of each simulation have been calibrated for the reaches and events in question. Despite the limitations discussed above the simulation

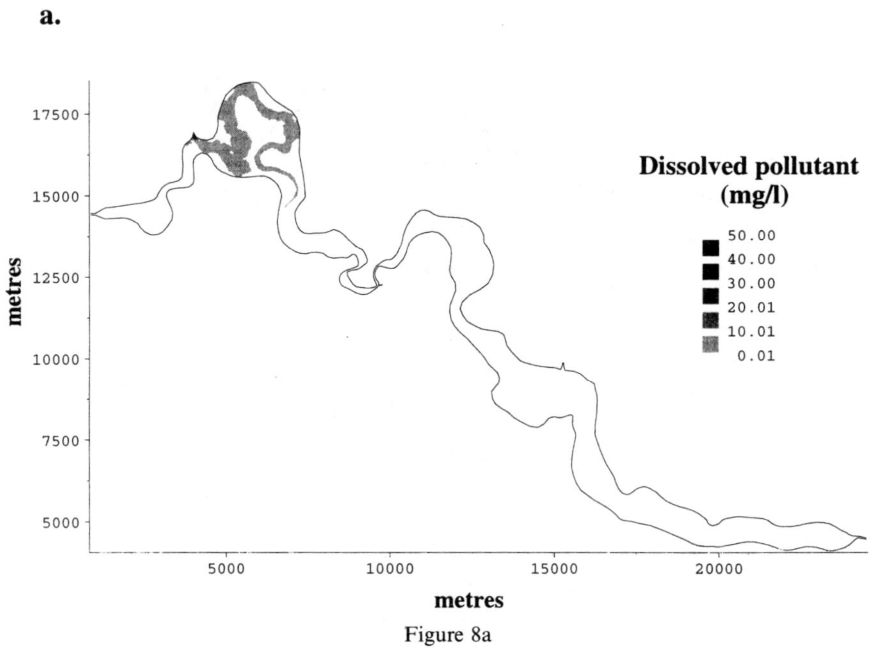

Figure 8a

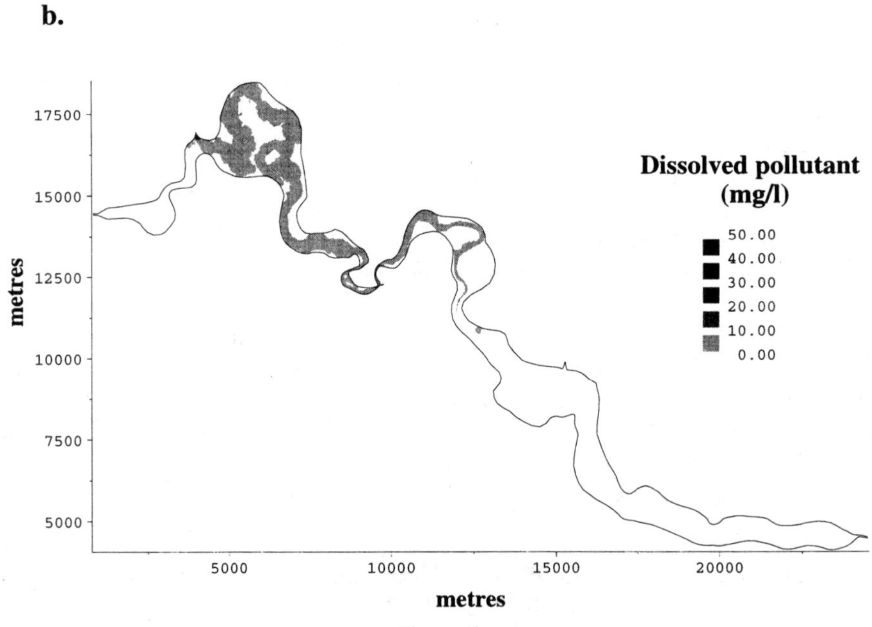

Figure 8b

examples were able to reproduce many of the features of floodplain pollution that have been noted in field studies. Notably, the model predicted the concentration of pollutants in small niches that were controlled by the physics of the flow, the floodplain environment and the nature of the pollutants.

The two simulation examples illustrate obvious applications of the modelling technique to chemical budget calculations. The model has the capability to provide reach-scale chemical budgets that are produced

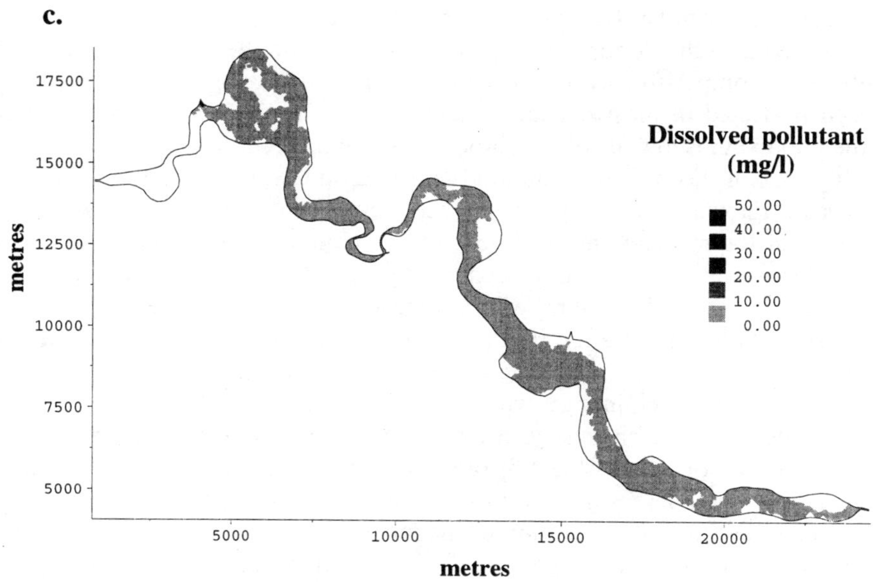

Figure 8c

Figure 8a–c. Evolution of the pollutant plume down the Severn reach

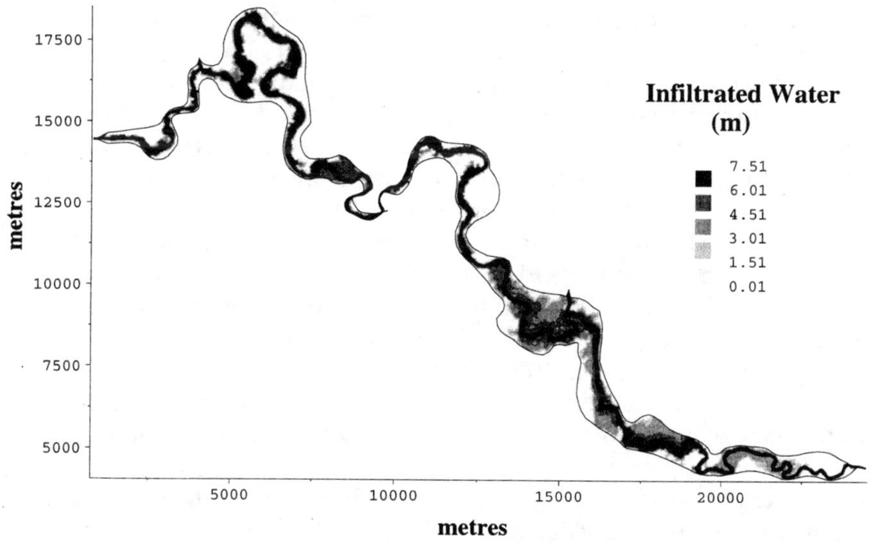

Figure 9. Cumulative infiltrated water along the Severn reach, following the flood event

by taking into account small-scale processes and local heterogeneity. This is potentially a powerful prediction tool for catchment management as well as for investigating the interactions of floodplain processes. In the River Culm example presented in this paper 18% of the polluted suspended sediment was deposited on the floodplain. This value is at the lower end of the range of values for floodplain accumulation of polluted suspended sediment calculated from chemical mass balance studies; for example, Berndtsson (1990) measured a loss of 11–47%; Imhoff *et al.* (1980), 31%; Marron (1992), 29–44%, and for studies measuring reach-scale suspended sediment deposition; for example, Walling *et al.* (1986) measured losses ranging from

8 to 54% for flood events down the River Culm. The Culm example also predicted that 3% of the dissolved pollutant was infiltrated into the floodplain and in the Severn example 2% was infiltrated. Field values of infiltrated pollution for comparison with model predictions are lacking, and the prediction of infiltrated pollutant has been neglected in most chemical mass balance studies. Although the percentage of water infiltrated into the floodplain is low, it still corresponds to a significant mass of pollutant, which is approximately equal to 10–15% of the mass of pollutant deposited on to the floodplain. In the Culm example the reach simulated was a known sink for suspended sediment deposition (Walling *et al.*, 1992) and this is reflected in the fact that the predicted mass of deposited pollutant is significantly greater than the mass of infiltrated pollutant. However, over a longer reach where regions of low deposition are also considered it is possible that the mass of infiltrated pollutant will become a much greater proportion of the overall pollutant budget. Hence consideration of infiltration into the floodplain is a vital component of the integrated model presented in this paper, even if a simple one-dimensional approach is used. There are many possible enhancements to this simple model linkage, with the consideration of two-dimensional subsurface flow, or the consideration of detailed soil chemistry, for example. However, the approach used in this paper is sufficient to illustrate the importance of the infiltration process and its inherent spatial variability, as well as forming a basis for further model testing, calibration and validation, using better representations of subsurface hydrology. Field testing and calibration of the model may, however, be more important than model linkage improvements if true chemical budgets are to be calculated. A field testing strategy is required that is consistent with the distributed nature of the model output, and that requires a finite number of field samples.

APPLICATIONS TO FIELD SAMPLING STRATEGIES

It is never possible to take enough field samples in order to express all the spatial variability in the distribution of pollutants within the floodplain environment (Walling *et al.*, 1986). It is therefore vital that sample locations are chosen that supply as accurate as possible a representation of this spatial variability, or at least provide an idea of the possible variability and the associated errors. Almost all of the studies of floodplain pollution have been field studies where the distribution of pollutants has been inferred from a limited number of at-a-point soil or water samples. Most commonly, the downstream variation of pollution is calculated from sample sites, from within the channel (Wolfenden and Lewin, 1978; Bird, 1987; Boult *et al.*, 1994) or from the floodplain near to the channel (Bradley and Cox, 1987; Leenaers, 1989; Macklin and Dowsett, 1989), spaced at intervals ranging from 100 m to 10 km. Other studies have measured pollutants on much smaller scale reaches, with sample sites being defined by local topography (Asselmann and Middelkoop, 1995; Walling *et al.*, 1986), local geomorphology (Lambert and Walling, 1987; Macklin and Lewin, 1989) and the age of floodplain deposits (Wolfenden and Lewin, 1977; Lewin and Macklin, 1987; Macklin and Lewin, 1989), or by more stratified approaches such as transects across the floodplain (Lewin and Macklin, 1987; Elder, 1985). In many of these strategies geomorphological analyses are used to identify regions of supposed hydraulic similarity.

Figure 10 shows a series of cross-sections taken across the floodplain of the River Culm reach, illustrating the predicted distribution of suspended sediment deposition and infiltrated pollutant relative to each other and the floodplain topography. The profiles illustrate that the variations in pollutant perpendicular to the channel can be as important as downstream variations, and that regions of highest pollutant concentration may not be predictable from the topography alone. It is unlikely that a random sampling procedure, transects or a topographically based sampling strategy would accurately express the spatial variability in pollution across the floodplain. It is proposed that model predictions from a series of hypothetical pollution and flood scenarios could be used as a starting point for the selection of a site-specific sampling strategy, which would not only reflect the spatial variation of pollution across the floodplain more accurately, but would also provide a calibration data set for the model that is consistent with the distributed output of the model. A possible sampling strategy for the Culm reach, which uses only a limited number of field samples and attempts to sample suspended sediment build up, is shown in Figure 11. Sites are chosen that sample

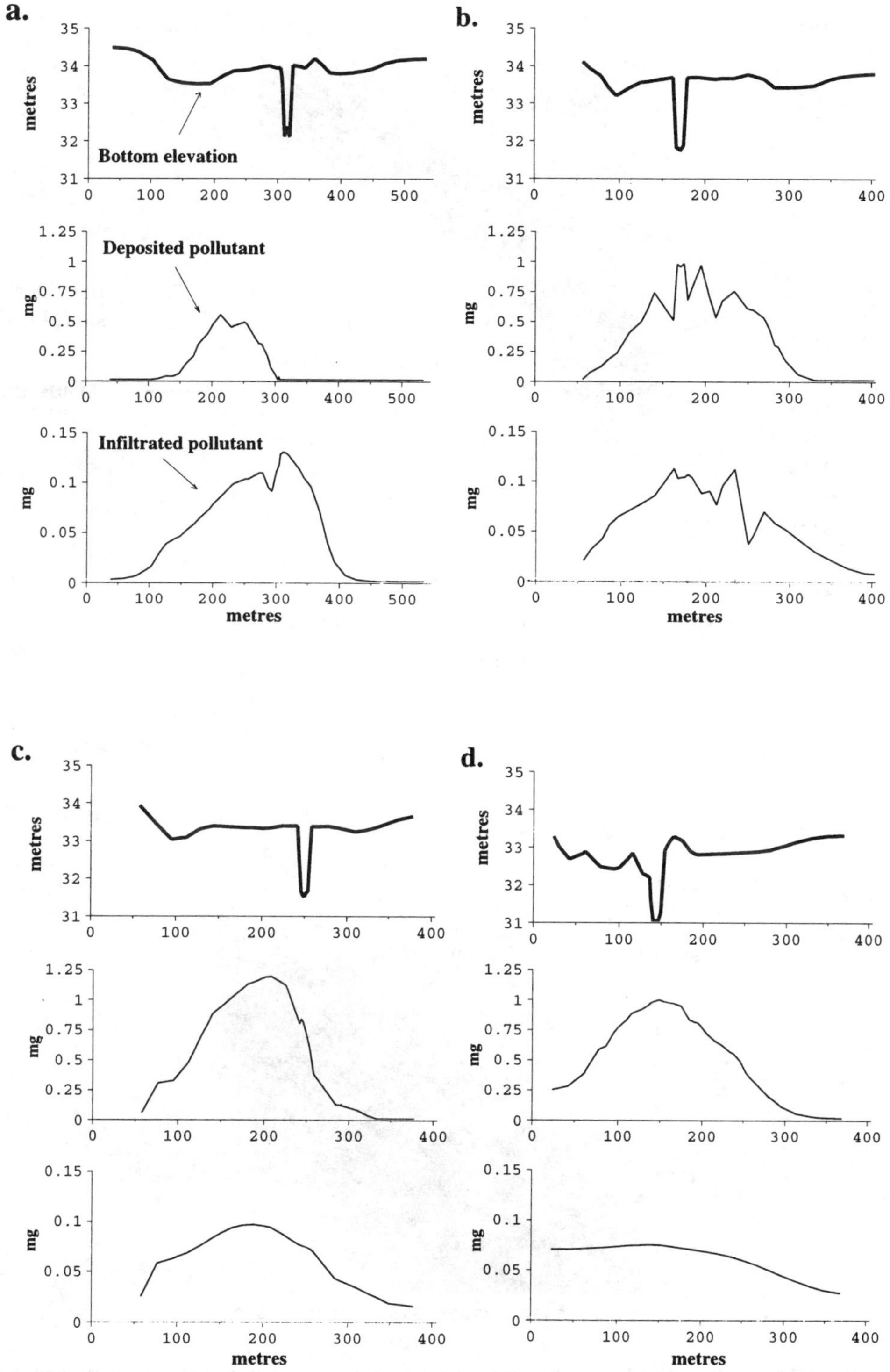

Figure 10. Cross-section profiles across the River Culm reach, comparing topography with the spatial build up of deposited and infiltrated pollutant. See Figure 5b for the locations of the cross-sections. Note the differences in the location of deposited suspended sediment and infiltrated pollutant, and the fact that variations in the distribution of pollutant cannot be described purely by variations in the floodplain topography

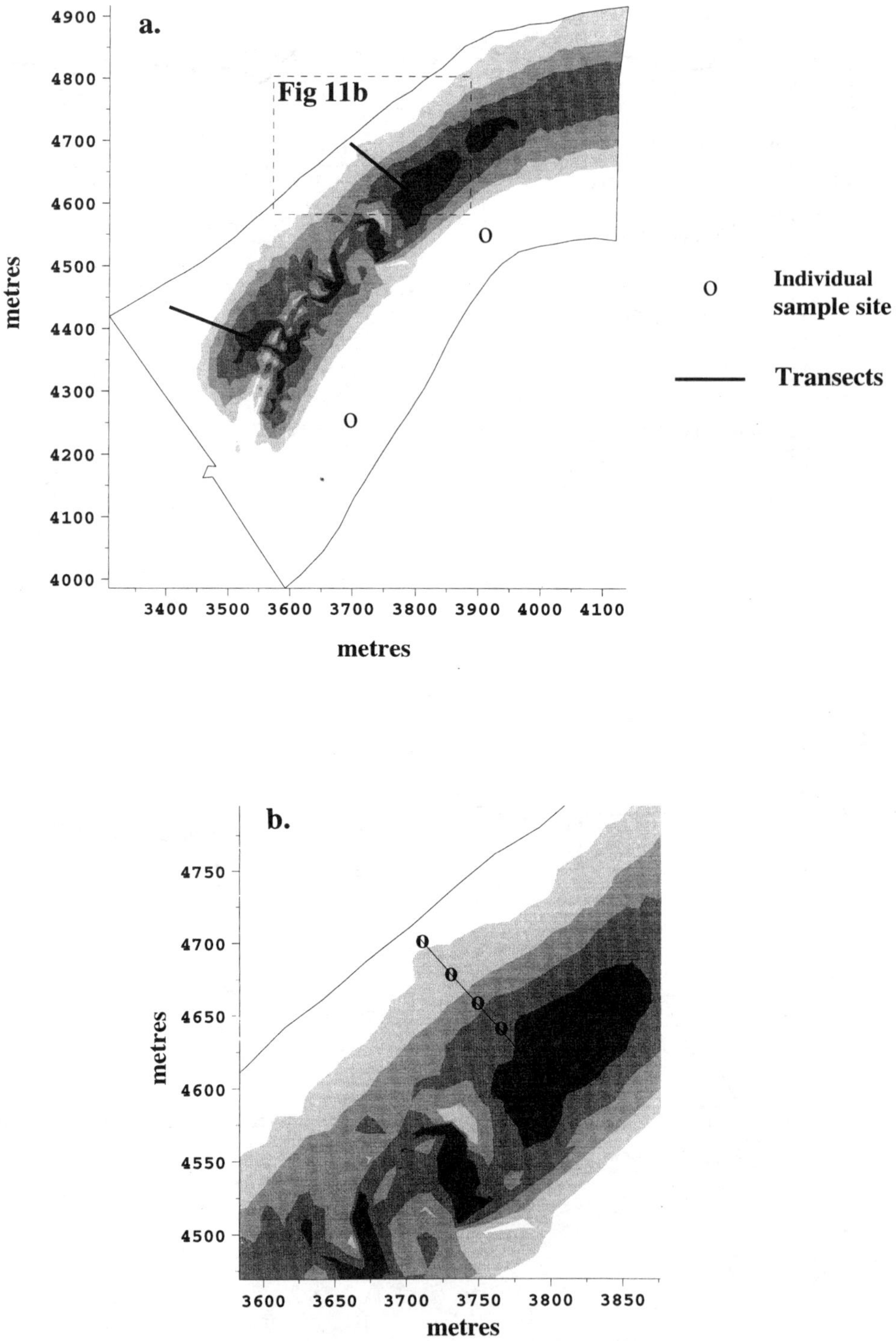

Figure 11. Proposed field sampling strategy for the River Culm example, using the results of the hypothetical simulation. (a) Sampling sites are chosen at locations of high and low pollutant build up, in order to calibrate the model predictions of extreme values. Sites are also chosen along transects across regions where there are high pollution gradients, in order to calibrate the accuracy of the spatial variability predicted by the model. It is thought that such an approach will produce a calibration that is consistent with the distributed nature of the model. (b) Detail of sampling sites along one of the transects

points of predicted high and low pollutant concentration and that also sample across regions where there is a significant predicted pollution gradient. This strategy attempts to calibrate the full range of absolute values predicted at points on the floodplain, as well as calibrating the degree of spatial variation. It is essential that the predicted pollutant gradients are sampled, since they provide the spatially integrated values of pollutant that are necessary for chemical budget calculations. Obviously, the sampling strategy in Figure 11 has been developed from the results of only one model simulation; however, it is easy to see how the results of a number of simulations, considering a range of flood events and testing a range of model parameters, could be integrated before a final sampling strategy is developed. Taylor (1996) stressed that detailed knowledge of any field site is required before a sampling strategy can be developed. This paper suggests that the results of hypothetical model simulations could become a further tool to be used in the development of sampling strategies, especially where the number of samples that can be taken is restricted and when the results of a local study have to be integrated over the reach scale. It is clear from this that closer ties between field and modelling studies; as well as intelligently thought out calibration strategies and model applications are required if the potential of such distributed models is to be realized.

SUMMARY

This paper has developed a new approach to modelling floodplain contamination that, for the first time, integrates two-dimensional hydraulic and sediment transport models with infiltration into the floodplain. Evidence is provided that demonstrates that the model is able to reproduce many of the features of pollutant build-up seen in the field. The model predicts that water and contaminants transferred to the floodplain during flood events can be significant in terms of reach-scale chemical budgets. Furthermore, the capabilities of distributed modelling approaches as tools in the planning of field sampling strategies is clearly illustrated. While there is considerable scope for refinement of the model components and their associated linkage mechanisms, as well as for research into appropriate parameterization, calibration and validation strategies, it is clear that the approach to modelling floodplain contamination taken in this paper represents the only current method for integrating small-scale process variability at the reach scale.

ACKNOWLEDGEMENTS

This research was supported by the UK Natural Environment Research Council (Grant no. GR3/09925) and Electricité de France Laboratoire National d'Hydraulique. David Price was supported by an NERC Studentship. Thanks go to Richard Hardy and Anne-Marie Bremner for advice, and to Stuart Lane and one anonymous reviewer for their constructive comments.

REFERENCES

Anderson, M. G. 1983. 'The prediction of soil suction for slopes in Hong Kong', *Final Report.* Geotechnical Control Office, Hong Kong.

Anderson, M. G. and Howes, S. 1985. 'Development and application of a combined soil water-slope stability model', *Q. J. Engng Geol.*, **18**, 225–236.

Asselman, N. E. and Middelkoop, H. 1995. 'Floodplain sedimentation: quantities, patterns and processes', *Earth Surf. Process. Landf.*, **20**, 481–499.

Bates, P. D. and Anderson, M. G. 1993. 'A two dimensional finite element model for river flow inundation', *Proc. R. Soc. London, Ser. A*, **440**, 481–491.

Bates, P. D., Anderson, M. G., and Hervouet, J.-M. 1994. 'Computation of a flood event using a two-dimensional finite element model', in Molinaro, P. and Natale, L. (Eds), *Modelling Flood Propagation over Initially Dry Areas.* American Society of Civil Engineers, New York. pp. 243–256.

Bates, P. D., Anderson, M. G., and Hervouet, J.-M. 1995. 'Initial comparison of two two-dimensional finite element codes for river flood simulation', *Proc. Inst. Civil Engrs, Water Maritime Energy*, **112**, 238–248.

Bates, P. D., Anderson, M. G., Price, D. A., Hardy, R. J., and Smith, C. N. 1996. 'Analysis and development of hydraulic models for floodplain flows', in Anderson, M. G., Walling, D. E., and Bates, P. D. (Eds), *Floodplain Processes.* John Wiley & Sons, Ltd., Chichester. pp. 215–254.

Berndtsson, R. 1990. 'Transport and sedimentation of pollutants in a river reach: a chemical mass balance approach', *Wat. Resour. Res.*, **26**, 1549–1558.
Beven, K. J. 1989. 'Changing ideas in hydrology: the case of physically based distributed models', *J. Hydrol.*, **105**, 79–102.
Bird, S. C. 1987. 'The effect of hydrological factors on trace metal contamination in the River Tawe, South Wales', *Environ. Pollut.*, **45**, 87–124.
Boult, S., Collins, D. N., White, K. N., and Curtis, C. D. 1994. 'Metal transport in a stream polluted by acid mine drainage — the Afon Goch, Anglesey, UK', *Environ. Pollut.*, **84**, 279–284.
Bradley, S. B. and Cox. J. J. 1987. 'Heavy metals in the Hamps and Manifold Valleys, north Staffordshire, UK: partitioning of metals in floodplain soils', *Sci. Total Environ.*, **65**, 135–153.
Bradley, S. B. and Lewis, J. 1982. 'Transport of heavy metals on suspended sediments under high flow conditions in mineralised region of Wales', *Environ. Pollut.*, **4**, 257–267.
Brooks, A. N. and Hughes, T. J. R. 1982. 'Streamline Upwind/Petrov–Galerkin formulations for convection dominated flows with particular emphasis on the incompressible Navier–Stokes equations', *Comp. Meth. Appl. Mech. Engng*, **32**, 199–259.
Brooks, S. M. and Anderson, M. G. 1995. 'The determination of suction-controlled slope stability in humid temperate environments', *Geograf. Annal.*, **77A**, 11–22.
Burt, T. P. 1992. 'The hydrology of headwater catchments', in Calow, P. and Petts, G. E. (Eds), *The Rivers Handbook*, Vol. 1. Blackwell, Oxford. pp. 3–28.
Chow, V. T. 1959. '*Open Channel Hydraulics*. McGraw-Hill Book Co, New York.
Christopherson, N. and Wright, R. F. 1981. 'Sulfate budget and a model for sulfate concentration in stream water at Birkenes: a small forested catchment in southernmost Norway', *Wat. Resour. Res.*, **17**, 377–389.
Elder, J. F. 1985. 'Nitrogen and phosphorus speciation and flux in a large Florida River wetland system', *Wat. Resour. Res.*, **21**, 724–732.
Falconer, R. A. 1984. 'A mathematical model study of the flushing characteristics of a shallow tidal bay', *Proc. Inst. Civil Engrs, part 2*, **77**, 311–332.
Falconer, R. A. 1986. 'A water quality simulation study of a natural harbour', *ASCE. J. Waterway, Port, Coastal Ocean Engng*, **112**, 15–34.
Falconer, R. A. and Liu, S. Q. 1994. 'Numerical modelling of hydrodynamic and water quality processes in an enclosed tidal wetland', in Falconer, R. A. and Goodwin, P. (Eds), *Wetland Management*. Thomas Telford, London. pp. 119–129.
Frenet, M. 1981. 'The distribution of mercury, cadmium, and lead between water and suspended matter in the Loire estuary as a function of the hydrological regime', *Wat. Res.*, **15**, 1343–1350.
Giesel, W., Renger, M., and Strebel, O. 1973. 'Numerical treatment of the unsaturated water flow equation: comparison of experimental and computed results', *Wat. Res.*, **9**, 174–177.
Graf, W. L. 1990. 'Fluvial dynamics of thorium-230 in the Church Rock event, Puerco River, New Mexico', *Ann. Assoc. Am. Geogr.*, **80**, 327–342.
Grimshaw, D. L., Lewin, J., and Fuge, R. 1976. 'Seasonal and short-term variations in the concentration and supply of dissolved zinc to polluted aquatic environments', *Environ. Pollut.*, **11**, 1–7.
Haverkamp, R., Vauclin, M., Touma, J., Wierenga, P. J., and Vachaud, G. 1977. 'A comparison of numerical simulation models for one-dimensional infiltration. *Soil Sci. Soc. Am. J.*, **41**, 285–293.
Haycock, N. E. and Burt, T. P. 1993. 'Role of floodplain sediments in reducing the nitrate concentration of subsurface run-off: a case study in the Cotswolds, UK', *Hydrol. Process.*, **7**, 287–295.
Hervouet, J.-M. 1989. 'Comparison of experimental data and laser measurements with the computational results of the TELEMAC-2D code (shallow water equations)', in Maksimovic, C. and Radojkovic, M. (Eds), *Computational and Experimental Methods in Hydraulics (HYDROCOMP '89)*. Elsevier, Amsterdam. pp. 237–242.
Hervouet, J.-M. 1993. 'Validating the numerical simulation of dam-breaks and floods', *Adv. Hydrosci. Engng*, **1A**, 754–761.
Hillel, D. 1980. *Fundamentals of Soil Physics*. Academic Press, New York. 413 pp.
Imhoff, K. R., Koppe, P., and Dietz, F. 1980. 'Heavy metals in the Ruhr River and their budget in the catchment area', *Prog. Wat. Technol.*, **12**, 735–749.
Jackson, R. D. 1972. 'On the calculation of hydraulic conductivity', *Proc. Soil Soc. Am.*, **36**, 380–382.
Jain, C. K. 1996. 'Application of chemical mass balance to upstream/downstream river monitoring data', *J. Hydrol.*, **182**, 105–115.
Lambert, C. P. and Walling, D. E. 1987. 'Floodplain sedimentation: a preliminary investigation of contemporary deposition within the lower reaches of the River Culm', Devon, UK', *Geograf. Annal.*, **69A**, 393–404.
Lane, S. N. 1998. 'Hydraulic modelling in hydrology and geomorphology: a review of high resolution approaches', *Hydrol. Process.*, **12**, 1131–1150 (this issue).
Leenaers, H. 1989. *The Dispersal of Metal Mining Wastes in the Catchment of the River Geul (Belgium — The Netherlands)*. Geografische Institut, Amsterdam, Rijksuniversitat Utrecht.
Lewin, J. and Macklin, M. G. 1987. 'Metal mining and floodplain sedimentation in Britain', in Gardiner, V. (Ed.), *International Geomorphology*, 1986, Part I. John Wiley & Sons Ltd., Chichester. pp. 1009–1027.
Lewin, J., Bradley, S. B., and Macklin, M. G. 1983. 'Historical valley alluviation in mid-Wales', *Geol. J.*, **18**, 331–350.
Macklin, M. G. 1996. 'Fluxes and storage of sediment-associated heavy metals in floodplain systems: assessment and river basin management issues at a time of rapid environmental change', in Anderson, M. G., Walling, D. E., and Bates, P. D. (Eds), *Floodplain Processes*. John Wiley & Sons, Ltd., Chichester. pp. 441–460.
Macklin, M. G. and Dowsett, R. B. 1989. 'The chemical and physical speciation of trace metals in fine grained overbank flood sediments in the Tyne Basin, north-east England', *Catena*, **16**, 135–151.
Macklin, M. G. and Lewin, J. 1989. 'Sediment transfer and transformation of an alluvial valley floor: the river South Tyne, Northumbria, UK', *Earth Surf. Process. Landf.*, **14**, 233–246.

Macklin, M. G., Ridgway, J., Passmore, D. G., and Rumsby, B. T. 1994. 'The use of overbank sediment for geochemical mapping and contamination assessment: results from selected English and Welsh floodplains', *Appl. Geochem.*, **9**, 689–700.

Marchuk, G. I. 1975. *Method of Numerical Mathematics*. Springer-Verlag, New York.

Marcus, W. A. 1987. 'Copper dispersion in ephemeral stream sediments', *Earth Surf. Process. Landf.*, **12**, 217–228.

Marron, D. C. 1992. 'Floodplain storage of mine tailings in the Belle Fourche river system: a sediment budget approach', *Earth Surf. Process. Landf.*, **17**, 675–685.

Millington, R. J. and Quirk, J. P. 1959. 'Permeability of porous media', *Nature*, **183**, 387–388.

Moulin, C. and Ben Slama, E. 1998. 'The two-dimensional transport module SUBIEF. Applications to sediment transport and water quality processes', *Hydrol. Process.*, **12**, 1183–1195 (this issue).

Moulin, C. and Gailhard, J. 1996. *SUBIEF V3P1 Users Manual, Report HE-43/95/074/B*. Laboratoire National d'Hydraulique, Direction des Etudes et Recherches, Electricite de France, Paris.

Pizzuto, J. E. 1987. 'Sediment diffusion during overbank flows', *Sedimentology*, **34**, 301–317.

Price, D. A. 1997. 'An integrated approach to modelling floodpain hydraulics, hydrology, and nitrate chemistry', *PhD Thesis*, University of Bristol, Bristol, 250 pp.

Ross, S. M. 1994. 'Sources and forms of potentially toxic soil–plant systems', in Ross, S. M. (Ed.), *Toxic Metals in Soil–Plant Systems*. John Wiley & Sons, Ltd., Chichester, U.K. pp. 3–25.

Taylor, M. P. 1996. 'The variability of heavy metals in floodplain sediment: a case study from mid Wales', *Catena*, **28**, 71–87.

Walling, D. E., Bradley, S. B., and Lambert, C. P. 1986. 'Conveyance losses of suspended sediment within a flood plain system', in Hadley, R. F. (Ed.), *Drainage Basin Sediment Delivery, IAHS Publ.*, **159**, 119–131.

Walling, D. E., Quine, T. A., and He, Q. 1992. 'Investigating contemporary rates of floodplain sedimentation', in Carling, P. A. and Petts, G. E. (Eds), *Lowland Floodplain Rivers*. John Wiley & Sons, Ltd., Chichester. pp. 165–184.

Webb, B. W. and Walling, D. E. 1985. 'Nitrate behaviour in streamflow from a grassland catchment in Devon, UK', *Wat. Res.*, **19**, 1005–1016.

Wolfenden, P. J. and Lewin, J. 1978. 'Distribution of metal pollutants in active stream sediments', *Catena*, **5**, 67–78.

Yalin, M. S. 1977. *Mechanics of Sediment Transport*, 2nd edn. Pergamon Press, Oxford, U.K.

13

KINEMATIC WAVE MODELLING OF VERTICAL MOVEMENT OF SNOWMELT WATER THROUGH A SNOWPACK

V. P. SINGH

Department of Civil and Environmental Engineering, Louisiana State University, Baton Rouge, LA 70803-6405, USA

L. BENGTSSON

Department of Water Resources Engineering, Lund University, Box 118, S-221 00 Lund, Sweden

AND

GÖRAN WESTERSTROM

Department of Water Resources Engineering, Luleå University of Technology, S-95187 Luleå, Sweden

ABSTRACT

Vertical movement of snowmelt water through snowpacks is modelled by applying the kinematic wave theory. Analytical solutions are obtained for moisture flux, particle velocity, time history and velocity of meltwater front and total moisture content for a single melt event assuming that the melt rate is constant. These solutions are extended to the case involving more than one event.

KEY WORDS kinematic wave theory; snowmelt movement; independent wetting events

INTRODUCTION

In northern and alpine environments, snowmelt runoff is a major contributer to both the annual maximum instantaneous discharge and the annual flow (Marsh and Woo, 1985). The main factors affecting snowmelt runoff include surface melt, snow metamorphosis, movement of water through the snow, interaction of the meltwater with the underlying soil and lateral flow at the base of the snowpack. Of these factors, the vertical movement of meltwater occupies an important position in the runoff-generation process.

In recent years, much attention has been focused on the various aspects of the vertical movement of snowmelt water. These comprise measurements of the rate of water movement through the snow cover (Wakahama, 1968a), the presence of flow channels in the snow (Wakahama, 1968b; Colbeck, 1978; Wankiewicz, 1978a, b) and the influence of ice layers on flow (Colbeck, 1973a; Langham, 1974a, b). Stratigraphic features of a snowpack vary greatly in space and time.

An accurate determination of the time distribution of spring runoff from snow-covered areas is needed for flood forecasting, water supply forecasting and the operation of regulatory hydraulic works. Fundamental to this determination is the prediction of the movement of snowmelt water. As a result, a number of prediction models have been developed.

Perhaps the earliest attempt to develop a theory of vertical percolation of water in snow was made by Colbeck (1971, 1972). By reviewing the general theory of Darcian flow, he showed that if water saturation were in the mid-range capillary forces were negligible (Colbeck, 1974). He then postulated that for snow the permeability to the water phase increased as a function of water saturation and the porosity was assumed to decrease linearly with depth. This formed the basis for application of the kinematic wave theory in subsequent studies. Colbeck (1973) investigated the effects of stratigraphic layers on water percolation and showed that during intense melting ice layers could be treated as porous media.

Colbeck and Davidson (1972, 1973) found a power-law dependence of relative permeability on liquid water saturation for describing water flow through homogeneous, repacked snow. The relative permeability of the snow to the liquid increased with the liquid saturation raised to the power of about 3. This relationship was investigated experimentally by Denoth *et al.* (1978) for snow having various grain sizes, densities and porosities, so determining the upper and lower limits for applicability of the kinematic wave theory. Colbeck and Anderson (1982) used data for lysimeters in California and Vermont to verify further the permeability–saturation relationship and the accuracy of the kinematic wave theory.

In a subsequent study, Colbeck (1974() constructed a two-layer model consisting of vertical flow through an unsaturated layer and flow along a boundary in a saturated layer. This was kinematic wave model. He also described the case of steady flow. Colbeck (1975) then developed a theory of water flow through a layered snowpack, which he represented in terms of an equivalent anisotropic porous medium. Ice layers lengthen the transit time for water movement by a factor equal to the ratio of the principal values of the permeability tensor. Ice layers rapidly disintegrate when significant quantities of liquid water are present (Gerdel, 1954), because of the thermodynamic interaction between water, air and ice (Colbeck, 1973b).

Colbeck (1975b, 1976) extended his kinematic wave treatment of ripe snow covers to dry snow covers. The flow through ripe snow covers is complicated by the diverting effects of individual layers, but simplified by the quasi-stable nature of uniform ice grains of 2 to 3 mm diameter at 0°C. Three requirements must be met for meltwater propagation. First, the snow cover temperature must be raised to 0°C through the supply of enough water. Secondly, the residual water requirement must be fulfilled. Thirdly, additional water must move. These requirements have the effect of delaying and reducing runoff due to rain on dry snow.

Colbeck (1977) incorporated some of the kinematic wave theory-based results in developing short-term forecasting of water runoff from snow and ice. In a similar vein, Dunne *et al.* (1976) employed these results to compute hydrographs generated by diurnal waves of snowmelt on the tundra and in the boreal forest of sub-arctic Labrador. In 1978 Colbeck provided an excellent review and discussion of the physical aspects of water flow through snow. Colbeck (1979) extended the kinematic wave theory for homogeneous and uniform snow to include the effects of ice layers and channels. His results suggested that most of the water moved down flow channels.

Jordan (1983a, b) developed a numerical model for describing the one-dimensional flow of meltwater in a deep snowpack based on unsaturated groundwater flow theory, and verified it using field observations. Although his model was a diffusion type, his results showed that the kinematic wave theory would be reasonably accurate. His observed kinematic wave speeds were similar to those reported by Wankiewicz (1978a) and Colbeck (1977). Akan (1984) simulated runoff from snow-covered hill slopes using a physically based model. He showed that springtime diurnal changes may influence the entire depth of a wet snowpack.

Bengtsson (1981) investigated snowmelt-generated runoff from small areas as a daily transient process using both the diffusion wave and kinematic wave theories. He found that the refreezing phenomena significantly influenced the saturation conditions in the top layer of the snow cover. Bengtsson (1982a, b) further analysed the hydraulics of meltwater percolation through a snowpack using the kinematic wave theory. He showed that even after moderately cold night it took several hours of snowmelt before the snowpack again reached its irreducible water content.

Bengtsson (1988) applied the kinematic wave theory to determine movement of meltwater in small basins. He computed moisture flux and particle velocities and compared them with observations of tracer-marked meltwater particles. He found that meltwater particles that reached the groundwater not very close to the ground surface did not contribute to runoff until several months after the melt event, and that the meltwater contributed significantly to the melt runoff only when the saturated overland flow occurred. Bengtsson *et al.* (1992) investigated the pathways of meltwater from a subdrained agricultural basin in southern Finland using oxygen-18 as tracer. They found that meltwater percolated through cracks in the ploughed, frozen soil and through regular spaced vertical sand-gravel drains. Bengtsson (1986) discussed the time resolution for snowmelt computations in order to calculate accurately runoff peaks and the resolution that can be obtained from different snowmelt models.

In a series of papers, Marsh and Woo (1984a, b, 1985) investigated the movement of meltwater in snowpacks. Using field observations they showed that in a naturally stratified snow cover the movement of meltwater was complicated by the interaction of the wetting front with stratigraphic horizons. When the wetting front from meltwater reached the premelt stratigraphic horizons, water ponded at the interface and flow fingers formed and penetrated the lower stratum. The flux in these fingers fed water to the impending horizons where it froze to form ice layers. These ice layers grew only at the stratigraphic boundaries and retarded the advance of the wetting front. Using field data, interactions between the processes of wetting front advance, freezing of meltwater, ice-layer growth and meltwater infiltration into the underlying frozen soil were incorporated into a simulation model (Kane, 1980). The variability of flow within a natural snow cover and its relationship with the physical properties of snow cover and melt rate was then described, and a multiple flow path model based on the kinematic wave theory was developed (Tucker and Colbeck, 1977). The model was successfully tested in the Arctic.

The above review shows that the kinematic wave theory has been applied and verified using field observations in numerous studies, and does seem to provide, under certain conditions, a sufficiently accurate representation of the movement of snowmelt water in snowpacks. However, a systematic, analytical treatment of snowmelt water movement using this theory is lacking. The aim of this paper is to develop this treatment and offer, in turn, additional insights into the process of water movement. The treatment is confined to one-dimensional vertical movement.

KINEMATIC WAVE FORMULATION

We consider vertical percolation in a homogeneous wet snowpack. The equations governing water percolation are the continuity equation and a flux law. By analogy with flow through unsaturated porous media, the continuity equation can be expressed as

$$\frac{\partial \theta}{\partial t} + \frac{\partial q}{\partial z} = 0 \tag{1}$$

where θ is the volumetric moisture (liquid water) content (dimensionless, volume of water per unit volume of snow), q is the volume flux of moisture (L/T), t is the time and z is the depth (L) measured as positive, downwards from the snow surface.

The flux law is given by the kinematic wave hypothesis:

$$q = q(\theta, z) \text{ or } \theta = \theta(q, z) \tag{2}$$

Based on the pioneering work of Colbeck (1972), Equation (2) can be written as

$$q = K \text{ or } u = \frac{q}{\theta} = \frac{K}{\theta} \tag{3}$$

where K is the unsaturated hydraulic conductivity and u is the average velocity of flow. For the liquid water content prevalent in snow under most natural conditions, the relationship between hydraulic conductivity and liquid water content can be expressed as a power law (Wankiewicz, 1976). It is, however, more convenient to employ relative saturation S_r, instead of the liquid water content θ, defined as

$$S_r = \frac{\theta - \theta_i}{n - \theta_i} \tag{4}$$

where θ_i is the irreducible water content, or water content held by capillary and surface forces even after snow has been allowed to drain for a prolonged period, and n is the porosity of snow. The term $n - \theta_i$ is frequently referred to as the effective porosity n_e. The value θ_i can be fixed as the origin for calculation of θ. Therefore, $\theta = 0$ will imply $\theta = \theta_i$ in the following discussion. Equation (4) expresses the proportion of the pore volume occupied by mobile water, and is also written (Colbeck and Anderson, 1982) as

$$S_e = \frac{S_w - S_i}{1 - S_i} \tag{5}$$

where $S_w = \theta/n$, $S_i = \theta_i/n$, and $S_e = S_r$ (also called the effective saturation).

The power law relation for hydraulic conductivity under drainage conditions can be written as

$$K = K_s \left(\frac{\theta - \theta_i}{n - \theta_i} \right)^a = K_s (S_r)^a, a \neq 0 \tag{6}$$

where K_s is the saturated hydraulic conductivity and a is an exponent. Wankiewicz (1976) suggested values of 3.3–3.5 for a, and Colbeck and Davidson (1973) used a value of 3. It may be pertinent to note that K can also be expressed in terms of the capillary pressure head, ψ, as

$$K = K_s \left(\frac{\psi_a}{\psi} \right)^b \tag{7}$$

where ψ_a is the air-entry capillary pressure for the boundary drying curve (which approximately equals the height of the capillary fringe above a water table) and b is an exponent. Wankiewicz (1976) suggested values of 13–14 for b. From Equations (4), (6) and (7) it follows that

$$S_r = \left(\frac{\psi_a}{\psi} \right)^c, c = b/a \tag{8}$$

Wankiewicz (1976) suggested a value of 4 for c, and his results lead to $\psi_a \approx -40$ to -30 mm (Jordan, 1983a). Thus, the kinematic flux law is prescribed by equating Equation (3) to Equation (6):

$$q = K_s \left(\frac{\theta - \theta_i}{n - \theta_i} \right)^a \tag{9}$$

Equations (1) and (9) are the kinematic wave equations whose solution is $q(z, t)$ or $\theta(z, t)$ in the z–t plane. In order to obtain a unique solution of these equations, initial and boundary conditions are specified. An initial condition is taken as one of uniform saturation throughout the snow (Colbeck, 1979). This can be expressed as θ_0 and will correspond to the initial flux of water q_0. Thus, the initial discharge from the column of snow is q_0. Expressed mathematically, the initial condition is

$$\theta(z, o) = \theta_0 \text{ or } q(z, o) = q_0 \tag{10}$$

The boundary condition is prescribed at the surface of a snowpack ($z = 0$) by

$$\theta(0, t) = N(t) \text{ or } q(0, t) = I(t), 0 < t \leqslant T \tag{11}$$

where T is some duration and $N(t)$ is an arbitrary moisture content as a function of time. Likewise, being related to $N(t)$ through Equation (9), $I(t)$ is an arbitrary time-dependent flux at the snow surface. These functions will correspond to the input of water at the snow surface in the form of rain or melt.

When water is introduced at the surface of a snowpack, it infiltrates the snowpack and begins to move vertically downwards, increasing the liquid water content. The downward movement of the liquid water content is given by Equation (1). If the snowpack is dry initially, the front of this downward movement defines the interface between the moist and dry parts of the snow matrix. This moving interface is the advance front (also called shock front or, simply, shock) and is not known *a priori*. This represents a discontinuity in the flow domain. Let the time history of this advance front, giving the space–time coordinates of the front, be $z = s(t)$ or $t = w(z)$. This front is a free boundary and has to be determined along with the solution. Following Sherman and Singh (1978), its equation can be deduced by observing from Equation (3) that $u(\theta) = u(\theta(z, t))$ and replacing z by $s(t)$ or t by $w(z)$. Accordingly,

$$\frac{ds(t)}{dt} = u(s(t), t) = \frac{K(\theta(s(t), t))}{\theta(s(t), t)}, s(0) = 0 \tag{12}$$

or

$$\frac{dw(z)}{dz} = [u(z, w, (t))]^{-1} = \frac{\theta(z, w(z))}{K(\theta(z, w(z)))}, w(0) = 0 \tag{13}$$

Equation (12) is valid for $\theta(z,t) > \theta_i$. Equation (12) or (13) requires the advance front to move with the same speed as that immediately behind the front. Note that this is an ordinary differential equation, whereas Equation (1) is a partial differential equation.

Thus, the problem of vertical movement of meltwater is formulated by Equations (1), (9) and (12), subject to Equations (10) and (11), and is well posed. This is a free boundary problem containing two unknowns, $\theta(z,t)$ and $s(t)$ or $w(z)$, that satisfy the partial differential Equation (1) and the ordinary differential Equation (12) or (13). Equation (1) is satisfied in the domain bounded by the positive t-axis and the curve $z = s(t)$ or $t = w(z)$.

CHARACTERISTIC SOLUTION FOR INDEPENDENT WETTING EVENTS

We will apply the method of characteristics to solve the problem formulated above. To that end, Equations (1) and (9) can be coupled as

$$\frac{\partial \theta}{\partial t} + \frac{\mathrm{d}K(\theta)}{\mathrm{d}\theta}\frac{\partial \theta}{\partial z} = 0 \tag{14}$$

where, for notational simplicity, we define

$$M = \frac{\mathrm{d}K}{\mathrm{d}\theta} = \frac{\partial\theta/\partial t}{\partial\theta/\partial z} = \frac{\partial z}{\partial t}\Big|_{\theta=\text{constant}} \tag{15}$$

The variable M defines the mobility of water in snow (Irmay, 1956) and is analogous to the wave speed in surface flow. Equation (14) can be expressed in characteristic form as

$$\frac{\mathrm{d}\theta}{\mathrm{d}t} = 0 \tag{16}$$

$$\frac{\mathrm{d}z}{\mathrm{d}t} = \frac{\mathrm{d}K(\theta)}{\mathrm{d}\theta} = M \tag{17}$$

According to Equation (16), θ is constant along the characteristics defined by Equation (17), which prescribes the characteristic speed or wave speed (or kinematic wave velocity). If Equation (6) is used for the $k - \theta$, function then Equation (17) takes the form

$$\frac{\mathrm{d}z}{\mathrm{d}t} = \frac{\mathrm{d}K(\theta)}{\mathrm{d}\theta} = M = \frac{aK_s}{n-\theta_i}\left(\frac{\theta-\theta_i}{n-\theta_i}\right)^{a-1} = \frac{aK_s}{n_e}(S_r)^{a-1} \tag{18}$$

This can be expressed in terms of the flux, q, as

$$\frac{\mathrm{d}z}{\mathrm{d}t} = M = \frac{a}{n_e}K_s^{1/a}q^{(a-1)/a} \tag{19}$$

which is the same as the expression derived by Colbeck and Davidson (1973) and is a direct consequence of applying the method of characteristics. In terms of the capillary pressure head, ψ,

$$\frac{\mathrm{d}z}{\mathrm{d}t} = M = \frac{aK_s}{n_e}\left(\frac{\psi_a}{\psi}\right)^{c(a-1)} \tag{20}$$

Wankiewicz (1976) introduced the concept of the fractional derivative of hydraulic conductivity with respect to the liquid water content, and Jordan (1983a, b) used this concept in his analysis. Accordingly,

$$r = \frac{1}{K}\frac{\mathrm{d}K}{\mathrm{d}\theta} = \frac{\alpha}{n_e}\left(\frac{K_s}{K}\right)^{1/a} \tag{21}$$

This is slowly varying function of K and is useful in that changes in liquid water content can be easily

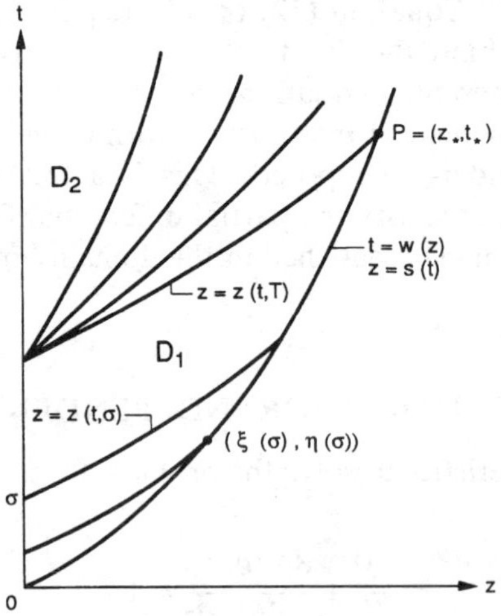

Figure 1. Solution domain for vertical flow through a snowpack

determined by measuring changes in density. Measurement of r for a given snow cover eliminates the need for direct measurement of the $K(S_r)$ relationship. Equation (18) can be written simply as

$$\frac{dz}{dt} = rq \tag{22}$$

Furthermore, Equations (5), (6) and (21) can be combined to produce

$$r = \frac{a}{n_e} = \frac{a}{\theta - \theta_i} \tag{23}$$

Thus, Equations (21)–(23) offer three different ways of determining r.

To solve Equations (16) and (17), we choose σ as the parameter on the t-axis and t as the parameter along the characteristics. The characteristics originate from the t-axis on the segment $0 \leqslant t \leqslant T$, as shown in figure 1. The parameter σ denotes the intersection of a characteristic $z(t, \sigma)$ with the t-axis. The solution domain can be divided into two subdomains D_1 and D_2, with $z(t, T)$ as the dividing line. In the (z, t, θ) space, the characteristics passing through the points $(0, \sigma, N(\sigma))$ satisfy (with initial and boundary conditions expressed in characteristic form)

$$\frac{d\theta(t,\sigma)}{dt} = 0, \theta(\sigma,\sigma) = N(\sigma) \tag{24}$$

$$\frac{dz(t,\sigma)}{dt} = \frac{dK(\theta(t,\sigma))}{d\theta(t,\sigma)} = M(\theta(t,\sigma)), z(\sigma,\sigma) = 0 \tag{25}$$

$z(t, \sigma)$ defines the position of water content θ at time t, which was at $z = 0$ and at $t = \sigma$; the corresponding water content at time t is $\theta(t, \sigma)$.

Solution for domain D_1

This domain is bounded by $0 \leqslant t \leqslant T$, $z(t, T)$ and $z = s(t)$. Solution of Equations (24) and (25) follows

$$\theta(t,\sigma) = N(\sigma) \tag{26}$$

$$z(t,\sigma) = M(\theta(t,\sigma))(t - \sigma) \tag{27}$$

Eliminating the parameter σ between Equations (26) and (27) gives

$$\theta(z,t) = N\left(t - \frac{z}{M}\right) \tag{28}$$

Because M is a function of θ, the solution given by Equation (28) is in implicit form. Alternatively, this equation can be expresses as

$$z - M(t - N^*) = 0, \sigma = N^*(\theta), N^* = N^{-1} \tag{29}$$

According to Equation (27), the characteristics are straight lines with the slope $M/\theta = \text{constant} = N(\sigma)^*$.

Depending upon the nature of the function $N(t)$, three possibilities can be identified: (1) $N(t) = N = \text{constant}$, (2) $\delta N/\delta t = N'(t) < 0$ and (3) $N'(t) > 0$. In the first case, the input due to rainfall or snowmelt as the snow surface is constant. In the second, it is a decreasing function of time, and it is an increasing function of time in the third case. To describe the variation of liquid water content in time t at a fixed z or in space at a fixed t, we may determine $\partial\theta/\partial z$ and $\partial\theta/\partial t$ from Equation (28). To that end, Equation (28) can be written as

$$F(z, t, \theta) = \theta - N\left(t - \frac{z}{M}\right) = 0 \tag{30}$$

By applying the implicit function theorem to Equation (30) we obtain

$$\frac{\delta\theta}{\delta z} = -\frac{\delta F/\delta z}{\delta F/\delta\theta} = \frac{-N'/M}{1 - N'(M'/M^2)z} \tag{31}$$

$$\frac{\delta\theta}{\delta t} = -\frac{\delta F/\delta t}{\delta F/\delta\theta} = \frac{-N}{1 - N'(M'/M^2)z} \tag{32}$$

where $N'(\sigma) = N' = \mathrm{d}N/\mathrm{d}\sigma$, and $M' = \mathrm{d}^2K/\mathrm{d}\theta^2 = \mathrm{d}M/\mathrm{d}\theta$. It can be shown that Equations (31) and (32) satisfy Equation (14) and can also be derived directly from Equations (26) and (27) as follows:

$$\frac{\delta\theta}{\delta z} = \frac{\delta\theta}{\delta\sigma}\cdot\frac{\delta\sigma}{\sigma z} = \frac{\delta\theta/\delta\sigma}{\delta z/\delta\sigma}; \frac{\delta\theta}{\delta t} = \frac{\delta\theta}{\delta t}\cdot\frac{\delta\sigma}{\delta t} = \frac{\delta\theta/\delta\sigma}{\delta t/\delta\sigma} \tag{33}$$

From Equation (26)

$$\frac{\delta\theta}{\delta\sigma} = N'(\sigma) \tag{34}$$

From Equation (27)

$$t = \sigma + \frac{z(t, \sigma|)}{M|_{\theta=N(\sigma)}} \tag{35}$$

$$\frac{\partial z}{\delta\sigma} = 1 - \frac{z}{M^2}M'N' \tag{36}$$

and

$$\frac{\delta z}{\delta\sigma} = -M + (t - \sigma)M'N' = -M + \frac{M'N'}{M}z \tag{37}$$

Therefore,

$$\frac{\delta\theta}{\delta z} = \frac{N'(\sigma)/M}{-1 + (M'/M^2)N'z}$$

and

$$\frac{\delta\theta}{\delta t} = \frac{N'(\sigma)}{1 - (M'/M^2)N'z}$$

which is the same as before (Equations (31) and (32)). We now examine the three cases.

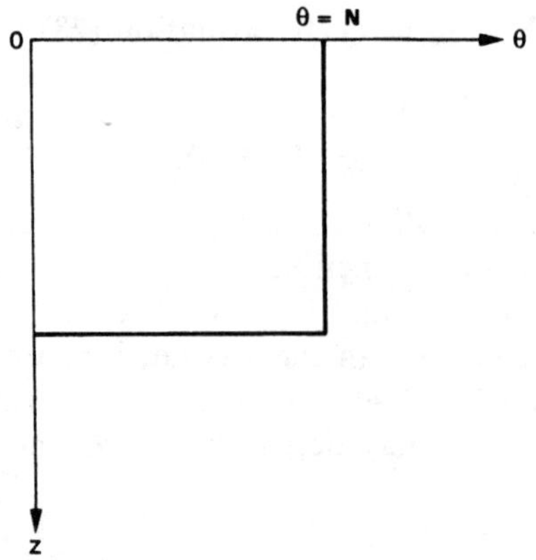

Figure 2. Moisture profile in domain D_1 for $N(t) = N$

(1) $N(t) = N = \text{constant}$

This gives $N(\sigma) = 0$, and $\delta\theta/\delta z = \delta\theta/\delta t = 0$. Therefore,

$$\theta(z, t) = \theta = N \tag{38}$$

The characteristics are parallel with common slope $S_0 = M(N)$ and constant moisture. The liquid water content will have the appearance shown in Figure 2. In this case the wave speed is a constant given by

$$M = \frac{aK_s}{n_e}\left(\frac{N - \theta_i}{n - \theta_i}\right)^a \tag{39}$$

and the flux is also a constant given by

$$q = K_s\left(\frac{N - \theta_i}{n - \theta_i}\right)^a \tag{40a}$$

The average velocity with which the liquid water moves is

$$u = \frac{K_s}{N - \theta_i}\left(\frac{N - \theta_i}{n - \theta_i}\right)^a = \frac{q}{N - \theta_i} \tag{40b}$$

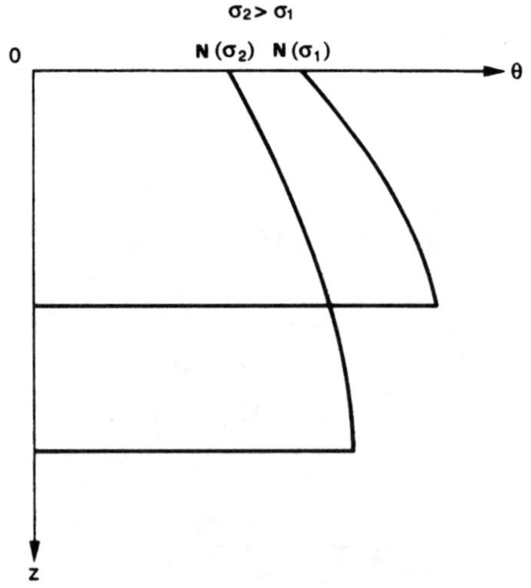

Figure 3. Moisture profile in domain D_1 for $N'(t) < 0$

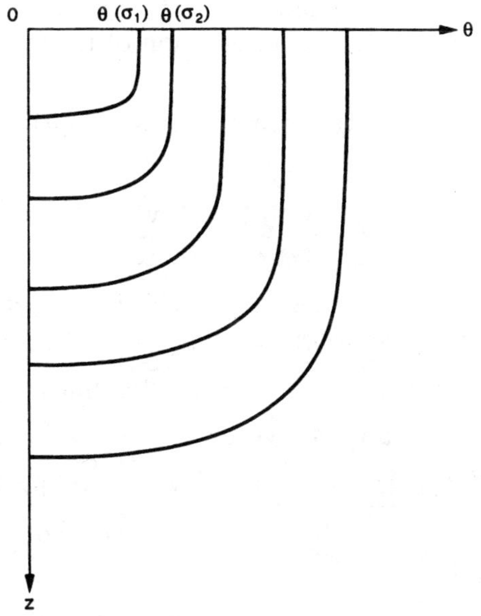

Figure 4. Moisture profile in domain D_1 for $N'(t) > 0$

(2) $N'(t) < 0$

This may correspond to the case of internal drainage following a wetting event. In this case, $\delta\theta/\delta z > 0$ and $\delta\theta/\delta t < 0$. The value of $\delta\theta/\delta z$ decreases with increasing z and the liquid water content spreads with depth, as shown in Figure 3. For a large value of z, Equation (31) asymptotically reduces to

$$\frac{\delta\theta}{\delta z} = \frac{M}{M'z}, \frac{\partial\theta}{\partial t} = -\frac{M^2}{M'z} \tag{41}$$

(3) $N'(t) > 0$

In this case, $\delta\theta/\delta z < 0$ and $\delta\theta/\delta t > 0$ for $z < M^2/(N'M')$. For fixed t, the moisture content decreases with depth, and for fixed z, t, increases with time as shown in Figure 4. When z increases, $\delta\theta/\delta z$ increases, approaching infinity at

$$z = \frac{M^2}{N'M'} \tag{42}$$

At this value of z, $\delta\theta/\delta t$ also approaches infinity. This marks the point of discontinuity and occurs at the location given by

$$z_{\mathrm{I}} = \frac{M^2(N(\sigma_{\mathrm{I}}))}{M'(N(\sigma_{\mathrm{I}}))N'(\sigma_{\mathrm{I}})} \tag{43}$$

$$t_{\mathrm{I}} = \sigma_{\mathrm{I}}\frac{M(N(\sigma_{\mathrm{I}}))}{M'(N(\sigma_{\mathrm{I}}))N'(\sigma_{\mathrm{I}})} \tag{44}$$

where σ_{I} is the time corresponding to the inflection point $(N(\sigma_{\mathrm{I}}), \sigma_{\mathrm{I}})$ of $N(\sigma)$. The discontinuity first occurs at the value of z corresponding to the maximum value of $N'(\sigma)$, which at the inflection point of $N(t)$. This discontinuity can also be called the wetting front.

Determination of the shock front (free boundary of domain D_1)

In order to determine the shock front, $t = w(z)$ or $z = s(t)$, it is convenient to express it in terms of the parameter σ introduced above. If we consider the characteristic curve $z(t, \sigma)$ in the z–t plane, then it intersects the free boundary at the point denoted by $(\xi(\sigma), \eta(\sigma))$ as shown in Figure 1. Therefore, $z = \xi(\sigma)$, $t = \eta(\sigma)$ constitutes the parametric representation of the free boundary in terms of the parameter σ. From Equation (27),

$$z(\eta(\sigma), \sigma) = \xi(\sigma) = M(N(\eta(\sigma), \sigma))[\eta(\sigma) - \sigma] \tag{45}$$

From Equations (12) and (26),

$$s'(t) = \frac{\xi'(\sigma)}{\eta'(\sigma)} = \frac{K(N(\sigma))}{N(\sigma)} \tag{46}$$

Equations (45) and (46) can be solved for $\xi(\sigma)$ and $\eta(\sigma)$, which, in turn, give the free boundary. To that end, Equation (45) is differentiated with respect to σ:

$$\xi'(\sigma) = M(\eta(\sigma), \sigma)[\eta'(\sigma) - 1] + [\eta(\sigma) - \sigma]M'(N(\sigma))N'(\sigma) \tag{47}$$

Substitution of Equation (46) into Equation (47) yields

$$\eta'(\sigma) - \frac{N(\sigma)M'(N(\sigma))N'(\sigma)}{K[N(\sigma)] - N(\sigma)M(\sigma)}\eta(\sigma) = -\frac{N(\sigma)[N(\sigma) + \sigma M'(N(\sigma))N'(\sigma)]}{K[N(\sigma)] - N(\sigma)M(\sigma)} \tag{48}$$

Solution of Equation (48) is given by

$$\eta(\sigma) = -\exp\left[\int_0^{\sigma} \frac{N(\sigma)M'(\sigma)}{K[N(\sigma)] - N(\sigma)M(\sigma)} \delta\sigma\right] \int_0^{\sigma} \frac{N(\sigma)[M(\sigma) - \sigma M'(\sigma)]}{K[N(\sigma)] - N(\sigma)M(\sigma)} \cdot IF_1(\sigma)\mathrm{d}\sigma \tag{49a}$$

where

$$IF_1(\sigma) = \exp\left[-\int \frac{N(\sigma)M'(\sigma)N'(\sigma)}{K[N(\sigma)] - N(\sigma)M(\sigma)} \mathrm{d}\sigma\right] \tag{49b}$$

Similarly, from equations (45) and (46), $\xi(\sigma)$ can be obtained. From Equation (46),

$$\eta'(\sigma) = \frac{N(\sigma)}{K(N'(\sigma))} \cdot \xi'(\sigma) \tag{50}$$

From Equation (45),

$$\eta'(\sigma) = \sigma + \frac{\xi(\sigma)}{M(N(\sigma))} \tag{51}$$

Differentiating Equation (51) gives

$$\eta'(\sigma) = 1 + \frac{\xi'(\sigma)}{M(N(\sigma))} - \frac{\xi(\sigma)}{M^2(N(\sigma))} \cdot M'(N(\sigma))N'(\sigma) \tag{52}$$

Eliminating $\eta'(\sigma)$ from Equations (50) and (52),

$$\xi'(\sigma) + \frac{M'(\sigma)}{M(\sigma)} \frac{K[N(\sigma)]}{N(\sigma)M(N(\sigma)) - K[N(\sigma)]}\xi(\sigma) = \frac{M(N(\sigma)K[N(\sigma)]}{N(\sigma)M(N(\sigma)) - K[N(\sigma)]} \tag{53}$$

Integration of Equation (53) yields

$$\xi(\sigma) = IF_2^{-1} \int_0^{\sigma} \frac{M[N(\sigma)K[N(\sigma)]}{N(\sigma)M(N(\sigma)) - K[N(\sigma)]} \cdot IF_2\mathrm{d}\sigma \tag{54a}$$

where

$$IF_2 = \exp\left[\int \frac{M'(N(\sigma))}{M(N(\sigma))} \cdot \frac{K[N(\sigma)]}{N(\sigma)M[N(\sigma)] - K[N(\sigma)]} \mathrm{d}\sigma\right] \tag{54b}$$

Equations (49) and (54) give a parametric representation of the free boundary. For the simple case, $N(t) = N =$ constant, $\theta =$ constant, which gives $K(\theta) = K =$ constant, $M(\theta) = M =$ constant, and

$M'(\theta) = 0$, Equation (49) yields

$$\eta(\sigma) = \frac{MN}{MN - K}\sigma \tag{55}$$

and Equation (54) yields

$$\xi(\sigma) = \frac{MK}{MN - K}\sigma \tag{56}$$

On eliminating σ between Equations (55) and (56), the equation for the free boundary becomes

$$z = \frac{K}{N}t \tag{57}$$

The liquid water content along the front is obtained by differentiating Equation (57) with respect to t and equating the derivative to Equation (12). Thus,

$$\frac{K}{N} = \frac{K}{\theta}, \text{or } \theta(s(t), t) = N \tag{58}$$

This shows that in the case $N(t) = N$, the liquid water along the front is the same as along the characteristics behind it.

To complete the solution for domain D_1, we have to determine the intersection of the bounding characteristic $z(t, T)$ and $z = s(t)$. Let that point of intersection be $p = (z^*, t^*)$. For the case $N(t) = N =$ constant, this point can be obtained explicitly. Combining Equation (26), with $\sigma = T$ and $z = z^*$, with Equation (57), with $z = z^*$, gives

$$\frac{Nz^*}{K} = T + \frac{z^*}{M} \tag{59}$$

which yields

$$z^* = \frac{KMT}{NM - K} \tag{60}$$

Similarly,

$$t^* = \frac{NMT}{NM - K} \tag{61}$$

In the present context, a simpler way, based on the volume balance, can be employed to determine the free boundary. Equation (28) gives

$$\theta(s(t), t) = N\left(t - \frac{s(t)}{M}\right) \tag{62}$$

Substituting Equation (62) into Equation (12) and integrating, gives

$$\int_0^{s(t)} \mathrm{d}s(t) = \int_0^t \frac{K(\theta(s(t), t))}{N\left(t - \frac{S(t)}{M}\right)} \mathrm{d}t \tag{63}$$

If $N(t) = N$, then

$$s(t) = z = \frac{K}{N}t \tag{64}$$

which is the same as Equation (57)

Shock velocity

The shock velocity is given by Equation (12) or (13) if the θ-profile is continuous in the z–t plane. In the simple case when $N(t) = N$, and $\theta =$ constant, the shock velocity is

$$u(s(t), t) = \frac{K(N)}{N - \theta_i} = \frac{K}{\theta - \theta_i} = \frac{q}{\theta - \theta_i} \tag{65}$$

which is the same as the velocity with which the liquid water content moves, as given by Equation (40b).

It may, however, be instructive to consider the nature of the shock when θ decreases with z and when θ increases with z and a discontinuity occurs in the θ-profile. We consider a region of snowpack with θ decreasing from θ_1 to θ_2, $\theta_1 < \theta_2$, with z; this corresponds to an advancing wave. For simplicity, let us take $\theta = \theta_1$ for $z < z_1$, and $\theta = \theta_2$ for $z > z_2$. The change in flux between θ_1 and θ_2 is

$$\Delta q = q(\theta_1) - q(\theta_2) \tag{66}$$

The θ-profile moves as a shock wave from z_1 to z_2 with velocity u_s equal to

$$u_s = \frac{q(\theta_1) - q(\theta_2)}{\theta_1 - \theta_2} \tag{67}$$

where $q(\theta_1)$ is a local maximum and $q(\theta_2)$ is a local minimum. Indeed, Equation (67) is a consequence of Equation (65). Because $q(\theta) = K(\theta)$ in our case, the shock front moves with velocity

$$u_s = \frac{K(\theta_1) - K(\theta_2)}{\theta_1 - \theta_2} \tag{68}$$

If, on the other hand, θ increases with z (trailing wave) then

$$u_s = \frac{dK(\theta)}{d\theta} = M(\theta) \tag{69}$$

will hold in the region $\theta_1 < \theta < \theta_2$. The net flux in this region is simply

$$\Delta q = q(\theta_2) - q(\theta_1) = K(\theta_2) - K(\theta_1) \tag{70}$$

Equation (67) is based on mass conservation. To that end, we integrate Equations (1) and (3) over z_1 to z_2 as applied to the front:

$$\frac{d}{dt}\int_{z_1}^{z_2} \theta dz + K|_{z_2} - K|_{z_1} = 0 \tag{71}$$

where $K|_{z_i}$ is the value of K at $z = z_i (i = 1, 2)$. To find the velocity of the front, the true moisture profile is replaced by a fictitious sharp front having the same total moisture, as shown in figure 5. Then the location

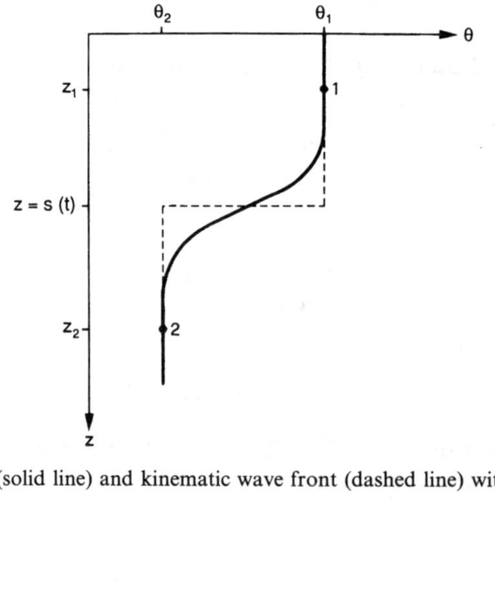

Figure 5. True wetting front (solid line) and kinematic wave front (dashed line) with the same moisture content

of sharp front $z = s(t)$ must satisfy

$$\int_{z_1}^{z_2} \theta \mathrm{d}z = \theta_1(s(t) - z_1) + \theta_2(s(t) - z_2) \tag{72}$$

Combining Equations (71) and (72), gives

$$\frac{\mathrm{d}s(t)}{\mathrm{d}t} = u_s = \frac{K_1 - K_2}{\theta_1 - \theta_2} \tag{73}$$

which is the same as Equation (68).

Solution for domain D_2

The domain D_2 is bounded by $T \leqslant t$, $z = z(t, T)$ and $z = (s(t)$ beyond the point $p = (z^*, t^*)$. In this case $N(t) = 0$, $t \geqslant T$. Let $N(t, \epsilon)$ be a continuous function of t coinciding with $N(t)$ on $0 \leqslant t \leqslant T$, $N(t, \epsilon) = 0$ on $t \geqslant T + \epsilon, \epsilon > 0$. It is seen from Equations (26) and (27) that the characteristics $z = z(t, \sigma, \epsilon)$, $T \leqslant \sigma \leqslant T + \epsilon$, cover the region above $z = z(t, T)$ completely. From these equations we get

$$M(t, \sigma, \epsilon) = \frac{z(t, \sigma, \epsilon)}{t - \sigma} \tag{74}$$

From equation (74) we derive, using $T \leqslant \sigma \leqslant T + \epsilon$,

$$\frac{z}{t - T} \leqslant M(z, t, \epsilon) \leqslant \frac{z}{t - T - \epsilon} \tag{75}$$

Letting $\epsilon \to 0$, we get, from Equation (75),

$$M(z, t) = \frac{z}{(t - T)} \quad \text{or } z = M(z, t)(t - T) \tag{76}$$

From Equation (18)

$$M(\theta) = \frac{aK_s}{n - \theta_i} \left(\frac{\theta - \theta_i}{n - \theta_i} \right)^{a-1}, \theta \leqslant \theta < N \tag{77}$$

Therefore,

$$z = \frac{aK_s}{n - \theta_i} \left(\frac{\theta - \theta_i}{n - \theta_i} \right)^{a-1} (t - T), \theta_i \leqslant \theta < N \tag{78}$$

or

$$\theta = \theta_i + (n - \theta_i) \left[\frac{(n - \theta_i)z}{aK_s(t - T)} \right]^{1/(a-1)}, \theta_i \leqslant \theta < N \tag{79}$$

The liquid water content flux, q, is given by

$$q(z, t) = \frac{z(\theta - \theta_i)}{a(t - T)}, t > T \tag{80a}$$

and the average flow velocity by

$$u(z, t) = \frac{z}{a}(t - T) \tag{80b}$$

In this domain, the characteristics all originate from the point $z = 0$ and $t = T$ and are straight lines with the slope $M(\theta(z, t))$. These characteristics form a fan shape. During the period $T \leqslant \sigma \leqslant T + \epsilon$, Equation (27), with θ replaced by N, still holds. Referring to Figure 6, when $\theta = N$, Equation (78) represents the line AB; when $\theta = \theta_1$, it represents the line AC. For $z < s(t)$, $\theta = \theta_i$, For $\theta_i < \theta < \theta = N$, the characteristics issue from within the angle ABC. These lines cover the domain D_2 in which the liquid water content

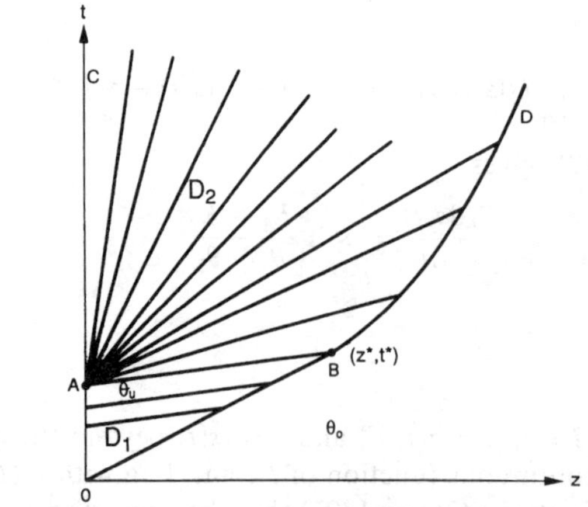

Figure 6. Solution domain or the case $\theta(0, t) = \theta,\ 0 < t < T$, showing wetting front, plateau and internal draining

depends on both z and t. The slope of these lines is given by Equation (77), with θ varying from one characteristic to the other. Along a given characteristic the liquid water content is constant. The line AB carries the liquid water content $\theta = N$. The value of θ decreases over to line AC, which carries the liquid water content $\theta = \theta_i$. This characteristic, corresponding to $\theta = \theta_i$, never leaves the snowpack surface, for $M(\theta_i) = 0$. The region BAC represents the area of internal drainage. The characteristic corresponding to $\theta_i + \Delta\theta$, where $\Delta\theta$ is a small quantity, does have a finite slope equal to $M = aK_s[\Delta\theta/(n - \theta_i]^{a-1}/(n - \theta_i)$, so the profile will eventually drain arbitrarily close to θ_i.

Free boundary of domain D_2

The curve OBD (Figure 6) marks the wetting front, part of which remains to be determined. This free boundary beyond the point p is obtained using Equations (49) and (54).

$$\frac{\xi(\sigma,\epsilon)IF_2(\sigma,\epsilon)}{\int_0^{\sigma} \frac{M(\sigma,\epsilon)K[N(\sigma,\epsilon)]}{N(\sigma,\epsilon)M(\sigma,\epsilon)-K[N(\sigma,\epsilon)]} IF_2(\sigma,\epsilon)\mathrm{d}\sigma} = \frac{\eta(\sigma,\epsilon)IF_1(\sigma,\epsilon)}{\int_0^{\sigma} \frac{N(\sigma,\epsilon)[M(\sigma,\epsilon)-\sigma M'(\sigma,\epsilon)]}{N(\sigma,\epsilon)M(\sigma,\epsilon)-K[N(\sigma,\epsilon)]} IF_1(\sigma\epsilon)\mathrm{d}\sigma} \tag{81a}$$

where $T \leqslant \sigma \leqslant T + \epsilon$. Letting $\epsilon \to 0$, Equation (81a) becomes

$$\frac{zIF_2(T)}{\int_0^{T} \frac{M(\sigma)K[N(\sigma)]}{N(\sigma)M(\sigma)-k[N(\sigma)]} IF_2(\sigma)\mathrm{d}\sigma} = \frac{tIF_1(T)}{\frac{\int_0^{T} N(\sigma)[M(\sigma)-\sigma M'(\sigma)]}{N(\sigma)M(\sigma)-K[N(\sigma)]}} IF_1(\sigma)\mathrm{d}\alpha \tag{81b}$$

this gives the free boundary for domain D_2.

Alternatively, the free boundary segment BD can be obtained using mass conservation. on each side of the line OB, given by Equation (57) if $N(t) = N$, both the liquid water fluxes and the liquid water contents are constant. Therefore, mass conservation leads to

$$\int_0^{s(t)} \theta - \theta_i)\mathrm{d}z = \int_0^{T} q(\theta(t))\mathrm{d}t + q(N)T \qquad (82$$

where θ is given by Equation (79). Solution of Equation (82) yields

$$s(t) = z = \frac{aK_s}{N - \theta_i}\left[\frac{N - \theta_i}{n - \theta_i}\right]^a \left[\frac{T}{a-1}\right](t - T)^{1/a} \tag{83}$$

which specifies the free boundary beyond the point p in the case $N(t) = N$.

Turning our attention to the liquid water profile during and following moisture application at the snowpack surface we refer to Figure 4. At $0 < t < T$, the moisture suddenly rises from θ_i to N. This wetting front

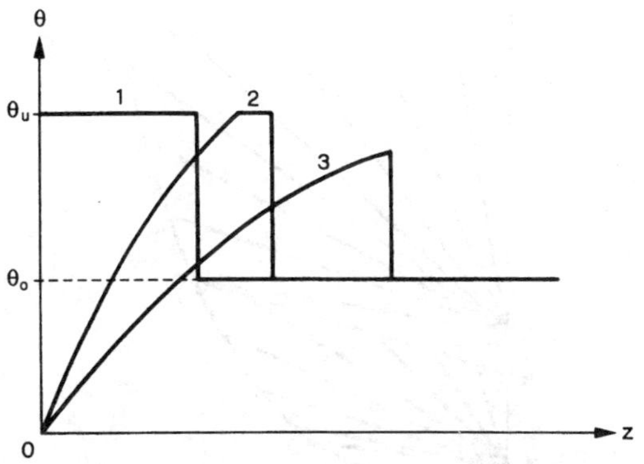

Figure 7. Moisture profiles during and following a wetting event, m

moves downwards separating the two regions of constant liquid water contents, with N behind and θ_i ahead of it. For $T < t < t^*$, the liquid water profile shows the region of internal drainage behind the wetting front and the zone of constant liquid water content. For $t > t^*$, the profile exhibits only the region of internal drainage behind the wetting front. Figure 7 shows these three profiles.

Total liquid water content

Since the liquid water content varies within the snowpack, it is useful to determine the total liquid water content as a function of time and depth. This can be achieved by integrating the moisture profile from $\theta = 0$ to a particular depth, say, L. Let $L \geqslant z^*$. At $t = 0$, the initial total liquid water content, W, is $W = \theta_i L$. For $0 < t < T$, say $t = t_1$, $W = Nz_1 + \theta_i(L - z_1)$, where z_1 is the intersection point of $t = t_1$, and $z = s(t)$. For $t > t^*$,

$$w(t, L) = \int_0^L \theta(z, t)\mathrm{d}z \tag{84}$$

where $\theta(z, t)$ is prescribed by Equation (79). Solution of this integral gives

$$W(t, L) = \theta_i L + \frac{a-1}{a}(n - \theta_i)\left[\frac{N - \theta_i}{aK_s(t - T)}\right]^{1/(a-1)} L^{a/(a-1)} \tag{85}$$

The above solutions for domains D_1 and D_2 determine completely the space–time history of liquid water content or liquid water flux. If a plane is cut across the t-axis in the characteristic diagram then we obtain the moisture content as a function of z at the time of cut. Likewise, taking a plane across the z-axis yields the temporal variation of the liquid water content at the position of cut. The characteristic diagram can also be transformed directly, by use of $K(\theta)$, to the moisture flux diagram, and, by dividing $q(\theta)$ by θ, to the

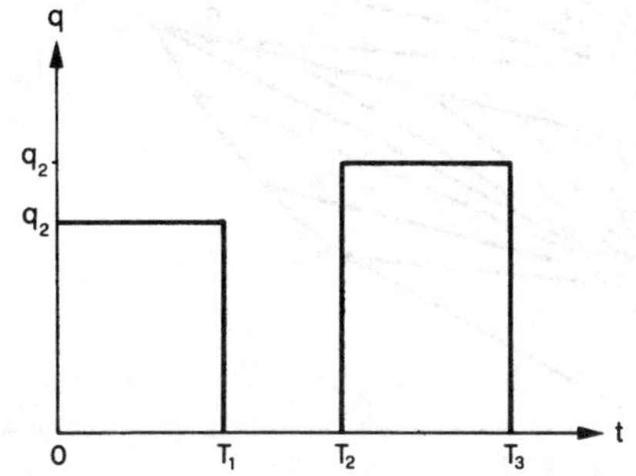

Figure 8. A sequence of two wetting events

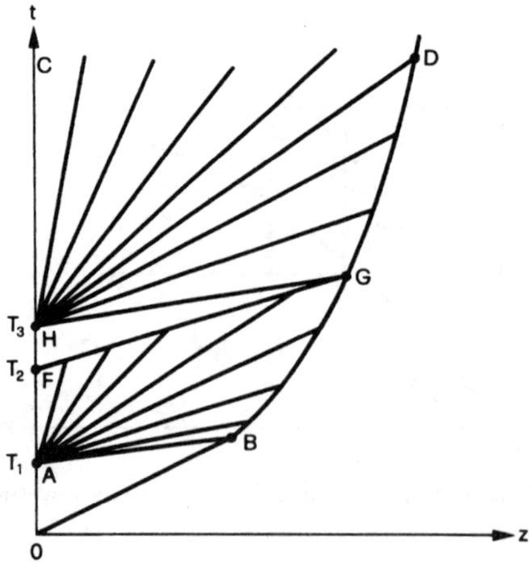

Figure 9. Solution domain for a sequence of two wetting events: one possibility

percolation velocity diagram. Integration of the moisture profiles from the snow cover surface to a particular depth yields the total water content within the cover up to that depth. When this is done at different times, temporal variation of total moisture content is obtained.

FLOW DUE TO COMPLEX SURFACE WETTING

The preceding discussion, developed for independent wetting event is extended to dependent wetting events, with the stipulation that during each event the liquid water flux at the snowpack surface is constant. The liquid water content at the surface is an arbitrary sequence of step functions. Whenever the flux increases at the snow surface, a wetting front originates there. When the surface flux decreases, a region of internal drainage occurs.

Two cases are distinguished for complex surface wetting due to dependent wetting events. These are illustrated by consideration of a sequence of two dependent events as shown in Figure 8, where the duration of

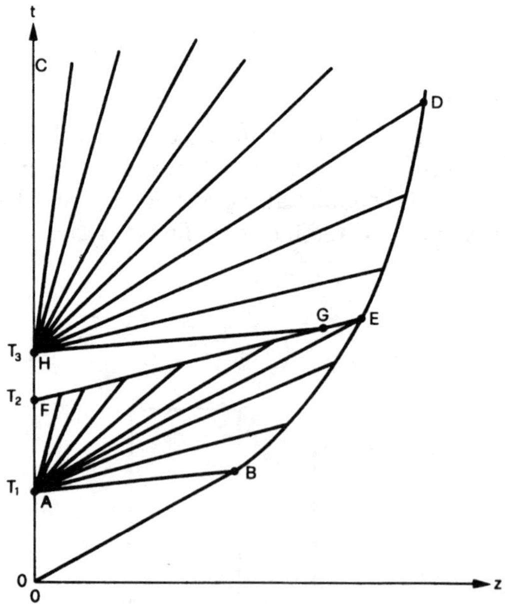

Figure 10. Solution domain for a sequence of two wetting events: another possibility

the first event is T, and the duration of the second event is $T_3 - T_2$. The two events are separated by a dry period of $T_2 - T_1$. The liquid water content is considered to be θ_i initially. Depending upon the length of the dry period and the duration of the second event, two cases can arise, as shown in Figures 9 and 10. In the first case, the initial wetting front is OBGD, the moisture plateau is OAB and the zone of internal drainage is BAFG for the first event. The second event begins at point F and ceases at point H. The second wetting front leaves the surface at $t = T_3$ and intersects the first wetting front at point G. The region FHG is a second region of constant liquid water content corresponding to the second event (F to H), overlapping a region of internal drainage associated with the first event. The region GHC is a region of internal drainage.

The second case is shown in Figure 10. The difference between this case and the first case lies in the structure of the second wetting front. If the duration of the second event is quite small, the second wetting front FG, as shown in the figure, may intersect the characteristics originating at $t = T_3$ (at point H). In that case, G to E represents a common internal drainage boundary for the two events. This discussion, although for two dependent events, does show all the features of the most general case and can, therefore, be easily extended to a sequence of any number of dependent events.

The solution for vertical flow due to complex wetting can be derived as follows. Referring to figure 9, Equation (26) gives the liquid water content in the region OAB, Equation (57) specifies the wetting from OB and Equation (83) gives the wetting front BG or BE. For the second event, the wetting front FG is obtained by applying Equation (67). To that end, we invoke the hydraulic conductivity relation defined by Equation (6). For notational simplicity, let

$$A = \frac{K_s}{(N - \theta_i)^a}$$

so Equation (6) can be given as

$$K(\theta) = A(\theta - \theta_i)^a \tag{86}$$

In order to apply Equation (67) we need to specify θ_1 and θ_2. Here, $\theta_2 = \theta^*$, and θ_1 is given by Equation (79) which can be given as

$$\theta_1 = \theta_i + \left[\frac{z}{aA(t - T_1)}\right]^{1/(a-1)} \tag{87}$$

Hence,

$$K(\theta_1) = A\left[\frac{z}{aA(t - T_1)}\right]^{a/(a-1)} \tag{88}$$

$$K(\theta_2) = K(\theta^*) = A(\theta^* - \theta_i)^a \tag{89}$$

This leads to

$$\frac{dz}{dt} = \frac{A\left[\frac{z}{aA(t-T_1)}\right]^{a/(a-1)} - A(\theta^* - \theta_i)^a}{\left[\frac{z}{aA(t-T_1)}\right]^{1/(a-1)} - (\theta^* - \theta_1)} \tag{90}$$

Let $B = (Aa)^{1/(a-1)}(\theta^* - \theta_i)$ and $y = (z/(t - T_1))^{1/(a-1)}$. Then

$$\frac{dz}{dt} = y^{a-1} + (a-1)(t - T)y^{a-2}\frac{dy}{dt} \tag{91}$$

Introducing Equation (91) and the above quantities, Equation (90) is transformed into

$$\frac{dt}{t - T_1} = \frac{a(a-1)(y - B)dy}{aBy - (a-1)y^2 - B^a y^{a-2}} \tag{92}$$

This can be integrated to yield

$$t = T_1 + (T_2 - T_1)\exp\left[\int_0^Y \frac{a(a-1)(y-B)}{aBy - (a-1)y^2 - B^a y^{a-2}}\mathrm{d}y\right] \tag{93}$$

Equation (93) gives the wetting front FG. For integral values of a, Equation (93) may be solved analytically; otherwise, standard numerical techniques may be used.

The segment GE in Figure 10 marks a wetting front with the second region of internal drainage overlapping the first. This can be derived, again, by applying Equation (67). For the first region of internal drainage, θ_1 and $K(\theta_1)$ are given by Equations (87) and (88). For the second region of internal drainage,

$$\theta_2 = \theta_i + \left[\frac{z}{aA(t-T_3)}\right]^{1/(a-1)} \tag{94}$$

and

$$K(\theta_2) = A\left[\frac{z}{aA(t-T_3)}\right]^{1/(a-1)} \tag{95}$$

Therefore,

$$\frac{\mathrm{d}z}{\mathrm{d}t} = \frac{1}{a}\frac{\left[\frac{z}{(t-T_1)}\right]^{1/(a-1)} - \left[\frac{z}{(t-T_3)}\right]^{1/(a-1)}}{\left[\frac{z}{(t-T_1)}\right]^{1/(a-1)} - \left[\frac{z}{(t-T_3)}\right]^{1/(a-1)}}\mathrm{d}z \tag{96}$$

Its integration follows

$$z = z_G \exp\left[\frac{1}{a}\int_{t_G}^{t} \frac{\left(\frac{1}{t-T_1}\right)^{1/(a-1)} \left(\frac{1}{t-T_3}\right)^{1/(a-1)}}{\left(\frac{1}{t-T_1}\right)^{1/(a-1)} - \left(\frac{1}{t-T_3}\right)^{1/(a-1)}}\mathrm{d}t\right] \tag{97}$$

where $(z_G,\ t_G)$ define the coordinates of the point G, which marks the initial point on this segment of the wetting front. The liquid water content in the second region of internal drainage is given by Equation (79) with T replaced by T_3.

CONCLUDING REMARKS

Vertical percolation through a snowpack can be formulated as a free boundary problem. Analytical solutions of this problem can be derived under fairly general conditions, and closed form solutions under the assumption that the melt rate at the snowpack surface due to an independent melting event is constant. These encompass algebraic equations for moisture content, capillary potential, flux, velocity, front velocity and the space–time history of the front (or free boundary). These solutions can be gainfully applied to snowmelt runoff forecasting. For complex melting conditions, numerical solutions are the only option, but they are simple and efficient.

REFERENCES

Akan, A. O. 1984. 'Simulation of runoff from snow-covered hillslopes', *Wat. Resour. Res.*, **20** (6), 707–713.

Bengtsson, L. 1981. 'Snowmelt generated runoff from small areas as daily transient process', *Geophysica*, **17** (1–2), 109–121.

Bengtsson, L. 1982a. 'Percolation of meltwater through a snowpack', *Cold Regions Sci. Technol.* **6,** 73–81.

Bengtsson, L. 1982b. 'A laboratory experiment on the effect of refreezing on the percolation of meltwater through a snowpack', *Research Report Series A, No. 11*, Luleå University, Water Resources Engineering Luleå, Sweden 26 pp.

Bengtsson, L. 1986 'Snowmelt simulation models in relation to space and time', *Proceedings, Modelling Snowmelt-Induced Processes. IAHS Publication No. 155*, pp. 115–123.

Bengtsson, L. 1988. 'Movement of meltwater in small basins', *Nord. Hydrol.*, **19,** 237–244.
Bengtsson, L., Seuna, P., Lepisto, A., and Saxena, R. K. 1992. 'Particle movement of meltwater in a subdrained agricultural basin', *J. Hydrol.*, **135,** 383–398.
Colbeck, S. C. 1971. 'One dimensional water flow through snow', *Research Report 296*, U.S. Army Cold Regions Research and Engineering Laboratory, Hanover, New Hampshire.
Colbeck, S. C. 1972. 'A theory of water percolation in snow', *J. Glaciol.*, **11,** 369–385.
Colbeck, S. C. 1973a. 'Effects of stratigraphic layers on water flow through snow', *Research Report 311*, U.S. Army Cold Regions Research and Engineering Laboratory, Hanover, New Hampshire. 13 pp.
Colbeck, S. C. 1973b. 'A theory of metamorphism of wet snow', *Research Report 313*, U.S. Army Cold Regions Research and Engineering Laboratory, Hanover, New Hampshire.
Colbeck, S. C. 1974. 'The capillary effects of water percolation in homogeneous snow', *J. Glaciol.*, **13,** 85–97.
Colbeck, S. C. 1975a. 'A theory of water flow through a layered snowpack', *Wat. Resour. Res.*, **11,** 261–266.
Colbeck, S. C. 1975b. 'Analysis of hydrologic response to rain-on-snow', *Research Report 340*, U.S. Army Cold Research and Engineering Laboratory, Hanover, New Hampshire.
Colbeck, S. C. 1976. 'An analysis of water flow in dry snow', *Wat. Resour. Res.*, **12** (1), 523–527.
Colbeck, S. C. 1977. 'Short-term forecasting of water runoff from snow and ice', *J. Glaciol.*, **19,** 571–587.
Colbeck, S. C. 1978. 'The physical aspects of water flow through snow', in Chow, V. T. (Ed.), *Advances in Hydroscience*, Vol. 11. Academic Press, New York. pp. 165–206.
Colbeck, S. C. 1979. 'Water flow through heterogeneous snow', *Cold Regions Sci. Technol.*, **1,** 37–45.
Colbeck, S. C., and Anderson, E. A. 1982. 'The permeability of a melting snow cover', *Wat. Resour. Res.*, **18** (4), 904–908.
Colbeck, S. C., and Davidson, G. 1972. 'Water percolation through homogeneous snow', in *The Role of Snow and Ice in Hydrology*, IASH–AISH, Gentbrygge, Belgium. pp. 242–247.
Colbeck, S. C., and Davidson, G. 1973. 'Water percolation through homogeneous snow', in *The Role of Ice and Snow in Hydrology, Proceedings of the Banff Symposium*, IASH–AISH Publication 107, pp. 242–257.
Denoth, A., Seidenbusch, W., Blumthaler, M., and Kirchlechner. 1978. 'Some experimental data on water percolation through homogeneous snow', in Colbeck, S. C. and Ray, M. (Eds), *Proceedings, Modeling of Snow Cover Runoff*, U.S. Army Cold Regions Research and Engineering Laboratory, Hanover, New Hampshire. pp. 253–256.
Dunne, T., Price, A. G., and Colbeck, S. C. 1976. 'The generation of runoff from subarctic snowpacks', *Wat. Resour. Res.*, **12** (4), 677–685.
Gerdel, R. W. 1965. 'The transmission of water through snow', *Trans. Am. Geophys. Union*, **35,** 475–485.
Irmay, S. 1956. 'Extension of Darcy law to unsteady unsaturated flow through porous media', *Proceedings of the IASH–IUGG Darch Symposium*, Dijon, France.
Jordan, P. 1983a. 'Meltwater movement in a deep snowpack, 1. Field observations', *Wat. Resour. Res.*, **19** (4), 971–78.
Jordan, P. 1983b. 'Meltwater movement in a deep snowpack, 2. Simulation model', *Wat. Resour. Res.*, **19** (4), 979–985.
Kane, D. L. 1980. 'Snowmelt infiltration into seasonally frozen soils', *Cold Regions Sci. Technology*, **3,** 153–161.
Langham, E. J. 1974a. 'Phase equilibria of veins in polycrystalline ice', *Can. J. Earth Sci.*, **11,** 1280–1287.
Langham, E. J. 1974b. 'The mechanism of rotting of ice layers within a structural snowpack', *Snow Mechanics, Publication 114*, IASH–AISH, Gentbrygge, Belgium. pp. 773–81.
Marsh, P. and Woo, M. K. 1984a. 'Wetting front advance and freezing of meltwater within a snow cover, 1. Observations in the Canadian Arctic', *Wat. Resour. Res.*, **20**(12), 1853–1864.
Marsh, P., and Woo, M. K. 1984b. 'Wetting front advance and freezing of meltwater within a snow cover, 2. A simulating model', *Wat. Resour. Res.*, **20**(12), 1865–1874.
Marsh, P., and Woo, M. K. 1985. 'Meltwater movement in natural heterogeneous snow covers', *Wat. Resour. Res.*, **21**(11), 1710–1716.
Sherman, B., and Singh, V. P. 1978. 'A kinematic model for surface irrigation', *Wat. Resour. Res.*, **14**(2), 357–363.
Tucker, W. B., and Colbeck, S. C. 1977. 'A computer routing of unsaturated flow through snow', *Special Report 77–10*, U.S. Army Cold Regions Research and engineering Laboratory, Hanover, New Hampshire. 39 pp.
Wakahama, G. 1968a. 'Infiltration of meltwater into snow cover, III. Flowing down speed of meltwater via snow cover', *Low Temp. Sci., Ser A*, **26,** 85–86.
Wakahama, G. 1968b. 'The metamorphism of wet snow', *Publication 79*, IASH–AISH, General Assembly of Bern, Switzerland. pp. 370–379.
Wankiewicz, A. 1976. 'Water percolation within a deep snowpack: Field investigation at a site on Mt. Seymour, British Columbia. *Ph.D. thesis*, University of British Columbia, Vancouver. 177 pp.
Wankiewicz, A. 1978a. 'A review of water movement in snow', in Colbeck, S. C. and Ray, M. (Eds), *Proceedings, Modeling Snow Cover Runoff*, U.S. Army Cold Regions Research and Engineering Laboratory, Hanover, New Hampshire. pp. 222–252.
Wankiewicz, A. 1978b. 'Water pressure in ripe snow packs', *Wat. Resour. Res.*, **14,** 593–599.

14

KINEMATIC WAVE MODELLING OF SATURATED BASAL FLOW IN A SNOWPACK

V. P. SINGH

Department of Civil and Environmental Engineering, Louisiana State University, Baton Rouge, LA 70803-6405, USA

L. BENGTSSON

Department of Water Resources Engineering, Lund University, Box 118, S-221 00 Lund, Sweden

AND

GÖRAN WESTERSTROM

Department of Water Resources Engineering, Luleå University of Technology, S-95187 Luleå, Sweden

ABSTRACT

Movement of snowmelt water through a thin saturated layer at the infiltrating base of a snowpack is modelled by applying the kinematic wave theory. Analytical solutions are obtained for flow depth, velocity and discharge assuming that the rate of input to the saturated layer due to vertical percolation is constant. This assumption results in a linear rise and recession of the snowmelt hydrograph. The solutions are extended to the case of time-varying input. An explicit consideration of infiltration leads to a free boundary problem.

KEY WORDS kinematic wave theory; snowmelt hydrograph; snowpack basal flow

INTRODUCTION

When the meltwater reaches the base of the snowpack, a thin saturated layer forms over the ground. The runoff in the saturated layer takes place if the percolation rate exceeds the infiltration rate (Bengtsson, 1981). The vertical flow from the unsaturated zone provides the boundary condition for computation of flow along the saturated basal layer. Colbeck (1974) found that the ground slope was much larger than the slope of the thickness of the saturated layer.

Based on measurements in Luleå, Sweden, Bengtsson (1981) found that no runoff took place at the beginning of the melting period, but almost all the meltwater ran off along the ground during the peak of the snowmelt period, when the soil was saturated to the field capacity. At a field site, Fujino (1971) found the speed of diurnal meltwater waves moving through the unsaturated zone to be about one-eighth the speed of wave movement along the base of the snow cover. Thus, the meltwater may spend much less time in this layer than it does in the unsaturated zone. However, the distance to be travelled by the meltwater in this layer is usually much larger than the snowpack thickness.

Colbeck (1974) was probably the first to develop a theory of water flow through snow overlying an impermeable barrier. Based on hydraulic reasoning, he developed a kinematic wave formulation for basal saturated flow. Since the snowpack was resting on an impermeable base, infiltration was not an issue in his analysis. A detailed discussion of this formulation is given in Colbeck (1978). The same formulation has been included by Colbeck (1977) in a model for short-term forecasting of water runoff from snow and ice. Dunne *et al.* (1976) included this formulation in their physically based model of the movement of water through snowpacks. Bengtsson (1981) also employed a similar formulation in his study of snowmelt-generated runoff from small areas in Sweden. On the other hand, Akan (1984) employed a Darcian flow representation, including infiltration. Since this was a numerical formulation, the effect of infiltration on either the formulation or the results was not shown. Kirnbauer *et al.* (1994) presented a review of distributed snowmelt runoff models.

The aim of this paper is to develop a kinematic wave model incorporating infiltration explicitly. Explicit solutions will be derived for the case when infiltration is constant. An explicit inclusion of infiltration results in a free boundary problem that does not appear to have been addressed in the context of snowmelt runoff.

KINEMATIC WAVE FORMULATION FOR SATURATED FLOW ON AN INFILTRATING BASE

For saturated basal flow, with infiltration into the underlying soil, the continuity equation can be expressed as

$$n_e \frac{\partial h}{\partial t} + \frac{\partial}{\partial x}(uh) = I(x,t) - f(x,t) \tag{1}$$

where h is the thickness of the saturated layer measured perpendicular to the ground surface, n_e is the effective porosity, u is the volumetric flux in the layer per unit cross-sectional area, I is the lateral inflow rate to the layer at its top for a duration T, f is the lateral outflow rate or infiltration rate into the underlying surface, x is the coordinate along the base of the snowpack (downwards positive), and t is time. Following Colbeck (1974), the kinematic flux law can be written as

$$u = KS \tag{2}$$

where S is the angle of slope (assumed to be small) and K is the saturated hydraulic conductivity. The flux, Q, per unit width is

$$Q = uh = KSh \tag{3}$$

Equations (1) and (3) can be combined to yield a single equation that, for a constant slope, is:

$$\frac{\partial h}{\partial t} + \frac{KS}{n_e}\frac{\partial h}{\partial x} = \frac{1}{n_e}[I(x,t) - f(x,t)] \tag{4}$$

It should be noted that the saturated layer thickness, h, varies in space (with x) and time (with t) in response to the inflow of water coming from the unsaturated zone above.

The filtration function, f, is dependent on the thickness of the saturated layer in the following sense:

$$f(x,t) > 0 \text{ if } h(x,t) > 0$$
$$f(x,t) = p \text{ if } h(x,t) = 0 \tag{5}$$

When snowmelt begins, the rate of vertical percolation may be smaller than the rate at which the underlying soil layer can absorb water. As snowmelt continues, the infiltration rate declines, and eventually the melt rate exceeds the infiltration rate and generates runoff. This means that there is some time lag between the commencement of runoff and the arrival of snowmelt water at the layer. In this study, time is measured with respect to this time lag.

It is further assumed that

$$I(x,t) > f(x,t), 0 \le t \le T, 0 \le x \le L$$

where L is the length of the saturated layer.

The following boundary and initial conditions can be taken as:

$$h(0,t) = h_0(t) \tag{6}$$
$$h(x,0 = h_1(x) \tag{7}$$

The problem of saturated basal flow reduces to the solution of Equation (4) subject to Equations (6) and (7).

It is plausible, on physical grounds, that there will be a curve $t = t^0(x)$ in $\{t \le T, 0 \le x \le L\}$ starting at $x = 0, t = T$, and such that $h(x, t^0(x)) = 0$. This curve gives the time history of the water edge as it recedes from $x = 0$ to $x = L$. Equation (4) is satisfied in the region $R \equiv \{0 < t < t^0(x), 0 < x < L\}$. Thus, $t^0(x)$ is a free boundary, and Equations (4)–(7), and $h(x, t^0(x)) = 0$, constitute a free boundary problem. The free boundary partitions the layer into wetted and non-wetted portions. In the domain curve $t = t^0(x)$; therefore, $h(x, t) = 0$. The determination of the free boundary is relatively simple when I and f are constant.

CHARACTERISTIC SOLUTION

Equation (4), subject to Equations (6) and (7), can be solved by the method of characteristics. The solution is then the surface $h(x, t)$ formed by all the characteristic curves passing through the segment $t = 0, 0 \leq x \leq L$ and the segment $x = 0, 0 \leq t \leq T$. The free boundary $t = t^0(x)$ is the locus $h(x, t) = 0$ in the x–t plane. If we take x as parameter then the characteristic form of Equation (4) is expressed as

$$\frac{\mathrm{d}t}{\mathrm{d}x} = \frac{n_e}{KS} \tag{8}$$

$$\frac{\mathrm{d}h}{\mathrm{d}x} = \frac{I(x,t) - f(x,t)}{KS} \tag{9}$$

The initial and boundary conditions in characteristic form become

$$t(x_0) = o, h(x_0) = h_1(x_0), 0 \leq x_0 \leq L \text{ (initial)} \tag{10}$$

$$t(0) = t_0, h(t_0) = h_0(t_0), 0 \leq t_0 \leq T \text{ (boundary)} \tag{11}$$

in which x_0 and t_0 are the points of intersection of characteristics with the x-axis and t-axis, respectively. It is assumed that the curves $t = t(x, t_0)$, which are solutions of Equations (8), (9) and (11), do not intersect for distant values of t_0. Likewise, the curves $t = t(x, x_0)$, which are solutions of Equations (8), (9) and (10), do not intersect for distinct values of x_0. For the simple case $h_1(x_0) = 0 = h_0(t_0) = 0$, this can be shown to be true, the proof is omitted here.

Depending upon the relative disposition of the bounding characteristic $t = t(x, 0)$, the line $t = T$ and the curve $t = t^0(x)$, three cases can be distinguished: A, B_1 and B_2. Cases B_1 and B_2 are parts of case B. Case A corresponds to the equilibrium situation and cases B_1 and B_2 to the partial equilibrium situation. A detailed discussion of these cases is given in Singh (1990).

Case A. $t^0(x) > T > t(x, 0),\ 0 < x < L$. This is an equilibrium case and is shown in Figure 1.

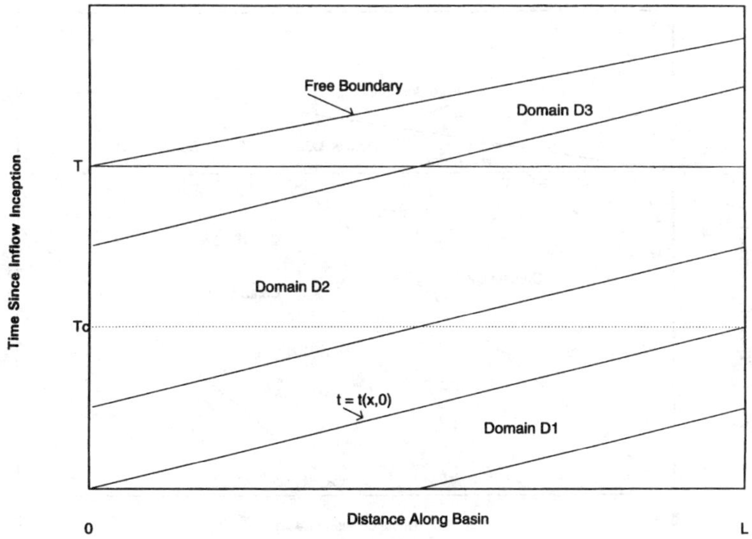

Figure 1. Solution domain for case A: equilibrium hydrograph

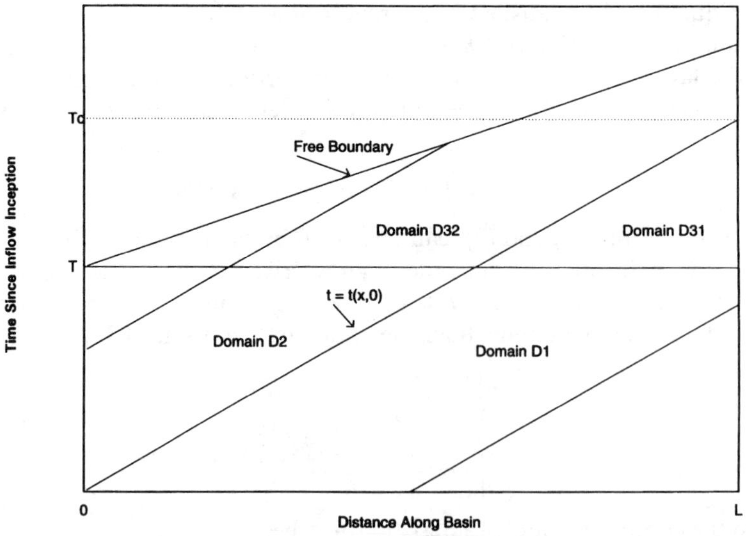

Figure 2. Solution domain for case B_1: partial equilibrium hydrograph

Case B_1. $t^0(x) > T$ and $t^0(x) > t(x, 0)$, but $t = T$ and $t = t(x, 0)$ intersect at $x = x^*$, i.e. $T = t(x, 0)$ and $0 < x^* < L$. This is a partial equilibrium case and is depicted in Figure 2.

Case B_2. $t^0(x) > T$, but $t = T$ and $t = t(x, 0)$ intersects at $x = x^*$ and $t = t^0(x)$ and $t = t(x, x^*)$ intersect at $x = \bar{x}$, i.e. $t^0(\bar{x}) = t(\bar{x}, x^*)$ and $0 < \bar{x} < L$. This is a partial equilibrium case and is shown in Figure 3.

Since $t^0(x)$ and $t(x, 0)$ are not known until we have solved the problem, it appears we cannot distinguish these cases beforehand. In the special case, however, where $I(x, t)$ and $f(x, t)$ are both constant, these cases can be distinguished beforehand.

SOLUTION FOR CASE A: EQUILIBRIUM HYDROGRAPH

The solution domain for this case, as shown in Figure 1, can be partitioned into domains D_1, D_2 and D_3. Domain D_1 is bounded by $t = 0$, $x = L$ and $t = t(x, 0)$; $t = t(x, 0)$ is the bounding characteristic. Domain

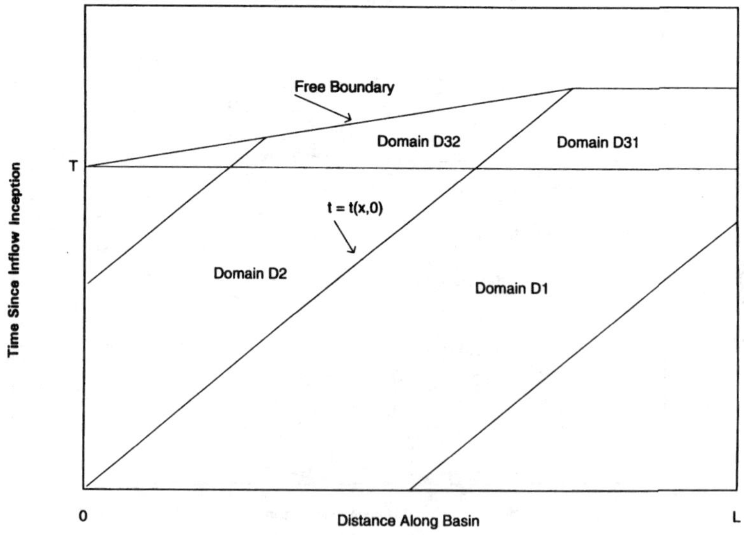

Figure 3. Solution domain for case B_2: partial equilibrium hydrograph

D_2 is bounded by $x = 0$, $t = T$, $x = L$, and $t(x, 0)$; and domain D_3 by $x = 0$, $t = T$, $x = L$ and $t = t^0(x)$. Domain D_1 corresponds to the rising limb, D_2 to the hydrograph crest segment and D_3 to the recession limb of the snowmelt-runoff hydrograph.

Domain D_1. Let $x_0, O \leq x_0 \leq L$, be the parameter representing the intersection of a characteristic curve with the x-axis. The solution surface can be expressed as $t = t(x, x_0), h = h(x, x_0)$, and $x = x$. The initial condition is given by Equation (10). Solution of Equation (9), subject to this condition, is

$$h(x, x_0) = h_1(x_0) + \frac{1}{KS}\int_{x_0}^{x} [I(x, t(x, x_0)) - f(x, t(x, x_0))]\mathrm{d}x \tag{12}$$

and that of Equation (8) is

$$t(x, x_0) = \frac{n_e}{KS}\int_{x_0}^{x} \mathrm{d}x = \frac{n_e}{KS}(x - x_0) \tag{13}$$

The parameter x_0 can be eliminated from Equations (12) and (13), and, in turn, h can be expressed in terms of x and t. The characteristics, given by Equation (13), are straight lines with a slope of $n_e/(KS)$. If I and f are approximated only as functions of time then Equation (12) becomes

$$h(x, x_0) = h_1(x_0) + \frac{1}{KS}\int_{x_0}^{x} [I(x, t(x, x_0)) - f(t(x, x_0))\mathrm{d}x \tag{14}$$

Under the assumption that I and f are constant, Equation (14) has an analytical solution:

$$h(x, x_0) = h_1(x_0) + \frac{1}{KS}(I - f)(x - x_0) \tag{15}$$

The assumption of constant I and f is not unduly restrictive during prolonged snowmelt periods. Equations (13) and (15) can be coupled to eliminate x_0:

$$h(x, t) = h_1(x_0) + \frac{1}{N_e}(I - f)t \tag{16}$$

If $h_1(x_0) = 0$,

$$h(x, t) = \frac{1}{n_e}(I - f)t \tag{17}$$

Equation (17) shows a linear dependence of the saturated layer thickness on time only. This condition assumes an initial depth of zero. The flow is uniform but unsteady, and changes with time as $(I - f)/n_e > 0$. The flux, Q, in this case is obtained using Equation (3) as

$$Q = \frac{KS}{n_e}(I - f)t \tag{18}$$

This states that Q varies linearly with time, with a slope of $KS(I - f)/n_e$.

Domain D_2. Let $t_0, 0 \leq t_0 \leq T$, be the parameter representing the intersection of a characteristic curve with the t-axis. The solution surface is then expressed in terms of x and t_0 as $h(x, t_0)$, $t = t(x, t_0)$ and $x = x$. The boundary condition is given by Equation (11). Then, the solution of Equation (9) is

$$h(x, t_0) = h_0(t_0) + \frac{1}{KS}\int_{0}^{x} [I(x, t(x, t_0)) - f(x, t(x, t_0))]\mathrm{d}x \tag{19}$$

and that of Equation (8) is

$$t(x, t_0) = t_0 + \frac{n_e}{KS}\int_{0}^{x} \mathrm{d}x = t_0 + \frac{n_e x}{KS} \tag{20}$$

By eliminating t_0 between Equations (19) and (20), h can be expressed as a function of x and t. The characteristics, given by Equation (20), are straight lines with a slope of $n_e/(KS)$. If I and f are t-dependent only, then Equation (19) takes the form

$$h(x, t_0) = h_0(t_0) + \frac{1}{KS}\int_0^x [I(t(x, t_0)) - f(t(x, t_0))]\mathrm{d}x \tag{21}$$

If I and f are constant, then

$$h(x, t_0) = h_0(t_0) + \frac{1}{KS}(I - f)x \tag{22}$$

If the upstream boundary is a vanishing one, then

$$h(x, t_0) = \frac{x}{KS}(I - f) \tag{23}$$

This condition holds when snowpack rests on a steeply sloping base. Equations (20) and (23) are uncoupled, and h is independent of t_0 and hence independent of t. The thickness varies linearly with a slope of $(I - f)/(KS)$. The flux in this domain, under this assumption, is

$$Q = (I - f)x \tag{24}$$

which shows a linear dependence of Q on x with a slope of $(I - f)$.

Domain D_3. In this domain, lateral inflow $I(x, t) = 0$ but lateral outflow $f(x, t) > 0$. Let $x_0^*, 0 \leq x_0^*, < L$, be the parameter representing the intersection of a characteristic $t = t(x, x_0^*)$ with the line $t = T$; $t = t(x, x_0^*)$ is an elongation of $t(x, t_0)$ beyond $t = T$. Therefore, Equations (8) and (9) are solved subject to

$$t(x_0^*) = T \tag{25}$$

$$h(x_0^*) = h(x_0^*, T) \tag{26}$$

The term $H(x_0^*)$ is obtained from Equations (19) and (20) with x replaced by x_0^*. Thus, the solution in domain D_3 can be expressed as $t = t(x, x_0^*)$, $h = h(x, x_0^*)$ and $x = x$. The solution of Equation (9) is

$$h(x, x_0^*) = h(x_0^*) - \frac{1}{KS}\int_{x_0^*}^x f(x, t(x, x_0^*))\mathrm{d}x \tag{27}$$

and that of Equation (8) is

$$t(x, x_0^*) = T + \frac{n_e}{KS}\int_{x_0}^x \mathrm{d}x = T + \frac{n_e}{KS}(x - x_0^*) \tag{28}$$

The parameter x_0^* can be eliminated between Equations (27) and (28), and h can be expressed as a function of x and t. Equation (28) shows that the characteristics are straight lines with a slope of $n_e/(KS)$. If f is only t-dependent then Equation (27) becomes

$$h(x, x_0^*) = h(x_0^*) - \frac{1}{KS}\int_{x_0^*}^x f(t(x, x_0^*))\mathrm{d}x \tag{29}$$

If f is a constant then Equation (29) has the solution:

$$h(x, x_0^*) = h(x_0^*) = \frac{f}{KS}(x - x_0^*) \tag{30}$$

where $h(x_0^*)$ from Equation (23) is

$$h(x_0^*) = \frac{x_0^*}{KS}\cdot(I - f) \tag{31}$$

Therefore, Equation (30) becomes

$$h(x, x_0^*) = \frac{x_0^*}{KS}(I - f) - \frac{f}{KS}(x - x_0^*) = \frac{x_0^* I - fx}{KS} \tag{32}$$

Eliminating x_0^* between Equations (28) and (32) yields h as a function of x and t:

$$h = \frac{IT}{n_e} + \frac{(I-f)}{KS}x - \frac{I}{n_e}t \tag{33}$$

The time at which the thickness h tends to vanish at any location is given, from Equation (33), as

$$t = T + \frac{n_e(I-f)}{KSI}x \tag{34}$$

The saturated layer at any location x decreases in thickness linearly in time with a slope of I/n_e. The flux in this domain is given by

$$Q = KS\frac{TI}{n_e} + (I-f)x - \frac{IKS}{n_e}t \tag{35}$$

which shows a linear dependence of Q in time at any location.

Maximum thickness and maximum discharge. For the case where I and f are constant, an analytical determination of these maxima is straightforward. The maxima occur at the end of domain D_1, or during domain D_2. To that end, the time to the maximum thickness for any value of x is given by Equation (13) with $x_0 = 0$:

$$t_{max}(x) = \frac{n_e}{KS}x \tag{36}$$

The maximum thickness, h_{max}, is obtained from Equation (15):

$$h_{max}(x) = h_1(0) + \frac{(I-f)}{KS}x \tag{37}$$

and the maximum flux is given as

$$Q_{max}(x) = KSh_1(0) + (I-f)x \tag{38}$$

Determination of free boundary $t = t^0(x)$. The free boundary $t = t^0(x)$ is determined by putting $h(x, x_0^*) = 0$ in domain D_3. Its determination is straightforward for the case where I and f are constant. To that end, Equation (32) gives

$$x_0^* I - fx = 0 \text{ or } x_0^* = \frac{fx}{I} = \psi(x) \tag{39}$$

Eliminating x_0^* between Equations (28) and (39) yields the free boundary equation:

$$t^0(x) = T + \frac{n_e(I-f)}{KSI}x \tag{40}$$

which shows that the curve $t = t^0(x)$ is a straight line with a slope of $n_e(I-f)/(KSI)$.

Summary of case A. For a fixed value of x, a thickness graph, showing the variation of h as a function of time, can be constructed, and by varying x a system of such hydrographs can be sketched. Figure 4 is a sketch of such a system for various values of x. In the figure, $\bar{t}(x_i)$ is the time taken by water to travel from the upper end of the layer to the location x_i. Figure 5 shows a three-dimensional view of the depth hydrograph.

Solution for cases B_1 and B_2. Partial equilibrium hydrograph. The solution domain for these cases, as shown in Figures 2 and 3, can be partitioned into domains D_1, D_2 and D_3, composed of D_{31} and D_{32}. Domains D_1 and D_2 are defined as before. Domain D_3 is partitioned into D_{31} and D_{32}. Domain D_{32} is bounded by $t = t(x, x^*)$, $t = T$, $t = t^0(x)$ and $x = L$; whereas domain D_{31} is bounded by $x = x^*$, $t = T$, $t = t(x, x^*)$ and $x = L$; the curve $t = t(x, x^*)$ is an elongation of the bounding water characteristic $t = t(x, 0)$ beyond $t = T$ until it intersects the line $x = L$.

Domain D_1. The solution is given by Equations (12) and (13).

Domain D_2. The solution is given by Equations (19) and (20).

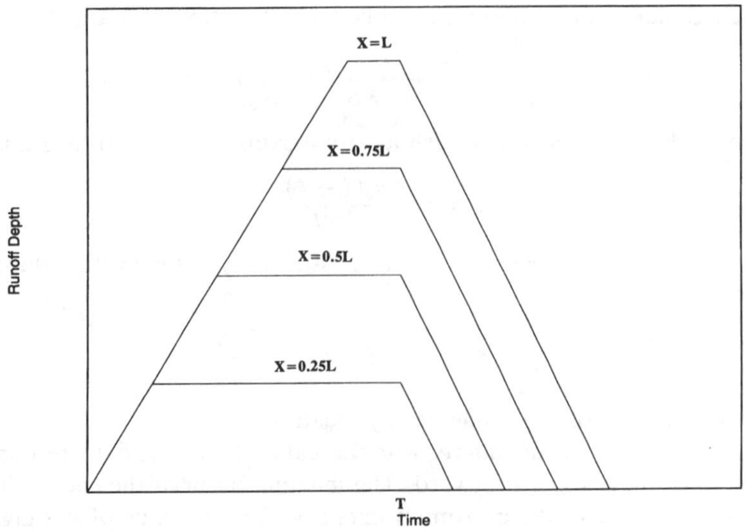

Figure 4. A system of hydrographs at various values of x in the equilibrium case, A

Domain D_3. The solution in domain D_{31} is given by Equations (26) and (27) for $0 < x < \bar{x}$. It only remains to determine the solution in Domain D_{32}. We define x_0^* by $T = t(x_0, x_0^*)$; here $x^* \leq x_0^* \leq L$. Thus, from Equation (13),

$$T(x_0, x_0^*) = \frac{n_e}{KS}(x_0^* - x_0) \quad (41)$$

The boundary condition for this subdomain is

$$h(x_0, x_0^*) = h(x_0^*) = h_1(x_0) + \frac{(I - f)}{KS}(x_0^* - x_0); t(x_0^*) = T \quad (42)$$

where $h(x_0^*)$ is obtained from Equation (12) with x replaced by x_0^*.

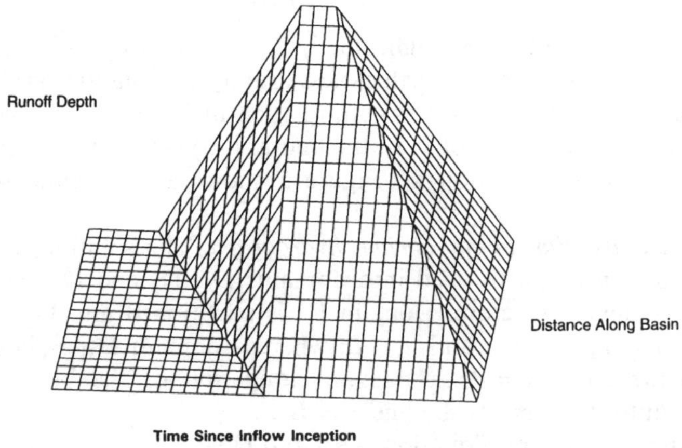

Figure 5. A three-dimensional view of the hydrograph in the equilibrium case, A

The solution of Equations (8) and (9) with $I = 0$, subject to Equation (42), is

$$h(x, x_0, x_0^*) = h(x_0, x_0^*) - \frac{1}{KS}\int_{x_0^*}^{x} f(x, t(x, t_0^*))\mathrm{d}x \tag{43}$$

$$t(x, x_0, x_0^*) = T + \frac{n_e}{KS}(x - x_0^*) \tag{44}$$

By eliminating the parameter x_0^* between Equations (43) and (44), h can be expressed as a function of x and t. Considering the simple case, f = constant, Equation (43) has the solution

$$h(x, x_0, x_0^*) = h(x_0, x_0^*) - \frac{f}{KS}(x - x_0^*) \tag{45}$$

where $h(x_0, x_0^*)$ is defined by Equation (15):

$$h(x_0, x_0^*) = h_1(x_0) + \frac{(I - f)}{KS}(x_0^* - x_0) \tag{46}$$

Using Equation (41), this reduces to

$$h(x_0, x_0^*) = h_1(x_0) + \frac{(I - f)}{n_e}T \tag{47}$$

Introducing this into Equation (45) gives

$$h(x, x_0, x_0^*) = h_1(x_0) + \frac{(I - f)}{n_e}T - \frac{f}{KS}(x - x_0^*) \tag{48}$$

Eliminating the parameter x_0^* between Equations (44) and (48), gives

$$h(x, x_0, t) = h_1(x_0) + \frac{1}{n_e}(IT - ft) \tag{49}$$

If the initial condition is a vanishing one, then

$$h(x, t) = \frac{IT - ft}{n_e} \tag{50}$$

showing the time-dependence of h only. The flux is given by

$$Q = \frac{KS}{n_e}(IT - ft) \tag{51}$$

In case B_2, the part of the boundary of D_{31} is $t = t^0(x)$. This is obtained by eliminating x_0^* between Equations (41) and (44), and from Equation (43). In the case f = constant and a vanishing initial condition, Equation (45), in conjunction with Equation (42), gives

$$I(x_0^* - x_0) + f(x_0 - x) = 0 \tag{52}$$

From Equation (52) one obtains

$$x_0^* = \psi(x, x_0) = x_0 - \frac{f}{I}(x_0 - x) \tag{53}$$

Thus, $t^0(x)$ is obtained as

$$T = \frac{n_e f}{KSI}(x - x_0) \tag{54}$$

$$t^0(x) = T + \frac{n_e}{KS}(x - x_0)\left(1 - \frac{f}{I}\right) \tag{55}$$

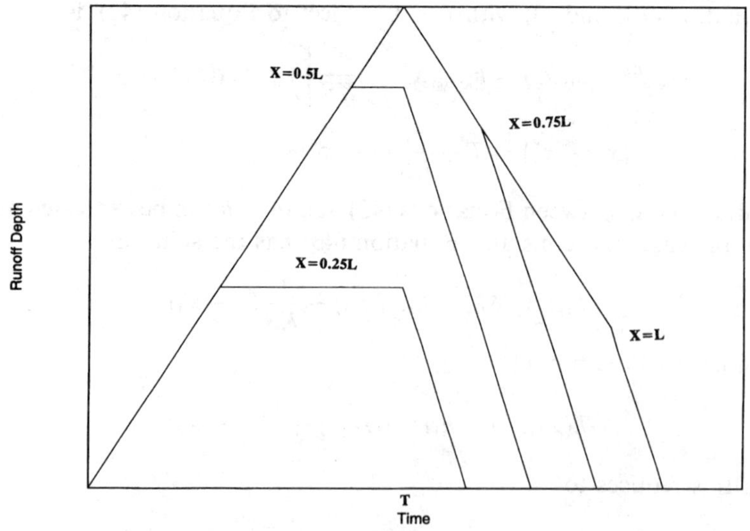

Figure 6. A system of hydrographs at various values of x in the partial equilibrium case, B_1

Combining Equations (54) and (55) yields

$$t^0(x) = \frac{I}{f} T \tag{56}$$

showing the x-independence of the free boundary. For a fixed value of T the free boundary depends only on the ratio of input to the infiltration rate.

Summary of cases B_1 and B_2. For a fixed value of x, a graph showing the variation of h with time can be constructed. Figures 6 and 7 sketch systems of such graphs for various values of x. In the figures $\bar{t}(x_i)$ represents the time taken by water to travel from the upper end of the saturated layer to the location x_i. Figures 8 and 9 illustrate the depth hydrograph in three dimensions.

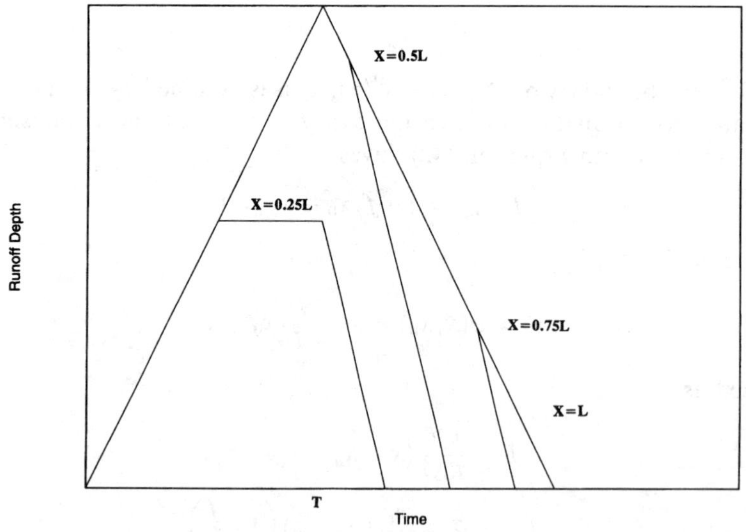

Figure 7. A system of hydrographs at various values of x in the partial equilibrium case, B_2

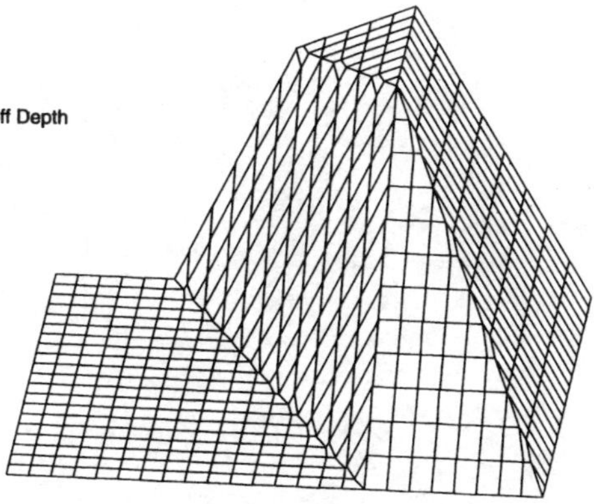

Figure 8. A three-dimensional view of the hydrograph in the partial equilibrium case, B_1

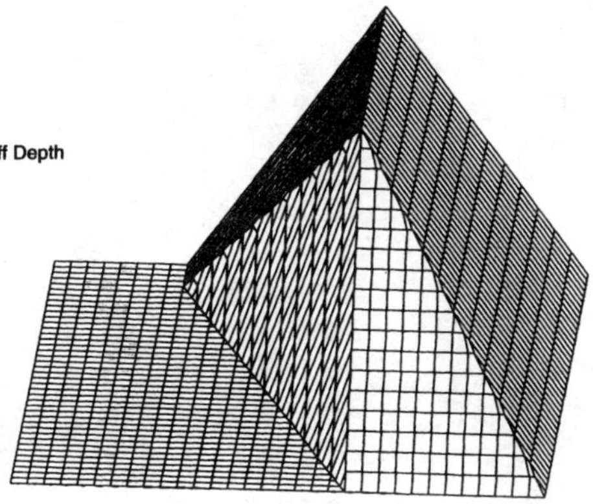

Figure 9. A three-dimensional view of the hydrograph in the partial equilibrium case, B_2

CONCLUSIONS

Analytical solutions are tractable for time-invariant inputs due to vertical percolation of snowmelt water and constant infiltration rate. These solutions exhibit essential features of a snowmelt-runoff hydrograph in terms of rise, recession, peak and time characteristics. The hydrograph rises linearly and recedes linearly in time. Infiltration, input and its duration, hydraulic conductivity and slope of the saturated layer exert significant influence on the hydrograph characteristics.

REFERENCES

Akan, A. O. 1984. 'Simulation of runoff from snow covered hillslopes', *Wat. Resour. Res.*, **29,** 707–713.

Bengtsson, L. 1981. 'Snowmelt generated runoff from small areas as a daily transient process', *Geophysica*, **17**(12), 109–121.

Colbeck, S. C. 1974. 'Water flow through snow overlying an impermeable boundary', *Wat. Resour. Res.*, **10**(1), 119–123.

Colbeck, S. C. 1977. 'Short-term forecasting of water runoff from snow and ice', *J. Glaciol.*, **19**(81), 571–588.

Colbeck, S. C. 1978. 'The physical aspects of water flow through snow, in Chow, V. T. (Ed.), *Advances in Hydroscience*, Vol 11, Academic Press, New York, pp. 165–201.

Dunne, T., Price, J. G. and Colbeck, S. C. 1976. 'The generation of runoff from subarctic snowpacks', *Wat. Resour. Res.*, **12**(4), 677–685.

Fujino, K. 1971. 'Measurement of flow down speed of meltwater in snow cover, 2. Flow of melt water in snow cover on a slope', *Low Temp. Sci., Ser. A*, **29,** 151–158.

Kirnbauer, R., Bloschi, G. and Gutknecht, D. 1994. 'Entering the era of distributed snow models', *Nord. Hydrol.*, **25,** 1–24.

Singh, V. P. 1990. 'Hydraulic considerations for water resources modeling. *V.U.B. Hydrologie 17*, Vrije Univerisiteit, Brussels, Belgium. 280 pp.

15

THE INTRODUCTION OF RUNOFF ROUTING INTO LARGE-SCALE HYDROLOGICAL MODELS

T. J. JOLLEY[1]* AND H. S. WHEATER[2]

[1]*Scottish Environment Protection Agency, East Kilbride G75 5PP, UK*
[2]*Department of Civil Engineering, Imperial College of Science, Technology and Medicine, London, SW7 2BU, UK*

ABSTRACT

Two alternative schemes are presented that are appropriate for the representation of runoff routing in large-scale grid-based hydrological models and atmospheric general circulation models (AGCMs). The first scheme characterizes routing processes as a single conceptual store. The second scheme, developed by Naden (1992), uses the normalized network width function to characterize the channel network form and a linear solution to the convective diffusion equation of one-dimensional flow to characterize the routing effect of a single channel. Both schemes are applied to the Severn catchment at the daily time-scale for the period 1981 to 1990 using a grid resolution of 40 km. Comparable results were obtained using both schemes (efficiencies were of the order of 80% in both cases). A combined model using a conceptual reservoir to represent hillslope routing and the network-based scheme to represent channel routing was developed to investigate the relative roles of hillslope and channel routing at the catchment scale. The application of this model demonstrated the important role of hillslope routing in reproducing the low frequency component of the catchment response. However, in terms of goodness-of-fit there was little to choose between the three schemes. Consequently, it is recommended that additional *a priori* knowledge of the routing processes should be used to condition the choice of model structure.

Hydrol. Process., Vol. 11, 1917–1926 (1997)

(No. of Figures: 6 No. of Tables: 2 No. of Refs: 31)

KEY WORDS runoff routing; large-scale hydrological models; AGCM; conceptual reservoir model; network routing model; combined model

INTRODUCTION

The development of land surface parameterization schemes for meso-scale meteorological models and atmospheric general circulation models (AGCMs) is generally divided into two components — the generation of runoff and the routing of runoff (e.g. Vorosmarty *et al.*, 1989; Becker and Pfuetzner, 1990; Naden, 1993). Runoff routing is required for two reasons. First, to ensure that the dynamics of freshwater inputs to the ocean and wetlands are realistically modelled, and, secondly, to contribute to the validation of land surface parameterization schemes through the use of observed river discharge data. The latter requirement is of particular importance as river discharge is perhaps the only truly large-scale observation available for model validation.

The decoupling of runoff generation from runoff routing is supported on the grounds that the routing of runoff through a channel system is perceived to be relatively straightforward and that the parameters for the routing scheme must be identifiable from the physical properties of the channel system. However, in practice, the division between the routing processes that occur in channels and in hillslopes is not a simple problem (Calver *et al.*, 1972; Messa and Mifflin, 1986). Consequently, there will be some dependency between the structures of the runoff generation and runoff routing models.

In this paper we present two alternative approaches to the problem of runoff routing for large-scale grid-based land surface parameterization schemes in meso-scale meterological models and AGCMs. They are

* Correspondence to: T. J. Jolley at Scottish Environment Protection Agency, Peel Park, East Kilbride G75 5PP, UK

considered appropriate in this context because they have been developed for application at the catchment scale, are computationally simple and have been demonstrated to provide reasonably accurate results within a hydrological context. Moreover, we investigate the conceptual disaggregation of runoff routing into the processes associated with channels and hillslopes.

RUNOFF ROUTING ALGORITHMS

Two approaches to the characterization of runoff routing were used in this study — conceptual reservoirs and a network-based routing algorithm.

Conceptual reservoirs

Conceptual reservoirs have commonly been used in hydrology to characterize the effects of storage on runoff dynamics both in black-box representations (e.g. Nash, 1958, 1960; Diskin, 1964; Muftuoglu, 1984) and conceptual representations (e.g. Horton, 1938, 1945; Crawford and Linsley, 1966). The difference between the two applications is whether or not a physical definition is given to the parameters of the reservoirs. These models provide a simple way of routing runoff from one grid square to another (or from grid square to outflow) and have proven to be reasonably accurate (Schultz, 1996).

The theory of conceptual reservoirs is well established (Dooge, 1973; Ibbitt and O'Donnell, 1974; Gill, 1976, 1977; Hughes, 1986) and will not be repeated in detail here. The relationship between the inflow (u), outflow (q) and storage state (s) of a reservoir is derived by combining the equation of mass continuity,

$$\frac{\mathrm{d}s}{\mathrm{d}t} = u - q \tag{1}$$

and the equation relating q and s,

$$q = ks^m \tag{2}$$

where k is called the reservoir constant and m is the reservoir exponent. Exact solutions have been derived for linear ($m = 1$) and second- ($m = 2$) and third- ($m = 3$) order non-linear reservoirs (Moore, 1992). The parameters m and k are generally replaced by a and b, where

$$a = mk^{\frac{1}{m}} \tag{3}$$

$$b = \frac{m-1}{m} \tag{4}$$

The reservoir constant, k, has dimensions T^{-1}, hence, an increase in k (or a) implies a decrease in the residence time of the reservoir. When applying conceptual reservoirs two parameters have to be identified, a and b. Although in general these parameters are calibrated using observations, they can be defined in terms of physical characteristics provided that a number of simplifying assumptions are made. For example, Horton (1938, 1945) related the parameters of a non-linear reservoir to turbulent overland flow, and Bras (1990) cites a number of papers that show that cascades of conceptual reservoirs are equivalent to specific solutions of the St Venant equations of one-dimensional gradually varied, open-channel flow. Hence, from the perspective of AGCMs, there would seem to be a reasonable possibility of identifying the reservoir parameters *a priori*.

Network-based routing algorithm

Rodriguez-Iturbe and Valdes (1979) employed a quantitative description of channel network form to establish a model of catchment response that was based on geometric parameters and exponential distribution of travel times within each link of the channel network. They called the model the geomorphological instantaneous unit hydrograph (GIH). The model was developed further by Gupta and Waymire (1983) who proposed that inaccuracies that arise owing to the use of Strahler ordering or Horton numbers to describe a channel network could be avoided if the network width function was used to describe the network form. The network width function describes the distribution of the number of channel links at any distance from the catchment outfall (Kirkby, 1976). Furthermore, the exponential distribution of travel times within a channel link was reviewed by Kirshen and Bras (1983) who proposed an alternative approach based on a simplification of the hydrodynamic equations of open-channel flow. Naden (1992, 1993) combined both of these developments into a network-based routing model. It is this model that was adopted in this study. This model is based on the physics of open channel flow and, as such, its parameters are directly related to the channel properties.

The model comprises two components, a normalized network width function and a routing function. The response function of the network is derived by convoluting these two functions and is, in effect, the instantaneous unit hydrograph of the contributing area. The routing function is a simple linear convective diffusion solution to the hydrodynamic equations of open-channel flow. The model has two parameters — a wave velocity (A) and a diffusion coefficient (D). Messa and Mifflin (1986) established that these parameters are related to the mean velocity (V_0), the Froude number (F_0), the bed slope (S_0) and the acceleration owing to gravity (g) through the expressions

$$A = \frac{3}{2} v_0 \tag{5}$$

$$D = \frac{1}{2} \frac{v_0^2}{g S_0 F_0^2} (1 - F_0^2) \tag{6}$$

Furthermore, Naden (1993) notes that because a linear solution is used the parameters are independent of flow, which has the advantage that runoff from different regions of a catchment can be routed independently and superimposed at the catchment outfall.

APPLICATION OF RUNOFF ROUTING MODELS

The performances of the models were assessed by comparing the simulated and observed daily runoff at the Haw Bridge river gauging station in the Severn catchment for the 10-year period 1981 to 1990. Daily runoff was generated by the 40 km grid-based water balance model developed by the authors (Jolley and Wheater, 1993, 1995). The river networks, location of Haw Bridge and the 40 km grid are shown in Figure 1. Analysis at the annual time-scale demonstrated that the water balance model was able to reproduce the mean annual runoff reasonably accurately [the model underestimated the mean annual runoff by 1% (Jolley and Wheater, 1995)], however, plots of the daily runoff (Figure 2) clearly demonstrated that the response is too dynamic, as runoff only occurs on those days when rainfall exceeds potential evapotranspiration.

The conceptual reservoir model

In the case of the conceptual reservoir model the total runoff from each contributing grid square was routed through a single reservoir. The advantage of this approach is that the number of model parameters is kept to a minimum; using a reservoir for each grid square would have introduced a large number of parameters creating a serious identification problem.

Jolley and Wheater (1993, 1995) demonstrated that the introduction of a linear reservoir leads to a significant improvement in both the monthly distribution of runoff and the timing of runoff peaks. However, it is not obvious that a linear reservoir would be appropriate at the daily scale. Consequently, linear and non-linear reservoirs were tested.

The choice of reservoir type fixed the value of the reservoir exponent (m), but the reservoir parameter (a) remained to be identified. The 10-year discharge series at Haw Bridge was used to identify the

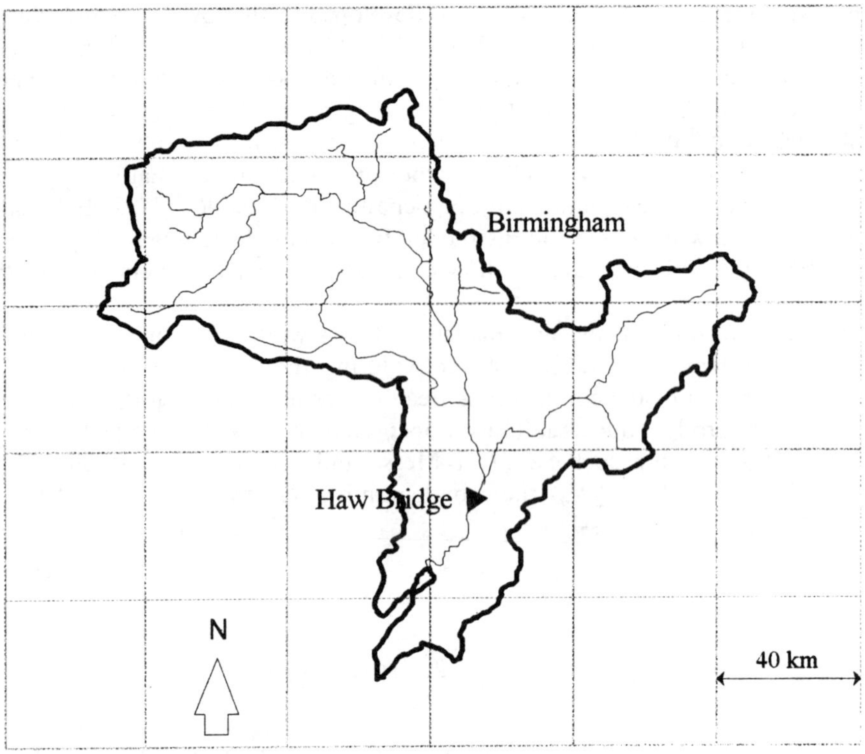

Figure 1. The Severn catchment showing the main channel network and the location of Haw Bridge

optimal value of a for the reservoir using a root mean squared error (RMSE) objective function. A split sample test was not carried out at this stage since the purpose was to investigate the ability of the models to replicate the dynamic structure of the discharge series and not to ascertain their values as predictive tools.

The results of the three simulations are given in Table I. The RMSE is expressed as a percentage of the mean annual discharge (106·52 m^3 s^{-1}) and R^2 is the efficiency (Nash and Sutcliffe, 1970).

Plots of a typical year during these simulations (Figure 3a,b and c) show that despite the relatively high values of R, all of the simulations are relatively poor visual fits. However, the plots do reveal that the recession periods are better modelled by the non-linear models than the linear model. Furthermore, they show that most of the peaks that occur during the drier months (days 150 to 270) are either missed by the models or seriously underestimated. This is a clear indication that the poor simulations can partly be attributed to weaknesses in the water balance model at the daily time-scale. Nevertheless, given the simplicity of the water balance model and the runoff routing scheme these results are encouraging.

Table I. The optimal values of the reservoir parameters a and the root mean squared error expressed as a percentage of the mean annual daily discharge (RMSE) and the efficiency (R^2) for each reservoir model

m	a	RMSE	R^2
1	0·098	43	0·80
2	0·055	45	0·78
3	0·047	46	0·78

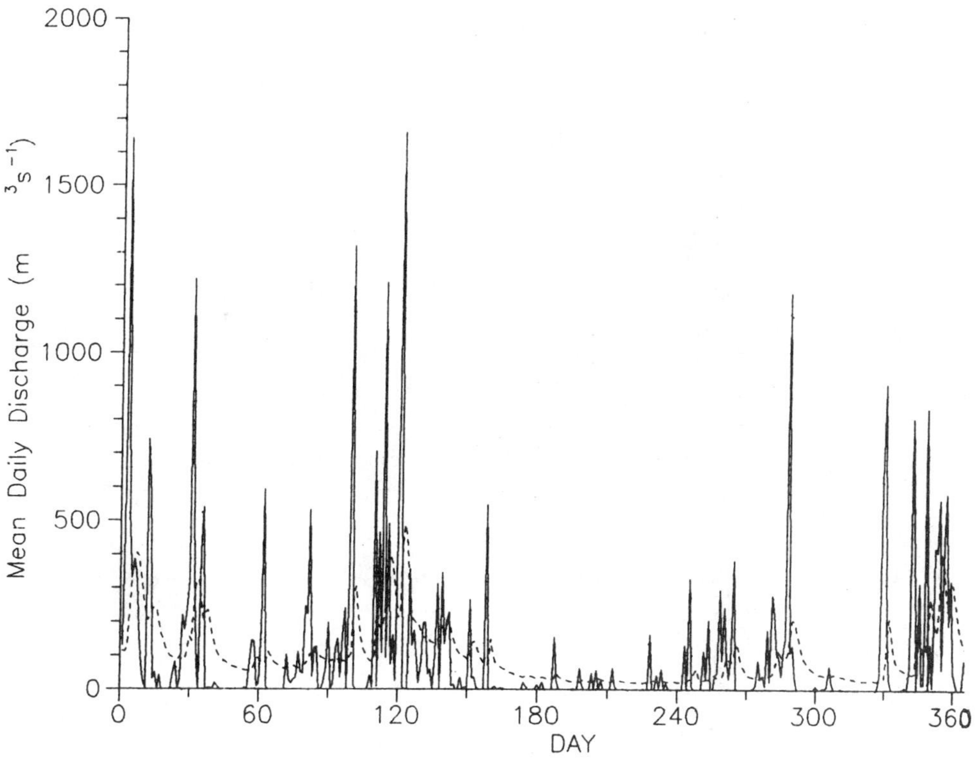

Figure 2. The observed (– – –) and simulated (——) mean daily discharge without routing

The network routing model

It was assumed that the runoff from each grid square could be routed independently to the catchment outfall. This is not an unreasonable assumption for large catchments (Beven and Wood, 1993) and is consistent with the need for computational efficiency. The network response function for each grid square was derived using the network width function that described the channel network of the contributing part of the grid square (including any channel links between the grid square boundaries and the catchment outfall). The grid square response was then derived by convoluting the grid square runoff with the network response function. The summation of the grid square responses gave the catchment response. It is noted that the need for convolution greatly increases the computation time compared with the conceptual reservoir method.

The network width functions were derived from the blue lines on the 1:50 000 scale Ordnance Survey maps of the region. The diffusion parameter (D) was fixed at 10 000 $m^2 s^{-1}$ in accordance with experience from previous applications (P. S. Naden, personal communication). The advection speed (A) can, in theory, be expressed in terms of the mean channel velocity [Equation (5)], or it can be treated as a calibration parameter. The optimal value based on the RMSE was 0·24 ms^{-1}. This is somewhat lower than the value of 0·6 ms^{-1} used by Naden (1993), which was based on an estimate of the mean velocity in the River Severn.

A typical year during the simulation is given in Figure 4. The RMSE was 59% of the mean annual daily discharge at Haw Bridge and the value of R^2 was 0·63. Visual inspection of Figure 4 reveals that the simulated runoff is too dynamic — rising limbs and recessions of hydrographs are too steep and low flows are underestimated. In addition, the simulated peaks tend to occur between one and three days after the observed peaks. Although the simulation is an improvement over the dynamics of the water balance model alone (Figure 2), Figure 4 suggests that the network routing model has not captured all of the routing effects and that the optimal value of A has been reduced in order to account for hillslope routing effects that have longer response times than the channel network.

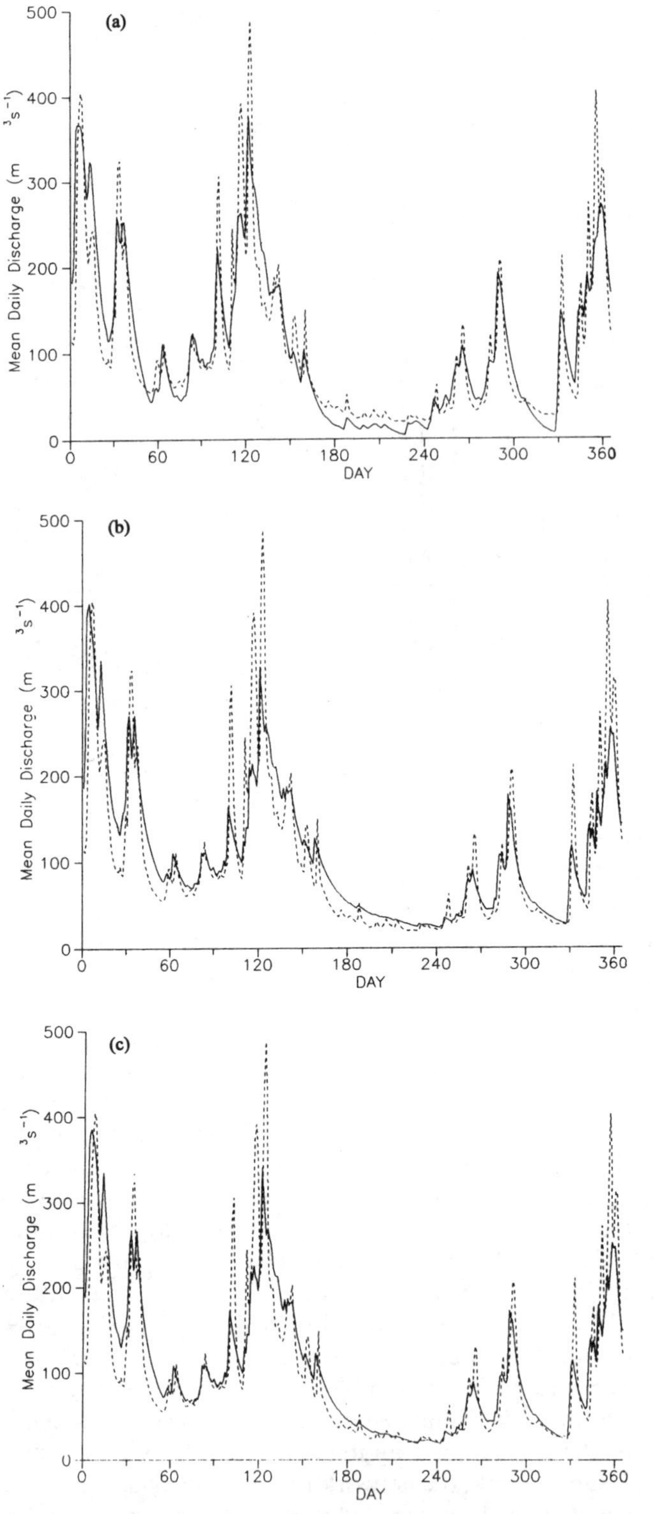

Figure 3. The observed (– – –) and simulated (——) mean daily discharge (a) with a linear reservoir to characterize routing; (b) with a 2nd order non-linear reservoir to characterize routing; (c) with a 3rd order non-linear reservoir to characterize routing

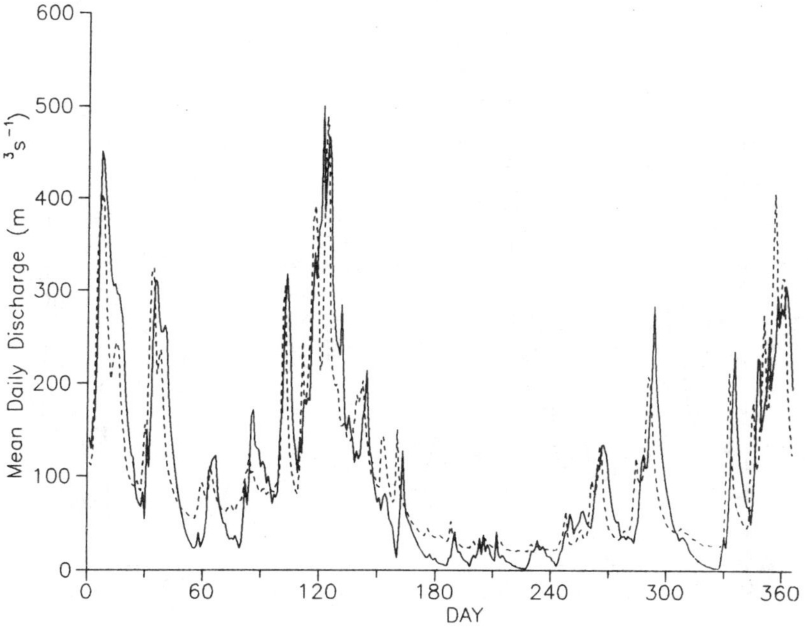

Figure 4. The observed (– – –) and simulated (——) mean daily discharge with the network routing model

COMBINED CONCEPTUAL RESERVOIR AND NETWORK ROUTING MODEL

In order to investigate the relative roles of channel and hillslope routing processes, a combined model was developed that used a conceptual reservoir to characterize hillslope routing processes and the network routing model to characterize channel routing processes. Runoff from each grid square was routed through a conceptual reservoir to represent hillslope routing processes, the outputs from these reservoirs were then convoluted with the grid square network routing functions (as described in the previous section). This approach is consistent with the conceptual distinctions made by Calver *et al.* (1972) and Messa and Mifflin (1986) between hillslope and channel response functions.

The increase in model complexity leads to an additional parameter, hence there are now two parameters (a and A) to identify. In the first instance, both parameters were optimized using the RMSE in order to investigate the ability of the model to reproduce the shape of the simulated runoff and the dependency between the two parameters. Two conceptual reservoirs were tested, a linear reservoir ($m = 1$) and a third-order non-linear reservoir ($m = 3$). The results are summarized in Table II together with the results of the third-order reservoir alone and the network routing function alone.

Table II. The calibrated parameters of alternative routing schemes using combinations of conceptual reservoirs (CR) and the network routing model (NRM). The parameters A and a are optimized using the root mean squared error (RMSE), which is expressed as a percentage of the mean daily discharge

Model	A (ms^{-1})	a	RMSE (%)	R^2
Cr ($m = 3$)	–	0·047	46	0·78
NRM	0·24	–	59	0·63
NRM + CR ($m = 1$)	2·53	0·100	45	0·79
RMN + CR ($m = 3$)	1·25	0·057	43	0·80

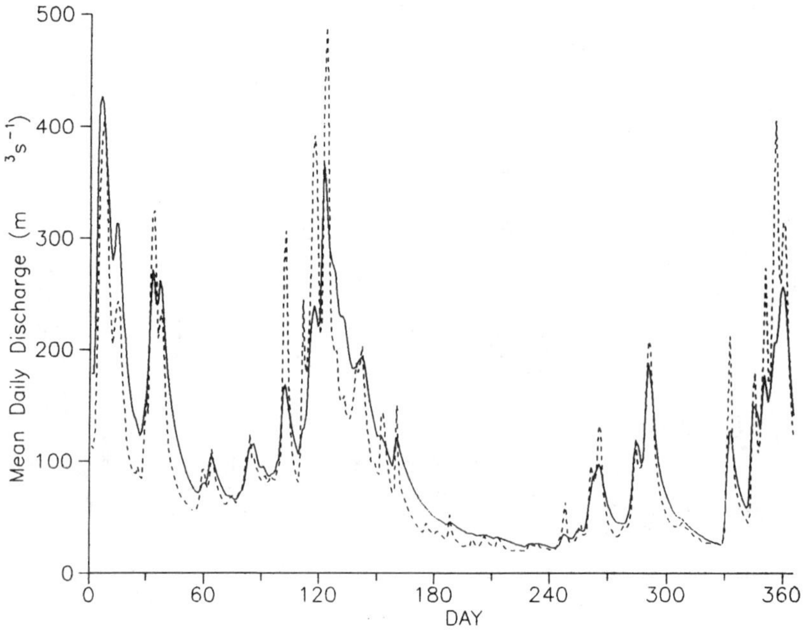

Figure 5. The observed (– – –) and simulated (——) mean daily discharge with the combined network routing model and 3rd order non-linear reservoir to characterize routing

The combination of the third-order non-linear reservoir gave the best visual fit, but, as before, even this model was far from a perfect fit (Figure 5), most likely because of the shortcomings in the water balance model. However, the simulation is a significant improvement on the use of the network routing model alone. A comparison of the model performance with and without the network routing model (Table II) suggests that the dynamic response is dominated by the characteristics of the conceptual reservoir as the introduction of the network routing model only led to a small improvement in performance and a slight increase in the optimal value of a. This supports the view that the channel network is not responsible for the low frequency component of the catchment response and this is corroborated by the work of Messa and Mifflin (1986) who state that it is the hillslope routing processes that control the initial rise and the recession of the catchment hydrograph. In addition, it is consistent with the conclusions of Wood *et al.* (1991), who recognized that the response of a catchment is controlled by the components with the longest residence time.

A plot of the RMSE surface (Figure 6) indicates that there is a strong dependency between a and A. The RMSE is almost completely insensitive to A for values of a less than 0·03, reflecting the long residence time of the conceptual reservoir. Furthermore, once A exceeds 0·9 ms^{-1} the network has little effect on the dynamics of runoff as the residence time in the network becomes insignificant relative to the daily time-scale. Hence, in practice it is more appropriate to identify A *a priori* using observations [or equation (5)] and to calibrate a. For example, when A was fixed *a priori* at 0·7 ms^{-1}, the optimal value of a was 0·067 and the RMSE and efficiency values were 48% and 0·76, respectively.

CONCLUSIONS

This study has demonstrated that the decoupling of channel and hillslope routing may not lead to improved routing schemes in meteorological models and AGCMs. Indeed, in this application a simple hydrological routing model performed equally well as a more sophisticated hydraulic model based on channel network morphology and the hydrodynamic equations of open channel flow. However, there are two important caveats. First, the lumping of runoff routing processes leads to problems with identifying the model

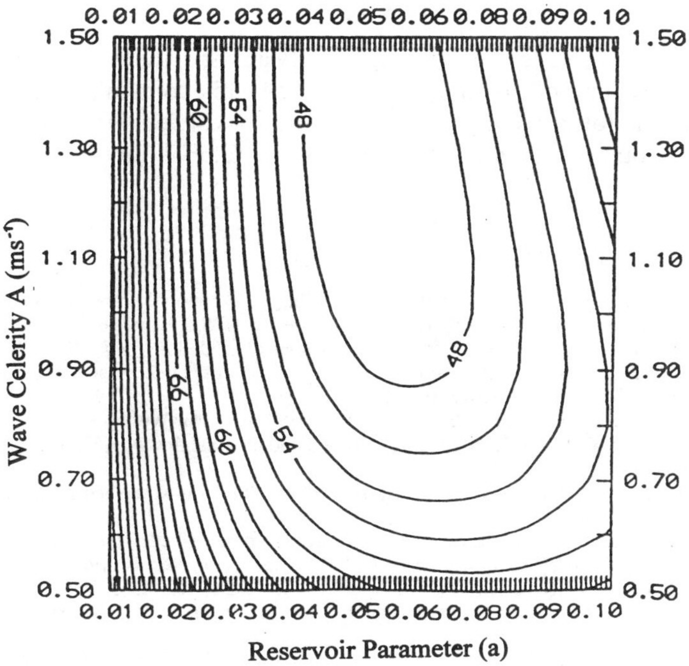

Figure 6. The RMSE plotted against the reservoir parameter a and the advection velocity A for the combined model

parameters *a priori*, and, secondly, the relative roles of channel routing and hillslope routing will not be independent of the climatic and physiographical characteristics of a catchment.

The first problem is associated with the recognized problem of equifinality (Beven, 1993) — the optimal identification of model structure and parameters depends on observations and the specification of a likelihood function. In order to disaggregate the lumped representation of runoff routing into its constituent parts, either we must obtain more complete sets of observations that can be used to identify those component parts, or we must weight the likelihoods of our model structures and parameters, being optimally based, on an *a priori* knowledge of the runoff processes. In an application of the network routing model to the River Severn using observed river flows, Naden (1993) showed that this model was a good representation of channel routing when the parameters A and D were set to 0·6 ms^{-1} and 10 000 ms^{-2}, respectively. Hence, we can use this *a priori* knowledge to say that the combined model (the conceptual reservoir and the network routing model) is more likely to be an optimal model than either of the separate models. However, Table II and Figure 5 both indicate that at the daily time-scale, hillslope routing is more significant than channel routing and, therefore, more attention should be given to the identification of this component.

The second problem can only be addressed through further applications to catchments with a wide range of climatic and physiographical characteristics. For example, it is suspected that in arid regions, where runoff predominantly occurs as overland flow, that the channel routing will be more important relative to hillslope routing. However, without further work, generalizations are not possible at this stage.

In conclusion, it is evident that the inclusion of routing will significantly improve the dynamic response of AGCMs. As with the hydrological modelling in general, the identification of an appropriate model structure depends on prior knowledge of the routing processes and the types and quantities of data available. In this study, as with most catchments, extensive data pertaining to the properties of the channel network are not known and therefore simple hydrological models are appropriate. The move to a more physically based model, such as the network routing model presented here, can only be justified where additional knowledge of the routing processes is available.

ACKNOWLEDGEMENTS

The authors wish to thank Pam Naden for providing software and the network width functions and for invaluable advise during the course of the study. We would also like to thank the Meteorological Office and the National Rivers Authority for providing data. This work was carried out with funding from the Natural Environmental Research Council's Terrestrial Initiative into Global Environmental Research (TIGER) programme.

REFERENCES

Becker, A. and Pfeutzner, B. 1990. 'Larger-scale hydrological modelling for regional transferring of hydrological information', in *Regionalization in Hydrology* (*Proc., Ljubljana Symposium, April 1990*), *IAHS*, **191**, 61–68.

Beven, K. 1993. 'Prophecy, reality and uncertainty in distributed hydrological modelling', *Adv. Wat. Resour.*, **16**, 41–51.

Beven, K., and Wood, E. F. 1993. 'Flow routing and the hydrological response of channel networks', in Beven, K. and Kirkby, M. J. (Eds), *Channel Network Hydrology*. Wiley, Chichester, pp. 99–128.

Bras, R. L. 1990. '*HYDROLOGY: An Introduction to Hydrological Science*', Addison-Wesley Series in Civil Engineering. p. 643.

Calver, A., Kirkby, M. J. and Weyman, D. R. 1972. 'Modelling hillslope and channel flows', in Chorley, R. J. (Ed.), *Spatial Analysis in Geomorphology*. pp 197–218.

Crawford, N. H. and Linsley, R. K. 1966. 'Digital simulation in hydrology: Stanford watershed model IV', *Technical Report No. 39*. Dept. Civ. Eng., Stanford University, Palo Alto.

Diskin, M. H. 1964. 'A basic study of the linearity of the rainfall-runoff process in watersheds', *PhD Thesis*, University of Illinois, Urbana.

Dooge, J. C. I. 1973. 'Linear theory of hydrologic systems', *Tech. Bull. 1468*, Agric. Res. Service. USDA, Washington, D.C. 327 pp.

Gill, M. A. 1976. 'Exact solution of gradually varied flow', *J. Hydrol. Div. ASCE*, **102**, 1353–1364.

Gill, M. A. 1977. 'Algebraic solution of the Horton–Izzard turbulent overland flow model of the rising hydrograph', *Nord. Hydrol.*, **8**, 249–256.

Gupta, V. K. and Waymire, E. 1983. 'On the formulation of an analytical approach to hydrologic response and similarity at the basin scale', *J. Hydrol.*, **65**, 95–123.

Horton, R. E. 1938. 'The interpretation and application of runoff plot experiments with reference to soil erosion problems', *Soil Sci. Soc. Am., Proc.*, **3**, 340–349.

Horton, R. E. 1945. 'Erosional development of streams and their drainage basins: hydrophysical approach to quantitative morphology', *Bull. Geol. Soc. Am.*, **56**, 275–370.

Hughes, D. A. 1986. 'Non-linear runoff-routing — a comparison of solution methods', *J. Hydrol.*, **85**, 339–347.

Ibbitt, R. P. and O'Donnell, T. 1974. 'Designing conceptual catchment models for automatic filtering methods', *IAHS Publ.*, **101**, 461–475.

Jolley, T. J. and Wheater, H. S. 1993. 'Macromodelling of the river Severn', in Wilkinson, W. B. (Ed.), *Macroscale Modelling of the Hydrosphere*, *IAHS Publ.*, **124**, 91–100.

Jolley, T. J. and Wheater, H. S. 1995. 'A large scale grid-based hydrological model of the Severn and Thames catchments', *J. Inst. Wat. Eng. Manage.*, **10**, 253–262.

Kirkby, M. J. 1976. *Hillslope Hydrology*. Wiley.

Kirshen, D. M. and Bras, R. L. 1983. 'The linear channel and its effect on the geomorphological instantaneous unit hydrograph', *J. Hydrol.*, **65**, 175–208.

Mesa, O. J. and Mifflin, E. R. 1986. 'On the relative role of hillslope and network geometry in hydrologic response', in Gupta, V. K., Rodriguez-Iturbe, I., and Wood, E. F. (Eds), *Scale Problems in Hydrology, Runoff Generation and Basin Response*. D. Reidel Publ. Co., Dordrecht.

Moore, R. J. 1982. 'Hydrological Models — Nonlinear storage models'. In: International Course for Hydrologists, International Institute for Hydraulic and Environmental Engineering. Institute of Hydrology, Wallingford.

Müftüoglu, R. F. 1984. 'New models for nonlinear catchment analysis', *J. Hydrol.*, **73**, 335–357.

Naden, P. S. 1992. 'Spatial variability in flood estimation for large catchments: the exploitation of channel network structure', *Hydrol. Sci. J.*, **37**, 53–71.

Naden, P. S. 1993. 'A routing model for continental-scale hydrology', in Wilkinson, W. B. (Ed.), *Macroscale Modelling of the Hydrosphere* (*Proc. International Symposium held at Yokohama, Japan, 21 July 1993*), *IAHS Publ.*, **214**, 67–80.

Nash, J. E. 1958. 'Determining run-off from rainfall', *Proc. Inst. Civ. Engrs*, **10**, 163–184.

Nash, J. E. 1960. 'A unit hydrograph study, with particular reference to British catchments', *Proc. Inst. Civ. Engrs.*, **17**, 249–282.

Nash, J. E. and Sutcliffe, J. V. 1970. 'River flow forecasting through conceptual models part I — a discussion of principles', *J. Hydrol.*, **10**, 282–290.

Rodriguez-Iturbe, I. and Valdes, J. B. 1979. 'The geomorphological structure of hydrological response', *Wat. Resour. Res.*, **15**, 1409–1420.

Schultz, G. A. 1996. 'Remote sensing applications to hydrology: runoff', *J. Hydrolog. sci.*, **4**, 453–476.

Vörösmarty, C. J., Moore, B., III, Grace, A. L., Gildea, M. P., Melillo, J. M., Peterson, B. J., Rastetter, E. B. and Steudler, P. A. 1989. 'Continental scale models of water balance and fluvial transport: an application to South America, *Glob. Biogeochem. Cycles*, **3**, 241–265.

Wood, E. F., Lettenmaier, D. P., and Wallis, J. R. 1991. 'Comparison of an alternative land surface parameterization with the GFDL high resolution climate model', in *Hydrological Interactions Between Atmosphere, Soil and Vegetation* (*Proc. Vienna Symposium, Aug. 1991*), *IAHS Publ.*, **204**, 53–74.

SECTION IV

GEOMORPHOLOGICAL MODELLING

16

NUMERICAL MODELLING OF FLOODPLAIN HYDRAULICS AND SUSPENDED SEDIMENT TRANSPORT AND DEPOSITION

A. P. NICHOLAS* AND D. E. WALLING

Department of Geography, University of Exeter, Exeter, Devon, EX4 4RJ, UK

ABSTRACT

A numerical model is presented for the prediction of floodplain inundation extents, flow depths and velocities, and patterns of suspended sediment dispersion and deposition. The model employs a simplified and novel treatment of fluid flow and a conventional convection–diffusion mass balance relationship for suspended sediment transport and deposition. These relationships are solved over a finite difference grid with a relatively fine nodal spacing, which enables a detailed representation of complex floodplain topography. The model is applied to a 600 m reach of the River Culm, Devon, UK. Continuous records of river stage and suspended sediment concentration monitored at the upstream boundary of the study reach provide the data needed to operate the model. These data are supplemented by measurements of the *in situ* settling characteristics of the suspended sediment load. Model predictions of overbank flow depths and velocities and patterns of suspended sediment concentrations and deposition amounts are presented. Predicted patterns of overbank deposition are described and compared with field measurements obtained with the aid of astroturf sedimentation traps. Grain size distributions obtained for samples of deposited sediment retrieved from the sedimentation traps are also examined and compared with model predictions. The results of both the field measurement programme and modelling work are shown to have significant implications for the representation of floodplain topography and suspended sediment particle size and settling properties in numerical simulations of overbank processes.

KEY WORDS floodplain; hydraulics; deposition; complex topography

INTRODUCTION

During periods when river discharge levels are high, floodplain regions adjacent to the channel may become inundated, acting either as additional flow conveyance zones or as floodwater storage locations. Such overbank flows are typically laden with high concentrations of fine-grained suspended sediment, which is transported away from the channel by convective and diffusive mechanisms and deposited in regions of reduced flow competence. These hydrological and geomorphological phenomena have significant social, economic and environmental implications, and investigations of out of channel environments and processes have attracted increased attention in recent years (cf. Carling and Petts, 1992; Anderson *et al.*, 1996).

Advances in numerical modelling of overbank processes have concentrated mainly upon the prediction of floodwater inundation depths and flow velocities. Numerical schemes have been developed that predict lateral and vertical distributions of hydraulic variables across compound channel cross-sections, for the simplified case of longitudinally uniform flow (Shiono and Knight, 1991). Alternatively, finite difference and finite element models have been presented that predict depth-averaged hydraulic variables in two horizontal dimensions (Wijbenga, 1985; Gee *et al.*, 1990). Hydraulic models of this type have also been coupled with two- and three-dimensional treatments of suspended sediment transport and deposition in estuarine and

* Correspondence to: A.P. Nicholas, Department of Geography, University of Exeter, Exeter, Devon, EX4 4RJ, UK.

coastal environments (Utnes, 1993; Lin and Falconer, 1995). A number of simulation models of suspended sediment transport and deposition in floodplain environments have been developed. These include the models of James (1985) and Pizzuto (1987), which predict suspended sediment concentrations, deposition rates and deposit grain size distributions across channel and floodplain cross-sections.

Field-based investigations of rates and patterns of floodplain sedimentation (cf. Walling *et al.*, 1992) have identified the central role played by floodplain topography in controlling floodwater flow hydraulics and thus suspended sediment transport and deposition. Spatial patterns of overbank deposition in these environments are strongly dependent upon small-scale topographic features, such as drainage ditches, bank breaches and depressions, which determine local floodplain inundation sequences and velocity vectors. Such field evidence strongly suggests that the realistic representation of these topographic elements must form the basis of any attempt to model overbank sedimentation.

This paper presents results from a recent study of overbank processes that has employed a combined field work and numerical modelling approach in order to predict realistic patterns of overbank deposition. The model has two main components; first, a simplified hydraulic scheme which predicts floodwater flow depths and velocities, and, secondly, a sediment transport element which employs a mass balance relationship describing suspended sediment dispersion by convective and diffusive mechanisms and sediment deposition as a function of particle settling rates. The model employs a two-dimensional finite difference grid with a relatively fine nodal spacing (5 m). Although a finite element approach would offer greater control over mesh geometry, the fine nodal spacing of the finite difference grid utilized here provides a detailed and accurate representation of the complex floodplain surface within the study reach. The use of such a grid has a number of implications for the selection of an appropriate modelling approach. These are discussed below.

MODEL DEVELOPMENT

As outlined above, geomorphological simulation models typically incorporate both hydraulic and sediment transport components. Commonly, of these two components, the non-linear equations of fluid motion are more difficult to solve than those describing sediment transport, erosion and deposition. The degree of difficulty involved in obtaining a stable solution to a given set of equations is dependent upon the complexity of both the process equations and the topographic boundary conditions employed. This has been illustrated by recent applications of state of the art hydraulic simulation models in floodplain environments (Bates *et al.*, 1992; Anderson and Bates, 1994). Bates *et al.* (1992) employed a two-dimensional finite element model for depth-averaged free surface flows to simulate floodplain hydraulics throughout an 11 km channel and floodplain reach of the River Culm, UK. The modelling scheme employed (RMA-2) solves the two-dimensional depth-averaged form of the Navier–Stokes equations. This model was successfully applied to simulate variations in flow depths and velocity vectors over spatial scales of the order of 10–100 m (the mean nodal spacing within the mesh employed was approximately 30–40 m). However, problems of numerical instability were observed in the most complex topographic situations, which necessitated the simplification or smoothing of the topography represented by the finite element mesh.

The main objective of the work presented here is to develop a numerical model capable of simulating realistic patterns of overbank sedimentation on floodplains. Field evidence has highlighted the central role played by small-scale topographic elements (e.g. depressions, drainage ditches, levees) in controlling such patterns of floodplain deposition, thus suggesting that any realistic sedimentation model must incorporate such features within its topographic mesh. However, the findings of the modelling work of Bates *et al.* (1992) illustrate that it may not always be possible to solve the Navier–Stokes equations in environments that are characterized by such complex topography. Accordingly, in this study the decision was made to employ a set of considerably simplified hydraulic equations so as to enable the retention of a detailed and accurate representation of the complicated floodplain topography within the study reach. The following equations of fluid motion were employed in the model described here.

Continuity of momentum:

$$u = \frac{h^{2/3} S_x^{1/2}}{n} \quad \text{where} \quad S_x = \frac{\partial h}{\partial x} + \frac{\partial z}{\partial x} \tag{1}$$

$$v = \frac{h^{2/3} S_y^{1/2}}{n} \quad \text{where} \quad S_y = \frac{\partial h}{\partial y} + \frac{\partial z}{\partial y} \tag{2}$$

Continuity of mass:

$$\frac{\partial(uh)}{\partial x} + \frac{\partial(vh)}{\partial y} = 0 \tag{3}$$

where h is the depth of flow, z is the bed elevation, n is the Manning roughness coefficient and u, v, S_x and S_y are, respectively, the velocity components and friction slopes in the x and y directions. Equations (1) and (2) are a pair of one-dimensional relationships equivalent to the diffusion wave form of the Saint-Venant momentum equation. The derivation of Equations (1)–(3) involves the neglect of both temporal derivatives and convective acceleration terms, so that friction slopes are approximated by water surface slopes.

The application of Equations (1)–(3) in floodplain environments has been proposed by Cunge *et al.* (1980) and presents a means of considerably reducing the complexity of the numerical problem. Implementation of this approach within the current study is justified by the complexity of topography represented by the finite difference grid and by the belief that the accurate representation of this topography holds the key to generating realistic predictions of sedimentation patterns. Despite this, Equations (1)–(3) must be applied with caution and their limitations recognized. The model is applied by dividing flood hydrographs into discrete 1-hour time intervals, during each of which the discharge crossing the upstream and downstream reach boundaries is assumed constant. The implementation of a steady-state approach will introduce errors into the equations of fluid flow, particularly for the case of high magnitude, short duration flood events. In addition, convective acceleration effects and momentum transfer owing to turbulence have also been neglected in the current study. These processes may be significant where flow depth and velocity gradients are high, as may be the case at the boundary between main channel and floodplain flow elements. In summary, the equations of fluid flow employed in this study are considerably less sophisticated than those used in certain other models of overbank hydraulics. However, this must be balanced against the highly complex nature of the floodplain topography within the model's finite difference grid, which represents a significant improvement over that utilized in many previous modelling studies.

A single-valued relationship between the water levels at the upstream and downstream boundaries of the floodplain reach is derived by the model (see below). These water levels provide the upstream and downstream boundary conditions employed by the model when solving the equations of fluid motion, Equations (1)–(3) above, throughout the finite difference grid. Rather than solving the fluid flow equations during each time step of each flood event, they are solved initially for a number of boundary water levels in order to generate a set of flow surfaces. These surfaces are then used, in conjunction with the relationship between the water levels at the upstream and downstream boundaries of the finite difference grid, to predict hydraulic patterns during all subsequent simulations.

In order to determine the relationship between the water levels at the upstream and downstream boundaries of the finite difference grid representing the study reach, a simple discharge calculation procedure is employed. This involves determining the discharge entering the reach, for a given upstream stage, and calculating the downstream stage required to yield the same discharge leaving the reach. Boundary cross-sections are divided into a number of elements of equal width, and discharge components acting normal to these sections are determined for each element from the following equations

$$h_i = w_b + S_y(y_b - y_i) - z_i \tag{4}$$

$$q_i = \frac{h_i^{5/3} S_x^{1/2}}{n} \Delta y \tag{5}$$

where h_i is the depth of flow over grid node i, z_i is the bed elevation, n is the Manning roughness coefficient, w_b is the water level at the boundary reference grid node (i.e. the grid node for which the boundary water level is specified), y_b and y_i are lateral distances, q_i is the discharge component normal to the boundary of the finite difference grid, Δy is the width of each grid cell and S_x and S_y are the components of the downstream friction slope in the x and y directions of the finite difference grid (the friction slope is assumed to be equal to the water surface slope). The total discharge crossing the reach boundary in the x-direction of the finite difference grid is given by the sum of the discharge components for each element across the relevant section. In Equation (4) it is assumed that floodplain flow moves in a direction parallel to the longitudinal axis of the valley floor and that the elevation of the water surface is constant along lines running perpendicular to this axis. This enables lateral variations in water levels and hence flow depths to be determined where the y-axis of the finite difference grid is not parallel to the cross-stream direction of the valley. As with certain other methods for estimating the relationship between stage and discharge (cf. Ervine and Ellis, 1987), Equations (4) and (5) take no account of the effect of momentum transfer between adjacent elements of differing depth and velocity; hence they are likely to overestimate the total discharge crossing the boundaries of the model grid. Wormleaton *et al.* (1982) present results from a series of flume experiments designed to assess the accuracy of a range of discharge calculation techniques. They report ratios of calculated to observed channel and floodplain discharges in the range 1·0–1·8 where lateral momentum transfer is neglected. Errors in discharge estimates are greatest where floodplain roughness is high or where the ratio of floodplain flow depth to main channel flow depth is minimized. The total discharge crossing the reach boundary for a given reference boundary water level (w_b) is also dependent upon the values of the roughness coefficients and friction slopes employed. In this study, friction slopes were approximated using surveyed floodplain and channel bed slopes at the two boundary cross-sections. Although this method will result in discrepancies between predicted discharges and discharges experienced in the field, it is reasonable to assume that the errors involved will be similar at both upstream and downstream reach boundaries. Furthermore, as it is only the boundary water levels that are carried forward to the next stage of the model, the errors in estimated discharges associated with the technique outlined above have no further effect upon the hydraulic calculations. Optimum floodplain and channel roughness coefficients are determined by calibrating the relationship between upstream and downstream water levels using field measurements of river stage. Manning roughness values of 0·05 and 0·04 were employed for the floodplain and channel regions, respectively. These values were also employed in the calculation of flow velocities throughout the finite difference grid, as outlined below. The division of grid nodes into just two roughness classes is consistent with previous attempts to model overbank flow hydraulics (e.g. Gee *et al.*, 1990). However, the use of additional classes, particularly at the channel margins and along field boundaries where vegetation is commonly more dense, might improve the prediction of flow velocities and depths in these areas. Whilst the relatively homogenous nature of the roughness surface represented here is acknowledged, small-scale topographic features are considered to be of far greater significance in controlling complex local sequences of floodplain inundation.

The next step in the modelling procedure is to generate the flow surfaces needed to make hydraulic predictions for any given upstream stage. These flow surfaces are termed water surface functions because they describe the water surface elevation at each node in the finite difference grid as a function of the specified upstream water level and the corresponding downstream water level determined using the discharge calculation technique. Water surface functions are generated for a number of upstream boundary water levels at 5-cm increments by solving the continuity of mass and momentum Equations (1)–(3). These differential equations are transformed to a set of difference approximations, which are then solved to yield the depth of the flow at each node in the finite difference grid for the given upstream and downstream boundary water levels. When a stable approximate solution is obtained the value of the water surface function (f_i) at each wet node is calculated as

$$f_i = \frac{w_u - h_i - z_i}{w_u - w_d} \qquad (6)$$

where w_u and w_d are the specified upstream water level and the corresponding downstream water level determined using the discharge calculation procedure.

Each calculated water surface function is used to predict flow depths and velocities throughout the grid over a specified interval of upstream water levels. This is achieved by rearranging Equation (6) and inserting the relevant boundary water levels, bed elevation and water surface function value. The patterns of flow depth generated in this way are then adjusted in certain places using a simple ponding algorithm, details of which are given in Nicholas (1994). This is necessary because certain floodplain regions containing flowing water at the upstream boundary water level for which the water surface function in use was generated, may be dry or contain ponded water at the slightly lower (i.e. 0–5 cm) upstream water level for which the hydraulic calculations are being made. Two types of ponding are identified by the model. First, backwater ponding, where a region is supplied with water from a point downstream, and, secondly, recession ponding, where floodplain depressions retain ponded water as the floodplain is drained on the falling limb of the flood hydrograph. Having determined the depth of flow at each node for the given upstream boundary water level, velocity components are calculated in the x and y directions using Equations (1) and (2).

Hydraulic predictions are carried forward to the sediment transport component of the model, which predicts patterns of suspended sediment dispersion and deposition for a range of sediment size fractions. This element of the model requires data quantifying the total suspended sediment concentration in the main channel flow and the particle size distribution of this sediment. The depth-integrated mass balance equation for suspended sediment transport by convective and diffusive mechanisms in two horizontal dimensions is then solved for each of these size fractions to allow the sediment concentration and amount of deposition of each fraction to be determined at each floodplain node within the solution grid

$$(uh)\frac{\partial c}{\partial x} - \frac{\partial}{\partial x}\left(\varepsilon h\frac{\partial c}{\partial x}\right) + (vh)\frac{\partial c}{\partial y} - \frac{\partial}{\partial y}\left(\varepsilon h\frac{\partial c}{\partial y}\right) + DR = 0 \tag{7}$$

where c is the depth-averaged suspended sediment concentration, ε is a horizontal mixing coefficient and DR is the net rate of deposition (i.e. deposition less erosion and/or resuspension). Horizontal mixing coefficients are determined using standard relationships (cf. Pizzuto, 1987)

$$\varepsilon = \lambda h U_* \tag{8}$$

where λ is a constant that was assigned a value of 0·13 after Fischer *et al.* (1979). A number of model runs were conducted using values of 0·013 and 1·3 for λ. The purpose of these additional simulations was to test the sensitivity of the model predictions to variations in this parameter (see below). In Equation (8), U_* is the shear velocity, which is given by

$$U_* = (ghS_f)^{1/2} \tag{9}$$

where g is the acceleration due to gravity and S_f is the water friction slope (assumed to be equal to the water surface slope). During the development of the model a number of equations predicting net deposition rates were tested (e.g. Engelund, 1970; Parker, 1978)

$$DR = c(V_s^2/\varepsilon_v) \tag{10}$$

where ε_v is a vertical mixing coefficient determined from a relationship of the same form as Equation (8). In Equation (10), net deposition rate is inversely related to flow depth, hence, where depths approach zero, the applicability of this relationship may be questionable. In view of this we employ a simplified deposition term

$$DR = kcV_s \tag{11}$$

where V_s is the sediment particle fall velocity and k is an empirical coefficient that controls the proportion of the sediment in suspension at a node that is deposited. This empirical parameter is used to calibrate the model with the aid of measured deposition amounts. It is clear from the form of Equation (11) that net deposition rates always exceed zero and that the model does not include an explicit treatment of floodplain erosion. In this respect, the model is similar to that proposed by Pizzuto (1987). This approach is adopted because there exists no practical means of monitoring floodplain erosion rates associated with individual flood events. In addition, medium-term (30–40 years) rates of floodplain sedimentation within the study reach, monitored using data derived from ^{137}Cs inventions of floodplain sediment cores (cf. Nicholas, 1994; Nicholas and Walling, 1997), support the hypothesis of positive net deposition over >95% of the floodplain surface.

In order to obtain an approximate solution to Equation (7) the terms of this differential equation are written as a series of difference approximations. Sediment fluxes across the boundary between wet and dry nodes are set equal to zero by adjusting mixing coefficients in these regions. When a stable approximate solution to Equation (7) is obtained, the total amount of deposition at each floodplain node during the given time step is determined from

$$D_i^t = k\Delta t \sum_{j=1}^{J} (V_{s_j} c_{i,j}^t) \tag{12}$$

where D_i^t is the amount of deposition at node i during time step t, Δt is the duration of the time step, V_{s_j} is the fall velocity of the jth size fraction, $c_{i,j}^t$ is the sediment concentration of the jth size fraction at node i during time step t, and J is the number of size fractions employed in the model. For most floodplain nodes this represents the total amount of deposition. However, for nodes in areas of the floodplain that are susceptible to recession ponding, there is the additional possibility of this deposition being supplemented by material that settles out in trapped stationary water. The extra sediment contributed in this way is given by

$$D_{a_i}^t = hp_i \sum_{j=1}^{J} c_{i,j}^{t-1} \tag{13}$$

where $D_{a_i}^t$ is the additional deposition at node i resulting from the settling of sediment from ponded water, hp_i is the depth of ponded water and $c_{i,j}^{t-1}$ is the sediment concentration of the jth size fraction at node i during the time step prior to the isolation of the ponded area from the main flow. Equation (13) assumes that 100% of the sediment contained in the ponded water is deposited. The validity of this assumption is dependent upon the time interval between flood events and the significance of flocculation processes within the ponded water.

THE STUDY REACH

Field data needed for model calibration and testing purposes were collected from a 600 m study reach of the River Culm at Rewe in south-east Devon. The River Culm has a drainage area of 276 km^2 above Stoke Canon where it enters the River Exe approximately 5 km north of Exeter. Its lower reaches are characterized by a meandering gravel-bed channel approximately 12 m wide, with fine-grained alluvial banks up to 1·5 m high. The floodplain here is relatively undisturbed (land use is predominantly rough pasture) and substantial inundation occurs on average seven times each year (Lambert and Walling, 1987).

The study reach (Figure 1) has an upstream boundary marked by the river monitoring site at Rewe. A second monitoring station was installed at the downstream reach boundary. The western edge of the reach is delimited by a steep hillslope which provides an excellent vantage point for observing floodwater inundation patterns. To the east, elevation increases more slowly up to the arable land above the valley floor. The 36-ha reach contains a wide variety of topographic features typical of lowland river floodplains. These include depressions, levees, bank breaches, abandoned channels, major and minor drainage ditches, and both

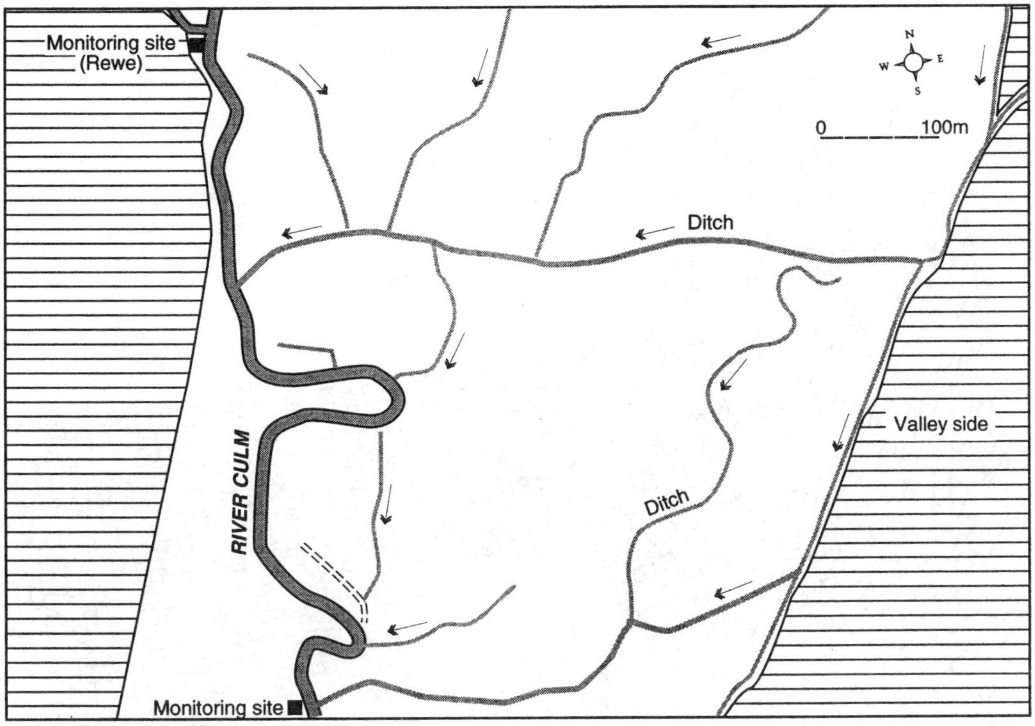

Figure 1. The study reach

straight and meandering channel sections. The major drainage ditch remains inundated at all times and can be seen (in Figures 3–5) running east to west across almost the full width of the floodplain.

The study reach was surveyed in detail using a Geodometer 400 integrated theodolite and Electronic Distance Measurement (EDM) unit. Over 3000 points were surveyed on the floodplain surface with the sampling density varying according to topographic complexity. Channel cross-sections were also surveyed at downstream intervals of 10 m along the reach. Interpolation of these data enabled the construction of a finite difference grid with a nodal spacing of 5 m. This scale of resolution allowed complicated topographic features to be represented by the grid with only minimal loss of detail. Figure 2 shows a simplified version of the grid, with a nodal spacing of 10 m for ease of viewing.

River stage and suspended sediment concentration were recorded continuously at the upstream and downstream boundary monitoring sites over a 16-month period from January 1992 to April 1993. Ten flood events occurred during this time period. Upstream stage measurements provided the input data required by the model, while downstream stage records were employed in the calibration of the discharge calculation procedure. These data were supplemented by information quantifying the particle size distribution and settling characteristics of the transported sediment.

FLOODPLAIN HYDRAULICS

The model predicts flow depths and two horizontal depth-averaged flow velocity components at each node. However, because of the difficulty of measuring those variables during flood periods, no data are available to assess the accuracy of these model predictions. Instead, they are compared with qualitative information obtained from ground photographs detailing the sequence of events by which the floodplain reach is inundated.

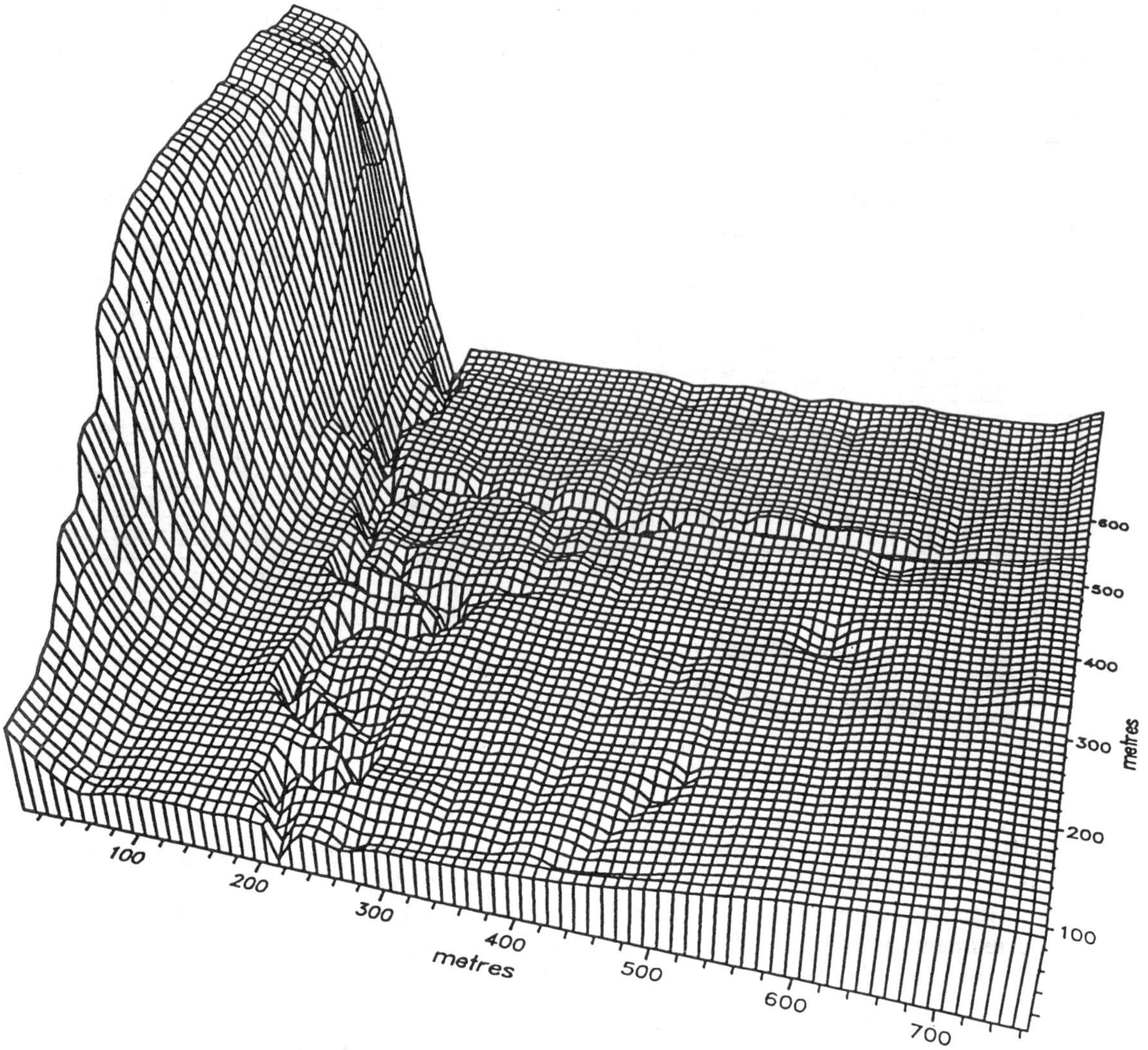

Figure 2. Reduced resolution image of the model's finite difference grid (nodal spacing is 10 m). Vertical exaggeration is 1 : 25

Figure 3 shows a plan view of predicted flow depths within the study reach for an upstream water level of 2 m (all water levels are relative to the mean channel bed elevation at the downstream reach boundary). This water level represents the earliest stage of flooding. Main channel flow depths typically exceed 1 m. Inundation of floodplain areas occurs mainly in the downstream portion of the reach, in the abandoned channel segments and minor drainage ditches on the east side of the channel, and inside the meander bends on the west side of the channel. Flow depths within these regions average 0·2–0·3 m while flow velocities are very low (typically 0·01–0·2 m s^{-1}). Model predictions compare favourably with field observations of inundation mechanisms. In particular, the importance of floodplain topography in controlling initial inundation patterns through the location of bank breaches and connected regions susceptible to backwater ponding is highlighted in both field observations and model predictions.

Figure 4 shows a plan view of predicted flow depths for an upstream stage of 2·35 m. This represents an intermediate flood stage. Although the extent of inundation is considerable (the average width of the water surface throughout the reach is approximately 400 m), numerous emergent floodplain areas remain and overbank flow is concentrated in a number of low-lying drainage channels and depressions. In the upstream

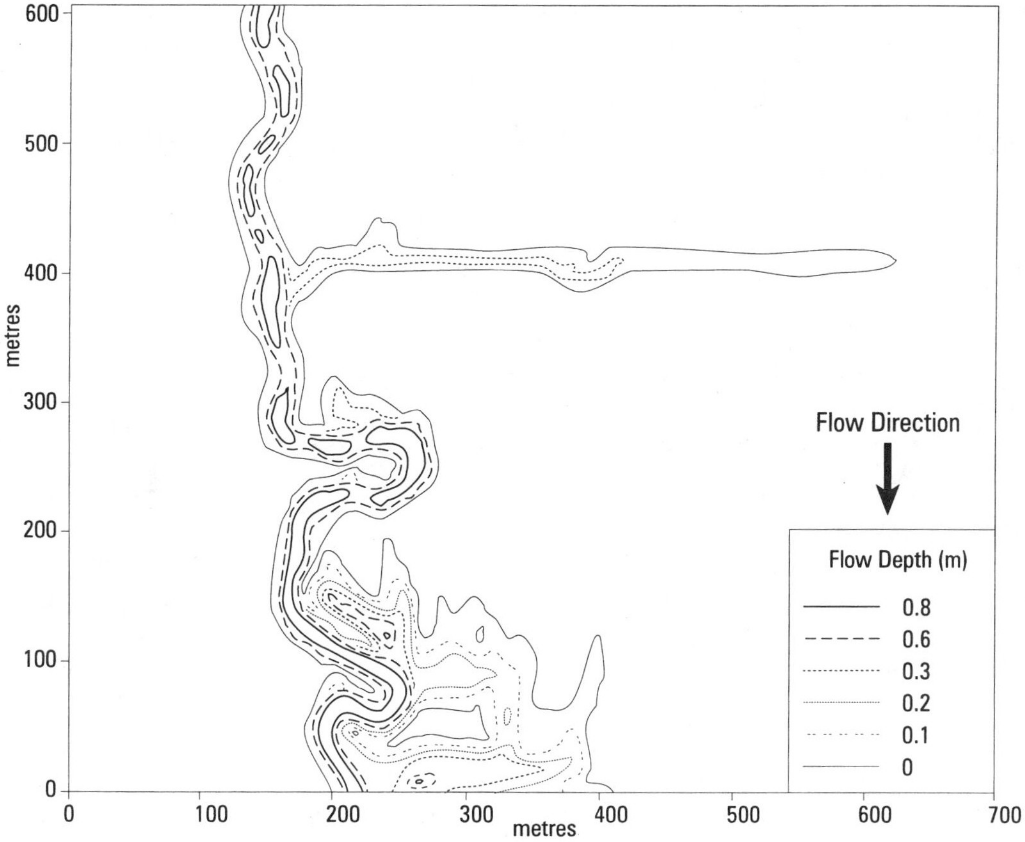

Figure 3. Predicted pattern of flow depth (m) within the study reach for an upstream water level of 2·0 m

part of the reach, inundation of the floodplain occurs initially by expansion of ponded water from the major drainage ditch running east to west across the reach. As the stage increases, the region of ponded water extends upstream until flow occupies the minor drainage ditches dissecting the upstream reach boundary. This occurs prior to the breaching of the levee bordering the east side of the main channel near the upstream reach boundary. The significant ponded areas in the downstream portion of the reach (see Figure 3) continue to expand as the stage rises. At varying threshold water levels these ponded regions are connected with upstream discharge sources and flow velocities across the floodplain increase.

Figure 5 shows a plan view of predicted flow velocity magnitudes for an upstream stage of 2·6 m. Predicted floodplain velocities typically range from 0·3 to 0·6 m s^{-1}. Main channel flow velocities generally exceed 0·6 m s^{-1} in downstream regions and 0·8 m s^{-1} in the upstream half of the reach. This difference in main channel flow velocities reflects the greater degree of flow confinement in the upstream part of the reach resulting from the existence of the well-defined levee in this area. Conversely, the topography in the downstream part of the reach is characterized by low-lying drainage channels which enable the channel flow to disperse across a wide area. A second factor promoting low main channel flow velocities and the dispersal of main channel discharge to floodplain regions is the interaction of main channel and floodplain flows. This is particularly evident in the tight meander bends, where main channel and floodplain flow directions are perpendicular to one another. At the water level shown in Figure 5, predicted flow directions in the main channel are approximately parallel to the channel banks while those on the floodplain are predominantly parallel to the longitudinal axis of the valley floor. At lower flow stages small-scale floodplain topographic

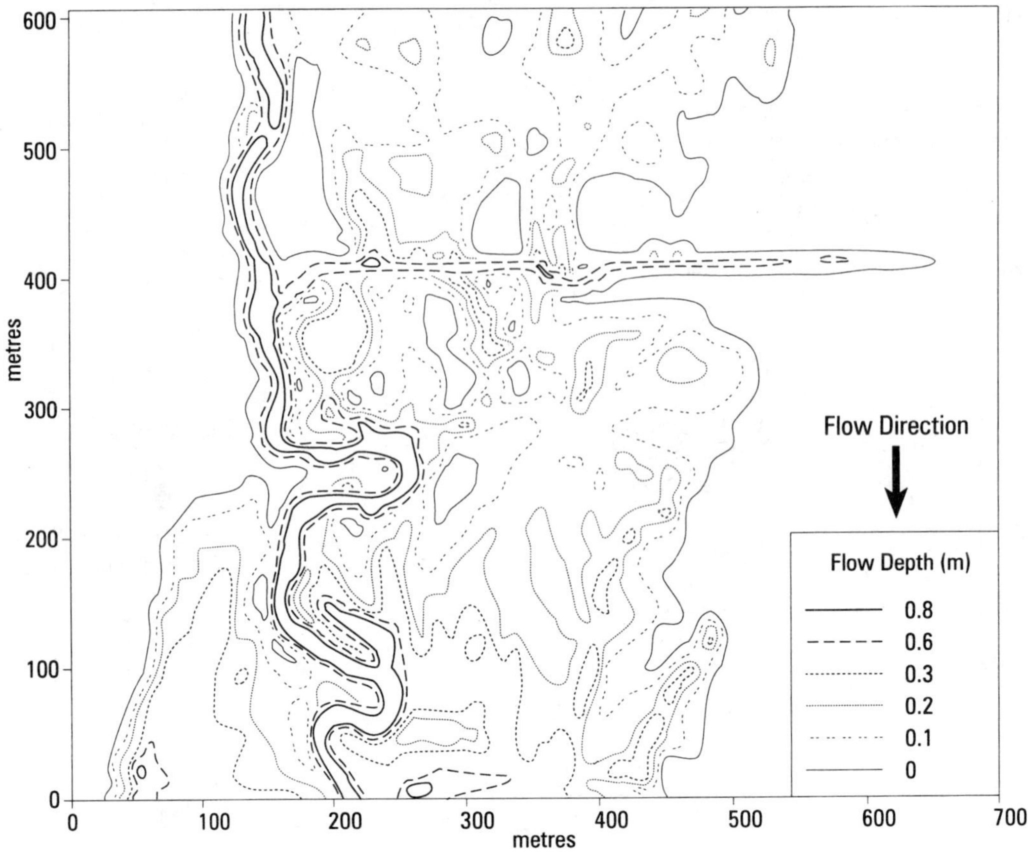

Figure 4. Predicted pattern of flow depth (m) within the study reach for an upstream water level of 2·35 m

features exert strong controls on predicted local flow directions, with velocity vectors being oriented along the network of shallow drainage ditches that dissect the floodplain reach.

In summary, although no field data are available to compare with model predictions of flow depths or velocities, observed and predicted inundation sequences compare favourably. In both instances the early stages of flooding are dominated by the ponding of floodwater in low-lying floodplain regions. As stage increases, floodplain flow is concentrated in a complex network of drainage ditches. Only at peak flood levels is the floodplain inundated by a single sheet of water moving approximately in a down-valley direction. The level of agreement between model predictions and field observations of inundation sequences can be attributed to the detailed representation of topography in the model's finite difference grid.

SUSPENDED SEDIMENT TRANSPORT AND DEPOSITION

Suspended sediment transport and deposition calculations are carried out using the predicted flow depth and velocity distributions for the given upstream stage, and field measurements of the concentration and grain size characteristics of the suspended sediment entering the reach at its upstream boundary. At this point it is important to identify the distinction between the *ultimate* and *effective* grain size characteristics of the suspended sediment load. In the past, fluvial suspended sediment has generally been characterized by its ultimate grain size distribution (i.e. the size distribution of the discrete primary particles of which the load is composed). However, a number of workers (e.g. Ongley *et al.*, 1981; Walling and Kane, 1984; Walling and Woodward, 1993) have highlighted the importance of aggregation in controlling the *in situ* physical

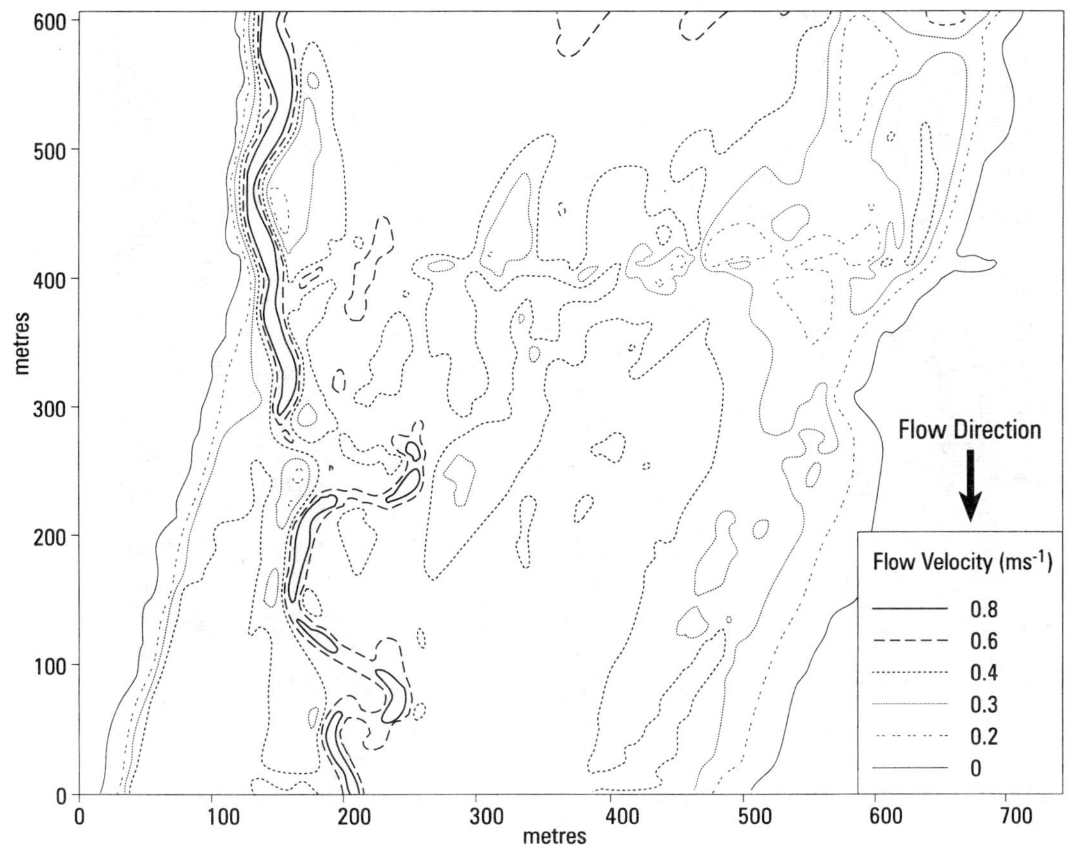

Figure 5. Predicted pattern of velocity magnitude (m s^{-1}) within the study reach for an upstream water level of 2·6 m

properties of suspended sediment. They have shown that the ultimate size distribution of suspended sediment may differ significantly from its effective size distribution, which includes composite particles composed of both mineral and organic matter. The settling velocities of composite particles may differ substantially from those of the individual constituent grains. The effective size distribution of the suspended sediment load was measured on 10 occasions over the study period using a custom-built water elutriation system. The principles of this system have been both documented and validated by several workers (Beavers and Jones, 1966; Umlauf and Bierl, 1987). The apparatus consists of a series of glass sedimentation chambers through which water is drawn during storm periods. Variations in flow velocities between each chamber, controlled by differences in their cross-sectional areas, cause each to retain sediment particles with a different settling velocity or equivalent spherical diameter. Further details of the system used are given by Walling and Woodward (1993). Using data derived from the elutriation apparatus, over the 10 flood events monitored during the study period, the mean contribution to the total load of each of five effective size classes was determined to be: <8 μm, 53%; 8–16 μm, 25%; 16–32 μm, 13%; 32–63 μm, 7%; and >63 μm, 2%. This average grain size distribution was employed in all model calculations. Because deposition rates are strongly controlled by particle fall velocity, the small amount of material in the coarsest size class makes a disproportionately large contribution to overbank deposits. As a result this size class was subdivided into two classes with different fall velocities: $>63^{A}$ μm, 1.9% and $>63^{B}$ μm, 0.1%. The latter class represents the very small proportion of large aggregates and coarse sand grains present in the load that may exceed several hundred micrometres in diameter (Woodward and Walling, 1992).

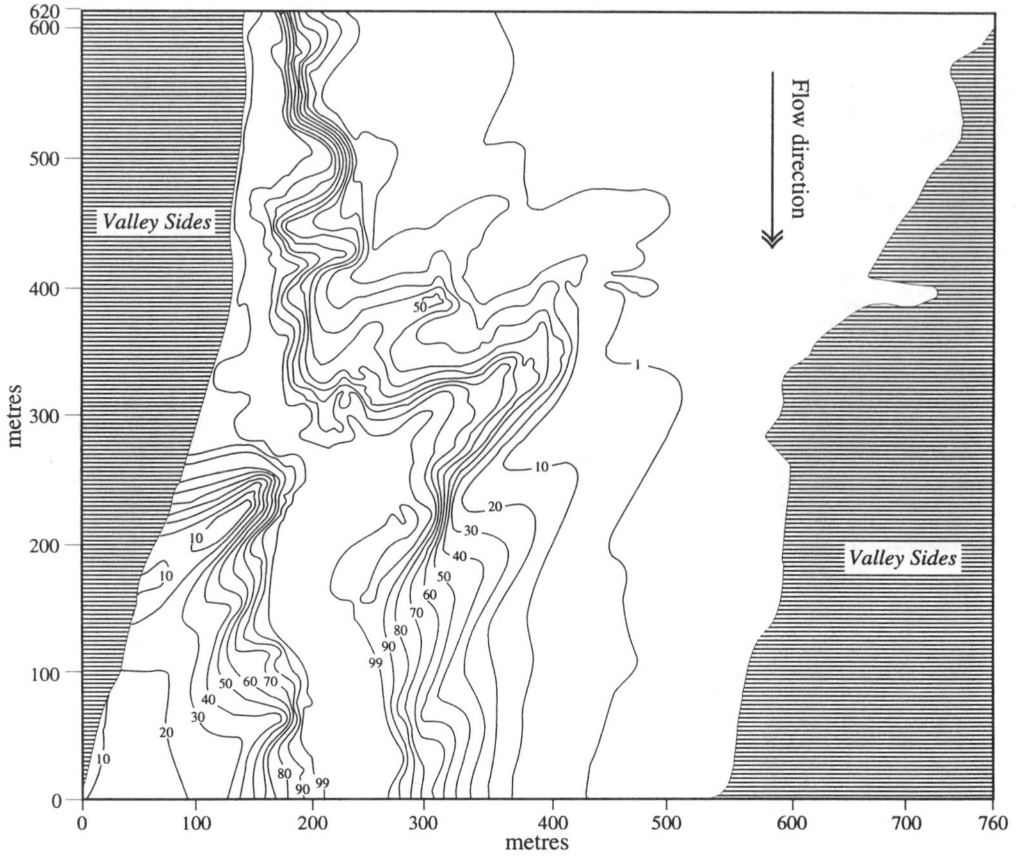

Figure 6. Predicted pattern of suspended sediment concentration (as a % of that in the main channel) for the peak of the flood event of 30 November 1992 to 5 December 1992

Equation (7) is solved separately for each of the six effective grain size classes employed in the model. Figure 6 shows the predicted pattern of total suspended sediment concentration (i.e. the sum of the six effective classes) as a % of the sediment concentration in the main channel, for the peak of the flood which occurred between 30 November 1992 and 5 December 1992. A number of trends are evident in this diagram. First, suspended sediment concentration decreases with distance from the main channel more rapidly in the upstream third of the reach than further downstream. This can be attributed to the effect of the levee bordering the channel in upstream areas, which confines the flow and impairs suspended sediment transport to the floodplain. Secondly, the width of the region of peak suspended sediment concentrations (i.e. 90–100% of main channel concentrations) is similarly greater in the downstream half of the reach than in the upstream half. This reflects the meandering nature of the channel and the existence of convective currents within the channel belt moving in a down-valley direction. These currents exit the main channel at one bend, move over the floodplain and then re-enter the channel at the bend downstream (cf. Willets and Hardwick, 1993). This mechanism stimulates suspended sediment transport and results in the high suspended sediment concentrations that characterize regions within the channel belt.

Outside the channel belt, rates of suspended sediment dispersion are highly variable spatially and are controlled by flow hydraulics and hence floodplain topography. The rate of decreasing suspended sediment concentration with distance from the main channel appears to be inversely related to flow depth (and also water surface slope). These variables control both flow velocities and mixing coefficients, hence rates of sediment dispersion increase where they are high (e.g. where bank breaches connect with floodplain drainage ditch networks). Conversely, sediment dispersion is suppressed across regions of shallow, slow moving flow over levees and other elevated parts of the floodplain.

As indicated above, Equation (11) includes an empirical parameter (k) that enables the model's deposition term to be calibrated. Calibration was carried out with the aid of direct measurements of overbank deposition, obtained for nine flood events using astroturf sedimentation traps (cf. Lambert and Walling, 1987). These consisted of 1·5-cm tufts of artificial grass attached to a pliable base and fixed to the floodplain surface using six-inch steel pins. Traps were deployed in closely spaced groups of nine to give an indication of the degree of measurement variability involved in their utilization. The parameter k was adjusted manually to a value of 1/30 so as to reduce the total error for the data set of 52 measured deposition amounts. Varying this value of k adjusts the deposition amounts over the whole floodplain by a factor that shows only very small local variations. Varying the value of λ was found to have a negligible effect upon predicted deposition amounts close to the main channel and within the channel belt. Outside the channel belt, increasing λ by a factor of 10 resulted in large relative increases in predicted deposition amounts, although these changes remained small in absolute terms (e.g. 0–10 g m^{-2}).

Figure 7 shows a plan view of the predicted pattern of deposition amounts (g m^{-2}) for the flood event of 10–11 April 1993. Also marked on this diagram are the sample positions of 16 sets of sedimentation traps

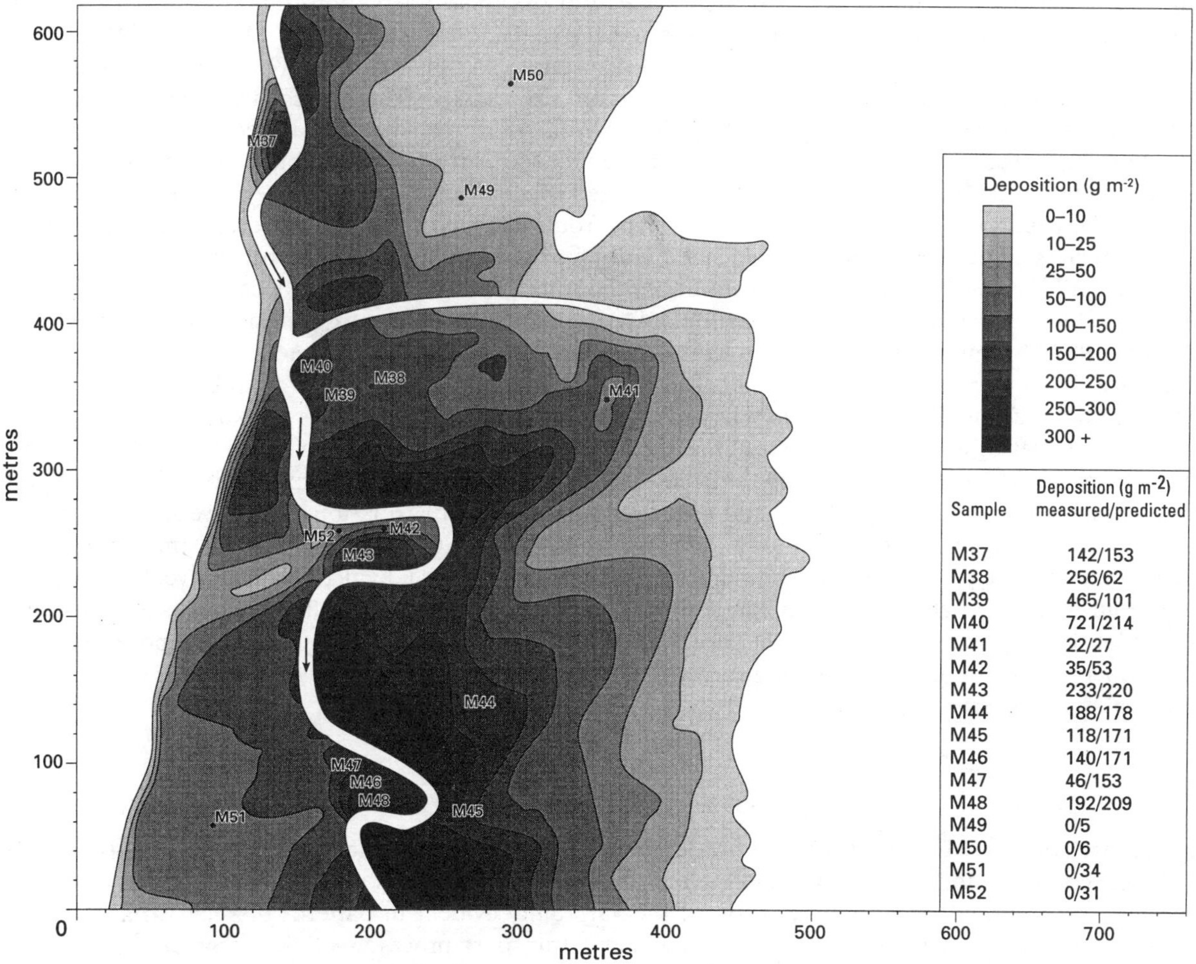

Figure 7. Predicted pattern of total deposition amounts (g m^{-2}) for the flood event of 10–11 April 1993. Also marked are measured and predicted deposition amounts for 16 sets of sedimentation traps

obtained for this flood, together with mean measured and predicted deposition amounts for each of these sets. The mean absolute error for the 16 sets is 88 g m^{-2}. However, three-quarters of the total error for this data set is associated with the three sets of sedimentation traps located close to the main channel on the downstream side of the major drainage ditch. Peak predicted deposition amounts for this flood event are 300 g m^{-2} (equivalent to 0·3 mm of deposition assuming a deposit density of 1 g cm^{-3}). These depositional highs occur in the downstream half of the reach on the inside of the meander bends and in the low-lying entrances to the abandoned channel elements. Maximum predicted deposition amounts in the upstream half of the reach are about half of this downstream value and occur on the inside of the meander bend and in the breach of the levee near the upstream reach boundary. In general, predicted deposition amounts decrease with distance from the channel at rates that reflect the patterns of suspended sediment dispersion shown in Figure 6 and the duration of inundation experienced at different locations (as indicated by Figures 3–5). Clearly, both of these factors are strongly influenced by the complex nature of the floodplain topography within the reach.

DEPOSIT GRAIN SIZE CHARACTERISTICS

Samples of deposited sediment were retrieved from 12 of the 16 sedimentation trap sets and their ultimate grain size characteristics were determined by laboratory analysis using a Malvern Mastersizer MS20 (after the removal of the organic fraction and chemical dispersion). As identified above, the model predicts the amount of deposition for each of six effective grain size classes. These predicted effective grain size distributions were converted to ultimate grain size distributions using measured ultimate grain size distributions for each of the effective classes determined for samples of material obtained from the elutriation apparatus described above. Figure 8 shows the predicted effective, predicted ultimate and measured ultimate grain size distributions for the 12 sedimentation traps from which sediment was retrieved (the $>63^{A}$ μm and $>63^{B}$ μm classes have been lumped into a single class for these purposes). A number of clear trends can be identified in Figure 8. First, all three distributions tend to be coarser for deposits located closer to the main channel. This is consistent with the predictions of other models of overbank deposition (e.g. James, 1985; Pizzuto, 1987). Secondly, there is a good level of agreement between the predicted and measured ultimate grain size distributions for the 12 samples. Despite this, there is clearly more variability in the measured ultimate grain size distributions than in the predicted ultimate distributions. This probably reflects the use of a limited number of effective size classes in the model, each of which are characterized using only a single ultimate grain size distribution. Finally, it is clear that the predicted effective size distributions are significantly coarser than both the predicted and measured ultimate size distributions. Effective size distributions are dominated by material >32 μm in diameter, while ultimate size distributions are composed of approximately 45% particles <8 μm in diameter with 55% of material divided relatively evenly between the other four grain size fractions. Clearly, the ultimate grain size distributions of flood deposits may bear little resemblance to the effective grain size distributions of the sediment immediately prior to deposition; hence, the former may be of limited value in any attempt to understand the influence of hydraulic conditions on sedimentation patterns.

SUMMARY

Spatial patterns of hydraulic variables (flow depth and velocity) and sedimentological variables (suspended sediment concentration, deposition amount and deposit grain size) associated with floodplain flows are extremely complicated. A large proportion of the variability evident in patterns of such variables can be attributed to the effect upon fluid flow and sediment transport processes of the complex topographical characteristic of river floodplains. This study has demonstrated the ability of a relatively simple numerical scheme to generate realistic predictions of hydraulic and sedimentological variables when coupled with a

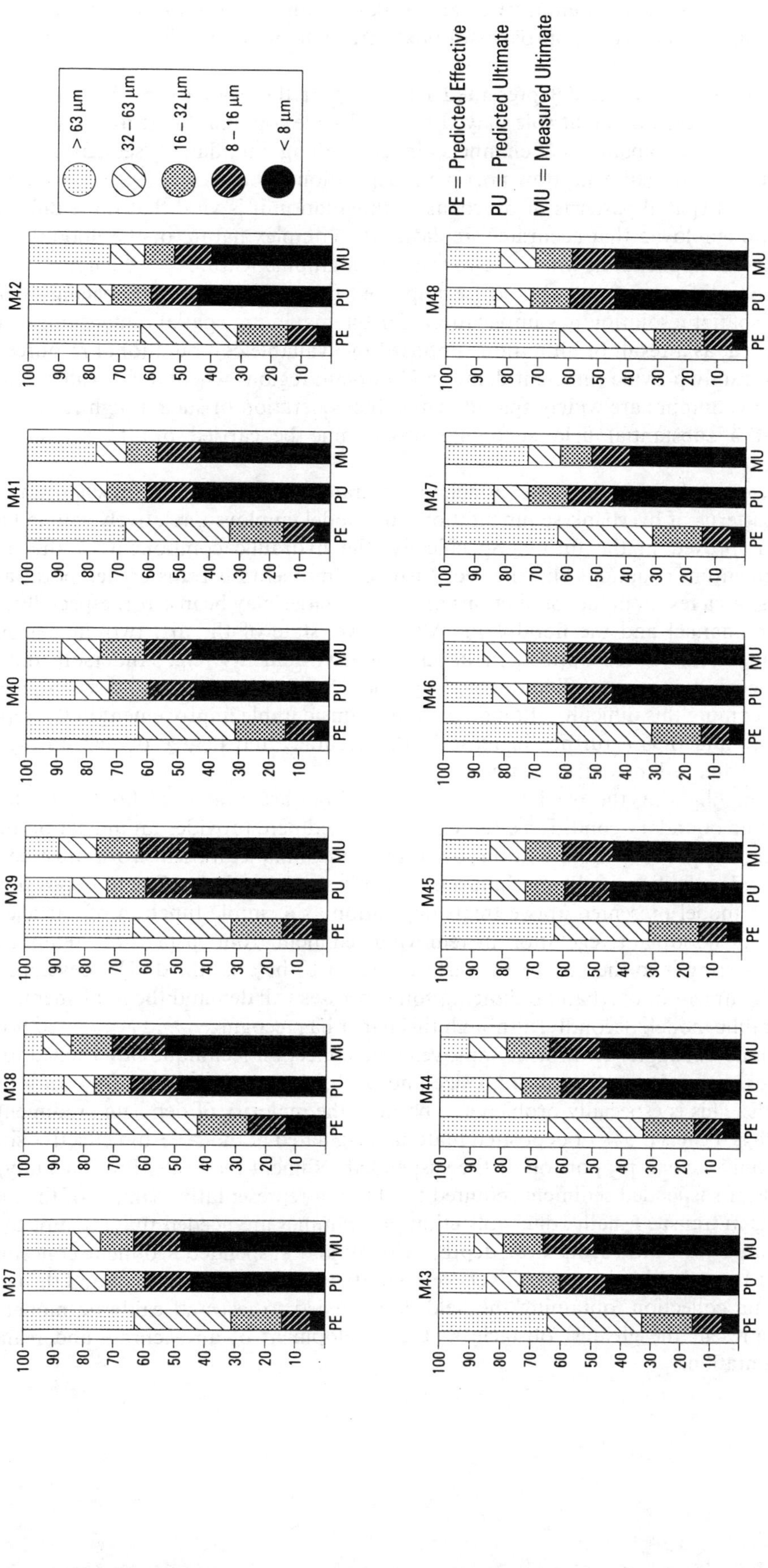

Figure 8. Predicted *effective*, predicted *ultimate* and measured *ultimate* grain size distributions for flood deposits from 12 sets of sedimentation traps

detailed finite difference grid representative of actual floodplain topography. Furthermore, the combined numerical modelling and field work approach adopted here highlights a number of important issues relevant to future investigations in this area.

The results of the numerical model presented here support the evidence of field studies (e.g. Walling *et al.*, 1992) in stressing the dominant role played by small-scale topographic features, such as levees, bank breaches, depressions and abandoned channels, in controlling inundation sequences, floodwater flow directions and suspended sediment transport and deposition processes. Clearly, in order to generate realistic predictions of spatial patterns of overbank sedimentation it is vital that finite difference and finite element meshes are employed that accurately replicate the complex nature of the floodplain geometry. In the case of the current application, where the relevant topographic features have length scales of the order of 10–50 m, a finite difference mesh should be employed that utilizes a nodal spacing of not more than 5 m. Such a high spatial resolution has implications for the maximum spatial dimensions of the floodplain reach to be modelled, as a result of limitations imposed by available computational resources. In addition, topographic maps rarely provide sufficient detail in floodplain regions where down-valley gradients may be very low and where contours are widely spaced. Thus, the generation of such a high resolution model grid will demand that a substantial field surveying programme be carried out to provide the necessary topographic data.

Despite the success of the numerical model in accurately simulating the mechanisms of floodplain inundation and patterns of overbank sedimentation, the model employs a relatively simple basis and could be substantially improved in the future. Specifically, the hydraulic component of the model employs continuity of momentum equations that neglect turbulent momentum transfer, temporal and convective acceleration terms. As a result, model predictions of flow velocities may be in error, especially at the interface between the main channel and the floodplain. Alternative, state of the art, two-dimensional models of floodwater hydraulics have been applied in floodplain environments (e.g. Anderson and Bates, 1994); however, such schemes may prove difficult to employ where floodplain topography is complicated (cf. Bates *et al.*, 1992). Overcoming this difficulty should result in a considerable improvement in the representation of hydraulic processes and thus a further increase in the accuracy and degree of realism evident in model predictions of overbank sedimentation patterns.

In addition to highlighting the need to couple state of the art models of floodwater hydraulics with detailed topographic boundary conditions, the work presented here provides an indication of several other important areas that require further attention. First, theoretical and empirical work must be carried out to improve current understanding of the mechanisms controlling suspended sediment deposition in overbank environments. The model presented above treats deposition as a simple function of particle settling rates; however, the role of floodplain vegetation in removing sediment from suspension, retaining it upon leaf surfaces and transferring it to the floodplain surface has yet to be quantified. The development of a fully physically based treatment of overbank sedimentation processes will demand the implementation of a three-dimensional hydraulic model. Secondly, although this paper has recognized and accounted for the transport of suspended sediment in aggregated form, improved data collection techniques are needed to provide more detailed records of suspended sediment grain size and settling characteristics as they vary over inter- and intra-event periods. This is especially problematic because the majority of deposited sediment is associated with the coarsest aggregates (35% of deposited material originated in the >63 μm effective size class) which in turn make up only a small proportion of the suspended sediment load (2% for this study). This greatly increases the mass of suspended sediment required to obtain a representative sample of the most important material while it is in transit. Finally, data collection programmes are needed that attempt to overcome the difficulties associated with measuring flow depth, velocity and suspended sediment concentration during flood periods over large floodplain areas. Given the spatial data resolution required and the labour intensive nature of such data collection, obtaining the field data needed to test and calibrate models of overbank processes may represent the greatest obstacle to the development of an accurate and realistic model of floodplain sedimentation.

ACKNOWLEDGEMENTS

The authors gratefully acknowledge the support of the UK Natural Environment Research Council in providing a postgraduate studentship for work on overbank sedimentation in Devon rivers. The cooperation of local landowners in permitting access to floodplain sites is also acknowledged with gratitude. This paper benefited substantially from the constructive comments of several anonymous reviewers.

REFERENCES

Anderson, M. G. and Bates, P. D. 1994. 'Evaluating data constrains on two dimensional finite element models of floodplain flow', *Catena*, **21**, 1–15.

Anderson, M. G., Walling, D. E., and Bates, P. D. (Eds), (1996). *Floodplain Processes*. John Wiley & Sons, Chichester.

Bates, P. D., Anderson, M. G., Baird, L., Walling, D. E., and Simm, D. J. 1992. 'On the potential for using two-dimensional finite element schemes in geomorphological investigations of floodplain environments', *Earth Surf. Process. Landf.*, **17**, 575–588.

Beavers, A. H. and Jones, R. L. 1966. 'Elutriation for fractionating silts', *Soil Sci. Soc. Am. Proc.*, **30**, 126–128.

Carling, P. A. and Petts, G. E. (Eds), 1992. *Lowland Floodplain Rivers: Geomorphological Perspectives*. John Wiley & Sons, Chichester.

Cunge, J. A., Holly, F. M., and Verwey, A. 1980. *Practical Aspects of Computational River Hydraulics*. Pitman, London. 420 pp.

Engelund, F. 1970. 'Instability of erodible beds', *J. Fluid Mech.*, **42**, 225–244.

Ervine, D. A. and Ellis, J. 1987. 'Experimental and computational aspects of overbank floodplain flow', *Trans. R. Soc. Edin. Earth Sci.*, **78**, 315–325.

Fischer, H. B., List, E. J., Koh, R. C. Y., Imberger, J., and Brooks, N. H. 1979. *Mixing in Inland and Coastal Waters*. Academic Press, London. 483 pp.

Gee, D. M., Anderson, M. G., and Baird, L. 1990. 'Large-scale floodplain modelling', *Earth Surf. Process. Landf.*, **15**, 513–523.

James, C. S. 1985. 'Sediment transfer to overbank sections', *J. Hydraul. Res.*, **23**, 435–452.

Lambert, C. P. and Walling, D. E. 1987. 'Floodplain sedimentation: a preliminary investigation of contemporary deposition within the lower reaches of the River Culm, Devon, UK', *Geograf. Annal.*, **69A**, 393–404.

Lin, B. and Falconer, R. A. 1995. 'Modelling sediment fluxes in estuarine waters using a curvilinear coordinate grid system', *Estuarine Coastal Shelf Sci.*, **41**, 413–428.

Nicholas, A. P. 1994. 'Modelling overbank deposition on floodplains: a case study of the River Culm, Devon, *PhD Thesis*, University of Exeter.

Nicholas, A. P. and Walling, D. E. 1997. 'Investigating spatial patterns of medium-term overbank sedimentation on floodplains: a combined numerical modelling and radiocaesium-based approach', *Geomorphology*, **19**, 133–150.

Ongley, E. D., Bynoe, M. C., and Percival, J. B. 1981. 'Physical and geochemical characteristics of suspended solids, Wilton Creek, Ontario', *Can. J. Earth Sci.*, **18**, 1365–1379.

Parker, G. 1978. 'Self-formed straight rivers with equilibrium banks and mobile bed: I — the sand–silt river', *J. Fluid Mech.*, **89**, 109–125.

Pizzuto, J. E. 1987. 'Sediment diffusion during overbank flows', *Sedimentology*, **34**, 301–317.

Shiono, K. and Knight, D. W. 1991. 'Turbulent open-channel flows with variable depth across the channel', *J. Fluid Mech.*, **222**, 617–646.

Umlauf, G. and Bierl, R. 1987. 'Distribution of organic micropollutants in different size fractions of sediment and suspended solid particles of the River Rotmain', *Z. Wass. Abwass.-Forsch.*, **20**, 203–209.

Utnes, T. 1993. 'Finite element current and sediment transport modelling', *J. Cont. Shelf Res.*, **13**, 891–902.

Walling, D. E. and Kane, P. 1984. 'Suspended sediment properties and their geomorphological significance', in Burt, T. P. and Walling, D. E. (Eds), *Catchment Experiments in Fluvial Geomorphology*. Geobooks, Norwich. pp. 311–334.

Walling, D. E. and Woodward, J. C. 1993. 'Use of a field-based water elutriation system for monitoring the *in situ* particle size characteristics of fluvial suspended sediment', *Wat. Res.*, **27**, 1413–1421.

Walling, D. E., Quine, T. A., and He, Q. 1992. 'Investigating contemporary rates of floodplain sedimentation', in Carling, P. A. and Petts, G. E. (Eds), *Lowland Floodplain Rivers: Geomorphological Perspectives*. Wiley, Chichester. pp. 165–184.

Wijbenga, J. H. A. 1985. 'Steady depth-averaged flow calculations on curvilinear grids', in *Second International Conference on the Hydraulics of Floods and Flood Control*. Cambridge. pp. 373–387.

Willets, B. B. and Hardwick, R. I. 1993. 'Stage dependency for overbank flows in meandering channels', *Proc. Inst. Civil Engrs. Wat. Marit. Energy*, **101**, 45–54.

Woodward, J. C. and Walling, D. E. 1992. 'A field sampling method for obtaining representative samples of composite fluvial suspended sediment particles for SEM analysis', *J. Sediment. Petrol.*, **62**, 742–744.

Wormleaton, P. R., Allen, J., and Hadjipanos, P. 1982. 'Discharge assessment in compound channel flow', *ASCE J. Hydraul. Div.*, **108**, 975–994.

17

MODELLING WIDTH ADJUSTMENT IN STRAIGHT ALLUVIAL CHANNELS

STEPHEN E. DARBY*

USDA-ARS National Sedimentation Laboratory, PO Box 1157, 598 McElroy Drive, Oxford, MS 38655, USA

ABSTRACT

Many existing numerical alluvial channel models are limited because they are unable to account for adjustments of channel width through time. In this paper, the basis for modelling width adjustment in straight river channels is discussed. Features and capabilities of selected models are discussed in terms of advances made in representing the physical processes and mechanisms involved in width adjustment, and in terms of approaches used to characterize initial and boundary conditions accurately. Validation and verification studies of existing width adjustment models are also considered. Existing models have, for the most part, not yet been subjected to rigorous evaluation of their numerical robustness, accuracy in replicating analytical solutions, accuracy in replicating laboratory data or their ability to simulate field situations. In part, these limitations reflect the lack of available detailed laboratory and field data sets required to evaluate complex models. The potential of width adjustment models for use as tools for the investigation of river channel dynamics is briefly described using two examples.

KEY WORDS numerical modelling; river width adjustment; morphological models; channel change; bank erosion; verification; validation

INTRODUCTION

Most existing deterministic alluvial channel models assume that the channel banks are rigid, or adjust bank geometry according to theoretical, but non-mechanistic, arguments regarding energy or stream power expenditure (Chang, 1988). Fixed width models are valid for many cases where the presence of natural or human controls, such as stream bank revetment or non-erodible bank materials, constrain width adjustment. Similarly, high resolution hydraulic models are often applied in simulations of rather small duration during which transient width adjustments may be negligible. However, river bank erosion and mass failure processes often act rapidly, and the validity of fixed width models in these circumstances is questionable. Simulations of the response of channel morphology through time can be obtained by solving a set of deterministic governing equations of flow resistance, flow momentum and continuity, sediment transport and conservation of sediment mass, in conjunction with a set of specific initial and boundary conditions. Initial conditions define the morphological and sedimentological characteristics of the channel prior to morphological adjustment. Boundary conditions specify the sediment load and water discharge at the upstream boundary of the reach as a function of time. Additional boundary conditions may be required for numerical solution of individual process equations. For example, flow momentum equations usually require specification of values of flow parameters at the interface of the water with the free surface and the banks.

In practice, the validity and accuracy of numerical modelling approaches are limited as a result of:

(1) difficulties in formulating complete and accurate mathematical models representing the governing processes, for the range of conditions encountered in nature;

* Present address: Stephen E. Darby, Department of Geography, University of Southampton, Southampton SO17 1BJ, UK.

(2) difficulties in accurately parameterizing the initial and boundary conditions; and

(3) inherent approximations associated with numerical techniques employed to solve the governing differential momentum and mass conservation equations. These approximations also lead to difficulties in choosing computational time step lengths, because the characteristic time-scales associated with bank erosion processes are often much longer than those associated with adjustment of the flow and the bed.

This paper reviews state of the art models of width adjustment for straight channels in the context of the first two issues above. Numerical solution techniques are not directly considered here because this topic has been covered in standard texts (Lapidus and Pinder, 1982). This paper complements a review recently undertaken by the American Society of Civil Engineers (ASCE) Task Committee on 'Hydraulics, Bank Mechanics, and Modelling of Width Adjustment' (ASCE, in press). The ASCE Task Committee identified the capabilities and limitations of existing approaches to predicting width adjustment for engineers and practitioners, covering empirical, analytical and computational methods. This paper focuses on providing detailed information about mechanistic numerical models only, with particular reference to flow in straight channel reaches. Mechanistic width adjustment models for irregular planforms have not yet been developed. Width adjustment models based on extremal hypotheses are not reviewed here as their features and capabilities are widely known and have been reported elsewhere (Ferguson, 1986; Chang, 1988; ASCE, in press). Models for non-cohesive and cohesive bank materials are discussed separately.

THE GEOMORPHOLOGICAL CONTEXT

River width adjustments can be initiated by a wide variety of causes which may act at a wide range of spatial and temporal scales. Any change in the flow or sediment regime can cause net erosion or deposition on individual channel banks that might lead to changes in width (ASCE, in press). Changes in channel width only occur when the direction and speed of migration of one bank is not equal to the direction and speed of migration of the opposite bank. For example, deposition on one bank can be matched by erosion on the opposite bank to maintain a steady width during lateral migration of a channel in dynamic equilibrium. For an individual bankline, migration rate is a function of the relative rates of supply and removal of sediment to and from the near-bank zone (Figure 1). Thorne (1982) outlined the concept of basal end-point control using a three point classification of bank migration conditions as they relate to sediment supply and transport processes in the near-bank environment:

(1) *Impeded removal.* Combinations of sediment transported by the flow, supplied by net fluvial entrainment of the bank material and supplied by geotechnical failures, supply bank material to the basal zone at a higher rate than it is removed down- and cross-stream by the flow. Net deposition results and bank heights and angles are reduced, helping to stabilize the bank with respect to geotechnical failure. This reduces the supply of failed bank materials, tending towards the second classification.

(2) *Unimpeded removal.* Delivery and removal of material to and from the basal zone are in approximate balance, no net change in the volume of sediment stored at the base occurs. The bank migrates at a constant rate determined by the sediment transport rate at the base of the bank. In essence, this is a state of dynamic equilibrium.

(3) *Excess basal capacity.* Sediment transport capacity is greater than the supply rate of sediment by bank erosion, bank collapse and fluvial transfer. Basal lowering occurs, increasing bank height and angle, and decreasing geotechnical stability. If a threshold of critical stability is crossed bank failure occurs, increasing the supply of sediment to the basal zone. The basal end-point status therefore tends back towards the second state.

The concept of basal end-point control illustrates how the rate of bank migration depends on a dynamic interaction between the flow, sediment transport, bank erosion and geotechnical bank stability in the near-bank zone (Figure 1). Focusing attention on the near-bank zone highlights the problems faced in

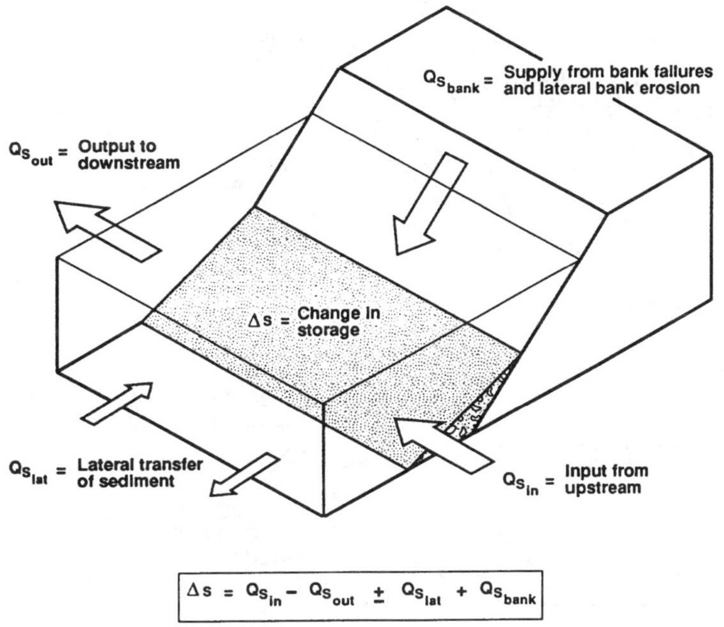

Figure 1. Sediment fluxes in the near-bank zone (from Thorne and Osman, 1988)

developing a model able to replicate the physics involved. For example, the distributions of boundary shear stress and sediment transport are difficult to predict in near-bank zones characterized by complex topography and roughness distributions (Shiono and Knight, 1990). In addition, a wide variety of bank erosion and deposition processes may be involved in the bank's contribution to this dynamic interaction (Thorne, 1982; ASCE, in press). Bank erosion processes include fluvial entrainment of bank material particles and mass failures. The effectiveness of fluvial entrainment processes is frequently influenced by subaerial or subaqueous weathering processes such as frost heave (Lawler, 1986) or desiccation. Depending on the geometry of the bank profile and the geotechnical properties of the bank materials, bank failure may occur by a variety of specific mechanisms, including planar (Lohnes and Handy, 1968; Osman and Thorne, 1988), rotational (Osman, 1985), cantilever (Thorne and Tovey, 1981) or piping/sapping (Hagerty, 1991) type failures (Figure 2). A separate analysis is required to represent each specific mechanism. Depositional processes are equally diverse. Alternate bars may become attached to stream banks, sediment may be deposited directly on sloping banks or islands may develop along the centre of the channel to create a divided flow with an accretionary inner bank (ASCE, in press).

The sheer number and diversity of specific processes and mechanisms involved means that, even assuming that the processes are fully understood and are correctly formulated mathematically, a single computational model is unlikely to have universal validity. Instead, computational models of river width adjustment have been developed to simulate specific, idealized, representations of natural fluvial systems.

NUMERICAL WIDTH ADJUSTMENT MODELS: REPRESENTATION OF PHYSICAL PROCESSES AND INITIAL AND BOUNDARY CONDITIONS

Alluvial channel modelling approaches reviewed here (Table I) are based on the sediment continuity equation for bed material load. In a Cartesian coordinate system this is expressed as

$$\frac{\partial z}{\partial t} + \frac{1}{1-\lambda}\left(\frac{\partial q_x}{\partial x} + \frac{\partial q_y}{\partial y}\right) = q_b \qquad (1)$$

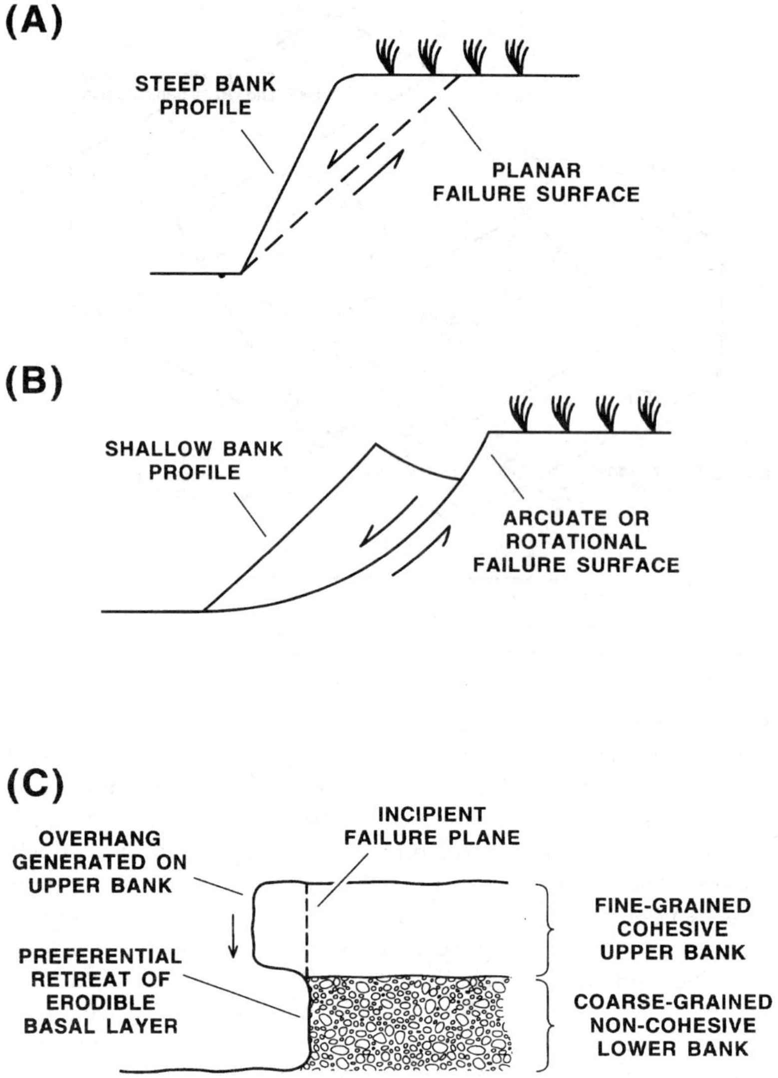

Figure 2. Specific mechanisms of bank collapse. (A) Planar failure; (B) rotational failure; (C) cantilever failure

where z is the local bed elevation (m), t is the time coordinate (s), q_b is the local volumetric inflow rate of bank material per unit width and reach length (m/s), λ is the sediment porosity, q_x and q_y are the local longitudinal and lateral sediment transport fluxes per unit channel width (m^2/s), respectively, x is the longitudinal coordinate (m) and y is the lateral coordinate (m). By specifying initial conditions at the computational nodes representing the prototype channel, evolution of the channel topography through time is simulated by solving Equation (1) in conjunction with a set of appropriate boundary conditions (see below). It is, therefore, often difficult to determine whether inaccuracies in model simulations are a result of limitations of the process models or errors in the characterization of the initial and/or boundary conditions.

Characterization of initial conditions

Initial conditions refer to the initial flow parameters, topography [$z(x, y)$] and boundary material characteristics (density and size distribution of bed material; cohesion, friction angle and density of

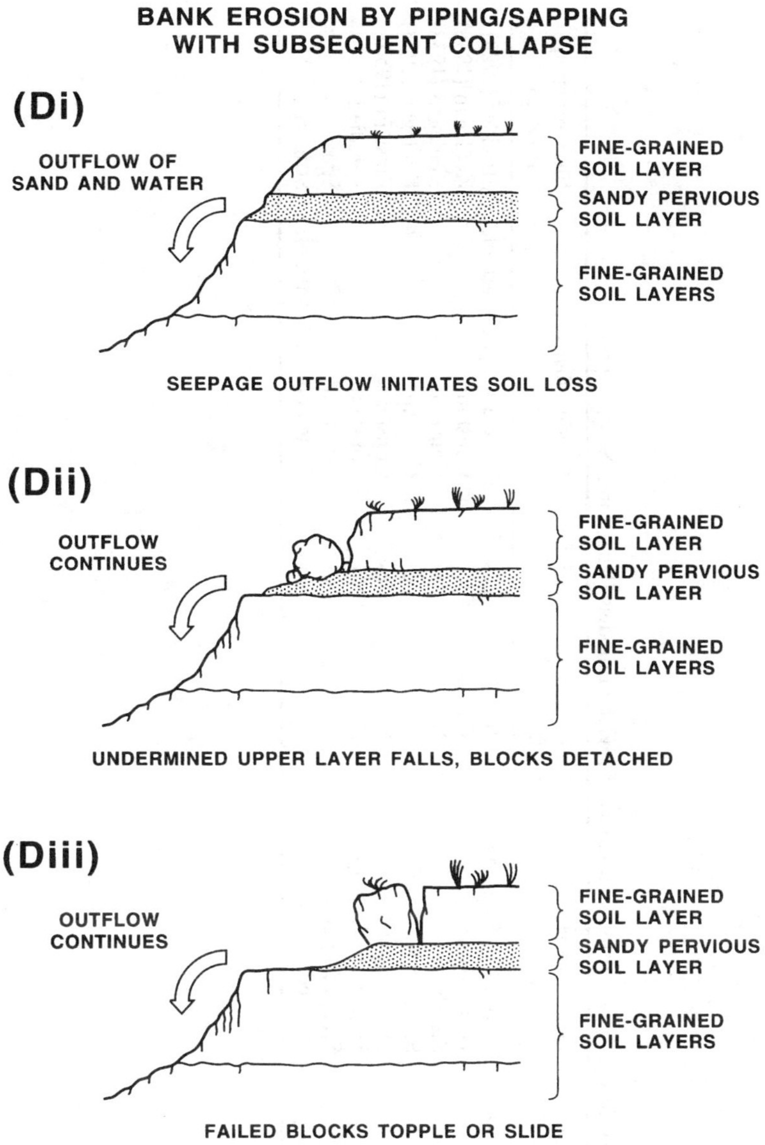

Figure 2. Specific mechanisms of bank collapse. (D) piping/sapping failure (from Hagerty, 1991)

bank material) at each node. Even for the simplest fluvial system, accurate representation of spatial non-uniformities of channel topography and boundary material characteristics using a finite number of computational nodes is difficult. For example, bank materials are usually non-uniform, and their geotechnical properties may vary laterally within the floodplain and/or longitudinally along the banklines. Even if the computational nodes adequately characterize the spatial domain of the channel, the number of measured data points is typically much smaller than the number of nodes, and the locations of the nodes and data points may not coincide, necessitating interpolation. Advances in field data collection techniques have recently been made, which, in some cases, considerably aid characterization of complex channel bathymetry. Specifically, photogrammetric survey techniques (Lane *et al.*, 1993) now enable very high resolution topographic data collection. These techniques are unsuitable for channel surfaces covered by the flow, or for river

Table I. Features of flow routing submodels of reviewed models (based on ASCE, in press)

Model	Dimension	Discharge variation over time	Secondary flow	Lateral shear	Friction factor	Flow resistance formulae
Darby and Thorne (1996a)	Quasi-2D	Stepped hydrograph	No	Yes	Time and space variable	Strickler (1923)
Li and Wang (1993, 1994)	3D	Unsteady flow	Yes	Yes	Constant	Keulegan (1938)
Kovacs and Parker (1994)	2D	Steady flow	No	Yes	Constant	Keulegan (1938)
Wiele (1992)	2D	Steady flow	No	Yes	Constant	Keulegan (1938)
Pizzuto (1990)	2D	Steady flow	No	Yes	Constant	Einstein (1950)
Borah and Bordoloi (1989)	1D	Stepped hydrograph	No	No	Constant	Specified
Alonso and Combs (1986)	1D	Stepped hydrograph	No	No	Constant	Specified
Osman (1985)	1D	Stepped hydrograph	No	No	Time and space variable	Stricker (1923)

Note: All these models neglect convective acceleration terms reaches longer than about 100 m. In any case, sampling problems and errors associated with characterization of boundary sediments still remain (Church *et al.*, 1987).

Specification of initial flow parameters is even more difficult than for initial topographic and sediment conditions. Accurate, synoptic, measurement of flow parameters at each node for the initial condition is impossible using either existing conventional measurement techniques, such as mechanical or electromagnetic current meters, or innovative devices such as acoustic doppler current profilers (ADCP). ADCP probes do enable simultaneous high resolution data sampling of flow parameters in a vertical column, but time delays are still inevitable during a synoptic sampling problem. It is these time delays during measurement that make it impossible to reference initial flow conditions to a common zero time datum. In the absence of accurate synoptic data for initial flow parameters, results of model simulations should be based on careful sensitivity analyses to investigate the uncertainty introduced into the solution through use of inaccurate initial conditions.

Characterization of boundary conditions

In time-based model validation and verification studies, simulations of stream gauged channel reaches are conducted so that model output can be compared with historical laboratory and/or field data. Time-dependent boundary conditions, such as discharge and sediment supplied from the upstream boundary, are based on measured data. The accuracy of these approaches is partially dependent on the quality of data used. Reviews of problems in collecting stream gauging data are available in standard texts such as Gregory and Walling (1973).

Prognostic studies require the user to predict unknown values of these time-dependent boundary conditions. These values depend on runoff and sediment yield processes controlled by meteorological events that are essentially unpredictable over the long time-scales (>1 year) required in many alluvial channel modelling studies. Most prognostic modelling studies use historically derived hydrographs and sediment rating curves and assume that runoff and sediment yield processes do not vary through time. Such approaches can only be applied on those streams where the requisite historical data are available and are not appropriate for investigations aimed at predicting channel response to environmental change.

Alternative approaches rely on the specification of steady design discharges and sediment loads to represent the integrated effects of the actual unsteady water and sediment discharges. This approach allows simulation of channel response to specific return period flows, which is useful in engineering design, but still leaves open the question: which steady flow represents the integrated effects on channel morphology of the naturally variable water and sediment discharges? Specification of this so-called 'dominant discharge' has been the subject of considerable controversy in the geomorphological literature for many years (Wolman and Miller, 1960; Biedenham and Thorne, 1994). For channels in dynamic equilibrium, the dominant discharge appears to be that which transports the most sediment, but the return period of that flow varies as a function of environmental conditions, such as runoff duration and boundary material characteristics. Further studies are required to refine the ways in which time dependent boundary conditions are specified in prognostic studies. Long-term channel monitoring programmes are also required to test predictions of channel behaviour, using the developed techniques.

Modelling the flow field

The three-dimensional flow of water in open channels is governed by the conservation of momentum and mass, expressed as a set of equations in Cartesian coordinates as

$$\frac{\partial u}{\partial t} + u\frac{\partial u}{\partial x} + v\frac{\partial u}{\partial y} + w\frac{\partial u}{\partial z} + u\frac{v}{r} = \frac{-1}{\rho}\frac{\partial p}{\partial x} + F_x \tag{2}$$

$$\frac{\partial v}{\partial t} + u\frac{\partial v}{\partial x} + v\frac{\partial v}{\partial y} + w\frac{\partial v}{\partial z} - \frac{u^2}{r} = \frac{-1}{\rho}\frac{\partial p}{\partial y} + F_y \tag{3}$$

$$\frac{\partial w}{\partial t} + u\frac{\partial w}{\partial x} + v\frac{\partial w}{\partial y} + w\frac{\partial w}{\partial z} + g = \frac{-1}{\rho}\frac{\partial p}{\partial z} + F_z \tag{4}$$

$$\frac{\partial u}{\partial x} + \frac{1}{r}\frac{\partial vr}{\partial y} + \frac{\partial w}{\partial z} = 0 \tag{5}$$

where u, v, w are the local velocity components (m/s) in the x, y and z directions, respectively, r is the local radius of curvature (m), ρ is the fluid density (kg/m^3), p is the pressure (N/m^2) and F_x, F_y and F_z are the local friction terms in the x, y and z directions, respectively. Although Equations (2)–(5) are exact descriptions of the flow field, inaccuracies in estimating the friction terms, and simplifying assumptions introduced to aid their numerical solution, introduce errors and uncertainty into estimates of the hydraulic parameters thus obtained. Detailed reviews of these limitations have been made by Knight and Shiono (1995) and ASCE (in press), but so that this paper can stand alone the main assumptions used in simplifying Equations (2)–(5) are listed here.

1. It is commonly assumed that the flow is steady ($dQ/dt = 0$) (Table I). The approach adopted is either to specify a steady discharge to represent the integrated effects of the hydrograph (see above), or to simulate unsteady flows through use of a stepped hydrograph. In the latter approach, discharge is assumed steady in any given discrete time step, but the value is allowed to vary between discrete time steps.

2. The channel morphology and flow field is frequently assumed to be spatially uniform, so that convective acceleration terms (derivatives of u, v or w velocity components with respect to the x-direction) in the momentum Equations (2)–(4) are often neglected. Experimental work has indicated that convective acceleration terms in Equations (2)–(4) are significant controls on the flow field in straight channels with natural bed topography (Whiting and Dietrich, 1991), so that the assumption of spatial uniformity is often unrealistic. In any case, such an assumption would in any case tend to defeat the point of mobile boundary modelling.

3. For straight river channels, secondary flows are commonly assumed to be negligible ($v = w = 0$) (Table I). Experimental data indicate that for straight channels with spatially non-uniform topography and variable flow depth, secondary flows significantly influence the flow field and boundary shear stress distribution (Knight and Sellin, 1987), especially close to steep banks. In natural straight channels, secondary flows are commonly generated by shear layer turbulence caused by large lateral variations of flow depth and/or friction factor (Shiono and Knight, 1990; also see below). These conditions occur frequently during overbank flows. Confinement of the model to straight river channels also tends to be self-defeating in the case of mobile boundary modelling: the channel ceases to be straight as soon as any bank erosion is allowed to occur.

4. In natural channels, and particularly during overbank flows, lateral gradients of flow depth and friction factor induce significant lateral velocity gradients. Lateral shear terms are frequently excluded from hydraulic submodels employed in width adjustment modelling (Table I). Although this assumption is appropriate for the central region of a wide channel (width to depth ratio > 20) which is not influenced by the banks, such terms are significant within the near-bank zone and are, therefore, important for width adjustment modelling (Shiono and Knight, 1990; ASCE, in press).

5. Accurate prediction of the friction factor is a problem. Some models simply allow the user to specify the Manning 'n' value, and assume this value is constant in space and/or time. More commonly, an empirically calibrated flow resistance equation is selected from the large number available in the literature and is used to relate flow and boundary material characteristics to the friction factor. To increase the potential for modelling a range of environments, many width adjustment models allow the user to select an equation from a menu (Table I). Usually, no selection criteria are provided for the user, and use of a single equation does not guarantee that it will be valid throughout the space–time domain of the model. Existing flow resistance

equations employed in width adjustment models do not account for the effects of bedforms, or in-channel and/or riparian vegetation.

Sediment flux field

Longitudinal and lateral sediment transport fluxes are estimated using specific empirically calibrated sediment transport equations. Often, the various models of width adjustment allow the user to select an appropriate equation from a menu (Table II). The capabilities and limitations of these equations have been reviewed elsewhere (Gomez and Church, 1989; Yang and Wan, 1991). All such sediment transport equations are valid only within the range of data used to calibrate them, and sediment transport flux estimates are also commonly subject to order of magnitude errors (Gomez and Church, 1989). The potential for errors in simulating sediment flux divergences in Equation (1) is, therefore, large.

Another important feature of the sediment transport submodel, in terms of influencing width adjustment simulations, is the capability to simulate transport of heterogeneous sediment (Borah *et al.*, 1982; Rahuel *et al.*, 1989; Bridge and Bennett, 1992; Table II). Width adjustment involves entrainment, transport and deposition of diverse sediment particles, especially for channels with non-cohesive bed materials and cohesive bank materials.

Fluvial entrainment and mechanisms of bank collapse

Fluvial entrainment of bank material results in incremental bank retreat. Increases of bank height owing to near-bank bed degradation, or increases in bank steepness owing to fluvial entrainment of the lower bank, may act alone or together to decrease the stability of the bank with respect to geotechnical failure (Osman and Thorne, 1988).

For non-cohesive banks, hydraulic shear erosion is simulated by applying Equation (1) across the flow width of the channel. Prediction of the time rate of cohesive bank erosion remains a difficult task (Partheniades, 1965; Grissinger, 1982; Lawler, 1986). Excess shear stress formulations are difficult to apply because the critical shear stress value is influenced by diverse processes (Partheniades, 1965; Grissinger, 1982). For example, weathering processes, such as frost heave or desiccation, may influence rates of fluvial erosion of exposed banks more than the intensity of the near-bank flow (Lawler, 1986). For undisturbed cohesive banks, the critical shear stress is related more closely to the electrochemical properties of the soil pore and eroding fluids (Arulanandan *et al.*, 1980) than readily measurable physical properties such as shear strength or cohesion.

Most cohesive bank width adjustment modelling approaches have been based solely on analysis of planar failures (Table III). Significant advances have been made in simulating this particular mechanism of failure. Analyses have been developed by Osman (1985), and reported in Osman and Thorne (1988), to account for the bank profile geometry associated with natural, eroding river banks that are destabilized through combinations of lateral erosion and bed degradation. Earlier stability analyses (e.g. Lohnes and Handy, 1968) were restricted to a simple bank geometry and excluded the effects of fluvial entrainment on the bank profile.

The Osman–Thorne stability analysis is subject to two main limitations (Simon *et al.*, 1991). First, it does not include the effects of either excess pore water or hydrostatic confining pressures. Secondly, the failure plane is assumed to pass through the toe of the bank. Simon *et al.* (1991) developed a bank stability analysis designed to account for hydrostatic and pore water pressure effects explicitly, and allowing the failure plane to intersect the bank profile at an arbitrary location. Recent research by Darby and Thorne (1996b) combines the features of the analyses of Simon *et al.* (1991) and Osman and Thorne (1988). This new approach accounts for failures on steep, eroding banks with arbitrary failure plane locations, accounting for excess pore water and hydrostatic confining pressures. Darby and Thorne (1996b) demonstrated that their analysis provided improved predictions of bank stability and failure plane geometry compared with previous analyses (Figure 3).

Table II. Features of sediment routing submodels of reviewed models (based on ASCE, in press)

Model	Routing method	Longitudinal flux divergence	Lateral flux divergence	Bed load	Suspended load	Transport equation(s)	Sorting	Bed material
Darby and Thorne (1996a)	Quasi-2D	Yes	Yes	Yes	Yes	Engelund and Hansen (1967)	Yes	Sand
Li and Wang (1993, 1994)	2D	Yes	Yes	Yes	No	Meyer-Peter and Muller (MPM) (1948)	No	Gravel
Kovacs and Parker (1994)	2D	No	Yes	Yes	No	Kovacs and Parker (1994)	No	Gravel
Wiele (1992)	2D	No	Yes	Yes	No	Parker (1979); Meyer-Peter and Muller (1948)	No	Sand gravel
Pizzuto (1990)	2D	No	Yes	Yes	No	Parker (1983)	No	Sand
Borah and Bordoloi (1989)	1D	Yes	No	Yes	Yes	Borah *et al.* (1982)	Yes	Sand gravel
Alonso and Combs (1986)	1D	Yes	No	Yes	Yes	Yang (1973); Graf (1971); Meyer-Peter and Muller (1948)	Yes	Sand gravel
Osman (1985)	1D	Yes	No	Yes	Yes	Borah *et al.* (1982)	Yes	Sand gravel
WIDTH	1D	Yes	No	Yes	Yes	Engelund and Hansen (1967)	No	Sand

Table III. Features of bank mechanics submodels of reviewed models (based on ASCE, in press)

Model	BANK PROCESS				BANK MATERIAL		
	Deposition	Fluvial erosion	Types of bank failure analysed	Longitudinal extent of failure included	Cohesive	Non-cohesive	Layered
Darby and Thorne (1996a)	No	Yes	Planar, curved	Yes	Yes	No	No
Li and Wang (1993; 1994)	Yes	Yes	None	No	No	Yes	No
Kovacs and Parker (1994)	No	Yes	None	No	No	Yes	No
Wiele (1992)	No	Yes	None	No	No	Yes	No
Pizzuto (1990)	No	Yes	None	No	No	Yes	No
Borah and Bordoloi (1989)	No	Yes	Planar	No	Yes	No	No
Alonso and Combs (1986)	No	No	Planar	No	Yes	No	No
Osman (1985)	No	Yes	Planar, curved	No	Yes	No	No

Note: Non-cohesive bank sediments are assumed uniform in size

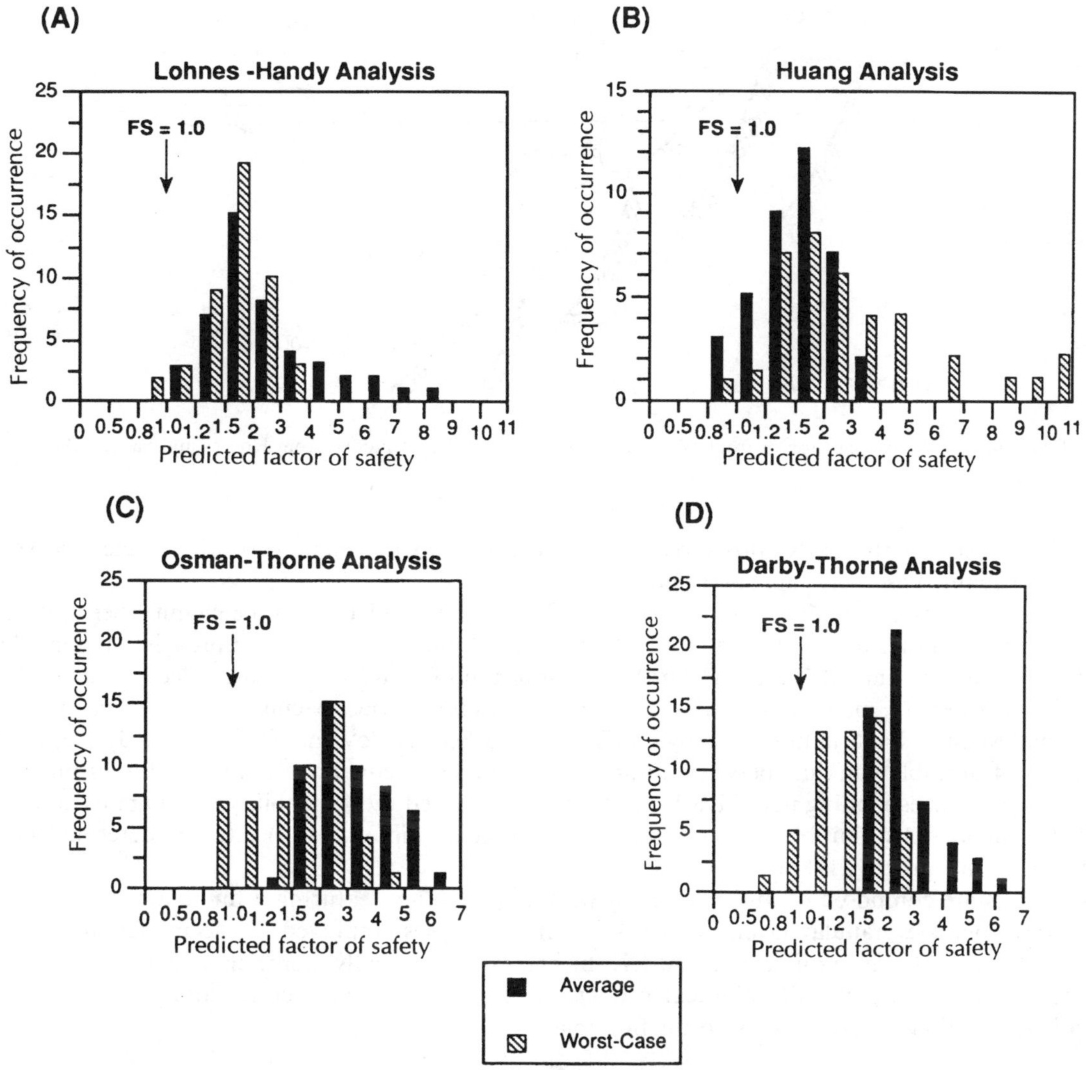

Figure 3. Comparison of predictive ability of (A) Lohnes and Handy (1968); (B) Huang (1983); (C) Osman and Thorne (1988) and (D) Darby and Thorne (1996b) bank stability analyses, for average and worst case soil parameters. FS = 1·0 indicates critical bank conditions (from Darby and Thorne, 1996b)

Pizzuto (1990), Wiele (1992) and Li and Wang (1993, 1994) simulated non-cohesive bank erosion using an heuristic model of slumping, wherein a failure surface inclined at the angle of repose is projected to the floodplain surface when bank slope exceeds the angle of repose of the boundary materials. Sediment above the failure plane is moved downslope, forming a deposit with a linear upper surface. One limitation of these analyses is that bedload formulations employed in the models are invalid at angles much less than the angle of repose, but bank materials subject to fluvial entrainment and bedload transport often stand on slopes inclined close to or at this angle.

Kovacs and Parker (1994) formulated a vectorial bedload transport equation for coarse sediment transport that is applicable to arbitrarily inclined slopes. According to their approach, widening is initiated when bank erosion along the lower part of the bank causes the local slope of the upper bank to exceed the angle of repose of the sediments. A discontinuity in slope is created between the over-steepened upper bank

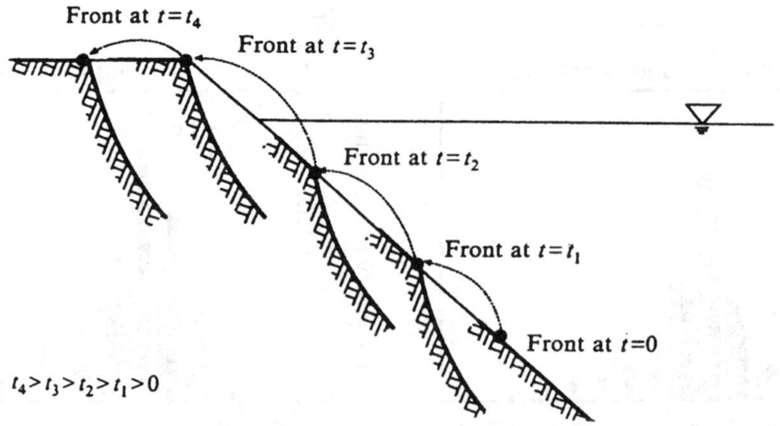

Figure 4. Upslope propagation of front of erosion in Kovacs and Parker model (from Kovacs and Parker, 1994)

and the lower part of the bank; this discontinuity migrates up the bank with a characteristic velocity, widening the channel as it propagates (Figure 4).

In all width adjustment models, the volume of bank sediments delivered to the channel per unit channel width and reach length is required as the source term, q_b, in Equation (1). Most approaches assume that the longitudinal extent of simulated mass failures is equal to the longitudinal spacing of computational nodes along the bankline. In many cases, however, the longitudinal grid spacing is much greater than the longitudinal extent of bank failures (Darby and Thorne, 1996a). Darby and Thorne (1996a) proposed that the fraction of unstable bankline between computational nodes is equal to the probability of mass failure computed at the corresponding node. Darby and Thorne obtained the probability of failure by substituting measured frequency distributions of bank material properties (Simon, 1989) into the deterministic bank stability equations (Huang, 1983).

Many banks are composed of multiple sediment horizons, often featuring a fine-grained cohesive layer above a non-cohesive granular layer. All bank stability analyses employed in the models reviewed here assume that the banks are non-layered (ASCE, in press). Additionally, some models (Table III) do not represent spatial variation in the physical properties of bank materials, either along the banks in the longitudinal direction, or extending into the floodplain.

Coupling flow, sediment transport and bank erosion submodels

Bank sediment from fluvial entrainment and bank collapse must be distributed laterally across the channel to determine the magnitude of the bank material inflow term (q_b) at each computational node. The physical properties (particle size distribution, density, cohesion) of the bank material deposited at each node must also be estimated, since these factors influence the transport dynamics of the bed and bank material mixture in the time steps following deposition of failed bank material on the bed (Darby and Thorne, 1996a; ASCE, in press).

For non-cohesive banks, the physical properties (particle size distribution, density) of disturbed bank materials can, as a first assumption, be taken to be identical to those of the undisturbed bank materials (ASCE, in press). In contrast, cohesive bank failure blocks may have physical properties quite distinct from those of intact bank materials (ASCE, in press).

Some mechanistic approaches (Osman, 1985; Alonso and Combs, 1986; Borah and Bordoloi, 1989) assume that the banks are composed of a fraction (ω) of cohesive material that becomes washload after failure, and a fraction ($1 - \omega$) of non-cohesive materials with the same physical properties as the bed material. In fact, failed bank materials become either bank material, bed material, bed material load or wash load, depending on the physical properties of the failed materials and the hydraulic properties of the flow. Darby and Thorne (1996a) noted that for undisturbed, cohesive, bank material, failure blocks sometimes disaggregate into a mass of particles with a measurable size range during mass failure. The disturbed particles, although composed of cohesive materials, are large enough to behave as non-cohesive sediment particles.

Darby and Thorne (1996a) suggested a criterion to discriminate whether or not the failure block would disaggregate. This criterion is based on the energy dissipated during mass failure and internal resistance of the failure block. They used the criterion to hypothesize that planar failures on steep banks would result in disaggregated blocks, while rotational failures on less steep slopes would remain as intact bank material. Knowledge of the size and density of deposited sediment assumed to behave as non-cohesive sediment clasts allows standard sediment transport analyses for mixed size sediment (Borah *et al.*, 1982; Rahuel *et al.*, 1989) to be applied to simulate transport of the bed and bank material mixture.

Various assumptions have been used to distribute eroded bank materials laterally across the channel. Pizzuto (1990) and Li and Wang (1993, 1994) translated bank material above the failure plane downslope, to form a deposit with a linear upper surface (Figure 5). The highest point of the deposit is the lowest point of the failure plane. The deposit extends downslope until its value equates with the volume eroded. Similar approaches could be used for cohesive bank failures (Figure 5). Wiele (1992) and Kovacs and Parker (1994) employed an approach in which the sediment continuity equation is manipulated to treat the bank erosion products as a transverse sediment flux, which is more realistic for grain by grain erosion of non-cohesive banks. The former approach is more consistent with the slumping or toppling mechanisms associated with cohesive bank failure (Wiele, 1992).

Darby and Thorne (1996a) distributed cohesive bank failure products uniformly across the near-bank zone, which they defined as extending into the channel for a distance equal to twice the bank height. Borah and Bordoloi (1989) used a linear distribution function based on local sediment transport capacity. Osman (1985) and Alonso and Combs (1986) do not describe how bank failure products are distributed in their models, other than stating that they are deposited close to the toe.

An important consideration in coupling flow, sediment transport and bank erosion submodels is related to the relative time-scales at which these processes operate. Typically, bank erosion occurs over a much longer time-scale than adjustment of flow processes. Care must therefore be taken to select the appropriate time step length. For example, selection of a relatively long time step length, required to characterize bank erosion processes, may lead to significant error in characterization of the flow and sediment transport field, and could even generate numerical instability. On the other hand, selection of a short time step length, required to characterize adequately rapidly adjusting flow parameters, will lead to increased computational time during simulations aimed at assessing longer term bankline response. Since selection of a short time step length will not affect the accuracy of bank erosion computations, this latter option is preferred for model accuracy. Use of computers with high processor speeds, which are now widely available, will help to counterbalance the additional computational time penalty incurred. Another approach (Darby and Thorne, 1996a) is to design computer simulation codes that allow the time step to vary according to whether or not bank erosion or deposition processes are active, or whether or not the flow is steady. During flows that are incapable of deforming the channel boundary, the time step can be maximized. In contrast, the time step is reduced under conditions of flow, sediment transport and bank erosion that could potentially lead to numerical instability.

Aquatic and riparian vegetation effects

Progress has been made in evaluating the effects of in-channel and riparian vegetation on stable channel forms (Hey and Thorne, 1986). Similar advances have been made in assessing the effects of vegetation on alluvial channel roughness (Kouwen and Li, 1980; Masterman and Thorne, 1994), and in experimental work aimed at quantifying the effects of vegetation on geotechnical stability (Gray *et al.*, 1991). Although no existing numerical width adjustment model accounts for the effect of vegetation on channel adjustment, the

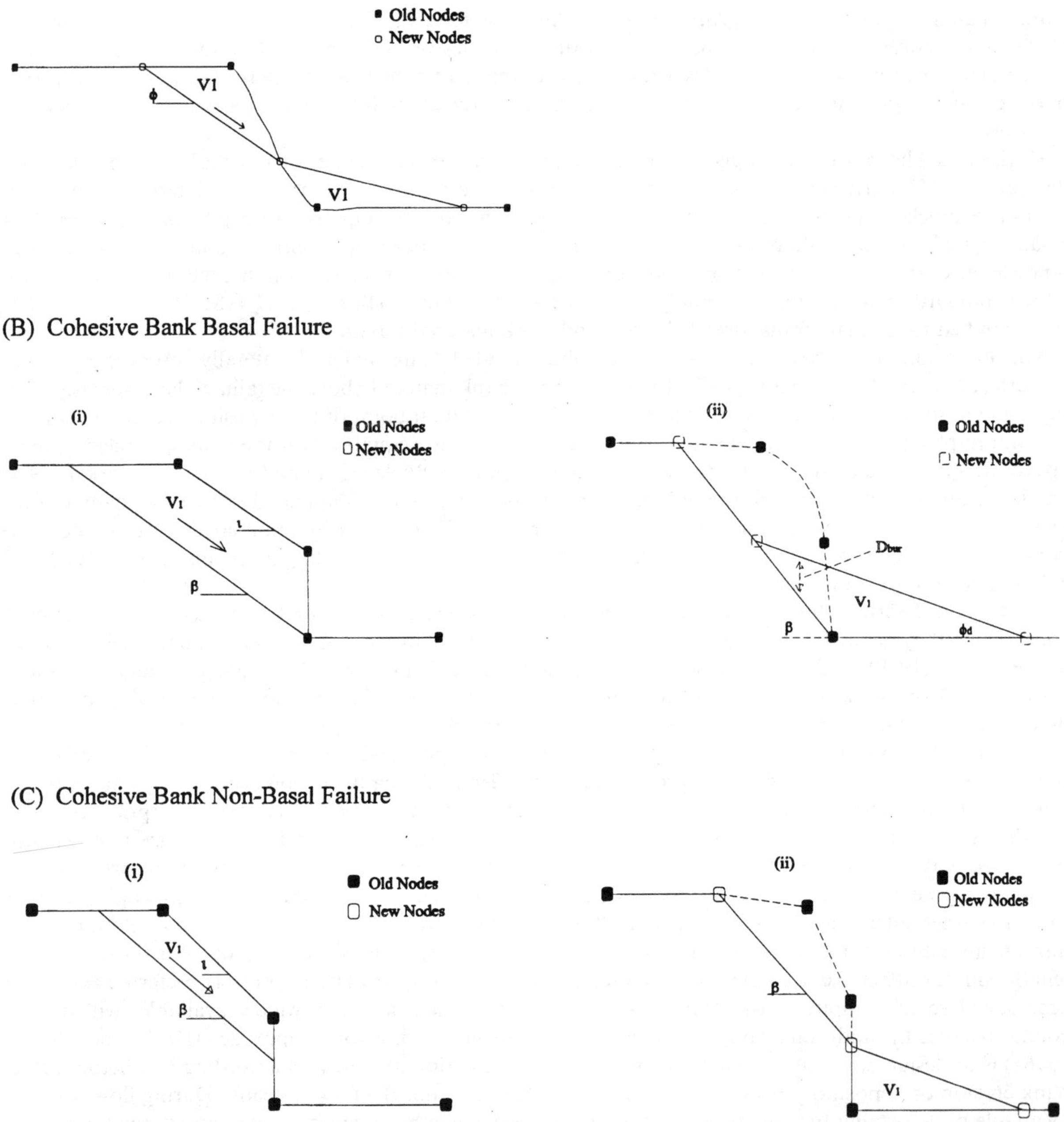

Figure 5. Illustration of methods of distributing bank failure material laterally across the channel. (A) Non-cohesive banks (modified from Pizzuto, 1990); (B) cohesive banks, basal failure; (C) cohesive banks, non-basal failure

results of these studies could quite easily be incorporated into existing flow resistance and bank erosion submodels. The way in which vegetation properties change through time as vegetation grows, decays or is disrupted or uprooted on dynamic geomorphological surfaces must also be accounted for. Field investigations of the establishment and growth of vegetation on eroding banklines (Hupp and Simon, 1991) might form the basis for approaches to modelling vegetation dynamics in alluvial channels.

VALIDATION AND VERIFICATION STUDIES

Evaluation and comparison of alluvial channel models is often difficult. Coincidence between model simulations and limited observed data is not surprising, nor difficult to achieve, particularly for models that have many adjustable parameters and coefficients.

Responding to the need to develop standard guidelines for model testing, the American Society of Civil Engineers has initiated a task force to develop standard procedures for verifying computational flow models. The proposed guidelines set out a four-point procedure to establish the reliability, computational integrity and accuracy of those computational models (S. Wang, personal communication, 1995).

(1) *Numerical convergence.* Convergence to an increasingly more precise solution as the resolution of the computational mesh is increased must be demonstrated. For morphological models, additional checks would be made on the degree to which water and sediment mass is conserved during the simulation. Such analyses would highlight any obvious numerical or coding errors, and demonstrate that the numerical solution of the mathematical model is robust. No information on the accuracy of the solution is provided at this stage.

(2) *Comparison with analytical solutions.* Numerical solutions should be compared with a range of standard, test case, analytical solutions to ensure that numerical solutions correctly approximate these exact analytical solutions. This demonstrates that the model faithfully replicates the physics of the processes involved.

(3) *Comparison of model-simulated data with benchmark laboratory data.* The capabilities and limitations of the model would next be evaluated through simulations of physical phenomena described by idealized standard laboratory data. Examples of such 'benchmark' data sets might include shear stress distributions in curved channels (Ippen and Drinker, 1962), overbank flow patterns in straight channels (Shiono and Knight, 1990) and flow over fixed bed dune fields (Nelson and Smith, 1989). The aim is not to reject or accept a particular model based on failure or success in simulating one or more particular test cases, but rather to use a variety of data sets, involving simulation of a range of idealized physical phenomena, to establish the capabilities of the model.

(4) *Comparison with field data.* Such comparisons are valid only if the data used are distinct from the data used during calibration. Data sets suitable for field evaluation of width adjustment models include those from the Upper Missouri River (Borah and Dashputre, 1994), the North Fork Toutle River, Washington (Simon, 1992), the South Fork of the Forked Deer River, Tennessee (Simon and Hupp, 1992), and Goodwin Creek, Mississippi (Murphey and Grissinger, 1985).

Evaluation of the models reviewed here is therefore conducted according to the above criteria.

Numerical robustness

None of the authors of the width adjustment models reviewed here report details of simulations designed explicitly to test their models for numerical stability and convergence criteria.

Comparisons with analytical solutions

The study by Wiele (1992) is the only one that explicitly includes a comparison of numerical and analytical solutions designed to check the accuracy of the numerical solution. This evaluation was, in fact, conducted only for the flow submodel of the bank erosion model. Most authors have tried to demonstrate that their respective models provide accurate simulations using parametric sensitivity analyses (Osman, 1985; Alonso and Combs, 1986; Borah and Bordoloi, 1989; Pizzuto, 1990; Li and Wang, 1993, 1994; Kovacs and Parker, 1994; Darby and Thorne, 1996c). In all cases, results indicate reasonable model behaviour.

Tests with laboratory data

The models by Pizzuto (1990), Wiele (1992), Li and Wang (1993, 1994) and Kovacs and Parker (1994) have been tested with common laboratory data (Ikeda, 1981). Results from these studies are shown in

Figure 6. Assessment of the relative merits of these models is difficult because there are some small, but significant, differences in the numerical values of coefficients used by each of the authors. Specifically, the critical dimensionless Shields stress is assumed to be 0·03 by Li and Wang (1993, 1994) and Pizzuto (1990), and 0·035 and 0·038 by Kovacs and Parker (1994) and Wiele (1992), respectively. The value of the internal angle of friction of the boundary material was assumed to be 33° by Pizzuto (1990) and 40° by the other authors (ASCE, in press).

Field testing

The models of Osman (1985), Alonso and Combs (1986), Pizzuto (1990), Li and Wang (1993, 1994) and Kovacs and Parker (1994) have not been subjected to field evaluation. However, as these models were specifically designed as research tools for idealized cases, there was no particular requirement to test with field data.

The models developed by Wiele (1992), Borah and Dashputre (1994) and Darby and Thorne (1996a) have been tested using field data independent of that used to determine the numerical values of the calibration parameters. Wiele (1992) compared simulated and measured cross-section profiles for a reach of the braided Sunwapta River in Jasper National Park, Alberta, Canada, over a periods of a few hours. Wiele (1992) suggested that model predictions were in 'reasonable agreement' with the observations.

Borah and Dashputre (1994) simulated a 312 km reach of the Upper Missouri River between Gavins Point Dam and Omaha, Nebraska, for the period 1957–1976. Comparisons between simulated and observed bank erosion rates are somewhat difficult to make, since observed bank erosion rates are based on a 20-year average for the simulation period, rather than rates for the shorter time spans over which event-driven erosion rates operate. Substantial discrepancies exist between simulated and observed rates of bank erosion, but the spatial distribution of erosion is successfully predicted (Borah and Dashputre, 1994).

Darby *et al.* (1996) used the Darby–Thorne model to simulate channel adjustments of a 13·5 km straightened reach of the South Fork of the Forked Deer River, Tennessee, during the period 1969–1993. The model appeared to be able to replicate the observed sequence of channel adjustments (Figure 7), and overall agreement between the magnitudes of simulated and observed widths, depths and deepening rates are within ±10%, but simulated widening rates are under-predicted by a factor of three (Darby *et al.*, 1996).

MODEL APPLICATIONS

Because of the many boundary conditions that exist in natural alluvial channels, and the difficulty in obtaining time-based data over the course of channel adjustment, the availability of comprehensive empirical data sets for analysis of channel dynamics over a range of environmental boundary conditions is limited. It is possible, however, to conduct numerical simulations to supplement limited empirical data. Two examples of this type of approach are outlined here. Both examples are based on use of the Darby–Thorne numerical model. Simulations were conducted for a hypothetical, straight, uniform channel reach, initially in a non-equilibrium condition, where sediment supply at the upstream boundary was half that of the transport capacity. The initial response of the channel was, therefore, incision, with potential for channel widening later as the bank height and instability increased owing to basal scour in the near-bank zones.

The results of parametric sensitivity tests conducted by Darby and Thorne (1996c) are illustrated in Figure 8. The results are useful in identifying the relative effects of change in independent variables on stable channel forms. For example, discharge (Q) and sediment load (Q_s) have high valued relative sensitivity parameters, for both width and depth. Relatively small changes in these variables, therefore, lead to relatively large changes in width and depth. For discharge, the relative sensitivity parameter for channel width is approximately twice the value of that for channel depth, suggesting that increases in discharge would lead to stable channel forms with increased width–depth ratios, a result consistent with empirical hydraulic geometry studies (Ferguson, 1986).

In contrast to morphological parameters such as width, depth or channel gradient, flow–energy parameters, such as total mechanical energy, boundary shear stress and energy dissipation rate (energy

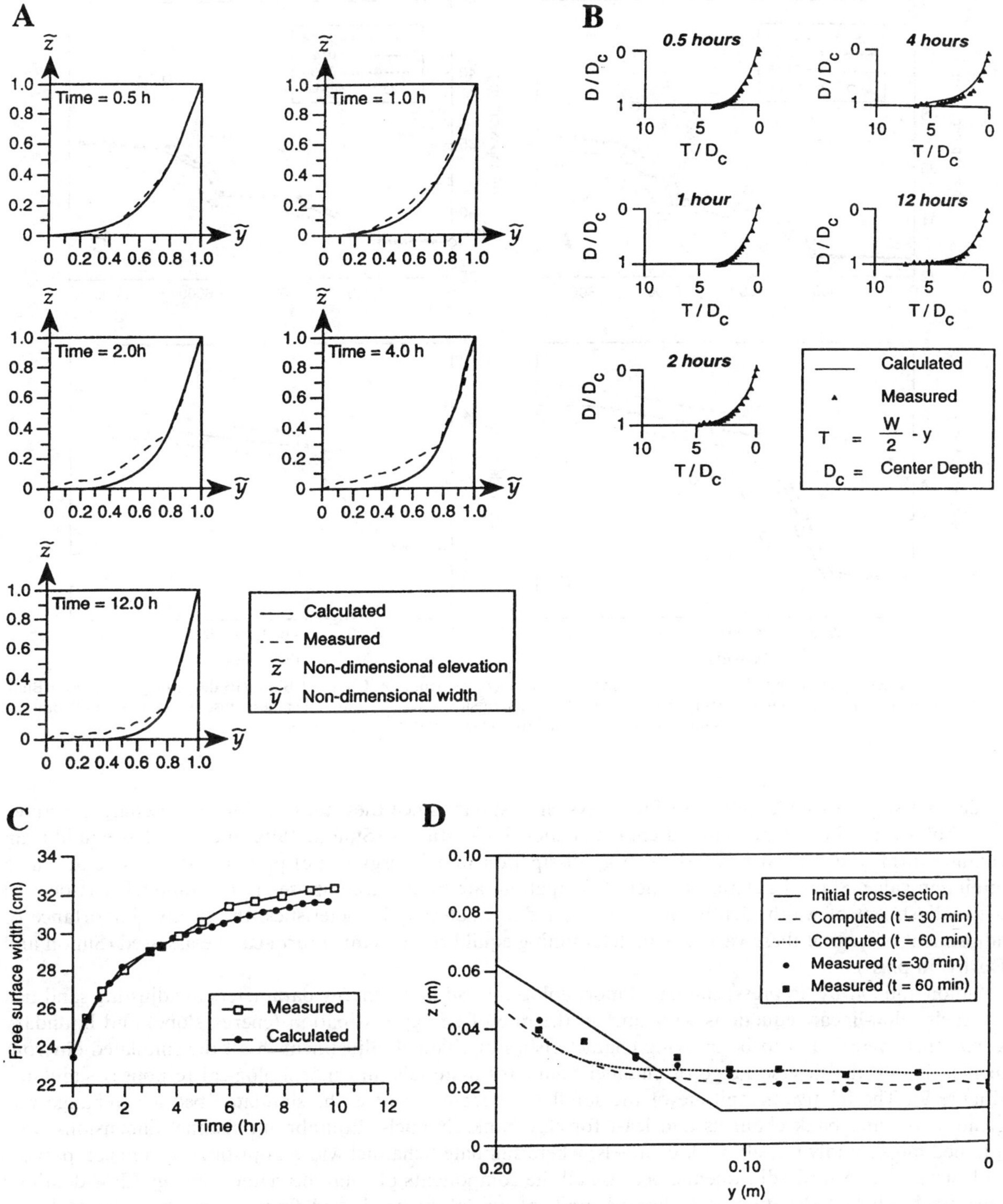

Figure 6. Comparison of Ikeda (1981) laboratory data and model simulations obtained by (A) Kovacs and Parker (1994); (B) Pizzuto (1990); (C) Wiele (1992) and (D) Li and Wang (1993) (from ASCE, in press)

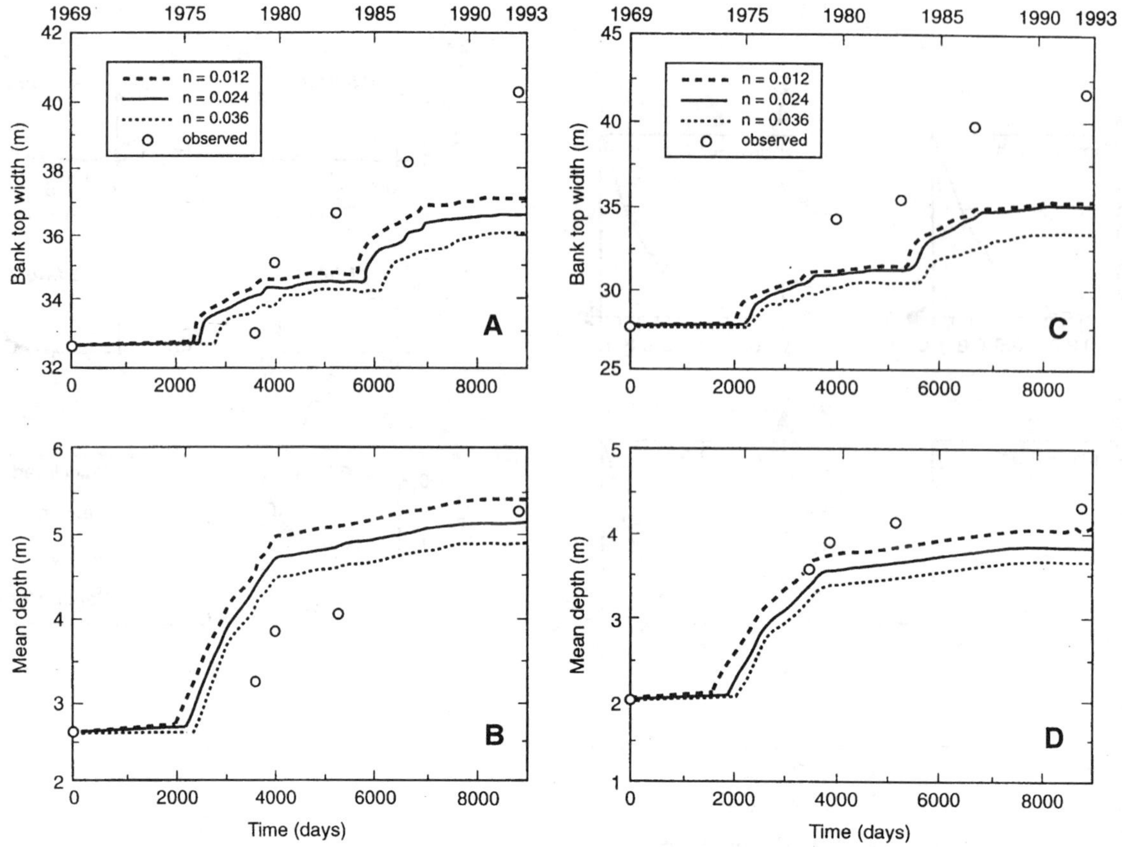

Figure 7. Field testing of Darby–Thorne channel widening model. (A) Widening at Chestnut Bluff; (B) deepening at Chestnut Bluff; (C) widening at Crossroads; (D) deepening at Crossroads for range of simulated Manning '*n*' values. Observed widths and depths are shown for comparison (from Darby and Thorne, 1996c)

slope), are single variable indices of fluvial system response since they are functions of discharge, channel morphology and boundary material characteristics. Field studies (Simon, 1992) indicate that equilibrium channel forms are defined when the energy dissipation rate (energy slope) plotted against time attains a minimum value. Since the change in energy dissipation rate can be decomposed into components associated with adjustment of width, depth, slope and boundary material characteristics, the relative importance of adjustment of each of these variables in determining equilibrium channel form can be assessed (Simon and Darby, in press).

Simon and Darby (in press) simulated morphological and flow–energy parameters in adjusting sand bed channels. Non-linear reductions with time in the rate of energy dissipation (energy slope) and boundary shear stress were found to be unifying characteristics of channel adjustment in all the simulated environments, despite the diverse range of channel boundary materials and morphological responses simulated (Figure 9). The relative magnitude of the simulated widening versus the simulated bed level change was greatest for sand bank channels and least for clay bank channels. Equilibrium channel dimensions were attained most rapidly for silt bank channels, where simulated channel width adjustments were comparable with simulated vertical adjustments, because all the components of total mechanical energy (flow depth or pressure head; bed elevation and channel gradient or datum head; and flow velocity or velocity head) decrease simultaneously (Figure 9). These simulations demonstrate how model studies could be used to investigate environmental controls on channel dynamics and stable channel forms.

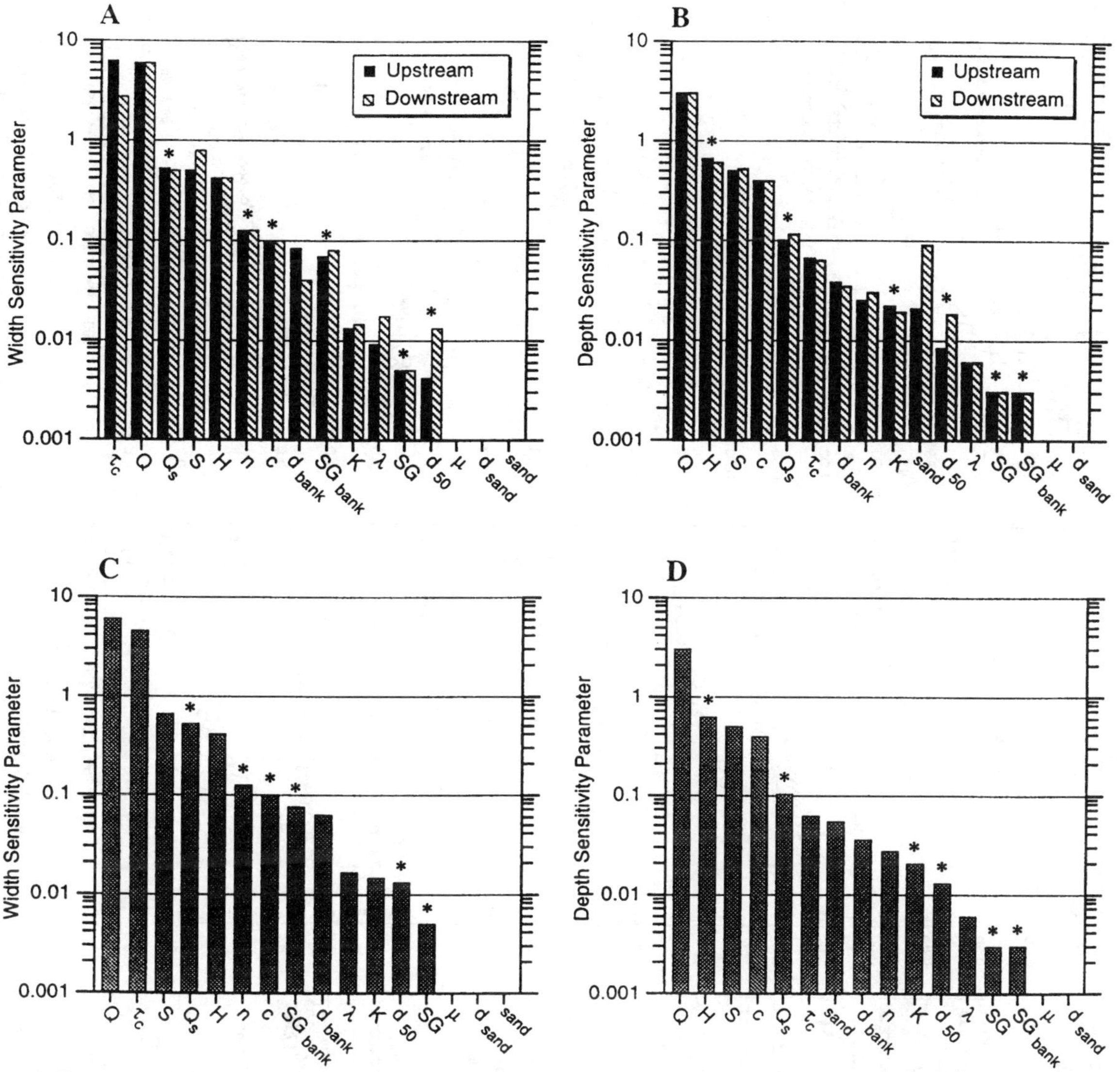

Figure 8. Simulated relative sensitivity parameters. (A) Width sensitivity parameters, upstream and downstream locations; (B) depth sensitivity parameters, upstream and downstream locations; (C) width sensitivity parameters, mean of locations; (D) depth sensitivity parameters, mean of locations (from Darby and Thorne, 1996c). Symbols: τ_c = critical shear stress for entrainment of bank material, Q = discharge, Q_s = sediment supply, S = channel gradient, H = initial bank height, n = Manning 'n', c = soil cohesion, d_{bank} = diameter of failed bank material particles, SG_{bank} = specific gravity of bank material, K = tension crack index, λ = bed material porosity, SG = specific gravity of bed material, d_{50} = median particle diameter of bed material, μ = Coulomb friction coefficient, d_{sand} = size of sand fraction of bank material, sand = percentage sand content of bank material

CONCLUSIONS

The following conclusions may be drawn.

1. Present knowledge of hydraulic, sediment transport and bank processes is sufficient to allow some tentative predictions of width adjustment to be made, but the mechanistic width adjustment models reviewed here should be regarded as research tools that cannot be used routinely in channel management applications (ASCE, in press).

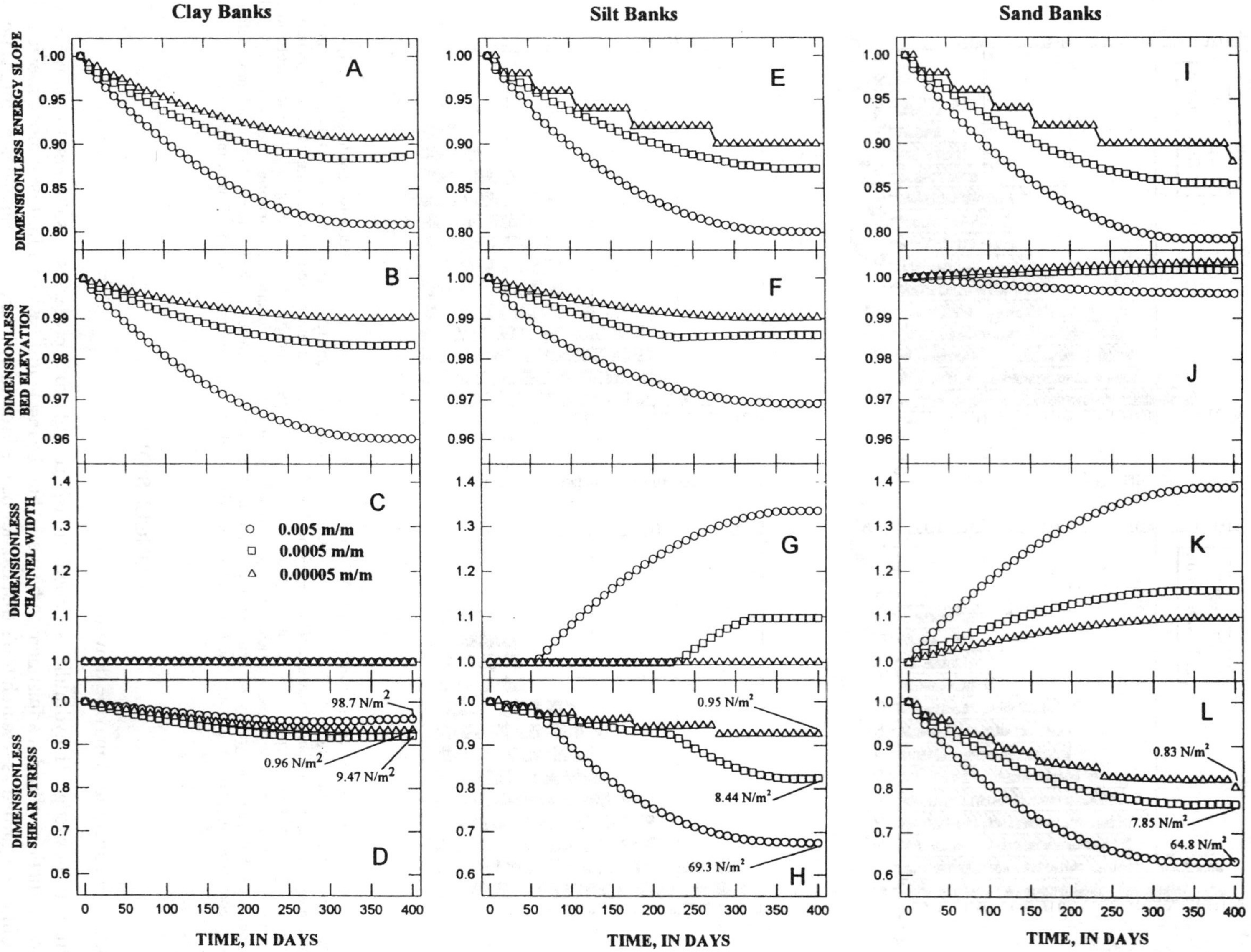

Figure 9. Simulated morphological and flow–energy parameters for a range of valley gradients and bank material types. For clay banks: (A) energy slope, (B) bed elevation, (C) channel width, (D) shear stress. For silt banks: (E) energy slope, (F) bed elevation, (G) channel width, (H) shear stress. For sand banks: (I) energy slope, (J) bed elevation, (K) channel width, (L) shear stress (from Simon and Darby, in press)

2. There are presently very few appropriate laboratory and field data sets suitable for testing width adjustment models. This has resulted in a lack of comprehensive testing and verification analyses of existing models using 'benchmark' field and laboratory data sets.

3. At present there is no universally applicable width adjustment model capable of dealing with all the circumstances under which width adjustments may occur.

4. Further research is required to advance the capabilities of width adjustment models. Improved methods of predicting boundary shear stress and sediment transport distributions in channels with complex, non-uniform bathymetry are required. The mechanics of cohesive bank erosion, failure and retreat are still poorly understood. Specifically, reliable methods of estimating fluvial entrainment of cohesive bank material, the longitudinal extent of mass failures and the physical characteristics of failed bank material are urgently required. Finally, comprehensive field and laboratory data are urgently required for testing and verification of existing and future width adjustment models (Table IV). These research topics represent necessary requirements for progress to be made in the medium to long term.

Table IV. Data required to apply numerical width adjustment models (from ASCE, in press)

Data item	Notes
Time-independent data (initial conditions)	
Cross-section surveys	Required to define initial channel morphology. Surveys are required at several sites along the prototype reach.
Bed material size distribution	Required to define the initial bed material characteristics. Data is required at each cross-section.
Bank material characteristics	Measurements of cohesion, friction angle and unit weight at left and right banks of each cross-section are required to define the bank material characteristics.
Time-dependent data (boundary conditions)	
Discharge	Value of discharge to be used in each discrete time step of the simulation.
Sediment supply	Value of sediment load at the upstream boundary of the prototype reach during each discrete time step of the simulation.

ACKNOWLEDGEMENTS

Some of the ideas presented here were developed or refined during meetings of the ASCE Task Committee on 'Hydraulics, Bank Mechanics and Modeling of Width Adjustment'; these discussions are greatly appreciated. The writer also expresses appreciation to Colin Thorne, Deva Borah and Sean Bennett for reviewing draft versions of this manuscript.

REFERENCES

Alonso, C. V. and Combs, S. T. 1986. 'Channel width adjustment in straight alluvial streams', in *Proceedings of the 4th Federal Interagency Sedimentation Conference, Las Vegas, Nevada, 24–27 March 1986*. pp. 5-31–5-40.

Arulanandan, K., Gillogley, E., and Tully, R. 1980. 'Development of a quantitative method to predict critical shear stress and rate of erosion of naturally undisturbed cohesive soils', *Report G1-80-5*. US Army Waterways Experiment Station, Vicksburg, Mississippi.

ASCE. 'Final report of the task committee on hydraulics, bank mechanics, and modeling of width adjustment', *J. Hydraul. Engng*, (in press).

Biedenharn, D. S. and Thorne, C. R. 1994. 'Magnitude–frequency analysis of sediment transport in the Lower Mississippi River', *Regul. Rivers: Res. Mgmt*, **9**, 237–251.

Borah, D. K. and Bordoloi, P. K. 1989. 'Stream bank erosion and bed evolution model', in Wang, S. (Ed.), *Sediment Transport Modeling*. American Society of Civil Engineers, New York. pp. 612–619.

Borah, D. K. and Dashputre, M. S. 1994. 'Field evaluation of the sediment transport model "STREAM" with a bank erosion component', in Cotroneo, G. V. and Rumer, R. R. (Eds), *Hydraulic Engineering '94*. American Society of Civil Engineers, New York. pp. 979–983.

Borah, D. K., Alonso, C. V., and Prasad, S. N. 1982. 'Routing graded sediments in streams: formulations', *ASCE J. Hydraul. Div.*, **108**, 1486–1503.

Bridge, J. S. and Bennett, S. J. 1992. 'A model for the entrainment and transport of sediment grains of mixed sizes, shapes, and densities', *Wat. Resour. Res.*, **28**, 337–363.

Chang, H. H. 1988. *Fluvial Processes in River Engineering*. John Wiley and Sons, New York.

Church, M. A., McLean, D. G., and Wolcott, J. F. 1987. 'River bed gravels: sampling and analysis', in Thorne, C. R., Bathurst, J. C., and Hey, R. D. (Eds), *Sediment Transport in Gravel-bed Rivers*. John Wiley and Sons, Chichester. pp. 43–88.

Darby, S. E. and Thorne, C. R. 1996a. 'Numerical simulation of bed deformation and channel widening in straight sand-bed rivers. I: model development', *J. Hydraul. Engng*, **122**(4), 184–193.

Darby, S. E. and Thorne, C. R. 1996b. 'Development and testing of riverbank stability analysis', *J. Hydraul. Engng*, **122**(8), 443–454.

Darby, S. E. and Thorne, C. R. 1996c. 'Modelling the sensitivity of channel adjustments in destabilized sand-bed rivers', *Earth Surf. Process. Landf.*, **21**, 1109–1125.

Darby, S. E., Thorne, C. R., and Simon, A. 1996. 'Numerical simulation of bed deformation and channel widening in straight sand-bed rivers. II: model evaluation', *J. Hydraul. Engng*, **122**(4), 194–202.

Einstein, H. A. 1950. 'The bedload function for sediment transportation in open channel flows', *Technical Bulletin*, 1026, United States Department of Agriculture, Washington, D.C.

Engelund, F. and Hansen, E. 1967. A monograph on sediment transport in alluvial streams, Teknisk Forlag, Copenhagen.

Ferguson, R. I. 1986. 'Hydraulics and hydraulic geometry', *Progr. Phys. Geog.*, **10**, 1–31.

Gomez, B. and Church, M. 1989. 'An assessment of bed load sediment transport formulae for gravel bed rivers', *Wat. Resour. Res.*, **25**, 1161–1186.

Graf, W. H. 1971. *Hydraulics of Sediment Transport*. McGraw-Hill, New York.

Gray, D. H., MacDonald, A., Thomann, T., Blatz, I., and Shields, F. D. 1991. 'The effects of vegetation on the structural integrity of sandy levees', Repair, Evaluation, Maintenance and Rehabilitation Research Program, *Technical Report REMR-El-5*. US Army Corps of Engineers, Washington, D.C..

Gregory, K. J. and Walling, D. E. 1973. *Drainage Basin Form and Process: A Geomorphological Approach*. Edward Arnold, London.

Grissinger, E. H. 1982. 'Bank erosion of cohesive materials', in Hey, R. D., Bathurst, J. C., and Thorne, C. R. (Eds), *Gravel-bed Rivers*. John Wiley and Sons, Chichester. pp. 273–287.

Hagerty, D. J. 1991. 'Piping/sapping erosion I: basic considerations', *J. Hydraul. Engng*, **117**, 991–1008.

Hey, R. D. and Thorne, C. R. 1986. 'Stable channels with mobile gravel beds', *J. Hydraul. Engng*, **112**, 671–689.

Huang, Y. H. 1983. *Stability Analysis of Earth Slopes*. Van Nostrand Reinhold, New York.

Hupp, C. R. and Simon, A. 1991. 'Bank accretion and the development of vegetated depositional surfaces along modified alluvial channels', *Geomorphology*, **4**, 111–124.

Ikeda, S. 1981. 'Self-formed straight channels in sandy beds', *ASCE J. Hydraul. Div.*, **107**, 389–406.

Ippen, A. T. and Drinker, P. A. 1962. 'Boundary shear stress in curved trapezoidal channels', *ASCE J. Hydraul. Div.*, **88**, 143–179.

Keulegan, G. H. 1938. 'Laws of turbulent flow in open channels', *J. Res. Natl Bureau Stand.*, **21**, 707–741.

Knight, D. W. and Sellin, R. H. J. 1987. 'The SERC flood channel facility', *J. Inst. Wat. Environ. Mgmt*, **1**, 198–204.

Knight, D. W. and Shiono, K. 1995. 'River channel and floodplain hydraulics', in Anderson, M., Walling, D., and Bates, P. (Eds), *Floodplain Processes*. John Wiley and Sons, Chichester. pp. 134–150.

Kouwen, N. and Li, R.-M. 1980. 'Biomechanics of vegetative channel linings', *ASCE J. Hydraul. Div.*, **106**, 1085–1103.

Kovacs, A. and Parker, G. 1994. 'A new vectorial bedload formulation and its application to the time evolution of straight river channels', *J. Fluid Mech.*, **267**, 153–183.

Lane, S. N., Richards, K. S., and Chandler, J. H. 1993. 'Developments in photogrammetry: the geomorphological potential', *Progr. Phys. Geog.*, **17**, 306–328.

Lapidus, L. and Pinder, G. F. 1982. *Numerical Solution of Partial Differential Equations in Science and Engineering*. John Wiley and Sons, New York.

Lawler, D. M. 1986. 'River bank erosion and the influence of frost: a statistical analysis', *Trans. Inst. Br. Geogr.*, **NS11**, 227–242.

Li, L. and Wang, S. S. Y. 1993. 'Numerical modeling of alluvial stream bank erosion', in Wang, S. S. Y. (Ed.), *Advances in Hydro-Science and Engineering*, Volume 1. pp. 2085–2090, University of Mississippi Press, Oxford, MS.

Li, L. and Wang, S. S. Y. 1994. 'Computational simulation of channel bank erosion and retreat', *Technical Report*. CCHE, The University of Mississippi.

Lohnes, R. and Handy, R. L. 1968. 'Slope angles in friable loess', *J. Geol.*, **76**, 247–258.

Masterman, R. and Thorne, C. R. 1994. 'Analytical approach to flow resistance in gravel-bed channels with vegetated banks', in Kirkby, M. J. (Ed.), *Process Models and Theoretical Geomorphology*. John Wiley and Sons, Chichester. pp. 201–218.

Meyer-Peter, E. and Muller, R. 1948. 'Formulas for bed load transport', *Proc. 2nd Congress of the International Association for Hydraulic Research, Stockholm, Sweden*. pp. 39–64.

Murphey, J. B. and Grissinger, E. H. 1985. 'Channel cross-section changes in Mississippi's Goodwin Creek', *J. Soil Wat. Conserv.*, **40**, 148–153.

Nelson, J. M. and Smith, J. D. 1989. 'Mechanics of flow over ripples and dunes', *J. Geophys. Res.*, **94**, 8146–8162.

Osman, A. M. 1985. 'Channel width response to changes in flow hydraulics and sediment load', *PhD Thesis*, Colorado State University, Fort Collins.

Osman, A. M. and Thorne, C. R. 1988. 'Riverbank stability analysis I: theory', *J. Hydraul. Engng*, **114**, 134–150.

Parker, G. 1979. 'Hydraulic geometry of active gravel rivers', *ASCE J. Hydraul. Div.*, **105**, 1185–1201.

Parker, G. 1983. 'Discussion of "Lateral bed load transport on side slopes" by S. Ikeda', *J. Hydraul. Engng*, **109**, 197–199.
Partheniades, E. 1965. 'Erosion and deposition of cohesive soils', *ASCE J. Hydraul. Div.*, **91**, 105–139.
Pizzuto, J. E. 1990. 'Numerical simulation of gravel river widening', *Wat. Resour. Res.*, **26**, 1971–1980.
Rahuel, J. L., Holly, F. M., Belleudy, P. J., and Yang, G. 1989. 'Modeling of riverbed evolution for bedload sediment mixtures', *J. Hydraul. Engng*, **115**, 1521–1542.
Shiono, K. and Knight, D. W. 1990. 'Turbulent open-channel flows with variable depth across the channel', *J. Fluid Mech.*, **222**, 617–646.
Simon, A. 1989. 'A model of channel response in disturbed alluvial channels', *Earth Surf. Process. Landf.*, **14**, 11–26.
Simon, A. 1992. 'Energy, time and channel evolution in catastrophically disturbed fluvial systems', *Geomorphology*, **5**, 345–372.
Simon, A. and Darby, S. E. 'Process-form interactions in unstable sand-bed streams: a numerical modelling approach', *Geomorphology*, (in press).
Simon, A. and Hupp, C. R. 1992. 'Geomorphic and vegetative recovery processes along modified stream channels of West Tennessee', *United States Geological Survey Open File Report, 91-502.* USGS Nashville.
Simon, A., Wolfe, W. J., and Molinas, A. 1991. 'Mass wasting algorithms in an alluvial channel model', in *Proceedings of the 5th Federal Interagency Sedimentation Conference, Las Vegas, Nevada.* pp. 8-22–8-29.
Strickler, A. 1923. 'Beitrage zur frage der geschwindigheitsformel und der rauhigkeitszahlen fur strome, kanale und geschlossene leitungen', *Mitteilungen des Eidgenossicher Amtes fur Wasserwirtschaft*, Bern, Switzerland, 16 pp.
Thorne, C. R. 1982. 'Processes and mechanisms of river bank erosion', in Hey, R. D., Bathurst, J. C., and Thorne, C. R. (Eds), *Gravel-bed Rivers.* John Wiley and Sons, Chichester. pp. 227–271.
Thorne, C. R. and Osman, M. A. 1988. 'The influence of bank stability on regime geometry of natural channels', in White, W. R. (Ed.), *River Regime.* John Wiley and Sons, Chichester. pp. 135–147.
Thorne, C. R. and Tovey, N. K. 1981. 'Stability of composite river banks', *Earth Surf. Process. Landf.*, **6**, 469–484.
Whiting, P. J. and Dietrich, W. E. 1991. 'Convective accelerations and boundary shear stress over a channel bar', *Wat. Resour. Res.*, **27**, 783–796.
Wiele, S. M. 1992. 'A computational investigation of bank erosion and midchannel bar formation in gravel-bed rivers', *PhD Thesis*, University of Minnesota, Minneapolis.
Wolman, M. G. and Miller, J. P. 1960. 'Magnitude and frequency of forces in geomorphic processes', *J. Geol.*, **68**, 54–74.
Yang, C. T. 1973. 'Incipient motion and sediment transport', *ASCE J. Hydraul. Div.*, **99**, 1679–1704.
Yang, C. T. and Wan, S. 1991. 'Comparison of selected bed-material formulas', *J. Hydraul. Engng*, **117**, 973–989.

18

MORPHOLOGICAL MODELLING OF RIVERS WITH ERODIBLE BANKS

ERIK MOSSELMAN*

Delft Hydraulics, PO Box 177, 2600 MH Delft, The Netherlands

ABSTRACT

A bank erosion mechanism and provisions to account for the associated planform changes and input of bank erosion products are added to a two-dimensional, depth-averaged model of river morphology. The model is applied to a reach of the meandering gravel-bed River Ohře (Eger) in the former state of Czechoslovakia. The agreement with observations is poor, but this can be ascribed to shortcomings in the flow and bed topography submodels rather than to shortcomings in the bank erosion submodel. Better results are expected when a three-dimensional flow model, equations for sediment mixtures and a bank accretion mechanism are included. This inclusion will have to be preceded by fundamental research on bank accretion mechanisms and on hiding and exposure effects in the relationship for the influence of gravity pull on sediment transport direction.

KEY WORDS bank erosion; mathematical modelling; river morphology

INTRODUCTION

A common class of high resolution models of river morphology is two-dimensional in the horizontal plane. These models are based on depth-averaged, steady-flow equations, a volumetric sediment balance and formulae relating magnitude and direction of sediment transport to local flow field and bed topography. The first two-dimensional model of this kind was developed by van Bendegom (1947), but its application without the modern computer devices of the present was very laborious. Subsequent work on helical flow in river bends (e.g. Kalkwijk and de Vriend, 1980) and forces on sediment grains on a transversely sloping bed (e.g. Engelund, 1974; Odgaard, 1981) resulted in the model of Struiksma (1985), Struiksma *et al.* (1985) and Olesen (1987). Similar models were developed by Shimizu and Itakura (1985) and Nelson and Smith (1989). The earlier flow and bed topography model by Kennedy *et al.* (1983) does not fall into this class of models since it does not use the sediment balance. Instead, it uses an axisymmetrical relationship to link the bed topography directly to the local flow field. All models compute the evolution of river bed topography for a given planform with fixed banks. In natural rivers, however, planforms do change, through bank erosion and bank accretion.

This paper presents the extension of the two-dimensional morphological model of Struiksma, Olesen and co-workers with a bank erosion mechanism and provisions to account for the resulting planform changes and input of bank erosion products. The resulting model is applied to the River Ohře in the former state of Czechoslovakia with two purposes. The first is to show the present predictive capabilities of the model. It is found that the agreement with observations is not good. Already, the reproduction of the bed topography is found to be poor, although basically the same model has given good results in applications to other rivers (Struiksma, 1985; Olesen, 1987). This leads to the second aim of the paper, to show how the construction and application of numerical models can guide further research. Better results

* Correspondence to: Erik Mosselman, Delft Hydraulics, PO Box 177, 2600 MH Delft, The Netherlands.

are expected with a three-dimensional flow model, with equations for sediment mixtures and with a bank accretion model. Equations and solution techniques for three-dimensional flow are readily available, but fundamental research in a laboratory or in the field is needed for a proper formulation of the equations for sediment mixtures and the bank accretion model.

MATHEMATICAL MODEL

Coordinate system

The mathematical model is formulated in a horizontal, channel-fitted coordinate system (s, n) which can be curvilinear and *non-orthogonal*. For clarity, however, the equations in this paper are written for a curvilinear, *orthogonal* coordinate system. Details on their non-orthogonal counterparts are given by Mosselman (1991, 1992). The s direction is taken to follow the streamwise direction of the river, i.e. more or less parallel to the streamlines, and boundary conforming at the banks. The n direction is the transverse direction. The coordinate increments are not dimensionless but represent metric distances. As a consequence, s and n are not independent, which implies that the order of cross-differentiation cannot simply be reversed but should be based on the relationship

$$\frac{\partial^2 f}{\partial s \partial n} - \frac{\partial^2 f}{\partial n \partial s} = \frac{1}{R_s}\frac{\partial f}{\partial s} - \frac{1}{R_n}\frac{\partial f}{\partial n} \tag{1}$$

in which f is a function of s and n, and R_s and R_n are the radii of curvature of the s and n coordinate lines, respectively. The right-hand terms can vanish in a numerical implementation if the difference equations are properly discretized.

Flow model

The depth-averaged, steady-flow equations read as

$$u\frac{\partial u}{\partial s} + v\frac{\partial u}{\partial n} + g\frac{\partial z_b}{\partial s} + g\frac{\partial h}{\partial s} + \frac{uv}{R_s} - \frac{v^2}{R_n} + \frac{gu^2}{hC^2} = 0 \tag{2}$$

$$u\frac{\partial v}{\partial s} + v\frac{\partial v}{\partial n} + g\frac{\partial z_b}{\partial n} + g\frac{\partial h}{\partial n} + \frac{uv}{R_n} - \frac{u^2}{R_s} + \frac{guv}{hC^2} = 0 \tag{3}$$

$$\frac{\partial(hu)}{\partial s} + \frac{\partial(hv)}{\partial n} + \frac{hu}{R_n} + \frac{hv}{R_s} = 0 \tag{4}$$

in which u and v are the flow velocities in the s and n directions, respectively, z_b denotes bed level, h is the water depth, g is the acceleration due to gravity and C is the Chézy coefficient for hydraulic roughness. The streamwise coordinate lines and the streamlines of the flow are more or less parallel, but generally they will not match exactly. Hence, v cannot simply be neglected, and the streamline curvature is in general not equal to the curvature of the coordinate lines. By simplifying an expression for the streamline curvature derived by De Vriend (1976), Olesen (1982, 1987) obtains

$$\frac{1}{R_f} = \frac{1}{R_s} - \frac{1}{u}\frac{\partial v}{\partial s} \tag{5}$$

where $1/R_f$ denotes the streamline curvature and $1/R_s$ denotes the curvature of the streamwise coordinate line.

Curved streamlines induce secondary flow. The development of the secondary flow intensity is described by

$$\lambda_r \frac{\partial}{\partial s}\left(\frac{Ah}{R*}\right) + \frac{Ah}{R*} = \frac{Ah}{R_f} \tag{6}$$

in which λ_r is the adaptation length of secondary flow, A is a coefficient for the influence of secondary flow on the direction of the bed shear stress, depending on the hydraulic roughness and the eddy viscosity model applied, and $R*$ is an 'effective' local radius of streamline curvature, which represents the intensity of the secondary flow. The secondary flow close to the bed adapts much faster to changing conditions than the higher parts, so that the value of λ_r for the development of the bed shear stress owing to secondary flow is smaller than the one for the development of secondary flow as a whole.

Equation (6) has been taken from Olesen (1987) and differs slightly from the relationships used by Rozovskii (1957) and De Vriend (1981). The differences do not matter because all these relationships for secondary flow adaptation are approximations instead of exact expressions anyway.

The length scale of secondary flow adaptation is small compared with other length scales in the model. It is therefore omitted in the flow computation, thus assuming an immediate adaptation of secondary flow. The adaptation length is retained in the calculation of the direction of sediment transport (see section on the sediment model), because omission there would cause stability problems in the numerical model (Olesen, 1987).

The convective influence of the secondary flow deforms the horizontal velocity distribution of the main flow. Kalkwijk and De Vriend (1980) show that this effect can be represented in depth-averaged models by an additional acceleration or friction term. The streamwise friction term is then replaced as follows

$$\frac{gu^2}{hC^2} \rightarrow \frac{gu^2}{hC^2} + \frac{k_{sn}u^2}{h}\frac{\partial}{\partial n}\left(\frac{h^2}{R*}\right) \tag{7}$$

in which k_{sn} is the secondary flow convection factor. Formally, the equation thus obtained holds along characteristics that do not coincide with the streamlines but gradually shift towards the concave bank. The error in applying the equation along streamwise coordinate lines is negligible in mildly curved channels.

The depth-averaged flow model is not very accurate in regions close to steep banks, because lateral friction is neglected and the similarity hypothesis for vertical profiles used to obtain the depth-averaged equations is not valid here. An erroneous growth of the main velocity at concave banks is prevented by an artifice for the near-bank decay of the secondary flow intensity (Olesen, 1987).

Bank erosion model

Bank erosion can result from a variety of mechanisms in which many factors play a role. It can be a result of discharge-induced flow and sediment transport, but also of processes beyond the basic system of river morphology. Ikeda *et al.* (1981) present a model in which the bank erosion rate is proportional to the excess flow velocity above a certain critical value. They take this critical value to be equal to the reach-averaged flow velocity. Osman and Thorne (1988) model the retreat of cohesive banks as a sequence of erosion cycles in which toe erosion decreases bank stability by increasing the slope and the height of the bank, leading to the more dramatic erosion of the upper bank by mass failure. Bank stability characteristics are important for the modelling of mass failures, but can be ignored in the time-averaged description of the sequence, which Osman and Thorne (1988) term 'parallel bank retreat'. The time-averaged behaviour can be described as a proportionality between bank erosion rate and excess shear stress, which, in a linear model, becomes equivalent to the excess flow velocity model of Ikeda *et al.* (1981).

During some time interval within the erosion cycle, higher and steeper banks are more likely to collapse than lower and less steep ones. The rate of bank retreat can hence be expected to be correlated with bank height and slope when looking at time-scales smaller than the period of an erosion cycle. For a uniform freeboard, that is a uniform difference between the water surface elevation and the top of the bank, correlations with bank height are equivalent to correlations with near-bank water depth. Models in which the bank erosion rate is proportional to excess near-bank water depth are proposed by Crosato (1989) and Odgaard (1989). Also, Hasegawa (1989) presents a relationship between near-bank water depth and bank erosion rate, but with opposite behaviour: larger near-bank water depths reduce the bank erosion rates. This is because Hasegawa's relation is an integrated description with information on flow field and bed topography of the full cross-section, instead of an equation for an underlying mechanism (Mosselman and Crosato, 1991).

Both the excess shear stress mechanism and the excess bank height mechanism have been implemented in the numerical model (Mosselman, 1992, 1995). In this paper only the excess bank height mechanism is used.

$$\frac{\partial n_B}{\partial t} = G\left(\frac{H - H_c}{H_c}\right) \quad \text{for } H \geqslant H_c$$
$$\frac{\partial n_B}{\partial t} = 0 \quad \text{for } H < H_c \tag{8}$$

in which $\partial n_B/\partial t$ denotes the rate of bank retreat, G is an erodibility coefficient, H is the total bank height and H_c is a critical bank height below which no bank erosion occurs. The total bank height is given by

$$H = h_w + H_{fb} \tag{9}$$

in which h_w is the near-bank water depth and H_{fb} is the freeboard. The mechanism thus formulated is equivalent to a mechanism in which bank retreat is a function of near-bank water depth when the freeboard is the same everywhere.

Sediment model

The magnitude of the sediment transport vector, $|\underline{s}|$, is taken to be a function of the magnitude of the flow velocity vector, $|\underline{u}|$, the water depth, h, the Chézy coefficient, C, the sediment grain size, D, and the streamwise bed slope, $\partial z_b/\partial s$, according to

$$|\underline{s}| = \left(1 - e\frac{\partial z_b}{\partial s}\right) f_s(|\underline{u}|, h, C, D) \tag{10}$$

in which e is a coefficient for the effect of streamwise bed slopes on sediment transport, and f_s denotes the sediment transport rate predicted by a transport formula. The relationship implies that the sediment transport rate is determined by the local flow field and bed topography only, which is appropriate for bedload and situations in which the adaptation length of the suspended load is short compared with other length scales in the model. The sediment transport formulae of Meyer-Peter and Müller (1948) and Engelund and Hansen (1967) have been implemented.

The direction of the sediment transport vector is defined as the angle, ψ, between the vector and the s direction. It is determined by the depth-averaged flow direction, the deviation of the near-bed flow direction from the depth-averaged flow direction owing to secondary flow and the effect of gravity on sediment grains on a sloping bed (Van Bendegom, 1947; Engelund, 1974). In the form used by Struiksma *et al.* (1985), ψ is given by

$$\tan \psi = \frac{v}{u} - A\frac{h}{R_*} - \frac{1}{f(\theta)}\frac{\partial z_b}{\partial n} \tag{11}$$

in which $f(\theta)$ is a function weighing the influence of gravity pull along a transverse bed slope. From experimental data presented by Zimmermann and Kennedy (1978), and using experience from the calibration of numerical models for river morphology, Struiksma and Crosato (1989) arrive at

$$f(\theta) = 0{\cdot}85\sqrt{\theta} \tag{12}$$

in which θ denotes the Shields parameter.

Bank erosion products contribute to the sediment balance. Treating the input of bank erosion products as a boundary condition for the transverse sediment transport, s_n, at the banks gives problems in mathematical analyses because it is not compatible with the notion that the sediment transport is determined by the local flow field and bed topography. The bank erosion products are therefore incorporated as a source term in the continuity equation for sediment. A distribution function, $\delta(s, n)$, specifies how the supply of eroded bank material, $s_{\mathrm{B}}(s)$, is distributed over the cross-section of the river. Then the full sediment balance reads

$$\frac{\partial z_{\mathrm{b}}}{\partial t} + \frac{\partial s_s}{\partial s} + \frac{\partial s_n}{\partial n} + \frac{s_s}{R_n} + \frac{s_n}{R_s} = \sum_{i=1}^{2} s_{\mathrm{B}_i}(s)\delta_i(s, n) \tag{13}$$

in which $i = 1$ denotes the right bank where $n = n_{\mathrm{B}_1}$, and $i = 2$ denotes the left bank where $n = n_{\mathrm{B}_2}$. The distribution function satisfies

$$\int_{n_{\mathrm{B}_1}}^{n_{\mathrm{B}_2}} \delta_i(s, n)\, \mathrm{d}n = 1 \tag{14}$$

A distribution over a narrow zone along the eroding bank is obtained by defining $\delta_i(s, n)$ as a Dirac delta function

$$\begin{aligned} \delta_1(s, n) &= \lim_{\Delta n \to 0} \frac{1}{\Delta n} \quad &&\text{for} \quad n_{\mathrm{B}_1}(s) < n < n_{\mathrm{B}_1}(s) + \Delta n \\ \delta_1(s, n) &= 0 &&\text{for} \quad n < n_{\mathrm{B}_1}(s) \vee n > n_{\mathrm{B}_1}(s) + \Delta n \end{aligned} \tag{15}$$

An analogous definition holds for $\delta_2(s, n)$. The corresponding difference equation used in the numerical model implies that bank erosion products are distributed evenly over the area of computational cells next to the banks. The input of bank erosion products, s_{B_i}, is equal to

$$s_{\mathrm{B}_i} = (1 - \omega)H\left|\frac{\partial n_{\mathrm{B}_i}}{\partial t}\right| \tag{16}$$

in which ω is a washload factor.

Computational procedure

The computation of river planimetry can be decoupled from the computation of bed topography when bank retreat is much slower than degradation or aggradation of the bed. This is a reasonable assumption for cohesive banks. Furthermore, the flow is assumed to be quasi-steady, which means that the computation of bed topography can be decoupled from the flow computation as well. The equations are hence separated into a steady-flow model, a bed deformation model and a bank migration model. The corresponding computational procedure consists of three steps. In the first step, the flow field is computed while keeping the bed and bank configuration fixed. Sediment transport rates and bank migration rates are calculated from the flow field. In the second step, bed level changes are computed from the sediment transport gradients and the input of bank erosion products. Finally, bankline changes are calculated from the bank migration rates.

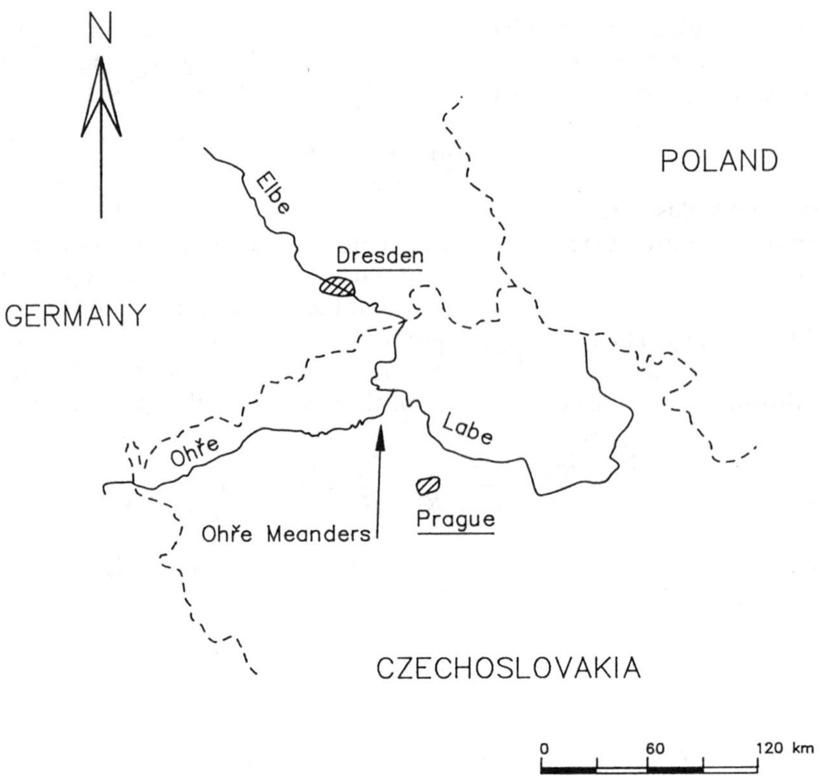

Figure 1. River Ohře

APPLICATION TO OHŘE RIVER, BOHEMIA, CZECHOSLOVAKIA

Introduction

The numerical model was applied to a reach of the meandering gravel-bed River Ohře (Eger) in the former state of Czechoslovakia. This river springs in the Fichtel Mountains in Germany and is a tributary to the River Labe (Elbe), see Figure 1. Its total length is about 300 km. The reach under consideration is about 15 km from the confluence, at the Ohře Meanders near Hostěnice, shown in Figure 2. The Písty Woods are a nature reserve where the meander bends are allowed to migrate freely. The shape of the river in Figure 2 is interesting in view of a discussion on the extent to which the skewing of river bend planforms can be ascribed to geometric non-linearities or to lateral confinements. The coexistence here of sharp skewed bends on the left, where the river impinges on hills, and round bends on the right supports Ikeda's (1989) opinion that the sharp skewed bends of the Beaver River in Alberta, Canada, shown by Parker *et al.* (1982, 1983), are primarily a result of the confinement within a narrow valley. Similar phenomena can be observed in the numerical simulations by Sun *et al.* (1996), where bends touching the meander belt borders are generally sharper and more skewed than bends completely inside the meander belt.

Schematization and data

The computational grid consisted of 25 × 69 points and is shown in Figure 3. Grid point numbers are indicated with ξ in the streamwise and η in the transverse direction. Only two bends of the Ohře Meanders were modelled. The average longitudinal slope of the last 100 km of the River Ohře is 1·3 m/km. The reach under consideration at about 15 km from the confluence is less steep, however, which might be a result of an overall concave shape of the longitudinal profile as well as to sedimentation upstream of the Hostěnice weir. A longitudinal slope of 0·14 m/km was used in the computations. The average river width is 54 m in this reach.

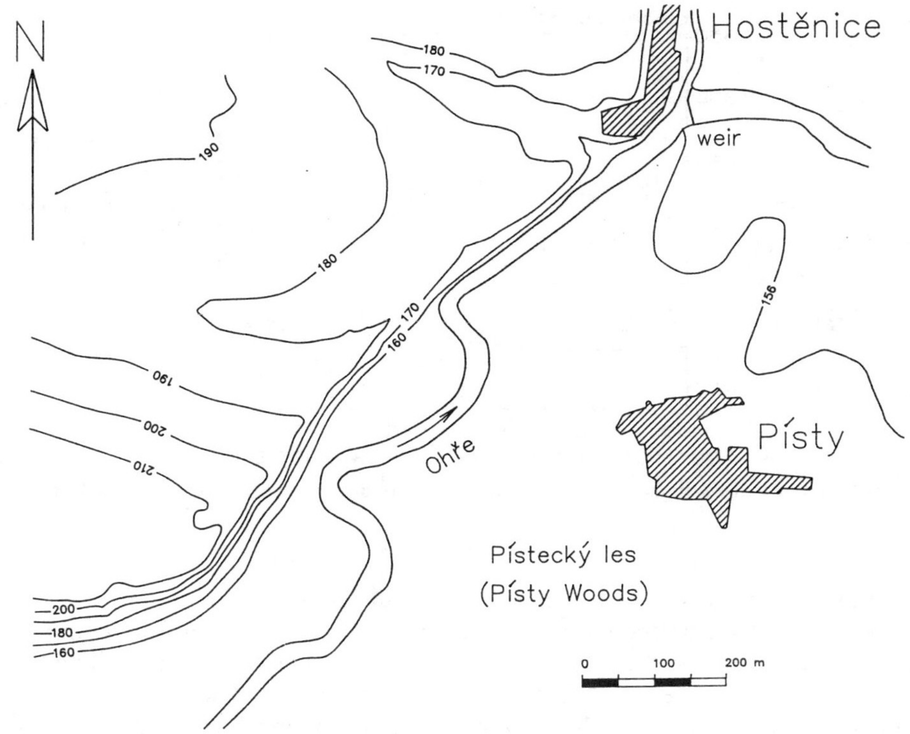

Figure 2. The Ohře Meanders near Hostěnice. Contour lines indicate heights in m above sea level

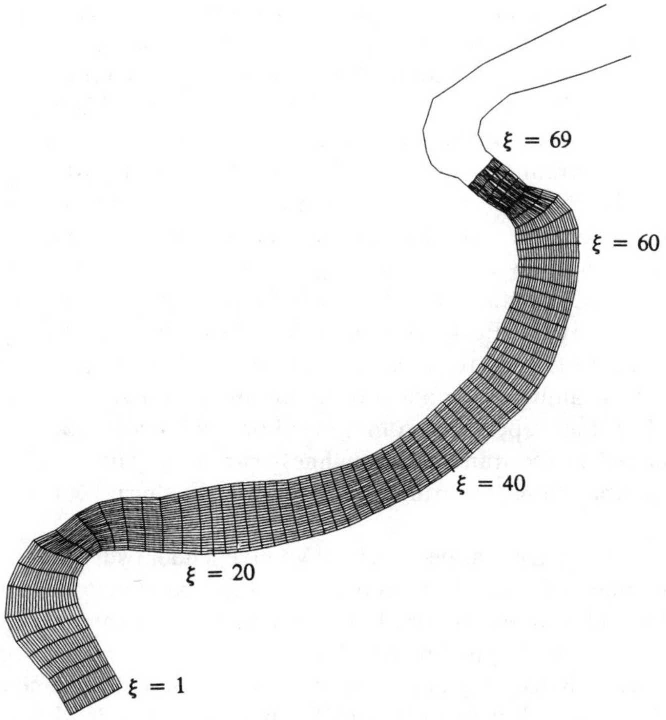

Figure 3. Computational grid for River Ohře with streamwise grid point numbers, ξ

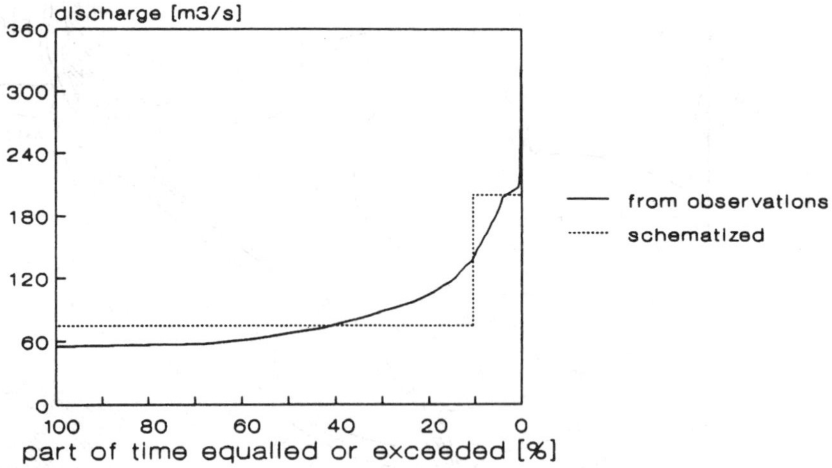

Figure 4. Observed and schematized discharge duration curves for periods in which the discharge equalled or exceeded 55 m^3/s

A discharge duration curve was determined from a data file containing processed discharge data. Discharges below 55 m^3/s have been omitted because they did not produce any substantial sediment transport. Discharges above the bank-full discharge of 200 m^3/s have been reduced according to estimates of the distribution of discharges over the main channel and the floodplains. The duration curve with processed discharge data is shown in Figure 4, together with a schematization in which the discharge is 75 m^3/s during 90% of the time and 200 m^3/s during 10% of the time. The suitability of this schematization is discussed by Mosselman (1992).

Despite the quasi-steady flow assumption, modelling a varying discharge is still possible by using different discharges in different computational steps, as long as the discharges are kept constant within each of the steps. Here the discharges immediately preceding the bed topography survey were not known, so computations were made for the two discharges of the schematization, 75 m^3/s and 200 m^3/s, separately. The results were expected to be envelopes of the possible outcomes from computations with a varying discharge.

The Chézy coefficient for hydraulic roughness, C, varies from 65 to 70 $m^{1/2}/s$. A constant value of 67·5 $m^{1/2}/s$ was chosen for the computations. Assuming logarithmic vertical velocity profiles, the corresponding coefficient for the influence of secondary flow on the direction of the bed shear stress, A, is 11 and the corresponding secondary flow convection factor, k_{sn}, is 0·4 (see Olesen, 1987). In sharp bends, however, k_{sn} can be neglected in order to reduce the errors that arise from not reproducing secondary flow cells in the model (De Vriend and Geldof, 1983). Furthermore, Olesen (1987) finds that calibration in mildly curved bends sometimes requires values of k_{sn} up to twice the theoretical value. He is not sure whether this is caused by a transverse variation of the alluvial roughness or by an underestimation of the influence of secondary flow convection, but A. M. Talmon (personal communication, 1991) concludes from curved flume experiments that it must be ascribed to the transverse roughness variation. The distribution of k_{sn} displayed in Figure 5 was used in the computations, ensuring that $k_{sn} = 0$ in the sharp bend and $k_{sn} = 0{\cdot}8$ in the mildly curved reach.

The locations of the banks had been mapped during two independent surveys, one in 1975 and 1976, and one in 1989. The banks did migrate in the period in between. A few bank material samples were taken during a visit in 1990. Nearly all samples were cohesive, but their composition varied considerably. Most samples consisted of a mixture of sand, clay, organic material and vegetation roots, but sometimes one or more of these components were absent. Bank erodibility parameters were not determined from the samples, but assigned uniform values by trial and error during calibration. The washload factor was chosen to be 1, because a mathematical analysis and trial computations indicated that the influence of bank erosion

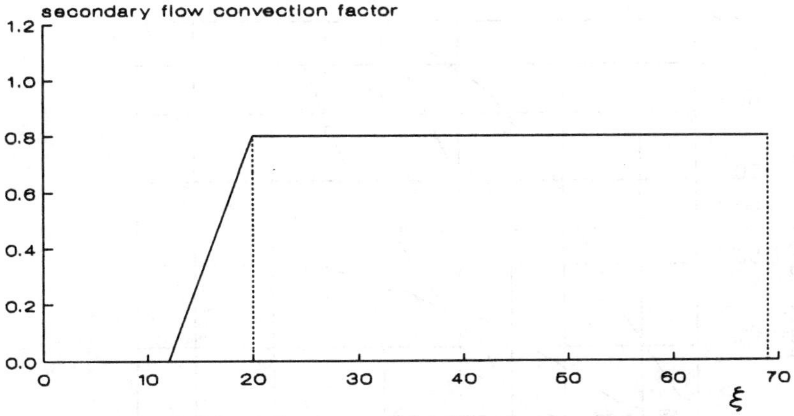

Figure 5. Secondary flow convection factor, k_{sn}, as a function of the streamwise grid point number, ξ

products on two-dimensional bed topography patterns can be neglected in rivers with low banks and only minor width changes (Mosselman, 1992). The maximum sediment production from bank erosion per unit channel length in this river reach is of the order of 10^{-3} times the sediment transport per unit river width.

Detailed bank elevation data were not available, but it was assumed that the freeboard was more or less uniform in the reach. This assumption may be questionable in view of the suggestion in the previous section that the sharp skewed bends on the left side of the river are a result of the impingement on valley walls. The question is, however, whether the top of the bank should be defined as the pivot point of the outer bend cut in the valley wall or as the top of the hillslope. The pivot point might be a fair definition for relatively short time spans, whereas inclusion of the hillslope might be necessary when studying the long-term evolution of the river.

Because of the assumed uniform freeboard, the dependence of bank erosion on bank height could be formulated as a dependence on near-bank water depth. Expecting inaccuracies in the computed flow velocities and water depths close to the banks, it was assumed that it was better to correlate bank erosion with values computed further away from the banks, modifying G and the critical near-bank water depth accordingly. Bank erosion was therefore related to water depths at $\eta = 5$ instead of $\eta = 1$ and at $\eta = 21$ instead of $\eta = 25$. Furthermore, bank erosion was assumed to take place during high discharge only.

Bed topography data stemmed from the survey in 1975 and 1976. The bed consists of sand and gravel with a wide distribution of grain sizes. Figure 6 shows a few grain size distribution curves. The grains were sorted in the transverse direction. Under a constant discharge, grain sorting is such that the sediment in the deeper pools is coarser than the sediment on the shallower bars (see Parker and Andrews, 1985), but here the opposite was observed. An explanation is that fine sediment brought into suspension during a flood had settled in the pools afterwards, whereas the smaller flow velocities at lower discharges could not remove the coarser sediment from the bars. Grain sorting was not taken into account, however. The computations were executed for uniformly distributed sediment with a median grain size, D_{50}, of 1·5 mm and a mean grain size, D_m, of 1·9 mm.

The sediment transport formulae of Engelund and Hansen (1967) and Ackers and White (1973) give the best results in one-dimensional morphological computations of a longer stretch of the river (E. Zeman, personal communication, 1990), but for the smaller flow velocities in the less steep reach considered here, the sediment transport formula of Meyer-Peter and Müller (1948) is more appropriate. The latter was used.

The unperturbed value of the function for the influence of gravity pull along a transverse slope, $f(\theta_0)$, was taken to be 0·22 for the discharge of 75 m^3/s and 0·30 for the discharge of 200 m^3/s.

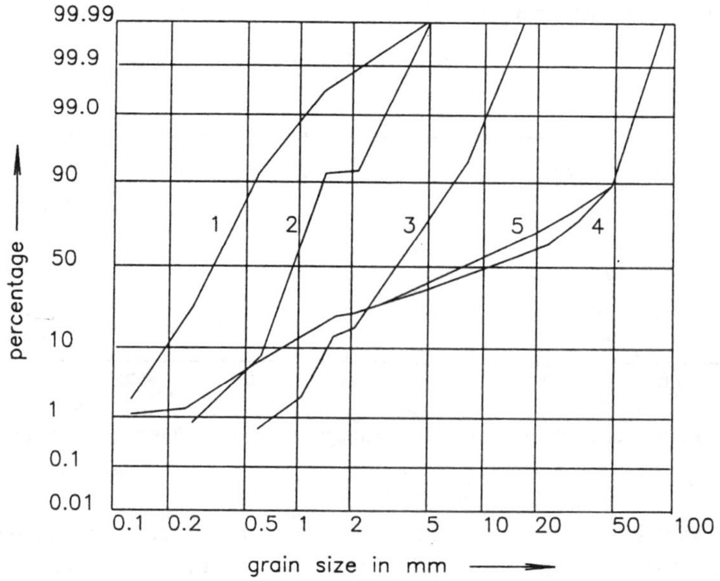

Figure 6. Grain size distribution curves. The curves 1–3 refer to three cross-sections of the Ohře Meanders, each curve being based on five samples from a survey in 1975 and 1976. Curve 4 represents an earlier sample. Curve 5 is used by the Technical University of Prague for one-dimensional morphological computations of the River Ohře

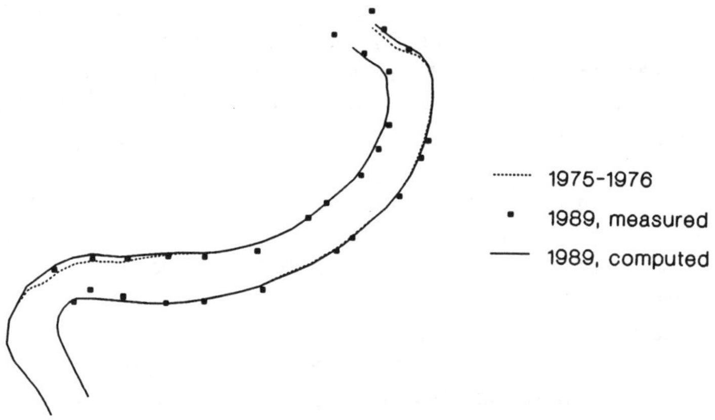

Figure 7. Measured and computed planform evolution over 14 years, the computation being based on only one step of bank migration. Bank erosion was taken to depend on excess near-bank water depth with $G/H_c = 0{\cdot}1 \times 10^{-6}\ \mathrm{s}^{-1}$ and $H_c - H_{fb} = 3{\cdot}1$ m

Computational results and discussion

Computed planform evolutions are shown in Figures 7 and 8. Only one step of bank migration was computed, instead of a series of bank migration steps with intermediate equilibrium flow fields and bed topographies. For the larger bank erodibility in Figure 8, this computation with only one bank migration step led to incorrect overmigration, but provided a clear visualization of the initial rates of bank retreat. Both figures show only locally good agreement between computed and measured bank migration. It would have been easy to use locally different values for the bank erodibility parameters to obtain perfect agreement, but such a calibration would be misleading. It is more useful to identify the causes of the deviations and to formulate recommendations to improve the predictive capabilities of the model.

Apart from a lack of more data, three main causes for the deviations can be identified. First, bank erosion as a function of excess bank height or near-bank water depth may not be the proper mechanism. The actual

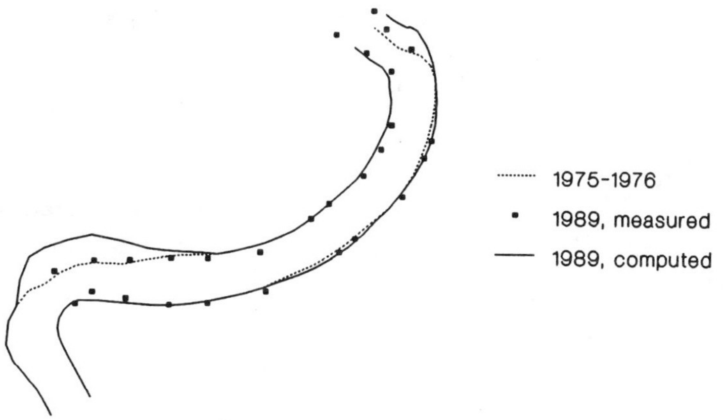

Figure 8. Measured and computed planform evolution over 14 years, the computation being based on only one step of bank migration. Bank erosion was taken to depend on excess near-bank water depth with $G/H_c = 0{\cdot}4 \times 10^{-6}\ s^{-1}$ and $H_c - H_{fb} = 3{\cdot}1$ m

bank retreat might have been caused by, for instance, excess near-bank flow velocities, ice effects or overbank flow. Mosselman (1992) does not find better results when bank erosion is related to near-bank flow velocities, but B. Jůza (personal communication, 1990) does find a correlation between the observed bank erosion rates and the computed cross-bank flow velocities in a two-dimensional depth-averaged simulation of the 1981 flood that severely inundated the village of Písty. Secondly, Figures 7 and 8 show that the banks advanced as well. The model does not contain any mechanism for bank accretion and, more generally, bank accretion has received much less attention in research than bank erosion. Thirdly, the poor agreement between the computed and the observed cross-sections shown in Figures 9 and 10 shows that the computation of flow field and bed topography in the river is inadequate, although previous computations of the bed topography of sand-bed rivers with basically the same model have been successful (Struiksma, 1985; Olesen, 1987). The present computations underestimated point bar heights and transverse bed slopes, and did not reproduce the reverse inclination of bed slopes near the concave bank. This third reason is the most important, because improved modelling of the mechanisms of bank advance and retreat does not make much sense if the variables that control these mechanisms are not computed correctly.

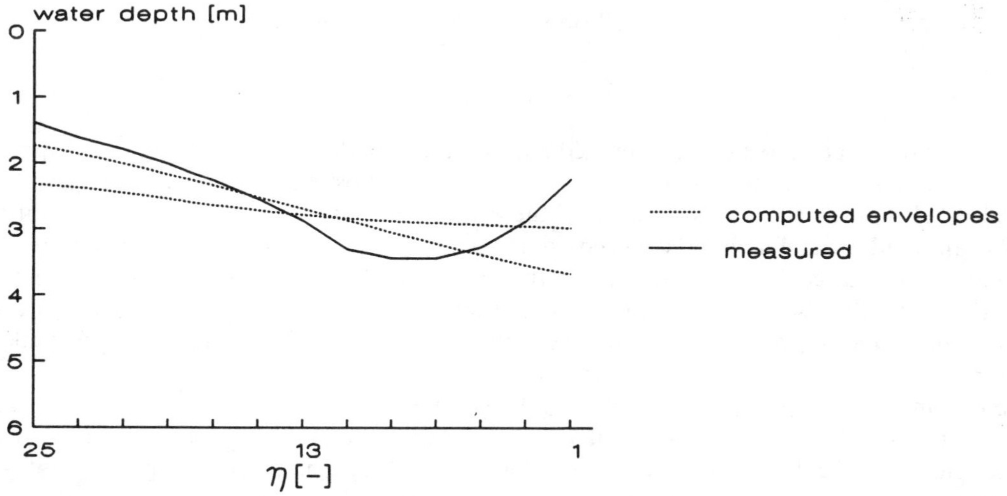

Figure 9. Computed water depths for $Q = 75$ and 200 m^3/s and measured water depths as a function of grid point number, η, along the transverse coordinate line $\xi = 40$

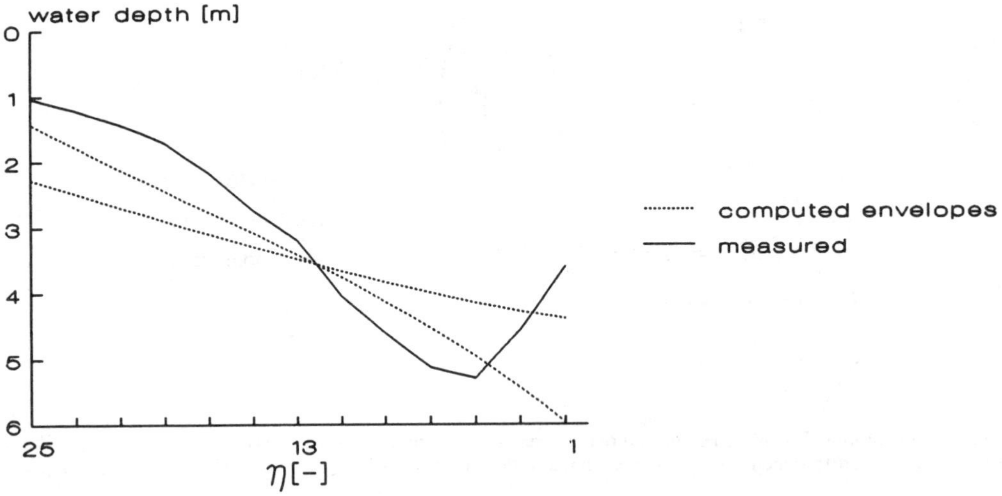

Figure 10. Computed water depths for $Q = 75$ and 200 m^3/s and measured water depths as a function of grid point number, η, along the transverse coordinate line $\xi = 65$

The deviations between the computed and the observed cross-sections in Figures 9 and 10 can be ascribed to the depth-averaged description of the flow field and neglect of grain sorting. The depth-averaged flow model yields inaccurate results in regions close to steep banks where the similarity hypothesis for vertical profiles of the velocity does not hold. Furthermore, the parameterization of secondary flow effects in Equations (7) and (11) does not reproduce the effects of the second, counter-rotating secondary flow cells that occur at sharp bends. It should also be noted that a different secondary circulation may occur when overbank flow crosses the main channel during a flood.

The reproduction of the observed grain sorting requires not only a morphological model for a mixture of different sediment size fractions, but also running this model with a varying discharge, because the observed coarser sediment on the bars and finer sediment in the pools is probably the result of bed evolution during a flood followed by a period of lower discharges. The high point bars might even be armoured remains of the bed formed during a large flood. The formulation of a two-dimensional morphological model for a mixture of different sediment size fractions would require experimental research on hiding and exposure effects in Equation (12) for the influence of gravity pull on the direction of sediment transport.

CONCLUSIONS

Bank erosion has been added to a depth-averaged mathematical model of river morphology. Application to a reach of the River Ohře (Eger) in the former state of Czechoslovakia yielded poor results, but an unequivocal conclusion that this was a result of the formulation of the bank erosion mechanism or the calibration of the bank erodibility parameters could not be drawn. Another reason for the poor results might be the absence of a bank accretion mechanism in the model. Bank accretion is a complex phenomenon in which point bar growth, discharge variations, channel incision, vegetation and washload play a role, but research on bank accretion has generally received much less attention than research on bank erosion. Therefore, more research on bank accretion is recommended.

More importantly, the reproduction of flow field and bed topography was also inadequate, despite successful application of basically the same model to sand-bed rivers. This can be ascribed to the depth-averaged description of the flow field and the neglect of grain sorting. The shortcomings of a depth-averaged description of the flow can be overcome when a three-dimensional flow model is used, although correct reproduction of two or more secondary flow cells in one cross-section will depend strongly on the turbulence model applied. Grain sorting can be dealt with by including equations for sediment mixtures and by applying the resulting model for a sequence of different discharges. The proper formulation of equations for sediment mixtures would require experimental research to examine hiding and exposure effects in the relationship for the influence of gravity pull on the direction of sediment transport.

ACKNOWLEDGEMENTS

This paper presents research carried out at Delft University of Technology between 1987 and 1991, under supervision of Prof. Dr M. de Vries and with financial support from the Centre for Civil Engineering Research, Codes and Specifications (CUR), the Netherlands Organization for Scientific Research (NWO) and the Technical University of Prague. The Ohře River Board and the Technical University of Prague provided data for the River Ohře.

REFERENCES

Ackers, P. and White, W. R. 1973. 'Sediment transport: new approach and analysis', *ASCE J. Hydraul. Div.*, **100**, 2041–2060.
Crosato, A. 1989. 'Meander migration prediction', *Excerpta*, GNI, 4. Libreria Progetto, Volume 4, published by GNI, Gruppo Nazionale Idraulica, Padova. pp. 169–198.
De Vriend, H. J. 1976. 'A mathematical model of steady flow in curved shallow channels', *Communications on Hydraulics*, 76. Delft University of Technology, Delft.
De Vriend, H. J. 1981. 'Steady flow in shallow channel bends', *Communications on Hydraulics*, 81. Delft University of Technology, Delft.
De Vriend, H. J. and Geldof, H. J. 1983. 'Main flow velocity in short river bends', *J. Hydraul. Engng*, **109**, 991–1011.
Engelund, F. 1974. 'Flow and bed topography in channel bends', *ASCE J. Hydraul. Div.*, **100**, 1631–1648.
Engelund, F. and Hansen, E. 1967. *A Monograph on Sediment Transport in Alluvial Streams*. Teknisk Forlag, Copenhagen.
Hasegawa, K. 1989. 'Universal bank erosion coefficient for meandering rivers', *J. Hydraul. Engng*, **115**, 744–765.
Ikeda, H. 1989. 'Sedimentary controls on channel migration and origin of point bars in sand-bedded meandering rivers', in Ikeda, S. and Parker, G. (Eds), *River Meandering*, AGU, Water Resources Monograph 12. American Geophysical Union, Washington, D.C. pp. 51–68.
Ikeda, S., Parker, G., and Sawai, K. 1981. 'Bend theory of river meanders, Part 1, linear development', *J. Fluid Mech.*, **112**, 363–377.
Kalkwijk, J. P. Th. and De Vriend, H. J. 1980. 'Computation of the flow in shallow river bends', *IAHR J. Hydraul. Res.*, **18**, 327–342.
Kennedy, J. F., Nakato, T., and Odgaard, A. J. 1983. 'Analysis, numerical modeling, and experimental investigation of flow in river bends', in Elliott, C. M. (Ed.), *River Meandering, Proceedings, Conference Rivers 1983, New Orleans*. ASCE, New York (1984). pp. 843–856.
Meyer-Peter, E. and Müller, R. 1948. 'Formulas for bed-load transport', *Proceedings 2nd Congress IAHR, Stockholm*. Paper No. 2, pp. 39–64.
Mosselman, E. 1991. 'Modelling of river morphology with non-orthogonal horizontal curvilinear coordinates', *Communications on Hydraulic and Geotechnical Engineering*, 91. Delft University of Technology, Delft.
Mosselman, E. 1992. 'Mathematical modelling of morphological processes in rivers with erodible cohesive banks', *Communications on Hydraulic and Geotechnical Engineering*, 92. Delft University of Technology, Delft.
Mosselman, E. 1995. 'A review of mathematical models of river planform changes', *Earth Surf. Process. Landf.*, **20**, 661–670.
Mosselman, E. and Crosato, A. 1991. 'Universal bank erosion coefficient for meandering rivers (discussion)', *J. Hydraul. Engng*, **117**, 942–943.
Nelson, J. M. and Smith, J. D. 1989. 'Evolution and stability of erodible channel beds', in Ikeda, S. and Parker, G. (Eds), *River Meandering*, AGU, Water Resources Monograph 12. American Geophysical Union, Washington, D.C. pp. 321–377.
Odgaard, A. J. 1981. 'Transverse bed slope in alluvial channel bends', *ASCE J. Hydraul. Div.*, **107**, 1677–1694.
Odgaard, A. J. 1989. 'River-meander model. I: development', *J. Hydraul. Engng*, **115**, 1433–1450.
Olesen, K. W. 1982. 'Introduction of streamline curvature into flow computation for shallow river bends', *Report 5-82*. Laboratory of Fluid Mechanics, Department of Civil Engineering, Delft University of Technology, Delft.
Olesen, K. W. 1987. 'Bed topography in shallow river bends', *Communications on Hydraulic and Geotechnical Engineering*, 87. Delft University of Technology, Delft.
Osman, A. M. and Thorne, C. R. 1988. 'Riverbank stability analysis. I: theory', *J. Hydraul. Engng*, **114**, 134–150.
Parker, G. and Andrews, E. D. 1985. 'Sorting of bed load sediments by flow in meander bends', *Wat. Resour. Res.*, **21**, 1361–1373.
Parker, G., Sawai, K., and Ikeda, S. 1982. 'Bend theory of river meanders, Part 2, nonlinear deformation of finite-amplitude bends', *J. Fluid Mech.*, **115**, 303–314.
Parker, G., Diplas, P., and Akiyama, J. 1983. 'Meander bends of high amplitude', *J. Hydraul. Engng*, **109**, 1323–1337.
Rozovskii, I. L. 1957. *Flow of Water in Bends of Open Channels*. Academy of Science of the Ukrainian SSR, Kiev. Translation 1961 by Y. Prushansky, Israel Program for Scientific Translations. S. Monson, Jerusalem, PST Cat. No. 363.
Shimizu, Y. and Itakura, T. 1985. 'Practical computation of two-dimensional flow and bed deformation in alluvial streams', Civil Engineering Research Institute Report. Hokkaido Development Bureau, Sapporo.

Struiksma, N. 1985. 'Prediction of 2-D bed topography in rivers', *J. Hydraul. Engng*, **111**, 1169–1182.
Struiksma, N. and Crosato, A. 1989. 'Analysis of a 2-D bed topography model for rivers', in Ikeda, S. and Parker, G. (Eds), *River Meandering*, AGU, Water Resources Monograph 12. American Geophysical Union, Washington, D.C. pp. 153–180.
Struiksma, N., Olesen, K. W., Flokstra, C., and De Vriend, H. J. 1985. 'Bed deformation in curved alluvial channels', *J. Hydraul. Res.*, **23**, 57–79.
Sun, T., Meakin, P., Jøssang, T., and Schwarz, K. 1996. 'A simulation model for meandering rivers', *Wat. Resour. Res.*, **32**, 2937–2954.
Van Bendegom, L. 1947. 'Some considerations on river morphology and river improvement', *De Ingenieur*, **59**, B1–B11 (in Dutch; English translation: *National Research Council Canada*, Technical Translation 1054, 1963).
Zimmermann, C. and Kennedy, J. F. 1978. 'Transverse bed slopes in curved alluvial streams', *ASCE J. Hydraul. Div.*, **104**, 33–48.

AUTHOR INDEX

SUBJECT INDEX